Musculoskeletal Ultrasonography in Rheumatic Diseases

Yasser El Miedany
Editor

Musculoskeletal Ultrasonography in Rheumatic Diseases

Springer

Editor
Yasser El Miedany, MD, FRCP
Department of Rheumatology
Darent Valley Hospital
Dartford, UK

ISBN 978-3-319-15722-1 ISBN 978-3-319-15723-8 (eBook)
DOI 10.1007/978-3-319-15723-8

Library of Congress Control Number: 2015933642

Springer Cham Heidelberg New York Dordrecht London

Printed on acid-free paper

Springer is part of Springer Science+Business Media (www.springer.com)

To my inspiring mom and dad for passing on the love of reading and the eagerness for learning

To my family for their love, care, and encouragement

Foreword

There are many challenges to providing a high quality musculoskeletal diagnostic and treatment service. Much of our diagnostic approach has relied traditionally on the skills of history and examination. These remain important, but are open to subjective interpretation. The same is true of assessing response to interventions, some of which are extremely expensive, and we need to justify their use with reliable methods of demonstrating improvement in the patient. We are working in health care systems where value (better outcomes at lower costs) is an increasingly important consideration.

Against this background, ultrasound has increasingly become a part of the armory of the musculoskeletal practitioner. It helps us

- in diagnosis, treatment, and assessing response to interventions;
- to make musculoskeletal disease management more selective and objective, and improves the chances of the patient getting the right treatment from the right person and at the right time;
- to enhance our expertise and differentiate ourselves from other musculoskeletal practitioners.

In summary, ultrasound helps musculoskeletal practitioners to improve value in health care, and possessing the skill of using this modality places the practitioner in a stronger position when it comes to the commissioning of high quality musculoskeletal services. All musculoskeletal practitioners should have the opportunity to train in developing expertise in this area.

Some people in musculoskeletal practice talk well about their service, but do not always deliver. Yasser El Miedany does not fit into this category. I have always been impressed by his ability to take great ideas and translate them into living practice in high quality services that we can all learn from. It comes as no surprise, therefore, that he has managed to assemble an international collection of ultrasound experts, encyclopedically covering all areas of musculoskeletal disease and practice. There is much we can all learn from the contents of these chapters. I commend this book to all musculoskeletal practitioners.

Chris Deighton BMedSci (Hons) MB BS (Hons) MD FRCP
President of the British Society for Rheumatology 2012–2014
Clinical Advisor National Institute of Health and Care Excellence Rheumatoid Arthritis Management Guidelines
Consultant Rheumatologist, Derby, UK
Associate Professor, Nottingham Medical School, UK

Preface

The use of musculoskeletal ultrasonography in rheumatology clinical practice has expanded swiftly over the past decade as an outcome of the technological developments and implementation of the window of opportunity concept in most of the inflammatory musculoskeletal conditions. Ultrasound booked its place as the "rheumatologist's stethoscope" as it enabled the treating health care professional to diagnose, prognosticate, monitor disease activity as well as response to management. In contrast with other imaging modalities such as X-ray and magnetic resonance, ultrasound has clear advantages, namely good tolerability, no radioactivity, ability to scan both joints and soft tissue at one sitting and its dynamic facility, which enables direct correlation between clinical and imaging outcomes. In spite of some concerns regarding standardization, the use of ultrasound in rheumatology is expected to grow further as prices of the machine fall and the opportunity to practice improves. This book will go beyond the traditional specific joint ultrasonography, which has already been discussed thoroughly in previous publications, to focus on ultrasonographic changes in different musculoskeletal disorders. The main purpose of this book is to serve as a guide for any physician who wishes to perform clinical musculoskeletal ultrasound regardless of experience. The aim is provide a practical approach to dealing with patients suffering from different rheumatic diseases in standard practice. It is therefore suitable to a wide range of audience from the novice embarking on providing a musculoskeletal ultrasonography service, to the experienced who wishes to develop their sonographic scanning skills.

The main theme of this book is to deliver a very practical and reader-friendly guide. On one hand it describes a "how to do it" approach, whereas in the meantime it delivers the evidence and advanced knowledge base of the relevant pathological sonographic appearances of the various musculoskeletal tissues. The illustrations and figures were meticulously selected to give the reader a clear guide towards implementation in real life practice. Focusing on the major values of ultrasound in rheumatologic diseases, this book with its 14 chapters is expected to fill an important void in the current literature. It represents what can be considered to be the best current thinking on role of ultrasonography in the assessment of pathology, diagnosis, and treatment of different rheumatic diseases. Therefore, this book can serve as both excellent introductory and a very good reference resource for future reading.

This work has been the outcome of cooperative effort of a large international group of leaders in musculoskeletal ultrasonography. They have done a superb job to produce authoritative chapters including vast amounts of scientific and clinical data to create state-of-the-art descriptions of sonographic changes encompassed by different rheumatic diseases. Special thanks to Dr. Chris Deighton who wrote the book's foreword and Dr. Adham Khalil for his support throughout the whole project which helped to make this book complete.

Personally, I feel privileged to have compiled this work and am enthusiastic about all that it offers our readers. I hope you too will find this edition a uniquely valuable educational resource.

Kent, England, UK

Yasser El Miedany, MD FRCP
Professor of Rheumatology and Rehabilitation (Egypt)/
Consultant Rheumatologist, UK

Contents

Contributors

Carlos Acebes, PhD Department of Rheumatology, Raigmore Hospital, Inverness, Scotland

Eva Álvarez Andrés, MD Department of Rheumatology, Paediatric Rheumatology Unit, Hospital Universitario Severo Ochoa, Madrid, Spain

Marina Backhaus, MD Department of Internal Medicine, Rheumatology and Clinical Immunology, Park-Klinik Weissensee Berlin, Academic Teaching Hospital of Charité, Berlin, Germany

Anna Ciechomska, MD, PhD Department of Rheumatology, Wishaw General Hospital, Wishaw, UK

Paz Collado Ramos, MD, PhD Department of Rheumatology, Paediatric Rheumatology Unit, Hospital Universitario Severo Ochoa, Madrid, Spain

Zbigniew Czyrny, MD Private Practice, Diagnostic Imaging, Warsaw, Poland

Hatem El-Azizi, MD Department of Radiodiagnosis, Cairo University Hospitals, Cairo, Egypt

Yasser El Miedany, MD, FRCP Department of Rheumatology, Darent Valley Hospital, Dartford, UK

Rheumatology and Rehabilitation Department, Ain Shams University, Cairo, Egypt

Daniela Fodor, MD, PhD Department of Internal Medicine, "Iuliu Hatieganu" University of Medicine and Pharmacy, Cluj-Napoca, Romania

John P. Harvie, MD Department of Rheumatology, Raigmore Hospital, Inverness, Scotland

Cristina Hernández-Díaz, MD, MS Laboratorio de Ultrasonido Musculoesquelético y Articular, Instituto Nacional de Rehabilitación, Mexico City, Mexico

Annamaria Iagnocco, MD Ultrasound Unit, Department of Rheumatology, Sapienza Università di Roma, Rome, Italy

Vasilia Iorgoveanu, MD Department of Internal Medicine and Rheumatology, University of Medicine and Pharmacy Carol Davila, Bucharest, Romania

Iustina Janţă, MD Department of Rheumatology, Hospital General Universitario Gregorio Marañon, Madrid, Spain

Carolina Botar Jid, MD, PhD Department of Radiology, Emergency Clinical County Hospital Cluj, "Iuliu Hatieganu" University of Medicine and Pharmacy, Cluj-Napoca, Romania

Adham Aboul Fotouh Khalil, MD, MSc New Kasr El Aini Teaching Hospital, Cairo University, Cairo, Egypt

Mihaela C. Micu, MD Department of Rehabilitation II, Rheumatology Division, Rehabilitation Clinical Hospital Cluj-Napoca, Cluj-Napoca, Romania

María Montoro, MD Department of Rheumatology, Hospital General Universitario Gregorio Marañon, Madrid, Spain

Esperanza Naredo, MD Department of Rheumatology, Hospital General Universitario Gregorio Marañon, Madrid, Spain

Rodina Nestorova, MD Department of Musculoskeletal Ultrasound, Rheumatology Centre "St. Irina", Sofia, Bulgaria

Margarida Alexandre Oliveira, MD Department of Rheumatology, Centro Hospitalar Cova da Beira, Covilhã, Portugal

Violeta Maria Vlad, MD, PhD Department of Rheumatology, Sf. Maria Hospital, Bucharest, Romania

Ximena Wortsman, MD Departments of Radiology and Dermatology, IDIEP-Institute for Diagnostic and Reasearch of the Skin and Soft Tissue, Clinica Servet, Faculty of Medicine, University of Chile, Santiago, Chile

Chapter 1
Fundamentals of Musculoskeletal Ultrasound

Hatem El-Azizi MD, PhD and Carolina Botar Jid MD

Introduction

Ultrasonography is a safe, well-tolerated, low-cost imaging technique with no ionizing radiation and few technical limitations. High-frequency ultrasound has better spatial resolution than magnetic resonance imaging (MRI), and ultrasonography really excels in its ability to perform real-time dynamic studies and interventions. Sonography also allows easy comparison with the contralateral side, which can help in identifying subtle abnormalities [1].

Basic Physics

Wavelength is the length of one complete cycle; it is the distance between two identical adjacent points in consecutive waves. Wavelength is typically measured between two similar points, such as two adjacent crests or troughs in a waveform. Wavelengths are most accurately measured in sinusoidal waves, which have a smooth and repetitive oscillation. It is normally given the symbol λ (lambda) and has units of meters or millimeters (Fig. 1.1).

Frequency is the number of waves per second. It is measured in hertz (Hz), with 1 Hz being one complete cycle per second. The audible sound for human has

H. El-Azizi (✉)
Department of Radiodiagnosis, Cairo University Hospitals,
Cairo, Egypt
e-mail: hatemazizi@hotmail.com

C. Botar Jid
Department of Radiology, Emergency Clinical County Hospital Cluj, "Iuliu Hatieganu" University of Medicine and Pharmacy, Cluj-Napoca, Romania
e-mail: inabotar@yahoo.com

Y. El Miedany (ed.), *Musculoskeletal Ultrasonography in Rheumatic Diseases*,
DOI 10.1007/978-3-319-15723-8_1

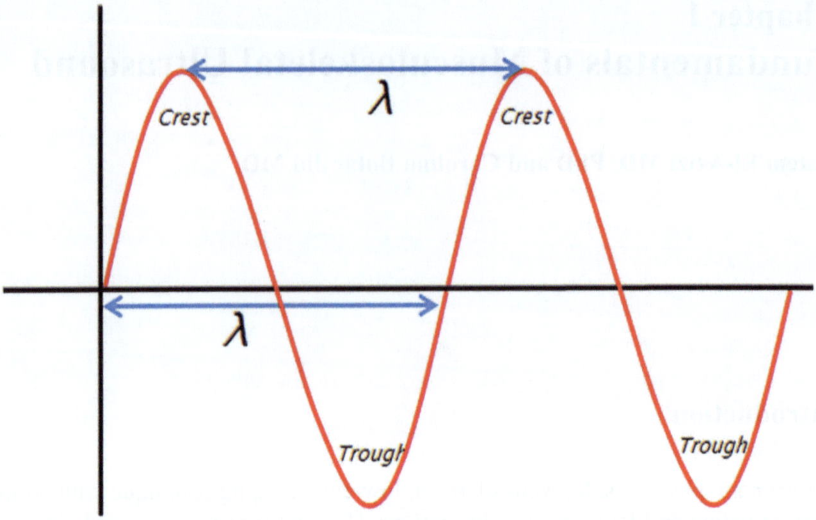

Fig. 1.1 Wavelength is the distance between two crests or any similar points in consecutive waves

the frequency of 20–20,000 Hz; any frequencies above this range are referred to as ultrasound. The frequencies used in diagnostic ultrasound typically ranges from 2 to 20 MHz (1 MHz = 1 million Hz). The frequency is inversely proportional to the wavelength; the higher the frequency, the shorter the wave length (Fig. 1.2). Higher frequency leads to a better resolution, while lower frequency provides better

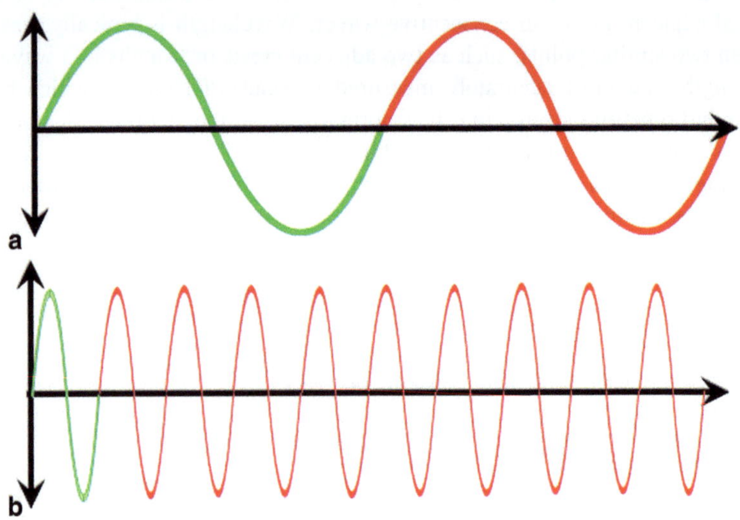

Fig. 1.2 Frequency inversely proportion to wavelength. **a** Low frequency and long wavelength. **b** High frequency and shorter wavelength

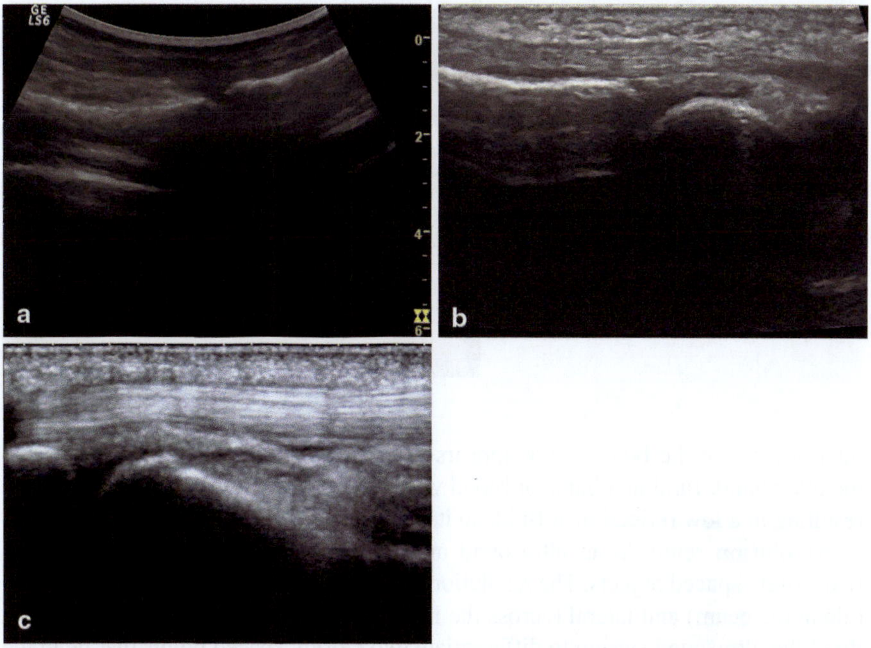

Fig. 1.3 Influence of different frequency on the resolution of flexor tendon at the level of MCP joint. **a** Very poor resolution at frequency 3.5 MHz. **b** Good resolution at frequency at 8 MHz. **c** High resolution at frequency 18 MHz. *MCP* metacarpophalangeal

penetration. To examine the musculoskeletal system, electronic high-frequency linear transducers are used, which provide uniform anatomical information on all images, have good resolution in both near and distance fields, and are easy to apply and maintain (Fig. 1.3). For overview and guidance, transducers with frequencies between 7.5 and 10 MHz are recommended, while for deeper-located structures, such as hip or shoulder joints, transducers with lower frequencies (2–6 MHz) are recommended. Detailed study of musculoskeletal structures requires the use of high-frequency transducers, 15–18 or 20 MHz [4, 10].

Acoustic Impedance is resistance at the interface between the two media or tissues. If the acoustic impedance between the two tissues is high, most of the sound beam will not be transmitted to the deeper tissues and it will be reflected back to the probe (air/soft tissue interface and soft tissue/bone interface). Minimal acoustic impedance between the tissues (subcutaneous/muscle interface) will lead to some of the sound beams being transmitted to the deeper tissue and some being reflected back. If the acoustic impedance between tissues is very low, almost all of the sound beam will be transmitted to the deeper tissue and minimal will be reflected back (soft tissue/fluid interface).

Reflection: Ultrasound images are created by the reflected ultrasound beam at the interface between tissues or media. The higher the reflection of sound beam at tissue interface, the higher the echogenicity in the image at the soft tissue/bone

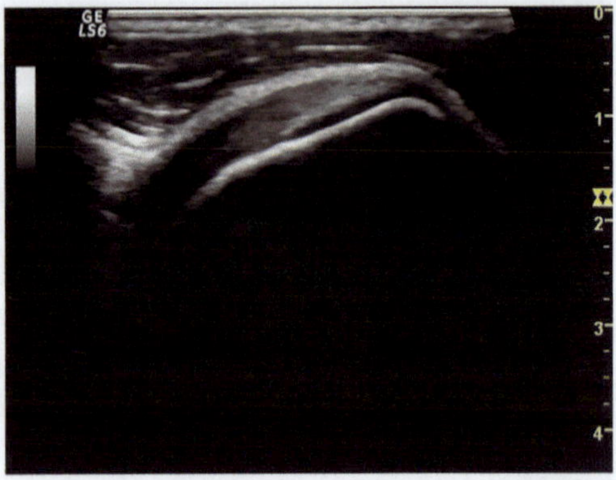

Fig. 1.4 High reflection of sound beam at soft tissue/bone interface with high echogenicity (*white*) of cortical outline

interface where the bone cortex appears highly echogenic (white) (Fig. 1.4). On the other hand, fluid in a bursa or blood vessels has a very low acoustic impedance resulting in a low reflection at fluid, so it appears black (Fig. 1.5).

Resolution refers to an ultrasound machine's ability to discriminate between two closely spaced objects. The resolution of an ultrasound image includes the axial (along the beam) and lateral (across the image). *Axial resolution* refers to the ability of the ultrasound system to differentiate two closely spaced points that lie in the plain parallel to the sound beam (Fig. 1.6). Each sound pulse is composed of 2–3 wavelengths emitted in longitudinal (axial) direction. The system can resolve two separate points in the image when the distance between the two points is equal to a single wavelength. Increase in the frequency will decrease the wavelength which will result in increase in the axial resolution of the ultrasound image (Fig. 1.3). *Lateral resolution* is the ability of the system to display small structures side by side (same depth) as separate from each other (Fig. 1.7). The ultrasound beam initially converges with increasing depth, and then widens out again with decreasing intensity and resolution. The focal zone of the beam is 3–4 wavelengths wide and is the area where lateral resolution is the highest. The ultrasound beam can be focused to improve image quality.

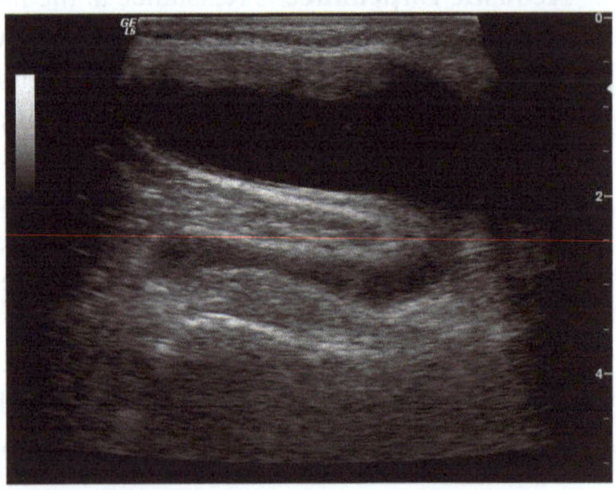

Fig. 1.5 Low reflection of sound beam at soft tissue/fluid interface with low echogenicity (*black*) of backer cyst fluid content

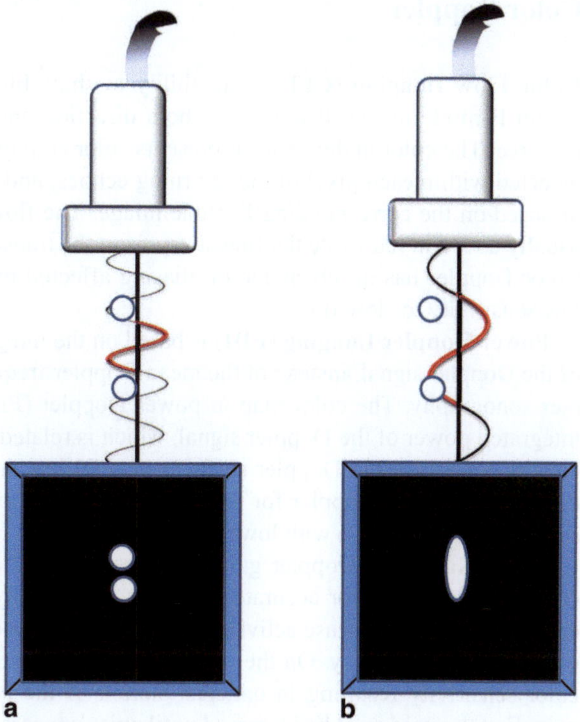

Fig. 1.6 Axial resolution. **a & c** Poor resolution at the near and far fields with both structures seen on the screen as single structure. **b** The best image seen at the level of the focus where both structures are seen as separate points on the image

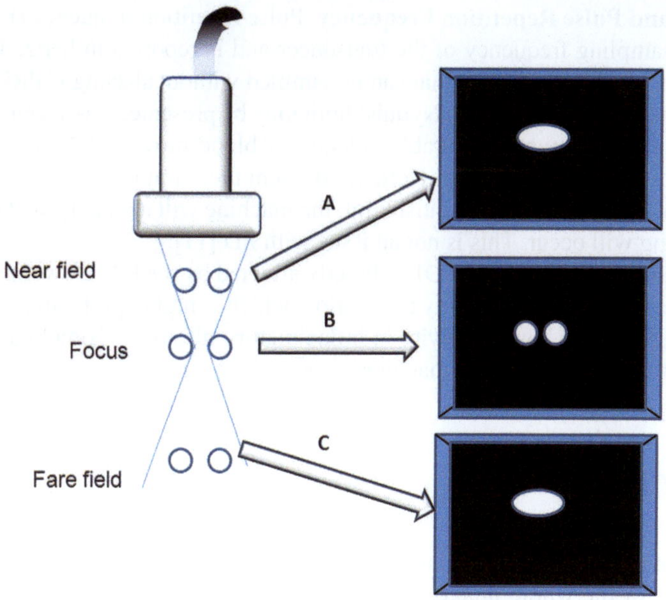

Fig. 1.7 Lateral resolution. **a & c** Poor resolution at the near and far fields with both structures seen on the screen as single structure. **b** The best image seen at the level of the focus where both structures are seen as separate points on the image

Color Doppler

Color Flow Imaging (CFI) is the ability to show blood flow in a selected area within B-mode image. It represents both direction and velocity of blood flow in this area. The color in the image represents color coding of Doppler frequency shift detected within each pixel of the returning echoes, and the detected color is superimposed on the corresponding B-mode image. The flow towards the transducer is usually coded in red while the flow away from the transducer is usually coded blue. Color Doppler has disadvantage of aliasing affected by the steering angle and its low sensitivity to slow flow.

Power Doppler Imaging (PDI) is based on the integrated power (or amplitude) of the Doppler signal, instead of the mean Doppler frequency shift as in color Doppler sonography. The color map in power Doppler (PD) sonography displays the integrated power of the Doppler signal, which is related to the number of red blood cells that produce the Doppler shift [3, 8]. PDI has three times the sensitivity of conventional color Doppler for the detection of flow and is particularly useful for small vessels and those with low-velocity flow [2].

Color Gain: The Doppler gain is different from the B-mode gain. Setting the color gain is crucial for accurate diagnosis of tissue hypervascularity which indicates the degree of disease activity. Increased gain causes noise and overestimation of the tissue vascularity. On the other hand, lowering the color gain decreases the color sensitivity resulting in underestimation of the activity. To adjust the color gain, first the gain must be increased until noise appears in the image then gradually decreased until noise disappears (Fig. 1.8).

Scale and Pulse Repetition Frequency: Pulse repetition frequency (PRF) is the Doppler sampling frequency of the transducer and is reported in hertz. The maximum Doppler shift frequency that can be sampled without aliasing is PRF/2, which is called the Nyquist limit. The Nyquist limit may be presented on-screen as a blood velocity (the maximum measurable velocity of blood moving directly towards or away from the transducer) or in Hertz (maximum measurable Doppler shift). If the blood velocity is above the Nyquist limit, the machine will misinterpret the velocity and aliasing will occur. This is not an issue with PD [11].

The sensitivity of CFI and PDI is directly affected by the PRF; lowering the PRF will increase the colur sensitivity to low flow which is highly preferable in rheumatology to detect the lowest activity in inflammatory diseases. Using high PRF will decrease the sensitivity of the machine to lower velocities.

Image Control

Gain: Gain correction is crucial to obtain an interpretable image as it affects the gray scale of the whole image. Decreased gain will give a black image, and details will be masked. Increased gain will give a white image, and details will be saturated (Fig. 1.9).

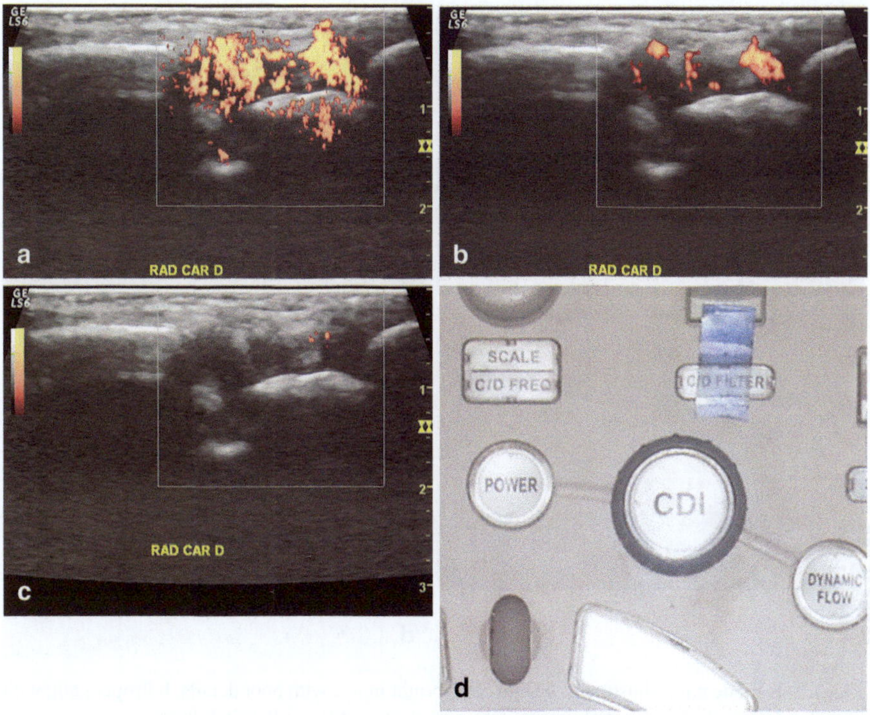

Fig. 1.8 Color gain adjustment. **a** High gain: color noise with overestimated hypervascularity. **b** Proper gain: with no color noise. **c** Poor color gain with false poor hypervascularity. **d** An example of color gain button

Time gain compensation (*TGC*): Echoes returning from deeper tissues within the body are weaker than those arising from structures closer to the transducer. Since the distance they have to travel is longer, they experience greater attenuation. Without TGC, the far field (bottom of the screen, deeper tissue) would always appear darker than the near field (top of the screen, tissue closest to probe); TGC correct the gain on the echoes returning from the far fields. Most of machines have multiple slider levers that allow you to control the gain throughout the entire scanning depth (Fig. 1.10).

Focus The focus of the image is usually marked on the side of the screen by a small arrowhead (Fig. 1.11). The ultrasound beam is narrowed at that depth revealing the best lateral resolution which improves the image quality and ensures high definition of tissue at that depth. The focus is usually adjusted by means of a knob or an up/down button on the control panel; the focus pointer should move to region of interest.

Depth adjustment increases or decreases the depth of the examined region on the image. It is best to have the structure that is being examined in the center of the screen.

Dynamic range (DR) at the receptor level has also modeled the relationship between the strongest and the weakest echoes by establishing a DR of echoes values.

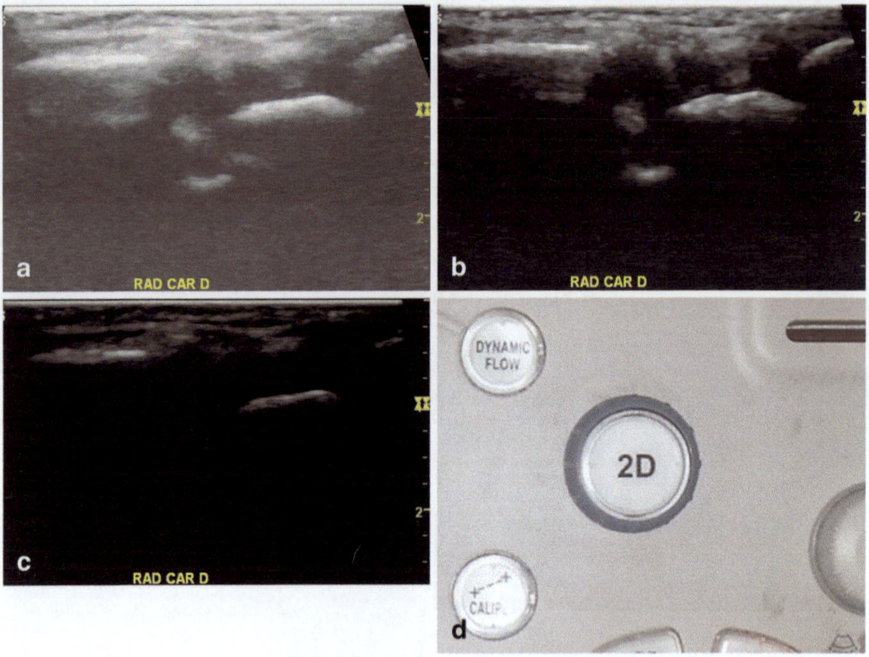

Fig. 1.9 B-mode gain adjustment. **a** High gain: bright image with poor details. **b** Proper gain: with no color noise. **c** Low gain: dark image with poor details. **d** B-mode gain button

The range of difference between echoes is 100 dB, which means ten billion times. The TGC can compensate 60 dB due to attenuation. It is necessary to compress the amplification elective of echoes; thus, weak echoes will be amplified, while strong echoes will not be amplified. Elective amplification is performed using a logarithmic curve. Another function of the receiver is the primary filtering of the electrical signal to eliminate very weak electrical signals corresponding to noise and multiple reflections.

Ultrasound Equipment

An ultrasound device consists of a console, comprising a computer, a monitor, a keyboard, and transducers.

The transducer or probe is the centerpiece of the equipment, with built-in piezoelectric crystals that emit and receive ultrasound. It contains a linear array of very thin crystals, characterized by a piezoelectric property. This property can be described as: the appearance of a difference in the electric potential between two surfaces of a piezoelectric crystal, when it is subjected to mechanical deformation. The phenomenon occurs inversely as well: A piezoelectric crystal subjected to a potential difference suffers a mechanical deformation, which generates ultrasound.

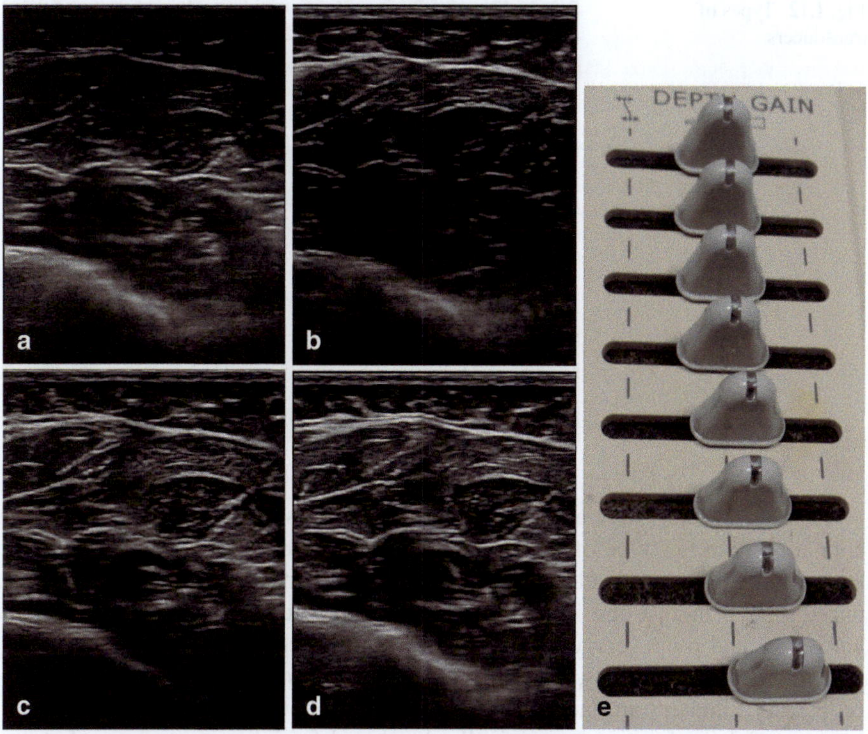

Fig. 1.10 TGC adjustment. **a** Too low gain in superficial part of the image. **b** Too low gain in the middle of the ultrasound image. **c** Too low gain in the deeper part of the image. **d** Proper gain in the entire image. **e** The TGC buttons. TGC time gain compensation

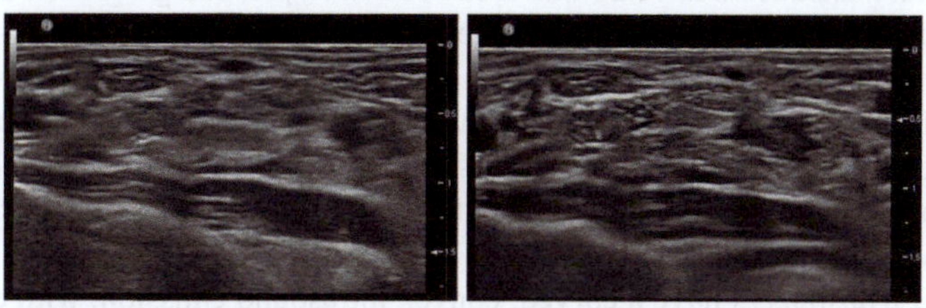

Fig. 1.11 Focus position is indicated by arrow head. **a** Positioning too low. **b** Proper positioning

The generated ultrasounds are characterized by frequency, amplitude, period, wavelength, and propagation speed [5, 6, 10]. Ultrasound frequency that can be generated and received by a transducer is determined by the resonance frequency and the thickness of the piezoelectric crystals. The nominal frequency of the transducer is predetermined by the construction unit.

Depending on the ultrasound transducers' emission and sequencing beam, they are classified into: linear, sectorial, and combined, with functions and/ or multiple frequencies (Fig. 1.12) [10].

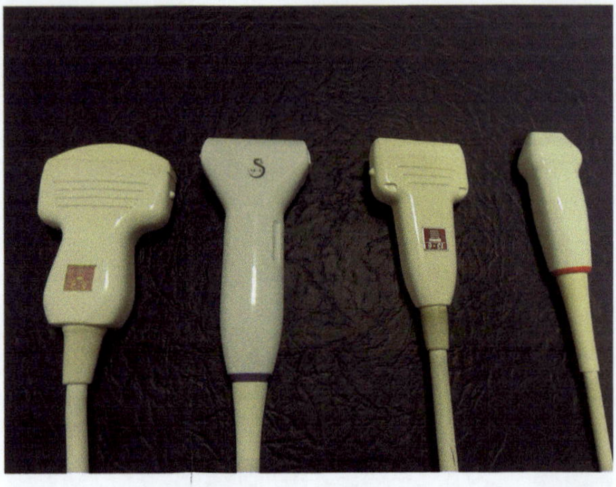

Fig. 1.12 Types of transducers

Linear transducers are able to cover almost all musculoskeletal ultrasound examinations. On this type of transducers, the ultrasound beam emerges parallel to each other and perpendicular to the surface of the transducer and the image produced is a rectangular one. These may have a contact surface length, with the skin, between 2 and 6 cm. Larger transducers provide a better overview and are used particularly for large joints, such as the hip joint, or for knee or shoulder joint stability tests. Smaller transducers, also known as "the hockey stick" or "fingerprint" because of their shape, were originally developed for intraoperative use, but are excellent for small and superficial structures and for the evaluation of inaccessible areas such as metacarpophalangeal (MCP), metatarsophalangeal (MTP), proximal interphalangeal (PIP), or distal interphalangeal (DIP) joints [4, 10].

Sectorial transducers have a small area of contact, providing a greater angle with good visualization of regions located deeper. They produce a triangular image on the screen (the area of a circle), the apex of the surface corresponding to the emission of ultrasound beam, because this emerges divergent from a common point on the transducer surface. These transducers are used to explore the musculoskeletal system for examination of meniscus, but they have a low resolution due to the low frequency.

Convex transducers have a curved, convex ultrasound emission, with electronic activation of the piezoelectric crystals and obtain a trapezoidal image. They are mainly used for abdominal ultrasound, whereas in musculoskeletal ultrasonography they are used to explore the hip joint, especially in overweight patients.

Combined transducers combine the several possibilities presented before. These include multifrequency transducers known as "broadband" (wide-band transducer), which include in a one-piece the necessary elements for an examination with a wide range of frequencies. Another type of combined transducers is transducers with multiple functions, which allow examination in several ways: bidimensional (2D) mode, M-mode, Doppler (continuous and pulsed), harmonic, three-dimensional elastography, etc.

Electronic memory is a fundamental component of any ultrasound machine. The electrical signal from the receiver is converted into binary data by an analog–digital converter. The information is stored and processed in the internal memory layers. Digital–analog convertor turns the processed information into an electrical signal. This electrical signal is demodulated and subjected to rectification and filtering processes to become video signal, which is send to the TV monitor.

The TV monitor consists of a cathode tube, whose electron beam is activated as horizontal lines. Each line corresponds to a row in digital memory. The ultrasound information is obtained and stored in columns, but is displayed in rows. An image displayed on the monitor is made up of 525 lines. In order to reduce the oscillation ("blinking") of image, each image is divided in two fields: field even-numbered lines and field odd-numbered lines, displayed alternately. During examination in real time, the monitor screen displays 30 images (60 fields) per second [10].

Probe Orientation and Handling

Probe Markers: Every probe has a mark on one of its sides (Fig. 1.13); it could be a raised marker or indentation or some other identifier that is correlated to a dot or the manufacturer's logo on the display screen. Structures on the same side of the probe marker will appear on the side of the screen mark on the display screen while structures on the opposite side of the probe marker will appear on the other side of the screen marker (Fig. 1.14). Most machines have a button that lets you flip the screen marker from right to left (Fig. 1.15).

Probe Handling When Scanning: To scan a structure in the longitudinal or sagittal view, the transducer is oriented along the long access of the body with the probe marker directed towards the patient head (Fig. 1.16), so the cephalad structures will appear on the side of the display screen with the marker. The transverse or axial is obtained by rotating the probe 90 ° from the long axis of the patient; the probe marker should be directed to the right side of the patient so that the right-side structures of the body will appear on the side of the display screen with the marker (Fig. 1.17).

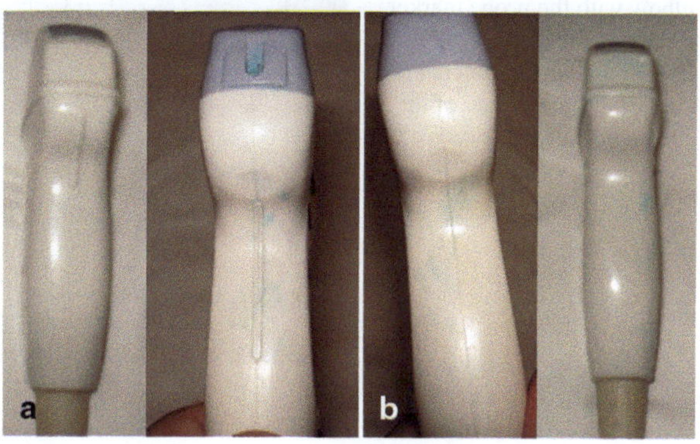

Fig. 1.13 **a** Probe side with marker. **b** Probe side with no marker

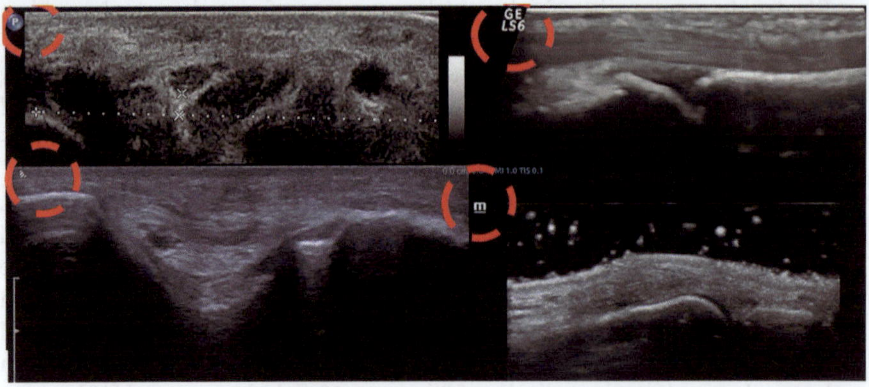

Fig. 1.14 Different screen marker showing either companies logo or initials' letter

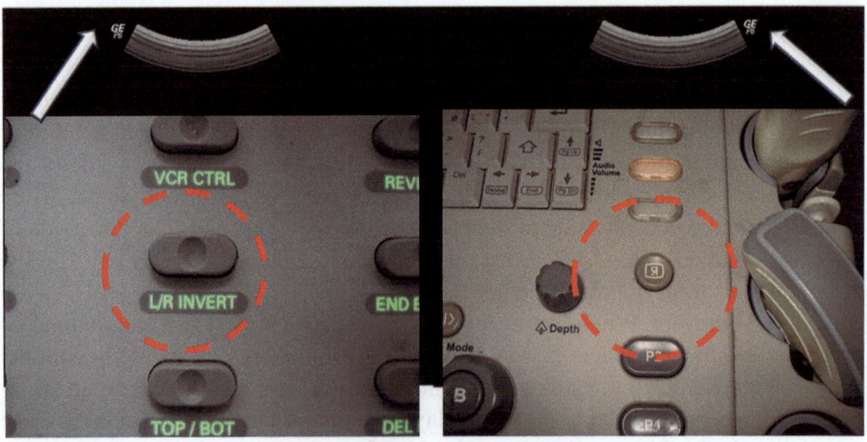

Fig. 1.15 Button for flipping the screen marker from right to left

Fig. 1.16 Longitudinal (*sagittal*) examination of the elbow with the probe marker directed to the head (*red arrow*) and the hummers (*H*) seen on the same side of the screen marker (*blue arrow*)

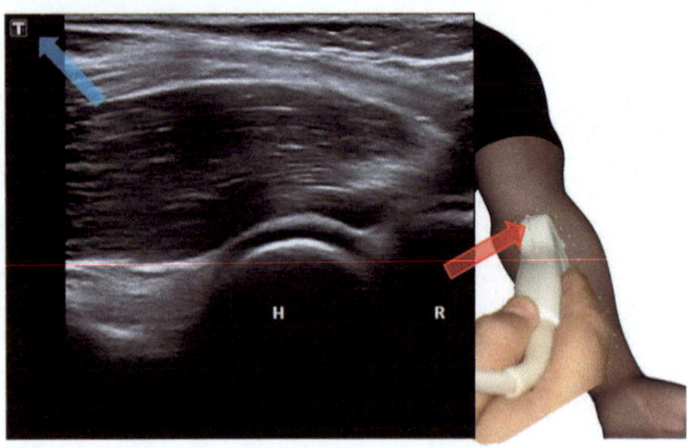

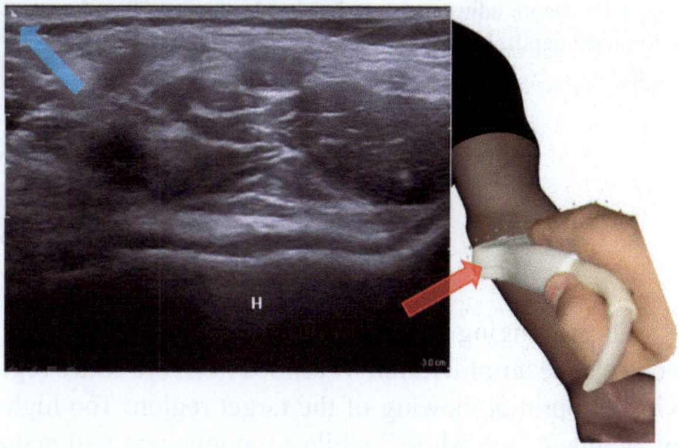

Fig. 1.17 Transverse examination of the elbow with the probe marker (*red* arrow) directed to the right side of the patient. The probe marker directed to the side of the display screen with the screen marker (*blue arrow*). Hummers (*H*)

Scanning Technique and Image Optimization

There are several notions of ultrasound techniques which are necessary to be known for achieving optimal ultrasound images. First of all, the examiner must choose the transducer (probe button) according to the region of interest. The capacity of penetration of the ultrasound beam is inversely proportional to the nominal frequency of the transducer. The frequency of the transducer is higher; the image resolution is better, but associated with the reduction of the penetration due to attenuation (which is proportional to the frequency of ultrasound). As a practical application, to obtain a detailed ultrasound image of superficial structures, transducers with high frequency and resolution are used, whereas for deeper structures, it is necessary to reduce the frequency and the resolution by default (Fig. 1.18).

Following transducer selection, ultrasound gel is placed on the transducer and the transducer is applied to the skin of the region of interest. Next step is to adjust the image depth (depth control on the console) so that the displayed image includes the region of interest without losing information and with no unused space (Fig. 1.19) [10].

The examiner must change the position (Fig. 1.11) and number of focuses so that the focal zone is located at the same position and depth as the targeted structure, in order to obtain an optimal lateral resolution [10].

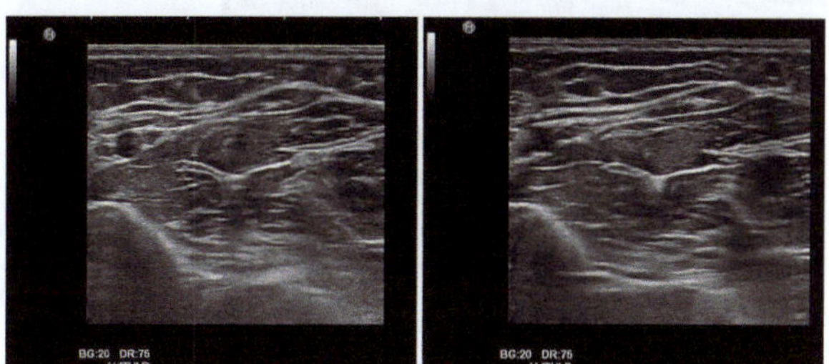

Fig. 1.18 Influence of frequency on the image with the same transducer. **a** Resolution frequency. **b** Penetration frequency

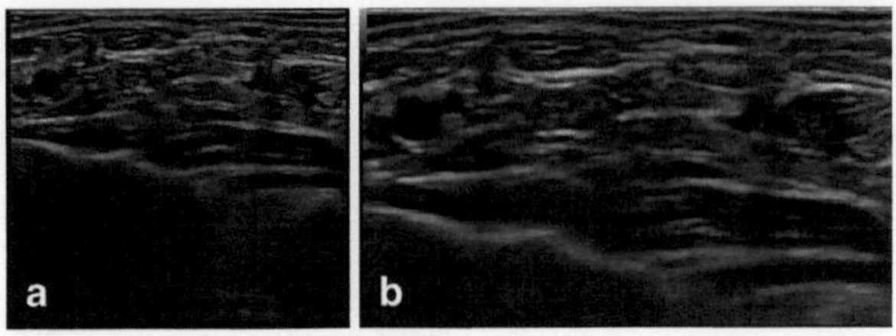

Fig. 1.19 Depth adjustment. a Reduced depth. b Increased depth

After changing the focal zone, it is necessary to adjust the overall gain which controls the amplification of echoes from the entire region scanned, in order to provide an optimal viewing of the target region. Too high gain will cause the display of an image "too white," while a too low gain will make the image too dark (black; Fig. 1.20). Although the optimal gain is subjective, it is adjusted properly when the sonographic picture of blood vessels/ pure fluid collections or the region located posterior to bone structures appears black or almost black on the ultrasound image.

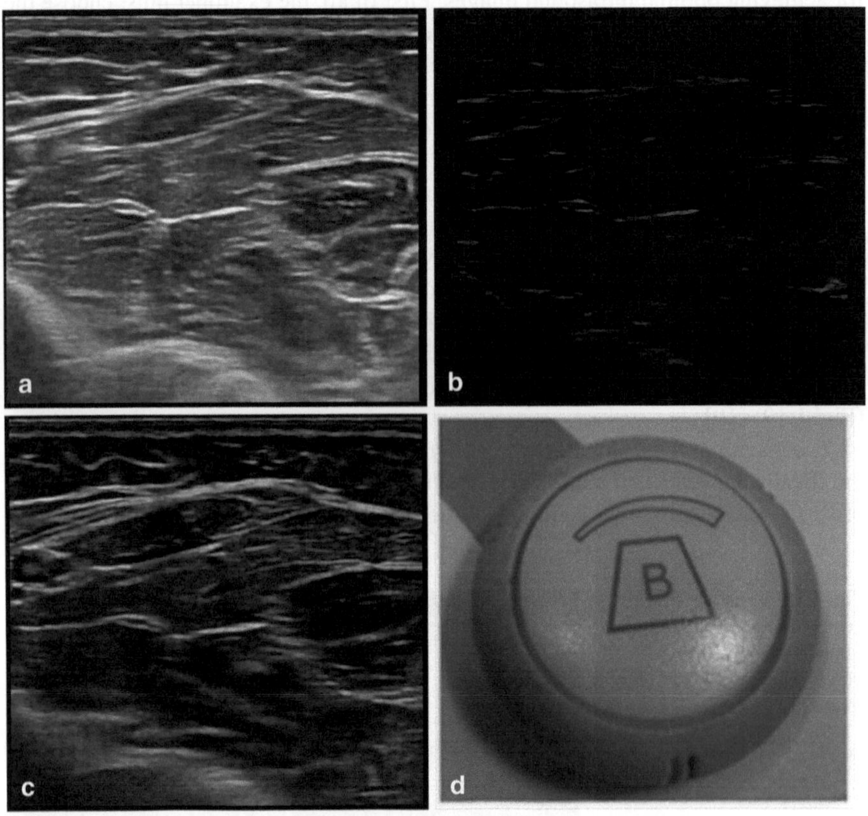

Fig. 1.20 Gain adjustment. a High gain: "too white image." b Too low gain: "too black image." c Proper gain. d An example of gain button

Subsequently, the examiner will adjust depth gain compensation (DGC) or TGC curve, which consists of multiple buttons, each representing the gain at a different depth in the image. TGC refers to ultrasound attenuation in the human body, attenuation being electronic compensated so that deep structures will produce similar echoes compared with the superficial ones. This compensation occurs exponentially, as well as attenuation: An echo coming from a structure located at a double distance from the transducer is four times amplified. DGC control can be adjusted by the examiner according to the structures of interest by manually sliding the buttons (Fig. 1.10) [10].

Ultrasound image can be freezed on the monitor using "freeze" button. The last image displayed on the TV monitor is still preserved indefinitely, until the cancellation of the order. Whereas with the picture freezing disappears the necessity of acquiring new information, in most applications the freezing transition is associated with the transducer at rest.

Magnification scale (scale, zoom) allows enlarging a particular region of interest image multiple times. Image magnification may occur on acquisition (write zoom) leading to increased resolution or on reading information from memory (read zoom), without increasing the resolution. Zoom image on acquisition can be applied only in real-time examination, while reading zoom can be applied to frozen images (Fig. 1.21).

All ultrasound machines are equipped with microprocessors that allow various facilities: measurement of distances, perimeters, areas, angles, and volumes; marking areas of interest; schematic display of the transducer position; enrolment of gen-

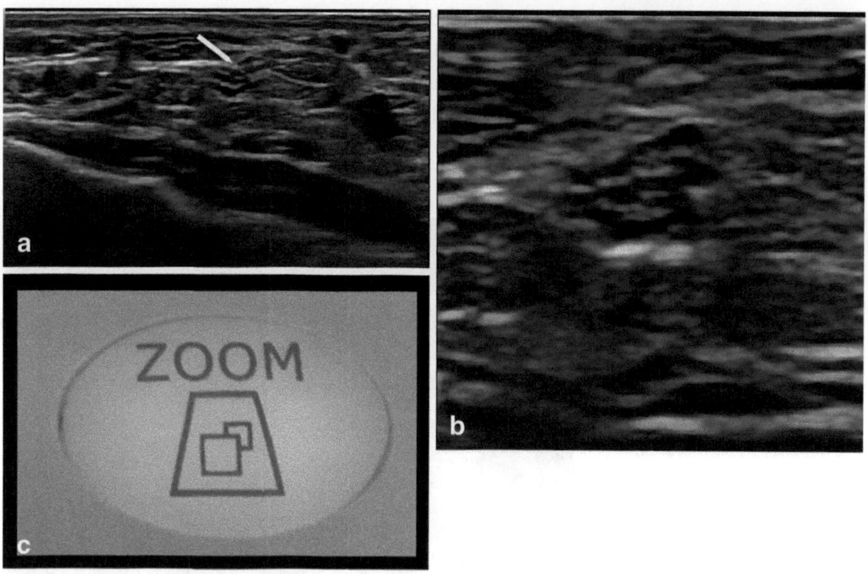

Fig. 1.21 Image magnification. **a** Non-enlarged image (*arrow* shows the region which is enlarged in *B*). **b** Magnified image. **c** Zoom button

eral information (date and time of the examination, transducer frequency and other data on the operating mode of the device, the institution where the examination is performed, etc.); and simultaneous display of multiple images in 2D ultrasound or in other different modes. Images are stored on sensitive papers or on devices with large capacity of digital memory. The technical features of the devices depend on the manufacturing company and are different from one machine to another [9, 10].

Terminology

Ultrasound, like other imaging methods, uses specific terminology due to ultrasound propagation and behavior in human body. The terms most frequently used are:

Anechoic or transonic is the term used for any structure that is entirely crossed by the ultrasound beam without any reflection on their route, resulting in a "lack of echoes," displayed as a black image on the monitor. This shows the fluid content of a structure, in most cases (Fig. 1.22).

Echoic or echogenic is the term used for the appearance of a structure that strongly reflects ultrasound, appearing white on the image. This can be produced in musculoskeletal ultrasound by connective structures (bone, calcifications) as well as other high-density structures (air, metal crystals, etc.; Fig. 1.23).

Hyper-, iso- , or hypoechoic are terms used when making comparisons between the intensities of echoes produced by the examined structure and the reference structure (hyperechoic—appears whiter, brighter than the reference structure; isoechoic—appears with the same echogenicity compared with the reference structure; hypoechoic—appears darker than the reference structure).

Impure liquid defines the aspect of a structure almost anechoic, containing inside scattered, fine echoes, often mobile under the pressure of the ultrasound beam (Fig. 1.24).

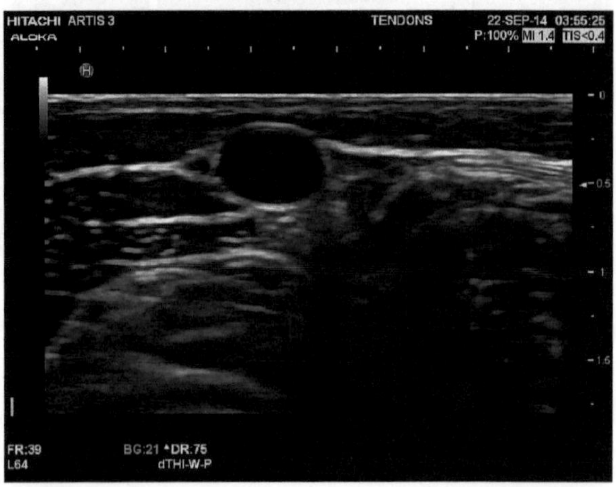

Fig. 1.22 Transonic image—a superficial vein

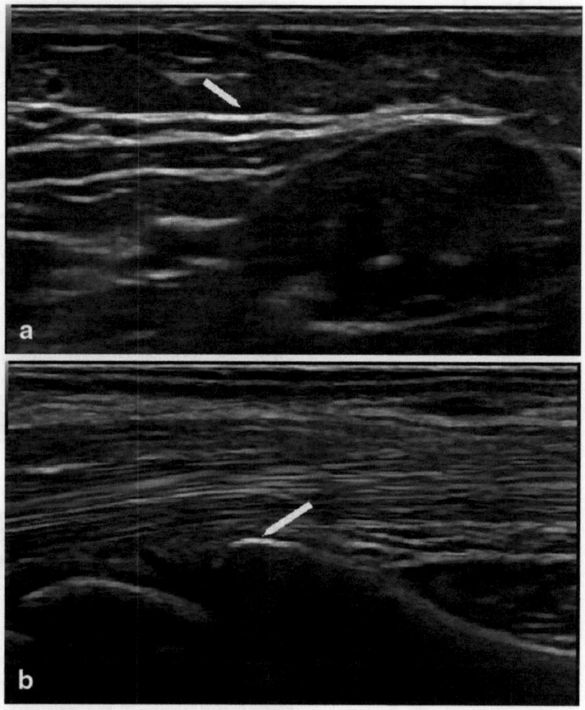

Fig. 1.23 Echoic structures—*arrows*. **a** Fascia. **b** Bone surface

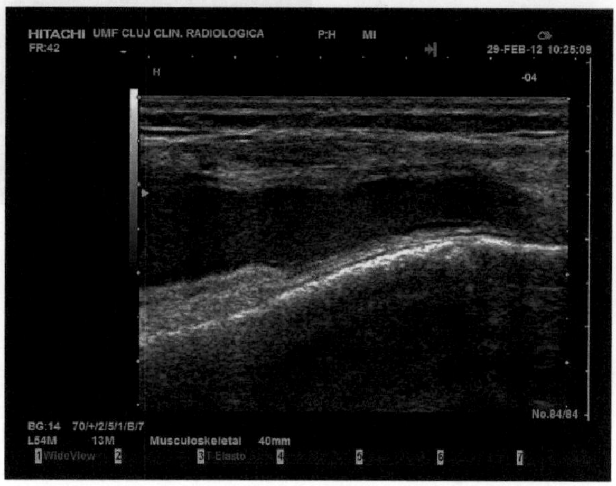

Fig. 1.24 Impure fluid—hemarthrosis on the suprapatellar recess

Echo structure defines all the echoes produced by solid-tissue elements constituted in a systematic structure that allows the identification and characterization of the examined elements (Fig. 1.25).

Sound attenuation defines the reduced intensity of echoes from the structures located deep, away from the transducer (Fig. 1.26).

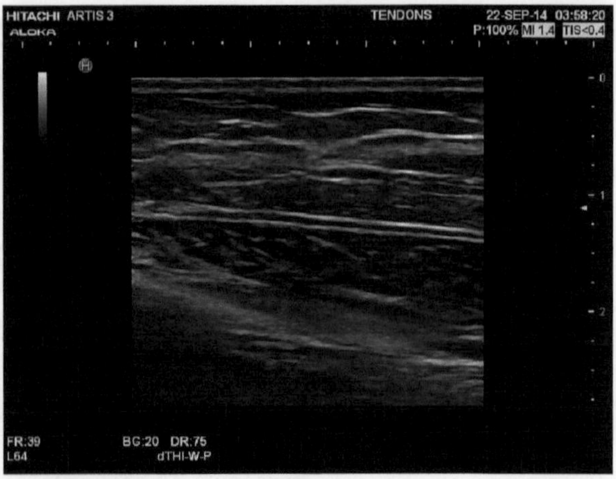

Fig. 1.25 Echo structure of the normal muscle—longitudinal view

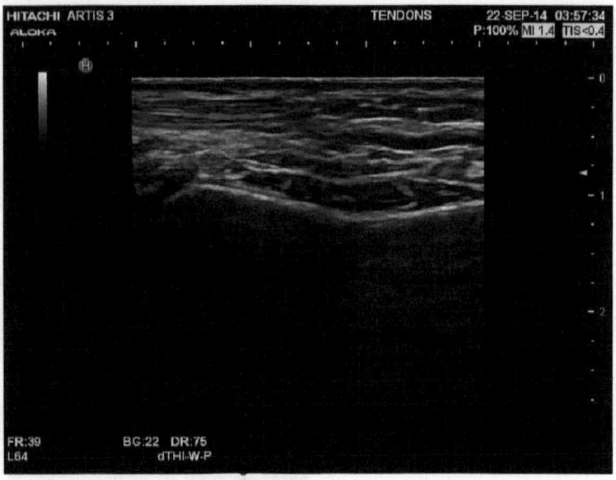

Fig. 1.26 Sound attenuation behind the bone surface—longitudinal view

Artifacts

Anisotropy: Anisotropy is an artifact produced by the linear configuration of tendons whereby hypoechoic change is seen if the transducer is slightly angulated. This artifact can mimic hypoechoic tendinopathy, but careful minor changes to transducer angulation make anisotropy disappear whereas true pathologic findings do not (Fig. 1.27). Anisotropy can be beneficial for confirming tendon position as the artifact can be produced in the linear tendon while the surrounding nonlinear echogenic fat is not affected; increasing contrast between the two structures [7] (Fig. 1.28).

Acoustic shadow defines important focal attenuation of ultrasound due to complete reflection. Distal to such structures ultrasound are no longer propagated, the appearance being of anechoic area "nonimage" (Fig. 1.29) that extends deep from

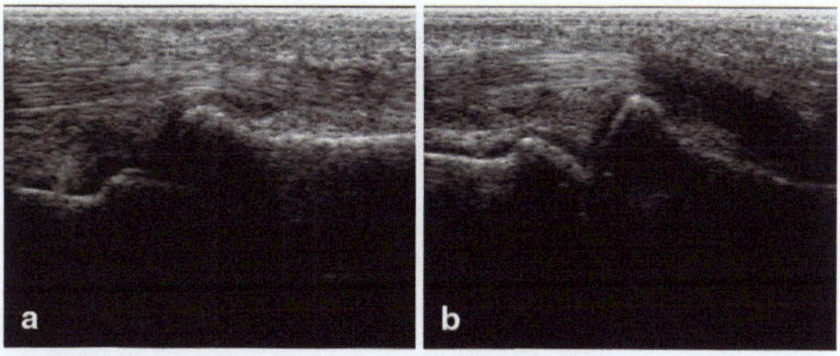

Fig. 1.27 **a** Tendon shows normal hyperechoic echopattern when the sound beam is perpendicular. **b** Anisotropy of the same tendon on probe angulation

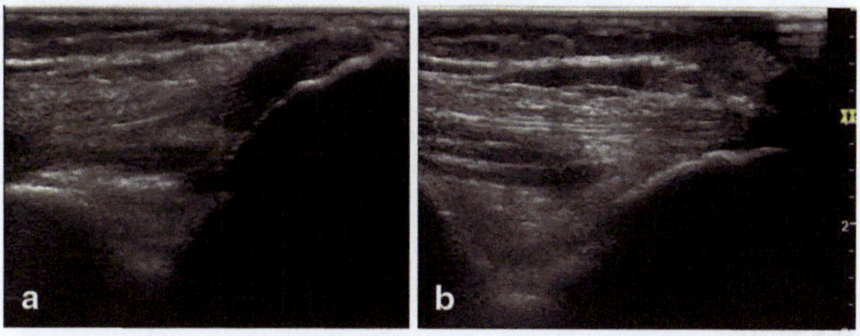

Fig. 1.28 Anisotropy of the triceps tendon, it shows a hypoechoic area at its distal end (**a**) which disappeared on correcting the probe angulation (**b**)

Fig. 1.29 Acoustic shadow extends posterior to intramuscular calcification (*arrow*)

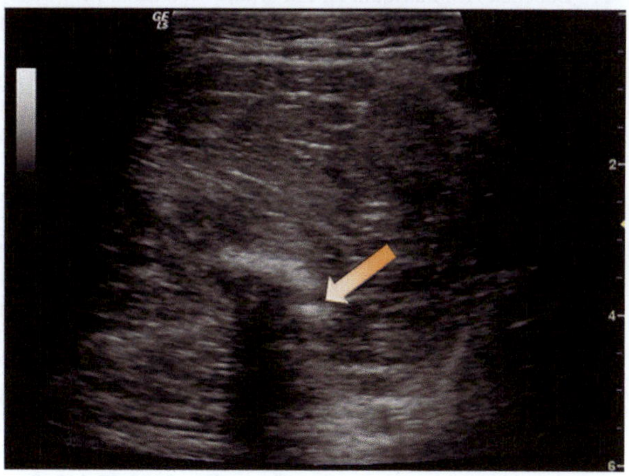

the involved interface. Examples of structures that produce shadowing include interfaces with bone or calcification and some foreign bodies

Acoustic enhancement (amplification) Apparent and selective increasing in the intensity of echoes beyond an anechoic structure [5, 6, 10]. (Commonly a cystic structure or blood vessel; Fig. 1.30) in these situations, the sound beam is less at-

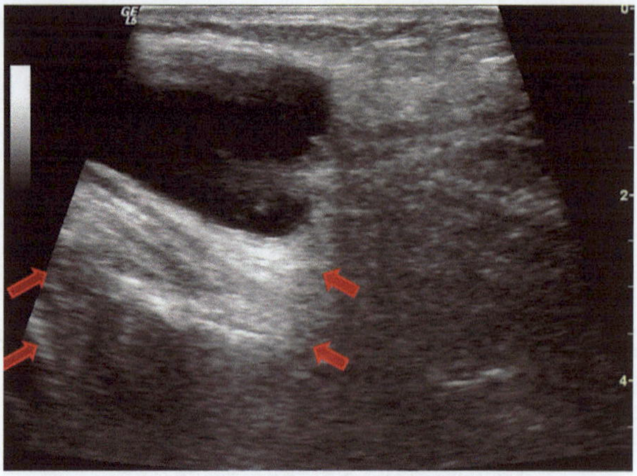

Fig. 1.30 Acoustic enhancement extends posterior to backer cyst (*arrow*)

tenuated compared with the adjacent tissues; therefore, the deeper soft tissues will appear relatively hyperechoic compared with the adjacent soft tissues

References

1. Ahmed R, Nazarian LN. Overview of musculoskeletal sonography. Ultrasound Q. 2010 Mar;26(1):27–35.
2. Babcock DS, Patriquin H, LaFortune M, Dauzat M. Power Doppler sonography: basic principles and clinical applications in children. Pediatr Radiol. 1996;26(2):109–15.
3. Bude RO, Rubin JM, Adler RS. Power versus conventional color Doppler sonography: comparison in the depiction of normal intrarenal vasculature. Radiology. 1994;192(3):777–80.
4. Colquhoun K, Alam A, Wilson D. Basic science: ultrasound. Curr Orthop. 2005;19:27–33.
5. Iagnocco A, Naredo E, Bijlsma JWJ. Becoming a musculoskeletal ultrasonographer. Best Pract Res Clin Rheumatol. 2013;27:271–81.
6. Joshua F. Ultrasound applications for the practicing rheumatologist. Best Pract Res Clin Rheumatol. 2012;26:853–67.
7. Robinson P. Sonography of common tendon injuries. Am J Rev. 2009;193:607–18.
8. Rubin JM, Bude RO, Carson PL, Bree RL, Adler RS. Power Doppler US: a potentially useful alternative to mean frequency-based color Doppler US. Radiology. 1994;190(3):853–6.
9. Schmidt WA. What the practising rheumatologist needs to know about the technical fundamentals of ultrasonography. Best Pract Res Clin Rheumatol. 2008;22(6):981–99.
10. Smith J, Finnoff JT. Diagnostic and interventional musculoskeletal ultrasound: part 1. Fundamentals. Clinical reviews: Current concepts. Phys Med Rehabil. 2009;1(1):64–75.
11. Torp-Pedersen ST, Terslev L. Settings and artefacts relevant in colour/power doppler ultrasound in rheumatology. Ann Rheum Dis. 2008;67:143–9.

Chapter 2
Pitfalls in Musculoskeletal Ultrasound

Violeta Maria Vlad MD, PhD

Introduction

Taking a good ultrasound (US) picture is an art. Interpreting it is a science. This is in fact everything US is about. Anatomic details should be carefully studied at the very beginning of US formation. However, good anatomy knowledge is not sufficient. The passage of US beam through human tissues, though so rewarding, is not without negative technical consequences. US images are the final result of multiple reflection processes combined with the machine technical capabilities and the knowledge of the performing physician. Any deviation of one of these three cornerstones leads to pitfalls—acknowledged as "difficulties/ traps"—and possible mistakes in images interpretation. Most of the pitfalls in US are represented by artifacts.

Oxford English dictionary [1] defines artifacts as "something observed in a scientific investigation or experiment that is not naturally present but occurs as a result of preparative or investigative procedure." Artifacts have been described since US was used for the first time as a diagnostic method in medical practice. The first musculoskeletal US report, published by T. Dussik in 1958 [2], led to the first description of anisotropy. Since then, the technique has known an exponential development and, consequently, a lot of artifacts have been described in literature, being defined, illustrated, and classified according to various criteria [3–13].

The first step for understanding artifacts is the admittance of the basic assumptions regarding the way an US image is obtained [3]:

1. Sound travels in straight lines.
2. Reflections occur from structures along the central axis of the beam.
3. Intensity of reflection corresponds to the reflector scattering strength.

V. M. Vlad (✉)
Department of Rheumatology, Sf. Maria Hospital, Bucharest, Romania
e-mail: vladvioleta1@gmail.com

Y. El Miedany (ed.), *Musculoskeletal Ultrasonography in Rheumatic Diseases*,
DOI 10.1007/978-3-319-15723-8_2

Table 2.1 Classification of US pitfalls

1. *Pitfalls due to GS artifacts*	Due to transducer position	Anisotropy Bone pseudodefect	–
	Due to interaction of US beam to the tissues	Attenuation	Shadowing
			Reverberation
			Mirror
			Side lobe/grating lobe
			Cartilage interface
		Enhancement through transmission	–
		Associated with velocity errors	
2. *Pitfalls due to Doppler artifacts*	Mirror	–	
	Reverberation		
3. *Pitfalls due to human error (GS and Doppler)*	Drop out		
	Machine incorrect adjustment		
	Transducer incorrect handling		
	Insufficient anatomy knowledge		
4. *Pitfalls due to machine error (GS and Doppler)*	Piezoelectric crystals damage		
	Transducer's fibers damage		

US ultrasound, *GS* grayscale

4. Sound travels at exactly 1540 m/s.
5. Sound travels directly to the reflector and back.

If any of these conditions is impaired, artifacts are generated.

This chapter focuses on the detailed description of all pitfalls that could lead to wrong acquisition/ interpretation of US images. As artifacts are connected to technical issues, it could be emphasized that pitfalls include artifacts and other possible sources of misinterpretation connected to human mistakes or machine inaccuracy. Both grayscale (GS) and Doppler pitfalls could be due to the technique itself, the particular type of machine used in the exploration, or the insufficient anatomic/ technical knowledge of the physician performing it. The chapter covers separately GS and Doppler artifacts, and then covers together the pitfalls due to the machine and human errors.

The classification of US pitfalls is given in Table 2.1.

Pitfalls Due to GS Artifacts

Artifacts Due to Transducer Position

Anisotropy, according to Wikipedia [14], is the property of being directionally dependent, as opposed to isotropy, which implies identical properties in all directions. US, especially musculoskeletal, is an anisotropic technique—in abdominal US the only tissue exhibiting anisotropy is the renal tissue [7]. Anisotropy is an artifact that appears in all the situations where the US beam is not strictly perpendicular to the examined tissue for ensuring maximal reflectivity. Anisotropy appears mostly in tendons; muscles and ligaments can exhibit it at a lesser extent. Nerves do not show anisotropic properties [9]. Any tissue containing parallel linear fibers can exhibit anisotropy. Due to highly organized structure of collagen fibers inside tendons, these are highly anisotropic structures. A tendon becomes anisotropic mainly at the point of its insertion where it follows the round shape of the bone epiphysis. Any angle deviation from 90° decreases the tendon echogenicity, making it resembling the muscle in appearance and obscures structural details; at higher deviation angles, the echogenicity will decrease even more [9, 11]. The presence of a hypo-echoic area near a tendon's insertion must lead to anisotropy considering before making a tendon tear diagnosis. The most important places to encounter anisotropy are: bicipital tendon inside the groove, quadriceps, and Achilles tendons at their insertion. Patellar tendon (also known as patellar ligament) has a straight direction, so it is usually visualized without anisotropy at both insertions (Figs. 2.1, 2.2, 2.3, 2.4, and 2.5).

Tips to Overcome Gently tilting the transducer until it becomes perpendicular to the structure makes anisotropy disappear, and the tendon's fibers appear intact. Sometimes it is enough to slightly move the patient's joint, like in hand flexors

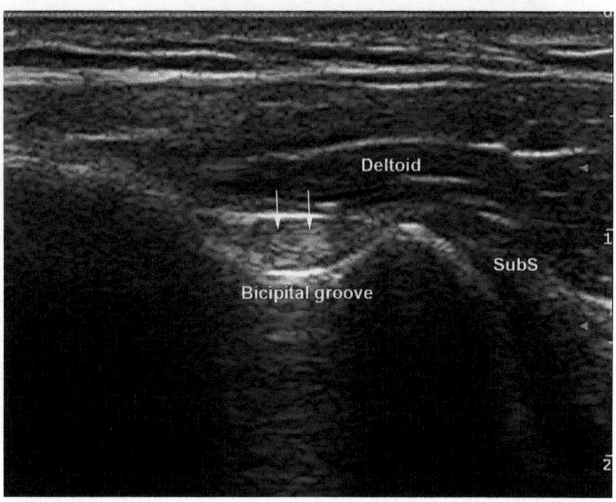

Fig. 2.1 Transverse view of bicipital tendon (*arrows*) inside the groove. The transducer is perpendicular to the tendon. *SubS* subscapularis tendon

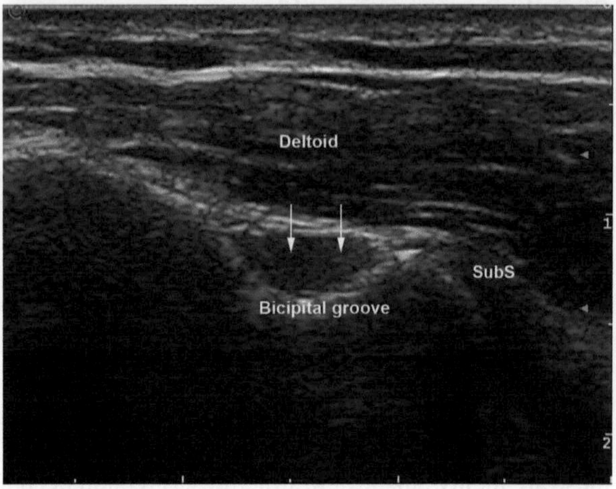

Fig. 2.2 Transverse view of "empty" bicipital groove (*arrows*). The transducer is not perpendicular to the tendon. *SubS* subscapularis tendon

evaluation. Knee flexion also makes anisotropy of quadriceps insertion disappear. Tendon tears should always be visualized in two perpendicular planes and they remain unchanged when tilting the probe. Dynamic evaluation of the tendon's movements can also help ruling out tears.

Bone pseudodefect is another artifact linked to the position of the probe in relation to the bone contour. When the beam touches the bone in a semi tangential way, the reflected echoes will have a different direction compared to the normal return and will not be detected on the screen. There will be no captured image in that area, so it will falsely look like a break in the bone contour. This artifact must be carefully differentiated from various pathologic cortical defects, which appear on two

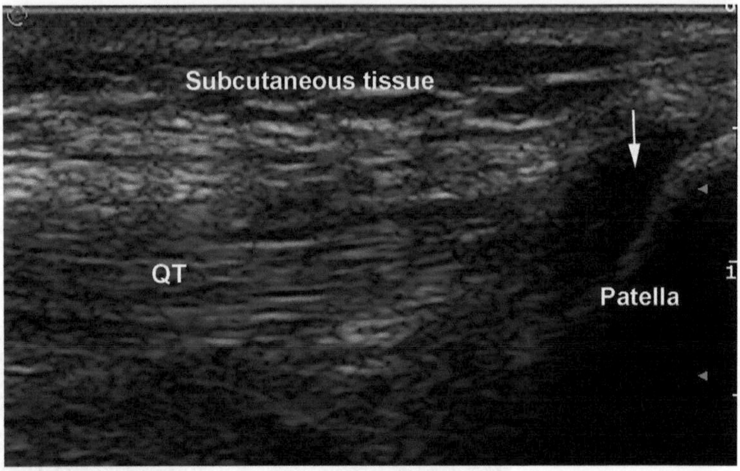

Fig. 2.3 Longitudinal view of quadriceps insertion on patella. The transducer is not parallel to the fibers at the insertion. *QT* quadricipital tendon. *Arrow:* anisotropy at insertion

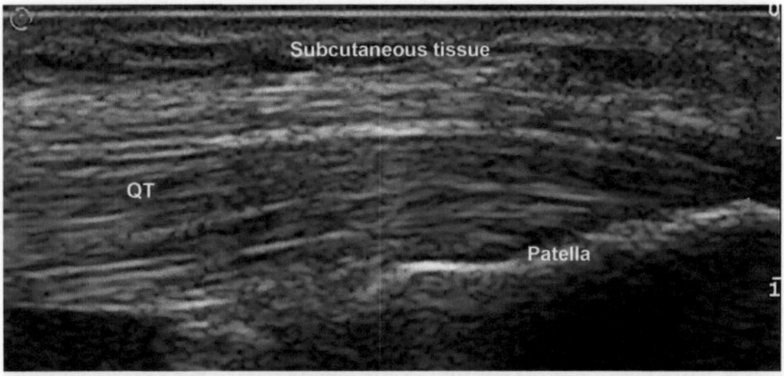

Fig. 2.4 Longitudinal view of quadriceps insertion on patella. When angulating the transducer it becomes parallel to the fibers at insertion and anisotropy disappears. *QT* quadricipital tendon

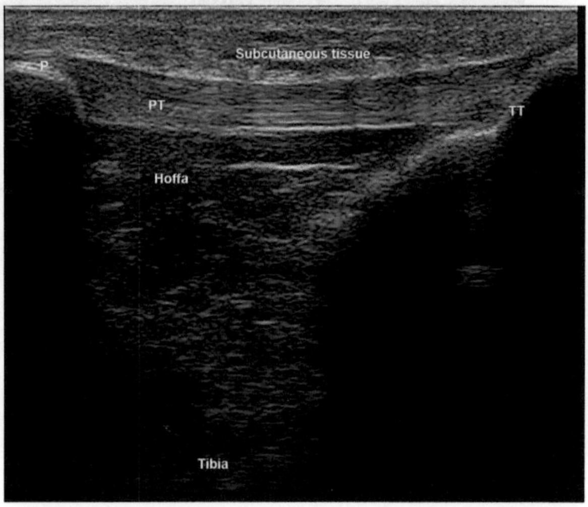

Fig. 2.5 Longitudinal view of patellar tendon. Due to its straight trajectory, no anisotropy appears at insertions. *P* patella, *TT* tibial tuberosity, *PT* patellar tendon

perpendicular planes and usually have a floor that may be visualized on the US image. Bone pseudodefect is frequently visible in the elbow area, at the coronoid fossa evaluation (Fig. 2.6).

Tips to Overcome Same as for anisotropy—gently tilting the probe for changing the angle of the beam toward the bone will make the defect disappear.

Artifacts Due to Interaction of Beam with Tissues

US beam interacts with tissues according to their specific characteristics. Generally, the waves are attenuated inside tissues because of reflection, scattering, refraction, or absorption processes, but sometimes their transmission might be enhanced. US

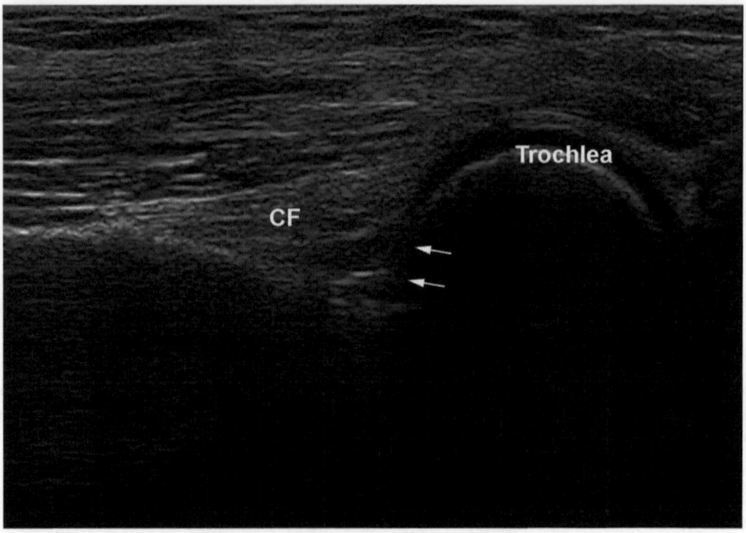

Fig. 2.6 Longitudinal view of the anterior elbow. Bone pseudodefect artifact is seen because of the transducer position, semitangential to the bone (*arrows*). *CF* coronoid fossa

Table 2.2 Attenuation coefficients and US speed for selected tissues at 1 MHz

Material	Attenuation coefficient	US speed (c)m/s
Water	0.0002	–
Air	40	330
Fat	0.5–1.8	1450
Soft tissue	0.3–0.8	1540
Bone	13–26	4080

US ultrasound

attenuation is 80% due to absorption, a small reminder of the beam being reduced by reflection, refraction, diffraction, and dispersion [15]. Different tissues have different attenuation coefficients (calculated at 1 MHz). The values are listed in Table 2.2 [4, 16].

Attenuation Artifacts

A brief description of the physical phenomena leading to beam attenuation would increase their understanding.

Reflection of the beam appears at the boundary between two types of tissues with different densities; as a consequence, the propagation speed of waves will be different too. The property of tissues that involves density and speed of waves is called acoustic impedance, marked as Z (Z=speed×density of the tissue). The final US image is the result of multiple reflection processes. At the boundary between two

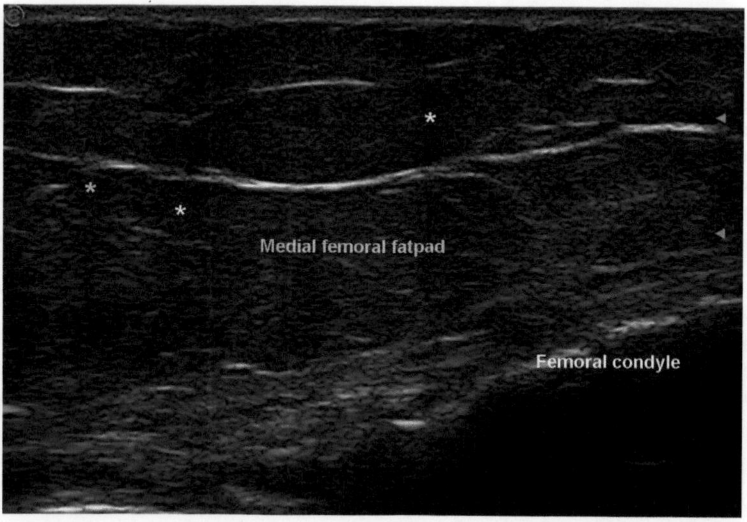

Fig. 2.7 Longitudinal view of the medial fat pad of the knee in an obese patient. Scattering and diffusion of US create shadows (*asterisks*) and allow a fair visualization of the structures below (femoral condyle). *US* ultrasound

tissues with highly different acoustic impedance, the reflection of the waves will be strong, so a strong echo will be generated. This situation is particularly encountered at the interface between air/bone and soft tissues. At the interface between soft tissue and air, 99 % of the waves are reflected, so no image will appear behind this. A particular situation illustrating this is the presence of an air gap between the transducer and the skin. To avoid that, a thick, uniform layer of gel must be applied on the transducer. Air is, though, a strong attenuator of the beam. Other strong attenuators are bone and calcifications. Behind these structures the attenuation will be so high that a shadow will appear, suggesting US waves do not penetrate that structure because they are all reflected. If the adjacent tissues have similar acoustic impedance, no echoes are generated. Most of the soft tissues have close values of acoustic impedance, so very few waves are reflected, most of the beam passing further to deeper structures [10].

Refraction is a change of US beam direction at interfaces between media with small differences in acoustic impedance (fat/muscle). Because of this phenomenon, real structures may appear in false locations. The degree of direction changing is influenced by the difference of impedance and also by the angle of the beam, which may easily be adjusted by tilting the transducer.

Scattering of the beam occurs when the reflecting surface is very small compared to the wavelength of US, and echoes are reflected through a wide range of angles, consequently reducing their detected intensity. When the beam encounters a number of small interfaces, comparable to the wavelength, a particular type of scattering appears, called *diffusion*. This is the situation of the fat or fibrous tissue (Fig. 2.7). Scattering and diffusion of the waves must be considered for correctly interpreting the echogenicity of small parts nodules.

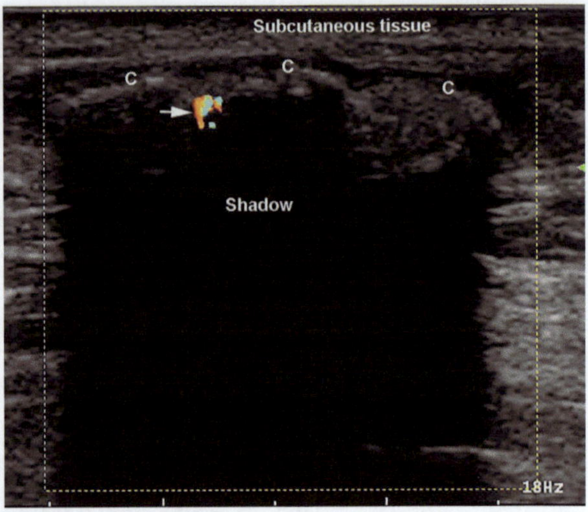

Fig. 2.8 Longitudinal view of the thigh of a young patient with juvenile dermatomyositis. A compact layer of calcifications are seen just under the skin. A "clean" shadow is covering all structures behind. The inflammatory aspect of calcifications, which were painful, is shown by the incomplete mirror Doppler artifact—the only two parts seen in the image are the bone and the mirror (*arrow*). *C* calcium deposits

Shadowing

Shadowing is the lack of US image behind structures like air or bone. Because of this artifact, bones and organs containing air like digestive tube cannot be visualized with US. Different types of shadows are produced by air and by bone—while shadow behind the bony structures and calcifications is uniform and homogeneous, called "clean," shadow behind the air is called "dirty." Dirty shadow has a heterogeneous aspect, with mixed areas of hypo- and anechogenicity. Sometimes dirty shadow may appear behind bony structures with a large radius of curvature, like humeral head [5]. A particular example of shadowing is represented by skin calcifications in dermatomyositis or scleroderma patients—the continuity of calcium deposits forms a continuous line immediately under the skin with no echoes behind it (Figs. 2.8 and 2.9). When the shadow is generated by a superficial structure, it may run over a deeper structure in the image and mimic pathology like tendon tear. A shadow may appear behind wood fragments as foreign bodies in subcutaneous tissue or deeper [5].

When a strong reflecting structure like a calcification is very small compared to the width of the main US beam, the posterior acoustic shadowing will be eliminated. This situation may occur in case of small calcifications inside a fluid collection. Awareness of this artifact, called *beam width,* will ensure proper diagnosis of calcium deposits, even without shadow (Figs. 2.10 and 2.11). Sometimes focus adjustment strictly in the area of interest will remove the artifact [5].

At the margins of a structure with different acoustic impedance compared to the tissue around it and with a highly curved surface (tendons' margins, cyst, etc.), an artifact called *lateral shadowing* (or refraction shadowing) appears. After reflection of the beam, marginal waves get also refracted at the edges of the structure, therefore no sound beam is returning to the probe; as a consequence, a shadow near

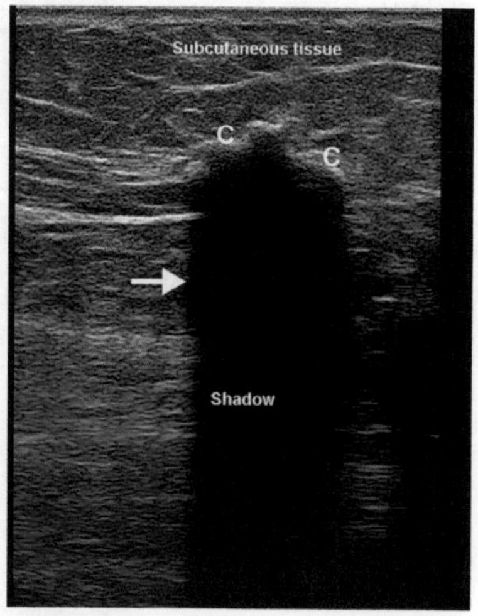

Fig. 2.9 Transverse view of the gluteal region in a scleroderma patient. A big calcification (*C*) is seen in the subcutaneous tissue. A "clean" shadow appears behind (*arrow*)

each lateral border of the structure may appear (Fig. 2.12). A particular case of refraction at the edges is represented by superficial varicosities (Fig. 2.13). Refractile shadowing also appears in case of tendon full-thickness tears, and sometimes it might be the only US sign of the tear.

Tips to Overcome Shadowing cannot be overcome, but it is of great help in interpreting US pathology. However, this type of artifact may look like another pitfall, caused by machine damage, presented below, so careful distinction is needed.

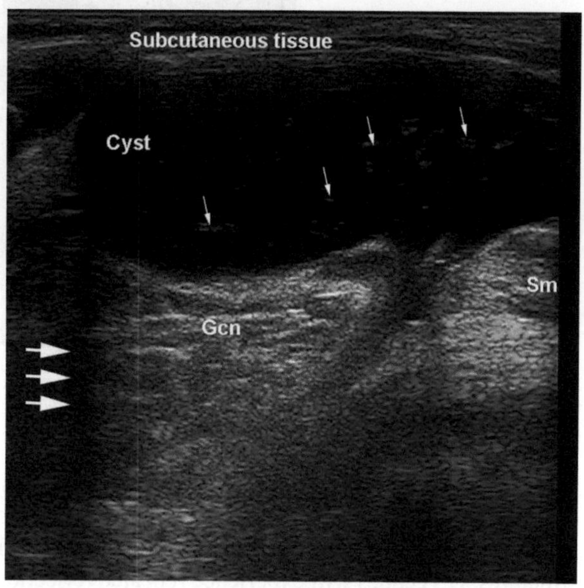

Fig. 2.10 Transverse view of the medial part of popliteal fossa. A Baker cyst is seen with small white deposits inside without shadow—calcifications exhibiting beam width artifact (*small arrows*). Also, a lateral shadowing phenomenon appears at the cyst margin (*large arrows*). *Gcn* gastrocnemius muscle and tendon, *Sm* semimembranosus tendon

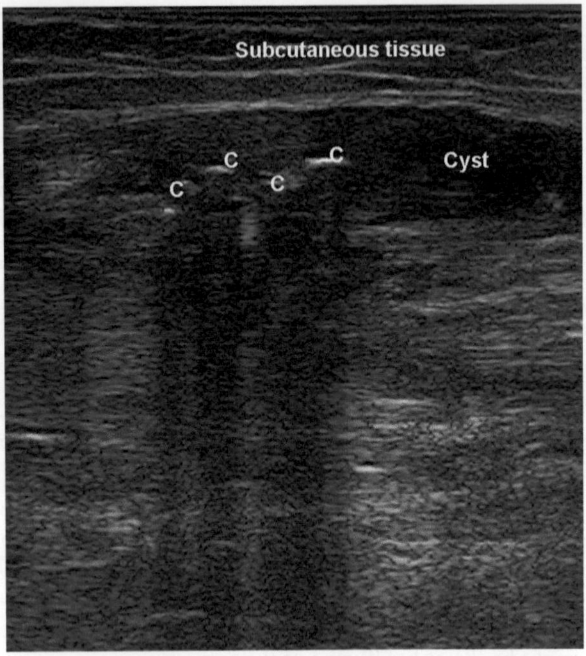

Fig. 2.11 Calcifications (*C*) in a small Baker's cyst with mixed echogenicity. Each calcification, though very small, has its own separate shadow, covering tissues behind

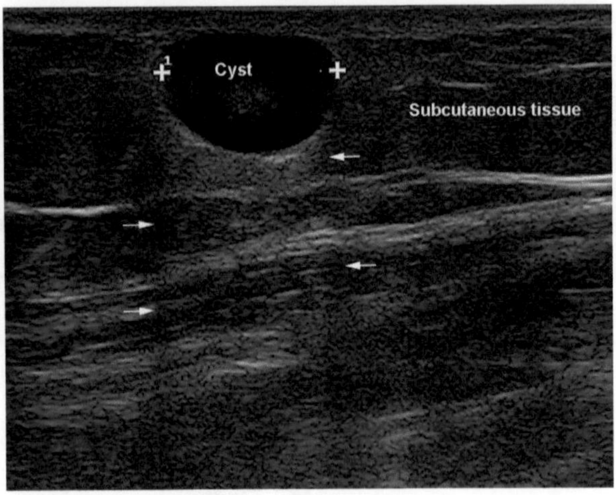

Fig. 2.12 Subcutaneous cyst, showing lateral shadows due to refraction (*arrows*)

Reverberation

Reverberation is an attenuation artifact generated by structures with two parallel surfaces, both highly reflective. Examples may be the prostheses used by orthopedists (shoulder, knee, hip) or a needle/biopsy device (Figs. 2.14 and 2.15). In such structures, US echoes will experience repeated reflections between the two surfaces

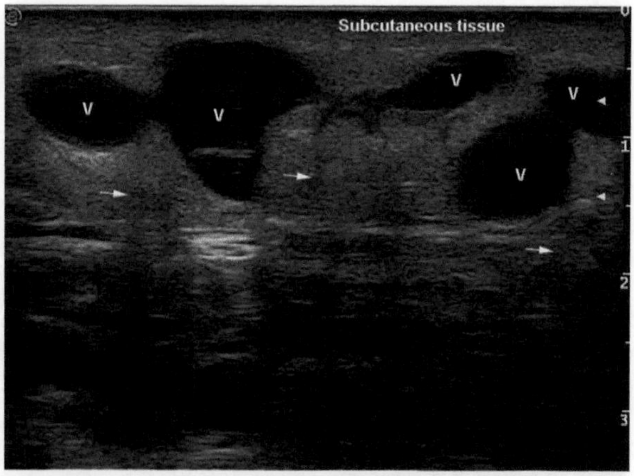

Fig. 2.13 Superficial varicosities (*v*) in subcutaneous tissue showing lateral shadowing (*arrows*)

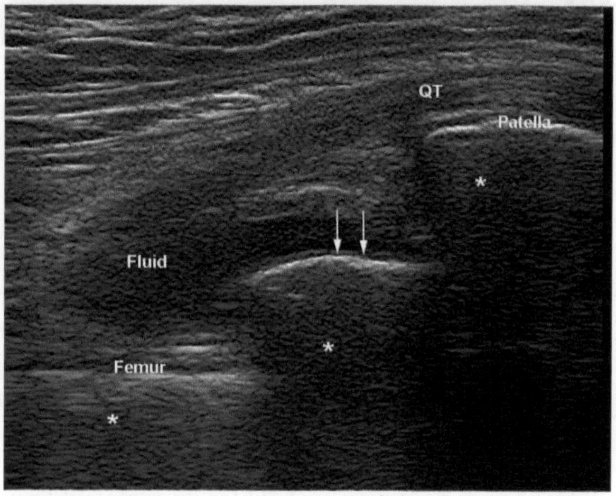

Fig. 2.14 Longitudinal view of the suprapatellar recess of a knee with prosthesis. Between patella and femoral cortical lines, another hyperechoic line is depicted—the prosthesis image (*arrows*). All three lines have reverberation artifacts behind them (*asterisks*). *QT* quadricipital tendon

of the prosthesis and the bone or between the two facets of the needle before going back to the transducer. Because of this mechanism, only the first echo will be properly located on the screen; the others will be displayed far away, and will appear deep in the image. Finally, the image on the screen will consist of multiple linear echoes spaced at equal distance, posterior to the first echo. Each echo will be weaker than the precedent one, due to attenuation. Bone exhibits reverberation artifact too; very often, bone contour appears again on the screen below the real one. Reverberation is also encountered in gynecologic US [17].

The dirty shadowing behind air bubbles is a particular type of reverberation. An example is the image representing a corticoid injection inside a cyst. Together with the corticoid, air from the syringe is injected, and it immediately rises in the upper part of the cyst, leaving a dirty shadow/reverberation artifact behind it.

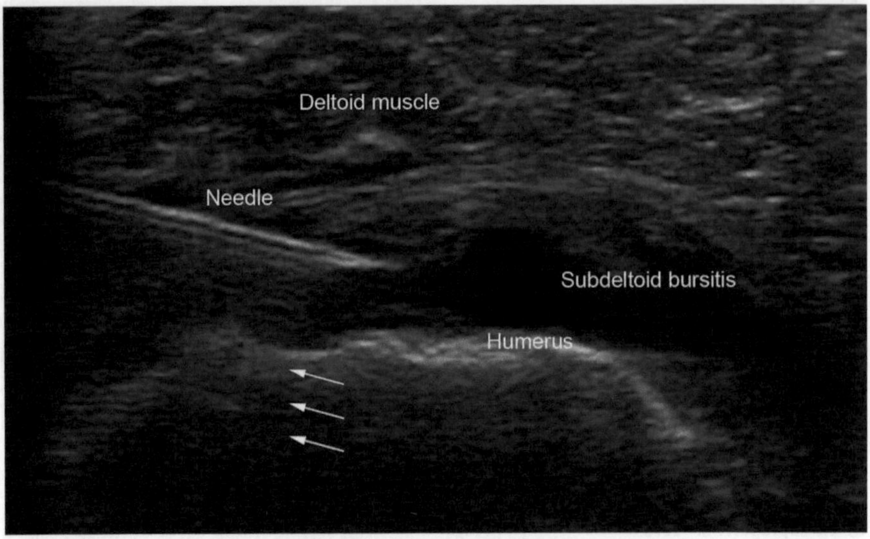

Fig. 2.15 Transverse view on the anterior shoulder. An anechoic image (subdeltoid bursitis) is shown under deltoid muscle with a needle inserted in it. Reverberation artifact of the needle is shown behind it (*arrows*)

A particular type of reverberation is *the comet tail artifact,* which accompanies the intrusion of a foreign object with very different acoustic impedance compared to the tissue around it. Comet tail artifact can appear behind a metallic prosthesis or behind glass. It has a characteristic shape, widening distally or proximally. Comet tail artifact is useful when marking the place of an injection by placing a needle between the skin and the transducer (Fig. 2.16). The artifact will indicate the precise location for the injection.

In the same category, *the ring-down artifact* appears when a very thin layer of fluid is trapped between air bubbles. In some publications, these last two types of artifacts are considered totally different [4], while in others they are considered the same [16].

Reverberation artifact may appear on the screen before even starting US examination because of the intense reflection of waves at the interface between the transducer and air. This artifact disappears when the transducer is in direct contact with the skin through a layer of gel (Fig. 2.17).

Tips to Overcome The last type is the only reverberation artifact that can be overcome. The others cannot be eliminated, as they are caused by the type of tissue involved. Sometimes this type of artifacts can be helpful, like in case of the characteristic appearance of air/gas present together. If not intraluminal, the aspect combining air and gas might be a characteristic US sign of an abscess collection [5]. Based only on reverberation artifacts, a whole new US area has developed—pulmonary US (details are given below). New type of machines with harmonic imaging can reduce a little reverberation artifacts [9].

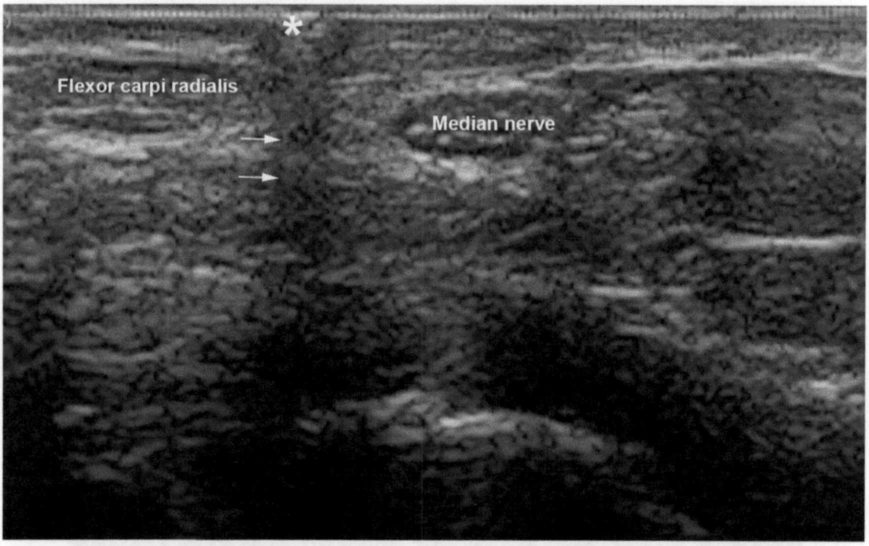

Fig. 2.16 Transverse view of the median nerve at the wrist. A needle (*asterisk*) is placed on the skin under the transducer. A comet tail artifact (*arrows*) is seen below the needle, indicating the injection place

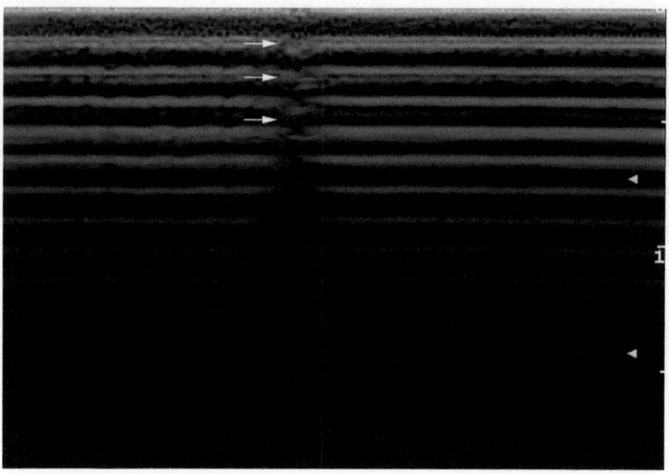

Fig. 2.17 Reverberation artifact before starting US examination. Many hyperechoic parallel lines equally spaced are seen. There is no gel on the transducer, except for a very small part, in the middle, which leaves a trace on the whole image (*arrows*). *US* ultrasound

Mirror Effect

A mirror-like image appears when the US beam is coming across a highly reflective and curved shape (curved margins of the long bones in longitudinal view). The main

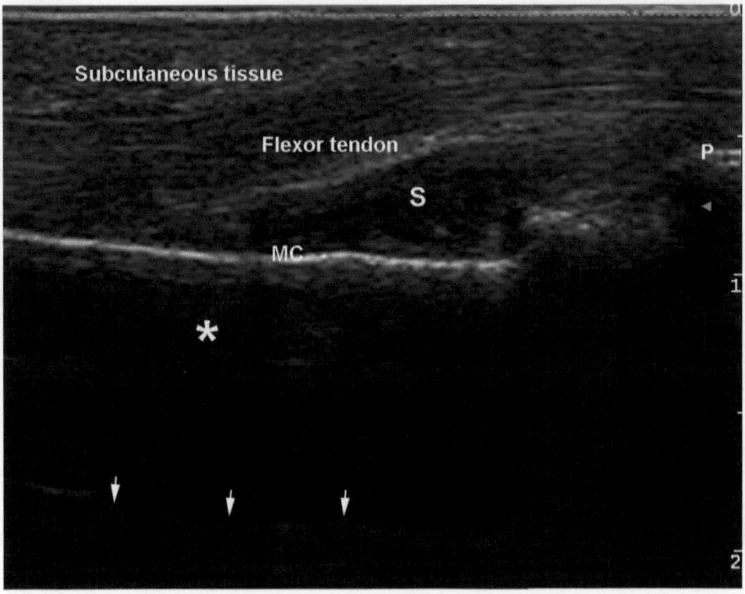

Fig. 2.18 Longitudinal view of a metacarpophalangeal joint volar side. A mirror image (*asterisks*) of the flexor tendon is seen below the bone contour. A mirror image of the bone contour is seen, too (*arrows*). *MC* metacarpal bone, *P* proximal phalanx, *S* synovial hypertrophy inside the joint

beam passes toward a deep reflecting interface, but after reflection, echoes from the deep structure first hit a more superficial interface, which is the other side of the same structure, then return to the profound one and then back to the transducer. As a result, they will be recorded later in time and will be misplaced on the screen [4]. In the final US image, the same structures will appear on both sides of the interface. The extra line appearing below bone contour could be interpreted as mirror or reverberation (Fig. 2.18). The mirror artifact is encountered in abdominal US too, where sometimes a tumor appears on both sides of a hepatic lobe, creating confusion.

Tips to Overcome It cannot be overcome, but it is essential to keep in mind that no structure should be interpreted below the bone hyperechoic sharp line.

Side-Lobe Artifact

Apart from the main beam, at the edges of the transducer, adjacent beams are generated from a radial expansion of the piezoelectric crystals inside its structure. These will have different directions for tissues penetration. Although they are usually weak, some of them manage to actually come back to the transducer (mainly if they encounter an enough reflecting interface), and furthermore can generate echoes. This is the artifact called *side lobe.* The structures appearing on the screen will then have a false position. Side-lobe artifact appears only with the use of linear transducers.

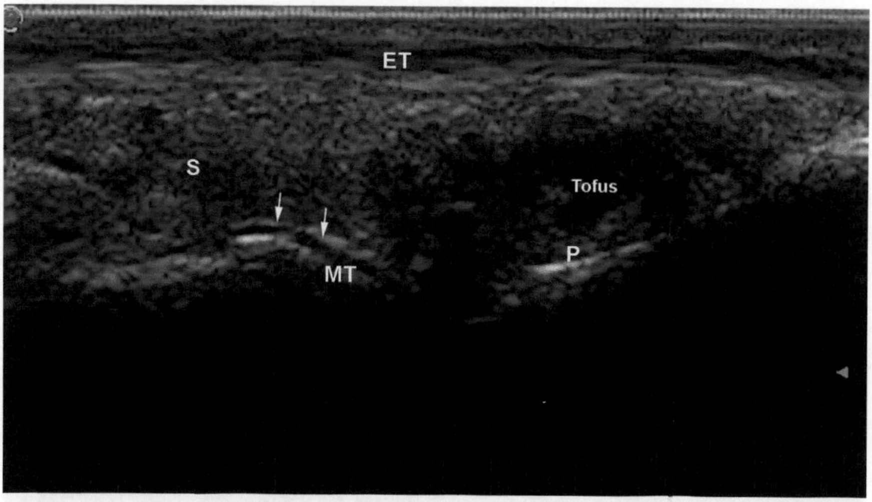

Fig. 2.19 Longitudinal view of a metatarsophalangeal joint. A double contour sign (*arrows*) suggesting urate deposit appears. A poorly delineated structure of mixed echogenicity is seen under the extensor tendon (*ET*), in the distal part of the joint, above the phalanx (*P*), interpreted as tophus. Synovitis is situated in the proximal part of joint (*S*).

Apart from that, some waves might have a much different direction than the main beam and still meet a reflective interface, this is called *grating-lobes* artifact. The falsely depicted structure may overlap an expected anechoic surface [6, 8].

Tips to Overcome The new US machines benefit from apodization, a technical characteristic that decreases the excitability of the crystals near the edge of the transducer, therefore reducing side-lobes and grating-lobes artifacts [8].

Cartilage Interface Sign

Also a consequence of multiple reflections at the interface between two distinct tissues with different acoustic impedances, the cartilage interface sign appears always accompanied by intra-articular fluid. Normally, the cartilage has an anechoic appearance without an upper hyperechoic limit. In contact with the fluid, a hyperechoic line can become visible at the interface. The "double contour" line must be differentiated from the urate deposition in gout, also a double contour appearance (Figs. 2.19 and 2.20). Interface cartilage sign is useful in differentiating calcifications inside fluid from osteophytes; these can appear like double lines due to the fact that they are covered by cartilage, like the rest of the bone (Figs. 2.21 and 2.22).

Tips to Overcome The artifact disappears when fluid is compressed with the probe, while the urate line remains unchanged during all kinds of joint movements. Gout also has other US-specific signs like the presence of tophi inside/outside of joint space.

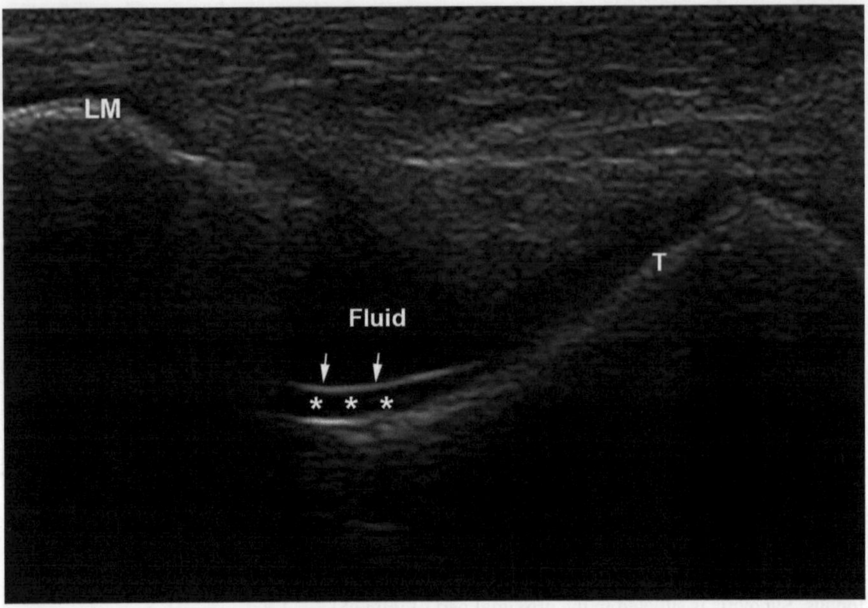

Fig. 2.20 Lateral view of tibiotalar joint with a little anechoic structure inside (possible fluid). A cartilage interface sign (*arrows*) is shown at the interface between the cartilage (*asterisks*) and the fluid. *T* talus bone, *LM* lateral malleola

Fig. 2.21 Longitudinal view of the anterior knee. A synovial hypertrophy image (*S*) is seen in the suprapatellar recess with a big calcification inside (*C*) with posterior shadow, covering the bone contour behind. *QT* quadriceps tendon, *Pat* patella

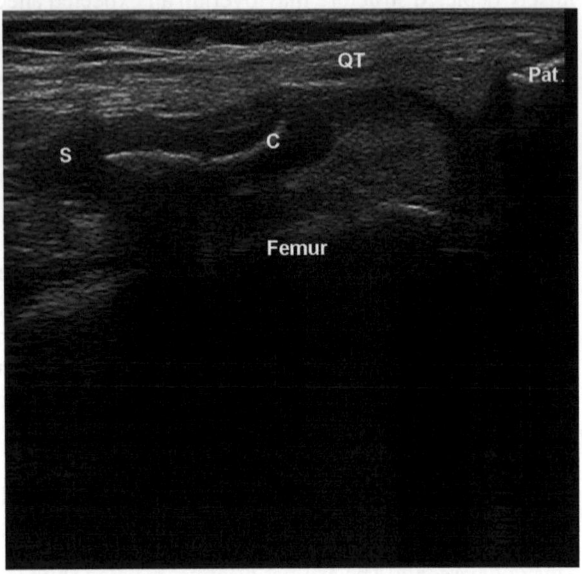

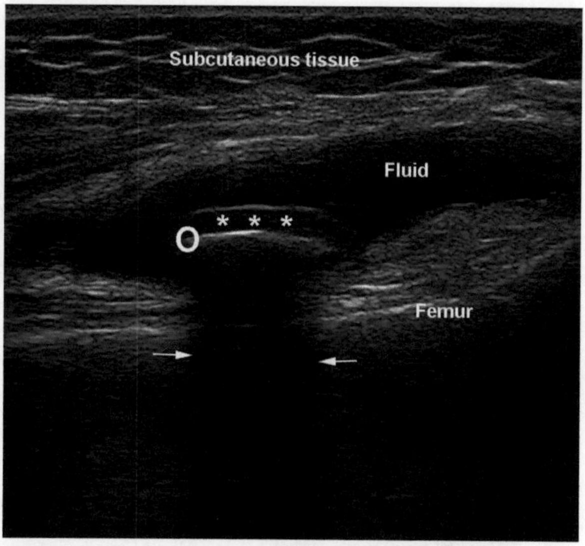

Fig. 2.22 Transverse view of the peripatellar recess of the knee. Inside knee collection (fluid), a hyperechoic line is seen with another line above it, representing the interface cartilage sign in an osteophyte (*O*). Differential diagnosis with a calcification—osteophytes are covered with cartilage (*asterisks*). A shadow is noticeable behind it (*arrows*)

Enhanced Through Transmission Artifact

Posterior enhancement appears when the beam is passing through a tissue with a low rate of attenuation, like the fluid. The echographic appearance of the fluid is black (anechoic), because fluid exhibits no reflection of the beam—as a result, the beam passes entirely through it. Behind the fluid collection, an area of higher brightness compared to the rest of the tissue is clearly depicted. Behind cystic lesions, this phenomenon is very clear (Fig. 2.23). The enhancement appears even in case of small fluid collections (Fig. 2.24). This artifact is actually a consequence of "time gain compensation (TGC)," a technical facility of US machines, which is meant to enhance the echoes coming from deeper structures based on a curve that increases exponentially with time [7]. This technical adjustment was necessary because the assumption that all tissues have the same degree of US attenuation is actually not true in real life.

Tips to Overcome Adjust the knobs from TGC panel as indicated in the drawing on the machine and in Fig. 2.25.

Artifacts Associated with Velocity Errors

Processing of an US image is based on the assumption that the sound travels inside the tissues of human body with a constant speed of 1540 m/s. In reality, the velocity of sound waves in different tissues is not the same [4, 15]. In tissues containing air or fat, the velocity of US is lower than the assumed one. As a consequence, the echoes' way back to the transducer will take longer and furthermore they will be displayed deeper on the image than they actually are. A fatty nodule or a cyst may have this effect. The different velocities inside human tissues are in Table 2.2.

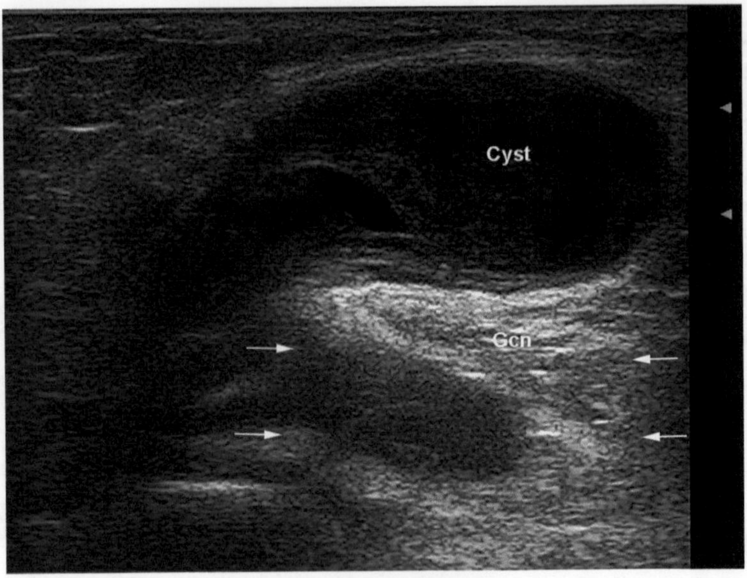

Fig. 2.23 Transverse view of a Baker cyst. Posterior enhancement artifact as brighter area behind it is shown (between *arrows*). *Gcn* gastrocnemius muscle and tendon

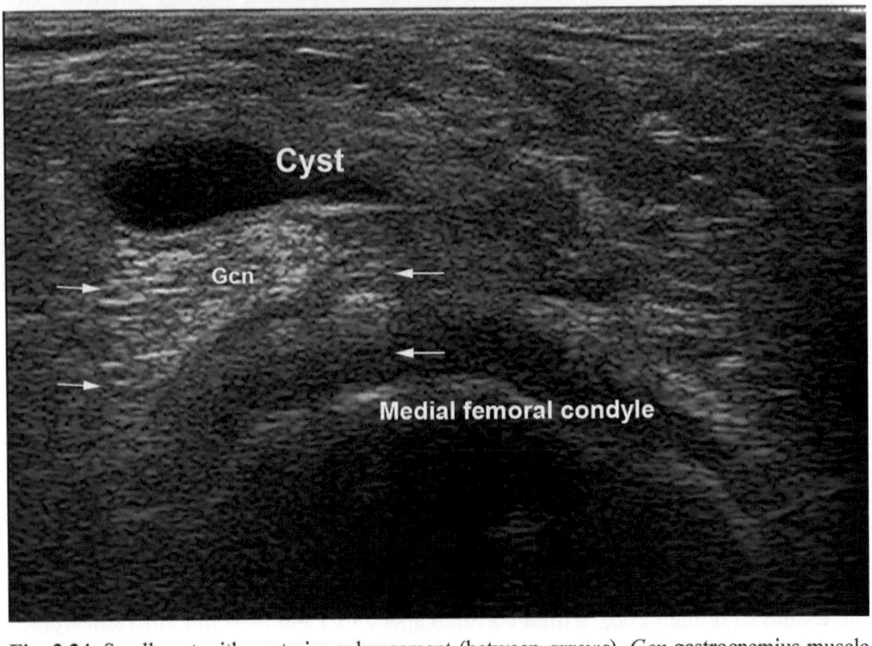

Fig. 2.24 Small cyst with posterior enhancement (between *arrows*). *Gcn* gastrocnemius muscle and tendon

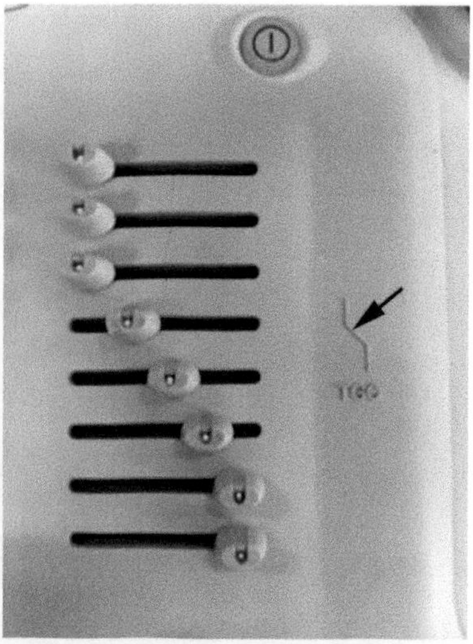

Fig. 2.25 Aspect of the TGC component of an US machine. Lateral to the knobs, an image is showing the modality to adjust it for accurate imaging (*arrow*). *US* ultrasound, *TGC* time gain compensation

Tips to Overcome It cannot be overcome, but awareness of this phenomenon is of major importance in case of a targeted local injection of a deep structure.

Pitfalls Due to Doppler Artifacts

In musculoskeletal US, Doppler investigation accompanies and completes GS evaluation, providing a "real-time" image of the blood flow inside the tissues. The choice of a Doppler mode evaluation depends mainly on the equipment used [12]. Power Doppler (PD) was initially described as the most sensitive type of evaluation for detection of slow flow inside tissues; however, new high-end machines provide color Doppler more sensitive than PD. A function of modern US equipment is the ability to prioritize the use of Doppler versus the use of GS—the machine is able to pick one of them and focus mainly upon it.

Even with all evidences for Doppler correlations to disease activity and C-reactive protein value, there are very few studies establishing the direct link to histological aspects obtained by direct biopsy of the same region examined with US [18, 19]. Anderson et al. [19] not only found color Doppler activity correlated with the extent of inflammation present in the synovial biopsies from rheumatoid arthritis (RA) patients but also found synovial pathology in biopsies taken from Doppler-negative sites. The absence of US Doppler signal in histologically affected areas could be in fact linked to pitfalls that hide colors. There is still a lot to be learned in that area, but separating artifacts from true Doppler signals on the screen is mandatory for an accurate examination [20]. US practitioners have to be aware of the fact that sometimes the two entities can overlap in the same picture. The signals appearance also depend on external factors—diurnal variation [21], the position of the joint, flexion

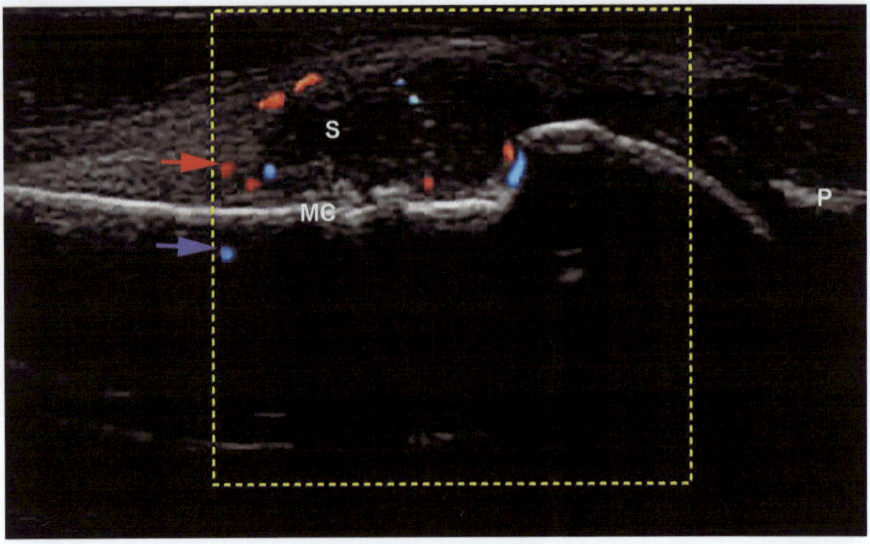

Fig. 2.26 Longitudinal view of an MCP joint dorsal side. A vessel (*red arrow*) and its mirror (*blue arrow*) are depicted symmetrically on both sides of metacarpal bone. *MC* metacarpal bone, *P* Phalanx, *S* synovitis inside the joint, *MCP* metacarpophalangeal

and extension [8]. The accuracy of establishing the true presence or absence of Doppler signals is mandatory for disease monitoring and treatment.

Standardization of Doppler use for rheumatologic US has been published [12]. However, although the authors recommend settings for the use of color Doppler and PD in rheumatology, they also conclude that the recommendations are in fact only suggestions. Perfect standardization for Doppler examination is not feasible due to the huge degree of variability implied in the exam. Most of Doppler artifacts are due to machine settings/misuse or human errors. The ones due to the technique are mirror and reverberation artifacts.

Mirror Doppler Artifact

This artifact has the same explanation with GS mirror effect—it is visible when the beam comes against a very highly reflective surface. In Doppler examination, the mirror effect is more frequently encountered. The mirror surface is usually the bone margin. A vessel localized near the bone may have a false mirror image on the other side in US image. Like in a real mirror, the symmetry of the real and fake images is maintained. The artifact is easily recognized when all three parts of the description above are present: the object, the mirror, and the false object. It must be distinguished from blood penetrating a bone erosion (Fig. 2.26). Sometimes only the mirror signal is exhibited on the image (Fig. 2.8).

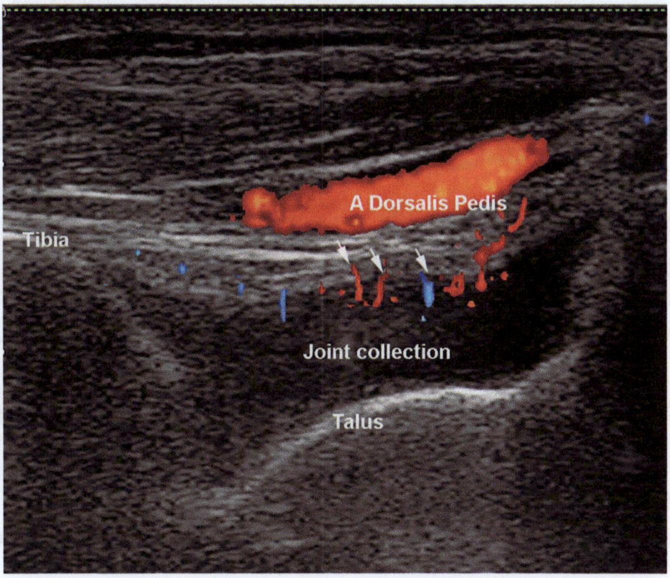

Fig. 2.27 Longitudinal view of the anterior recess of the tibiotalar joint with the Doppler box on. A hypoechoic joint collection is seen inside the joint. Above it, the dorsalis pedis artery crosses the image. Below the vessel reverberation Doppler artifacts (*arrows*) seem like Doppler signal inside the collection. The top of the Doppler box is at the top of the image

Tips to Overcome It is important to remember not to interpret images and "hot spots" below a bone with intact contour.

Reverberation Doppler Artifact

It is based on the same mechanism as its GS correspondent, with the distinction that in Doppler evaluation the two closely spaced interfaces conducting to repeated wave reflections are the vessel walls. As a consequence of reverberation, false vessels can appear on the screen. A superficial vessel can appear again deeper on the screen image either once (simple reverberation) or more times, with a showering of color behind it (complex reverberation) [8]. This can alter US image interpretation, as color seen on the screen is usually automatically connected to disease activity. However, it should be emphasized that Doppler reverberation artifacts are placed behind the superficial vessel (Fig. 2.27).

Tips to Overcome Standardization of the color box position in US Doppler evaluation helps overcome this pitfall. Always put the Doppler box up to the top of the image to capture any small superficial vessel that could cause reverberation and false vascularization below.

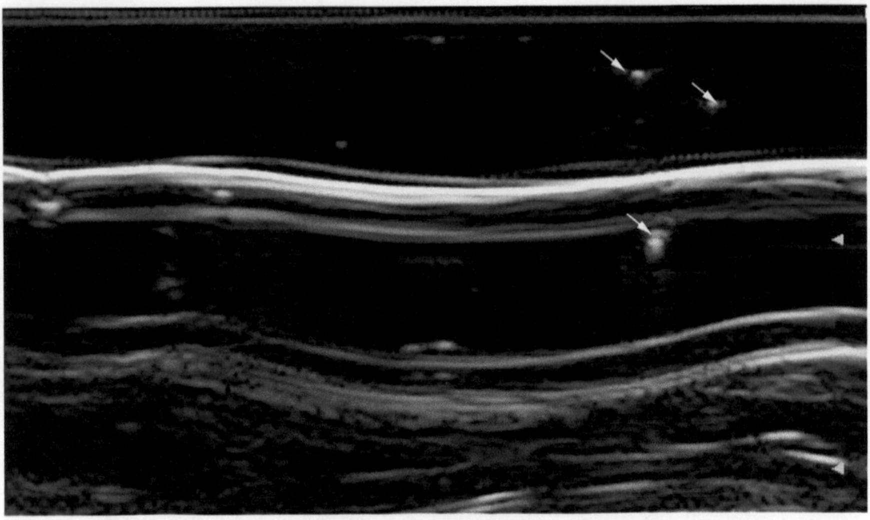

Fig. 2.28 Image on the screen before the exploration starts, with a huge layer of gel on the transducer. Gas bubbles are seen inside (*arrows*), with "dirty" shadows behind them

Pitfalls Due to Human Errors (GS and Doppler)

Drop-Out Pitfall

US implies multiple wave transmissions back and forth between the transducer and structures inside the human body. As already shown above, intense sound reflections may appear at the interface between air and any structure nearby. In order to avoid that, ultrasonic media represented by US gels are the perfect solution. The gel is in fact a liquid enriched with a thickening agent which is designed to improve its spreadability on the skin [22]. Hydrogels are in current use for US purposes, as opposed to lipogels, which have the disadvantage of fattening the skin and being hard to remove. Hydrogels contain mainly water, so they can be wiped off without any residues and, as a practical issue, they do not stain the clothes. US gel is highly important; in fact, it is mandatory for the examination. It provides optimal contact between the crystals of the transducer and the patient's skin. In fact, without the layer of gel, all the waves from the transducer would reflect and there would be no image on the screen because of the huge difference in acoustic impedance at the interface transducer/air. Gel should be completely free of air bubbles which are known to impair transmission. If present, these appear on the screen as white spots inside the layer of gel (Fig. 2.28).

The "drop-out" pitfall is due to the lack of conducting gel on a part/on the whole transducer surface. It is an air artifact appearing at the contact between the transducer and the skin. The part of the transducer not covered with gel will only show reverberation artifacts and no echoes (Fig. 2.29).

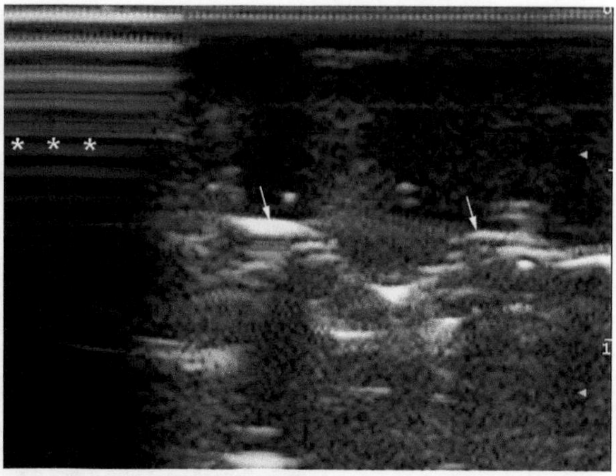

Fig. 2.29 Image on the screen before the exploration starts, with a part of the transducer missing gel (*asterisks*), and the other part exhibiting gas bubbles inside gel (*arrows*)

Tips to Overcome Make sure a generous and uniform layer of gel is correctly applied on the probe at the beginning of the procedure and during it.

Insufficient Knowledge of the Machine Settings: Gain–Focus–Depth Adjustment

In the matter of properly using the machine, US performing is very much similar to driving a car. At the first driving lesson, one has to learn about the classic triad, chair–mirror–seatbelt, before attempting to actually move the car. Any driving skills are useless if the elements from the triad are not accurately adjusted. The things are about the same when learning how to perform US. The classic triad here is gain–focus–depth. Any mistakes in correctly adjusting these settings will lead to pitfalls. The science of making perfect machine adjustments is known as "knobology" [20].

Gain is the artificial increase of signal strength and it is responsible for the image brightness [8]. A proper gain adjustment is essential for depicting the correct echogenicity of a structure and it should be reviewed at each examination. The "too low" or "too high" gain can lead to missing or, respectively, obscuring a structure (Figs. 2.30, 20.31 and 2.32). No standard values are available for GS gain, but according to the usual experience it is set at about 50–60 %. Modern US equipment has an automatic gain adjustment knob, which is very useful especially for beginners, until they learn to know the normal echogenicity of tissues.

A separate gain adjustment for Doppler evaluation is necessary. It is very important to keep in mind that Doppler gain is not the same as GS gain. Doppler gain is responsible for the machine's sensitivity to flow [8, 12]. Like in GS, no standard values are available—the adjustment has to be done for every examination. The best method to set the gain is to increase it till pitfalls appear—colored signals behind the bone contour on the screen, known as *random noise* (Fig. 2.33). After that, it is

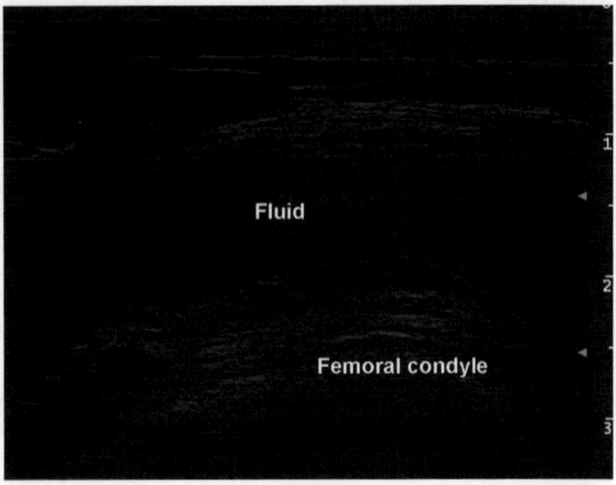

Fig. 2.30 Transverse view of the peripatellar lateral recess of a knee with fluid collection inside. Gain is too low (28 %), the image is too dark

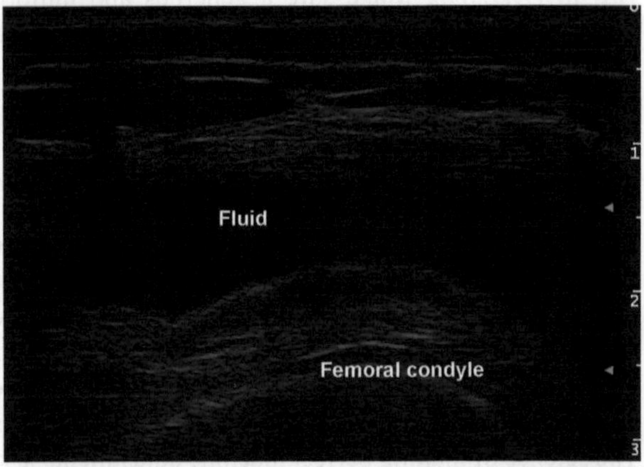

Fig. 2.31 Transverse view of the peripatellar lateral recess of a knee with fluid collection inside. Gain is set at 50 %

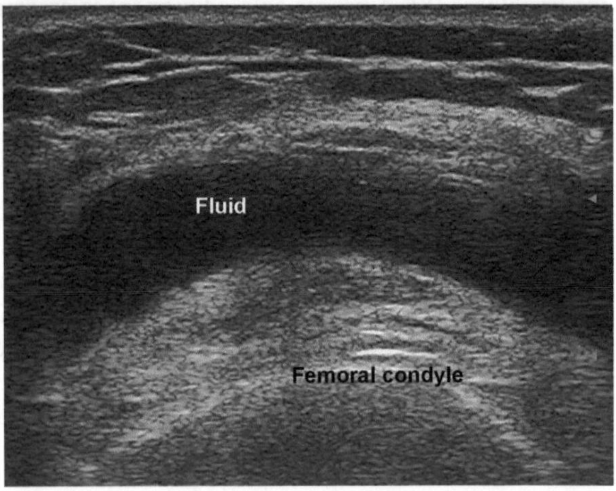

Fig. 2.32 Transverse view of the peripatellar lateral recess of a knee with fluid collection inside. Gain is too high (100 %), the image is too bright

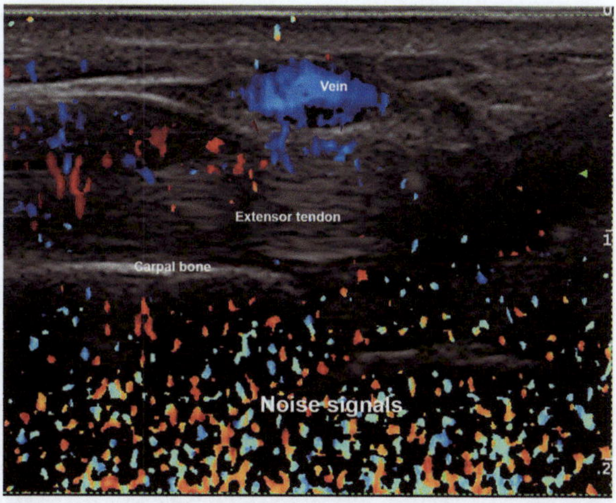

Fig. 2.33 Longitudinal view of the dorsal side of carpal joints. The Doppler gain is set "too high," so a lot of signal noise appears below the bone contour and below the vessel. To adjust the image, the Doppler gain should be lowered until all signals below the bone disappear

indicated to lower the Doppler gain until these completely disappear below the bone contour. A "too low" gain might lead to missing flow.

Another artifact due to gain incorrect adjustment is *blooming artifact.* In case of too high gain, the colors give the impression of reaching beyond the vessel walls, making these look much larger than they really are (Figs. 2.34 and 2.35). Lowering

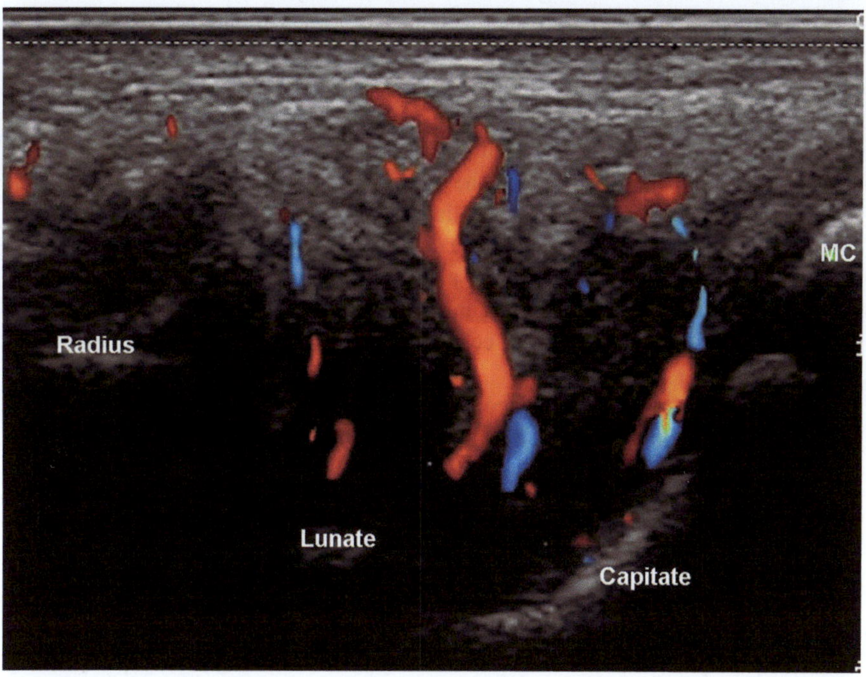

Fig. 2.34 Longitudinal view of the dorsal side of carpal joints in a rheumatoid arthritis patient. The signal is so high that the vessels seem a lot larger than they really are—blooming artifact. *MC* metacarpal bone

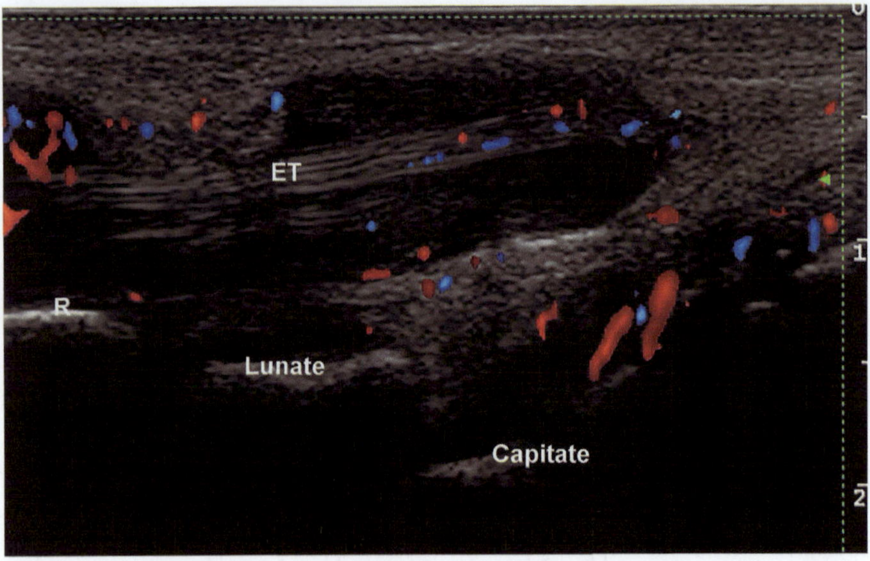

Fig. 2.35 Longitudinal view of the dorsal side of carpal joints in a rheumatoid arthritis patient. Normal size of Doppler vessels in a rheumatoid arthritis patient. *ET* extensor tendon

the gain would correct that impression, but might lead to missing vessels. As it indicates real blood flow, blooming artifact must not always be corrected.

An interesting concept regarding this issue has been postulated [23] by the American Institute of Ultrasound in Medicine (AIUM): as low as reasonably achievable (ALARA). The perfect balance in obtaining the most accurate images is of course a matter of personal experience of the practitioner.

In Doppler evaluation, other machine-important parameters to be considered and adjusted for pitfall elimination are the pulse repetition frequency (PRF) and the wall filters. Suggestions for accurately handling most US machines regarding these characteristics are available in literature [12].

Inappropriate *focus* adjustment can also lead to false brightness information regarding a structure in GS, and might under-evaluate the blood flow. The focus area is common for GS and PD, and the echoes in the focal area are higher in amplitude [12].

The *depth* should be adjusted according to the position of the target structure—if more profound, the depth should be increased and consequently the image will become smaller (as for hip evaluation). In case of inappropriate depth setting, a pitfall could be the omission of a deep structure from the screen. Decreasing the depth can perfectly depict a superficial structure, making all its details available for evaluation: tendon fibers, small bone erosions, etc. (Figs. 2.36, 2.37 and 2.38).

Apart from the three mentioned cornerstones, *frequency* of the transducer may be responsible for pitfalls, if inaccurately chosen. It is very important to be aware of the frequency recommended for GS in every joint according to its depth [24, 25].

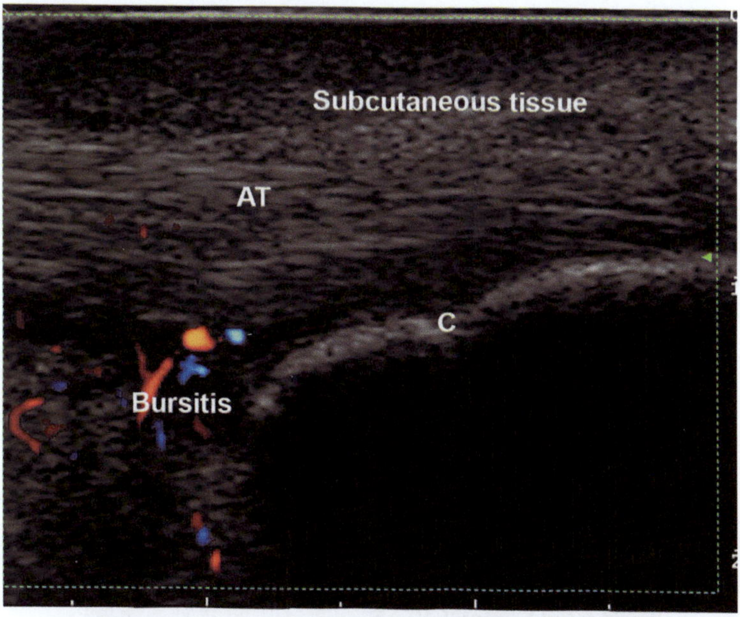

Fig. 2.36 Longitudinal view of posterior ankle compartment at various depth adjustment, the same transducer frequency (12 MHz). Lowering the depth will make tiny details of the superficial tissues visible: subcutaneous tissue, Achilles tendon parallel fibers (*AT*), the sharp hyperechoic line of calcaneus (*C*), and a hypoechoic area below the tendon with Doppler signal inside (retro-calcaneal bursitis). The focus is also set for superficial structures

Details regarding the frequency are not a matter of this chapter; however, a high frequency transducer is known to increase resolution, so that tiny details of local anatomy can be perfectly visualized as separate entities on the US screen.

Tips to Overcome Random noise signals have random appearances on the screen, as opposed to real flow signals, which appear in the same location. Remember to adjust gain, focus, depth, frequency at every examination. Remember Doppler frequency is different from GS frequency.

Inaccurate Handling of the Transducer

The pressure applied with the transducer on the skin should be kept under control in US examination. For superficial structures, ideally the transducer should not come in direct contact with the skin and a generous, uniform layer of gel should intermediate the contact. An easy method to accomplish that objective is defining a correct position of holding the probe: from its base (not from its top near the fiber) with the examiner's little finger resting on the patient's skin (Fig. 2.39). This position is meant to allow as much applied pressure as desired. *Compressing* on the skin may

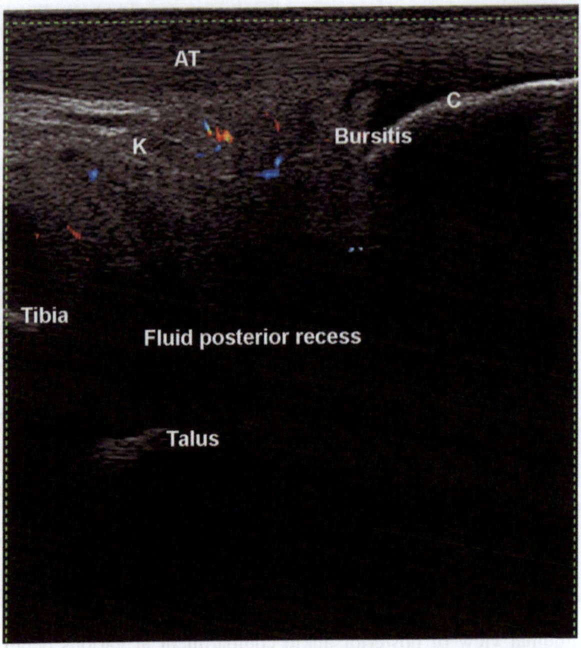

Fig. 2.37 Longitudinal view of posterior ankle compartment at various depth adjustment, the same transducer frequency (12 MHz). (Same patient as in Fig. 2.36.) Increasing the depth will allow visualization of the posterior recess of the tibiotalar joint, which in this case is filled with anechoic material, possible fluid. Setting the depth too small would lead to missing the collection. Achilles tendon (*AT*) is seen with little detail. *K* Kager fat pad, *C* calcaneus

lead to structures' disappearance (as fluid). On the contrary, when deep structures like hip are involved, the pressure applied should be higher. The same pressure is needed to help visualizing other profound structures, like the distal insertion of bicipital tendon. In vascular US, compression is useful for veins' patency check. If the vein is patent, its walls will collapse at compression and return to normal when the pressure is released. Compression is also useful in the differential diagnosis between fluid and synovial hypertrophy in rheumatic pathology.

For Doppler evaluation, a pitfall related to handling the probe is *motion artifact*. Regardless if it is connected to the probe, to the examiner, or to the patient (as in Parkinson's disease), motion can significantly alter the image on the screen (Fig. 2.40). A lot of colored pixels appear connected with movements, making the image difficult to evaluate. Highly like random noise artifact in appearance, motion artificial pixels are found in the part of the screen connected to the movement. When stopping the motion the colors disappear, and only real flow persists. Motion signals have a random distribution pattern. An interesting and, somehow, funny observation is that a motion pitfall can appear on the screen even when the examined patient starts speaking.

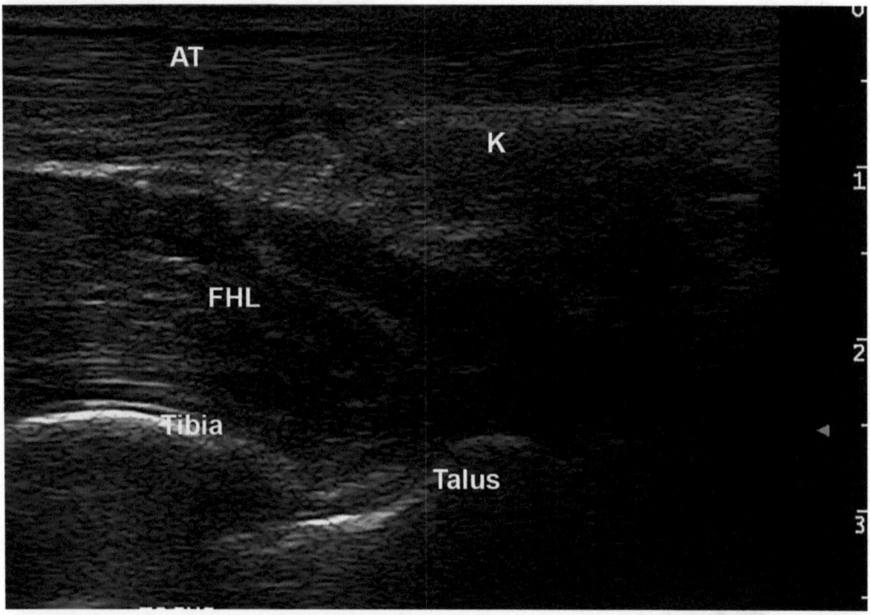

Fig. 2.38 Longitudinal view of posterior ankle compartment at various depth adjustment, the same transducer frequency (12 MHz). (Other patient, same area as in Fig. 2.36.) Increasing depth (3 cm) and adjusting focus will show the normal aspect of the posterior recess of the tibiotalar joint and the structures nearby. *FHL* flexor hallucis longus muscle, *AT* Achilles tendon, *K* Kager's fat pad

Tips to Overcome Remember to keep control over the pressure applied to the probe. Never move the hand and/or the transducer while the Doppler box is put. Remind the patients to keep relaxed and quiet during musculoskeletal US examination. Be relaxed yourself.

Insufficient Knowledge of Local Anatomy

This kind of pitfalls is mainly linked to image interpretation and not to its acquisition. Apart from solid knowledge of anatomic details of the region examined by US, the practitioner has to keep in mind some very frequent anatomic pitfalls that might alter the image.

Physiologic vessels near the bone contour can easily be interpreted as Doppler signals entering the cortical area, which could be an early erosion alarm sign. Knowing the normal sites of feeding bone vessels would increase the probability of an accurate diagnosis; however, nutrient vessels are not always found in the same areas. No article in the literature describes the normal position of all feeding vessels.

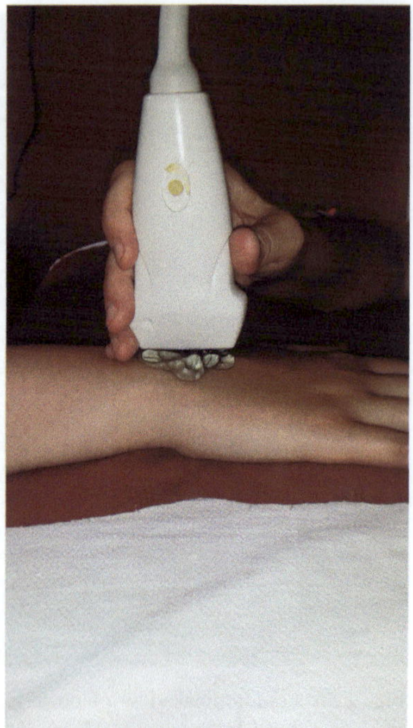

Fig. 2.39 Correct position of the hand when holding the probe. The little finger of the examiner is resting on the patient's skin

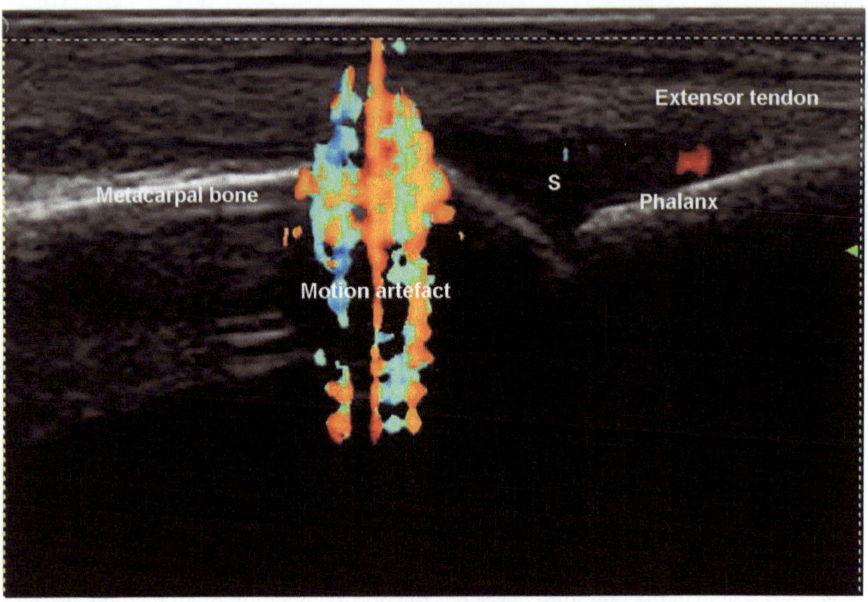

Fig. 2.40 Longitudinal view of the dorsal side third MCP joint. A hypoechoic area with Doppler signal inside is seen inside the joint—synovitis (*S*). A huge color mixture is seen in the middle of the image—motion Doppler artifact, caused by slight movements of the examiner or of the patient. *MCP* metacarpophalangeal

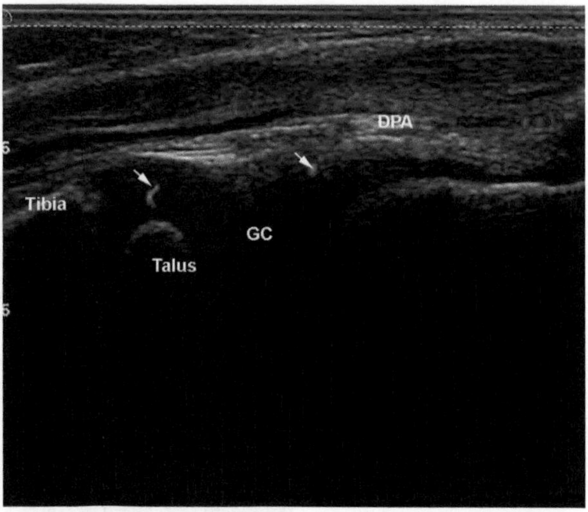

Fig. 2.41 Longitudinal view of the tibiotalar joint anterior recess in a 3-year-old healthy boy. Feeding vessels inside growing cartilage (*GC*) of the talus—*arrows*. *DPA* dorsalis pedis artery

Sometimes a radiologic confirmation of the vessel is possible [9]. According to personal experience, feeding vessels are usually unique, not accompanied by bone cortical visible lesions, and not accompanied by Doppler signal inside the joint. Nutritive vessels appear frequently in pediatric US as small vessels entering the ossification centers inside the hyaline growing cartilage (Fig. 2.41). Other normal vessels that could become pitfalls are the circumflex artery inside bicipital groove (Fig. 2.42)—prone for misinterpretation as bicipital inflammatory tenosynovitis, and the anatomic variant of the median artery of the forearm, visible right near median nerve.

Sesamoid bones may appear randomly inside the structures visible on the US screen. Very frequently, they can be seen as calcifications inside the tendons running above the bone, being falsely interpreted as calcific tendonitis. Their acoustic shadow may project over structures below and create a pitfall in image interpretation.

Tips to Overcome For avoiding this phenomenon, simple calcifications inside otherwise normal tendons should better not be interpreted as pathologic, especially if the patients are asymptomatic.

Other anatomic variants can cause interpretation pitfalls and sometimes they can even cause pathology. As an example, the presence of a supracondylar process on the anteromedial aspect of the distal humerus may become connected to the medial epicondyle by the ligament of Struthers. In case of muscular hypertrophy, the arteries/nerves of the region may become affected (median nerve entrapment or brachial artery compression) with clinical consequences.

Tips to Overcome Become aware, if possible, of most anatomic variants before becoming an US practitioner.

Pediatric US performance could be a huge source of pitfalls in images interpretation if not aware of the differences from adult sonoanatomy. As a result of

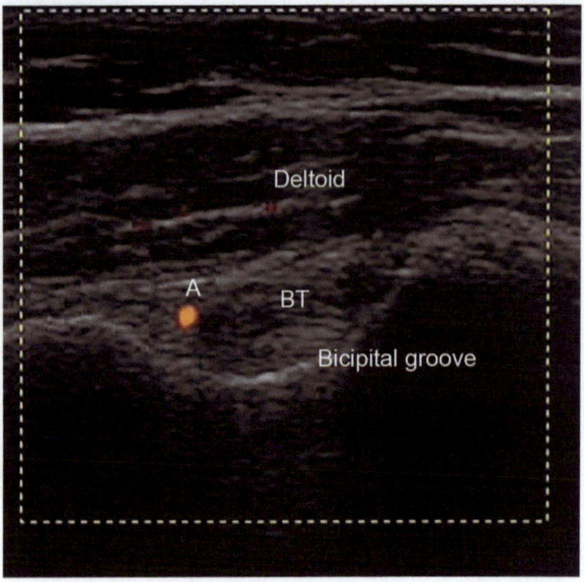

Fig. 2.42 Transverse view of bicipital groove. Circumflex artery (*A*) inside the groove, near the biceps tendon (*BT*)

incomplete ossification, the cartilage with its anechoic appearance will seem very large, resembling fluid inside the joint. Inside the cartilage, ossification centers appear later in the growing process, as hyperechoic foci with/without shadowing behind them (Fig. 2.43). Growing cartilage is particularly well vascularized— many feeding vessels can be seen inside it, with no pathologic significance.

Tips to Overcome Become aware of pediatric peculiar US aspect of joints before starting children evaluation with US on your own.

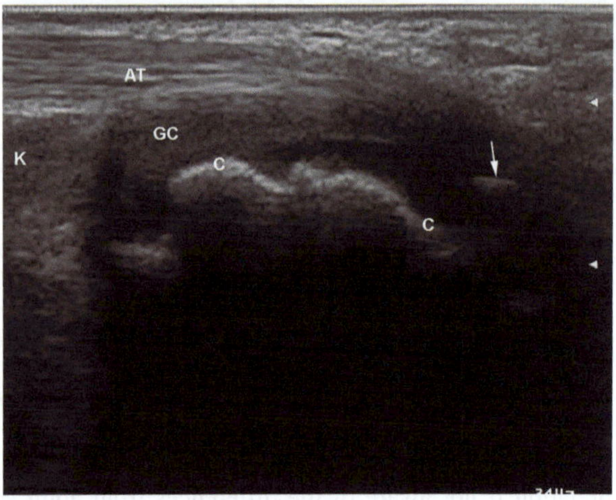

Fig. 2.43 Longitudinal view of the posterior ankle of a 3-year-old girl. Calcaneus bone (*C*) is covered by a hypoechoic layer of growing cartilage (*GC*). Ossification center (*arrow*) is seen inside the cartilage. *AT* Achilles tendon. *K* Kager's fat pad

Pitfalls Due to Machine Properties and Errors

The quality of US images depends on the machine image resolution. *Resolution* is the capacity of the beam to make a distinction between two structures [8] on a vertical line (axial resolution) or on a horizontal line (lateral resolution). Less performant machines lack accurate resolution, so close tiny structures can appear together on the screen as one structure. Adjusting the frequency of the probe to the examined area is mandatory for improving the resolution. Increasing the frequency will increase resolution. If that is impossible, a stand pad of gel interposed between the transducer and the skin may also contribute to resolution improvement, by increasing the depth. An important issue in ultrasonography is the purchase of a machine perfectly adjusted to the needs of the practitioner and the regular upgrade of its functions.

The early signs of US machine deterioration are always an alarm sign. A sign of *piezoelectric crystals and/or transducer's fibers damage* is the appearance of vertical black lines on the screen. Even if it is easily recognizable, this might be difficult to resolve, as it may have many causes, including the beginning of the machine end.

Tips to Overcome In this case, consultancy from the manufacturer of the machine is needed. Also, regular service and upgrades are mandatory.

Artifacts Useful for Medical Practice

Artifacts were classified as "good" and "bad" a long time ago [7]. Recently, they have helped toward the use of US in a field where it was never used before—in lung evaluation. Echography of the lungs was considered not possible for many years, given the fact that the beam would only encounter air in its path. With no mismatch in acoustic impedance no image will appear. But when the air is replaced for example by fluid, as in pulmonary edema from heart failure, a mismatch appears and creates some lines visible on the screen and called "B lines." These are vertical hyperechoic lines from the pleural line to the bottom of the screen, moving with respiration [26]. These are in fact *reverberation artifacts* with the origin in pulmonary septa filled with water. Their discovery opened a whole new perspective as they were proven correlated to CT findings. They are now included in the bedside differential diagnosis of dyspnea: the one of pulmonary origin (chronic obstructive pulmonary disease, COPD) does not exhibit B lines on the US screen as the cardiac one does.

Last but not least important, interventional sonography benefits a lot from artifacts. *Reverberation artifact* of the needle is helpful for following its entire trajectory during the procedure. Apart from that, even before the procedure artifacts can help "marking" the spot of a needle insertion, using *the comet tail* shape appearance behind a metallic object (a needle) placed between the probe and patient's skin. Same artifact can be used in the teaching process of interventional US, as students' direct approach on patients would be totally unethical and also dangerous.

Concluding Remarks

Good or bad, pitfalls appear on the screen at every US examination. Increasing their awareness among echographists is imperative for accurate evaluation and diagnosis of various rheumatology conditions. Echography development as a "friendly" technique, easily repeatable, not harming for the patient, allowing multi-planar and dynamic evaluation, has brought to attention two other fields medical practitioner is used to forget about: physics and anatomy. A good ultrasonographer is a combination between solid clinical background, detailed anatomy and physical principles knowledge, and technical skills in handling the machine [24]. The very fact that US is widely known as an "operator-dependent" technique is prone to create pitfalls. As a result, the images obtained by a less-skilled practitioner can be a lot different from expert images. Less performant (and less expensive) equipment is also prone to create pitfalls.

In the end, teaching the main pitfalls has some important purposes: an alert sign toward their existence, toward the possibilities of their overcoming, and to the modalities of making them useful in medical practice. Referring to the "operator dependency" of US, in fact this attribute charges the operator with a huge responsibility and gives them an opportunity to use theoretical knowledge combined with imagination and ability to discern between them.

References

1. Oxford Dictionary of English 3rd edition, Online version 2013, http://www.oxforddictionaries.com/definition/english/artefact.
2. Kane D, Grassi W, Sturrock R, Balint PV. A brief history of musculoskeletal ultrasound: from bats and ships to babies and hips. Rheumatology (Oxford);43(7):931–3.
3. Aldrich JE. Basic physics of ultrasound imaging. Crit Care Med. 2007; 35(5 Suppl):131-7.
4. Feldman MK, Katyal S, Blackwood MS. US artifacts. Radiographics. 2009;29:1179–89.
5. Taljanovic MS, Melville D, Scalcione LR, Gimber L, Lorenz E, Witte R. Artefacts in musculoskeletal ultrasonography. Semin Musculoskelet Radiol. 2014;18:3–11.
6. Nilsson A. Artefacts in sonography and Doppler. Eur Radiol. 2011;11:1308–15.
7. Holsbeeck MT, Introcaso JH. Musculoskeletal ultrasound, second edition. St Louis: Mosby; 2001. pp. 9–23.
8. Irsay L, Mandl P, Balint PV. Pitfalls in gray scale artifacts. In: Wakefield RJ, D'Agostino MA. Musculoskeletal ultrasound in rheumatology. Philadelphia: Saunders, Elsevier; 2010. pp. 29–40.
9. Bianchi S, Martinoli C. Ultrasound of the musculoskeletal system. Berlin: Springer; 2007.
10. Hagiandreou NJ. B mode ultrasound: basic concepts and new technology. Radiographics. 2003;23:1019–33.
11. Rutten GCM, Jagger G, Blickmann JG. US of the rotator cuff: pitfalls, limitations and artefacts. Radiographics. 2006;26:589–604.
12. Torp Pedersen ST, Terslev L. Settings and artefacts relevant in colour/power Doppler ultrasound in rheumatology. Ann Rheum Dis. 2008;67:143–9.
13. Joshua F. Ultrasound applications for the practicing rheumatologist. Best Pract Res Clin Rheumatol. 2012;26:853–67.

14. http://en.wikipedia.org/wiki/Anisotropy. Accessed 13 March 2015.
15. Badea M, Dudea S. Ecografie generala-prelegeri pentru examenul de competenta. Ed. Universitara Iuliu hateganu. Cluj Napoca; 1997. S. 5–29.
16. Bushberg JT, Seibert JA, Leidholdt EM, Boone JM. The essential physics of medical imaging. 2nd ed. Philadelphia: Lippincott Williams & Wilkins; 2002. pp. 469–553.
17. Zalud I, Rocha F. Artefacts, pitfalls and normal variants. Donald Sch Ultrasound Obstet Ginecol. 2012;6(1):1–8.
18. Koski JM, Saarakkala S, Helle M, Hakulinen U, Heikkinen JO, Herminen H. Power Doppler ultrasonography and synovitis: correlating ultrasound imaging with histopathological findings and evaluating the performance of ultrasound equipment. Ann Rheumdis. 2006;65:1590–5.
19. Andersen M, Ellegard K, Hebsgaard JB, Christensen R, Torp-Pedersen S, Kvist PH, Soe N, Romer G, Vendel N, Bartels EM, Samsoe BD, Bliddal H. Ultrasound colour Doppler is associated with synovial pathology in biopsies from hand joints in rheumatoid arthritis patients: a cross sectional study. Ann Rheum Dis. 2014;73:678–83.
20. Balint PV, Mandl P, Kane D. "All that glistens is not gold"- separating artefacts from true Doppler signals in rheumatological ultrasound. Ann Rheum Dis. 2008;67:141–2.
21. Semerano L, Gutierrez M, Falgarone G, Filippucci E, Guillot X, Boissier MC, Grassi W. Diurnal variation of Power Doppler in metacarpophalangeal joints of patients with rheumatoid arthritis: a preliminary study. Ann Rheum Dis. 2011;70:1699–70.
22. http://www.dermaviduals.de/english/publications/products/ultrasound-gels-effects-compositions-applications.html. Accessed 13 March 2015.
23. American Institute of Ultrasound in Medicine (AIUM). Medical Ultrasound Safety. ISBN1–930047-71–1, 1994;reapproved 2002:1–40.
24. Schmidt WA, Schmidt H, Schicke B, Gromnica-Isle L. Standard reference values for musculoskeletal ultrasonography. Ann Rheumdis. 2004;63:988–94.
25. Iagnocco A, Naredo E, Biljsma JWJ. Becoming a musculoskeletal ultrasonographer. Best Pract Res Clin Rheumatol. 2013;27:271–81.
26. Gargani L. Lung ultrasound: a new tool for the cardiologist. Cardiovasc Ultrasound. 2011;9(1):6.

Chapter 3
Rheumatoid Arthritis

Esperanza Naredo MD, María Montoro MD and Iustina Janţă MD

Abbreviations

ACPA	Anti-citrullinated protein antibodies
ACR	American College of Rheumatology
ADA	Adalimumab
b	Biologic
BM	B-mode
BME	Bone marrow edema
CD	Color Doppler
CR	Conventional radiography
CRP	C-reactive protein
cs	Conventional synthetic
CT	Computed tomography
DAS	Disease activity index
DMARDs	Disease-modifying antirheumatic drugs
ECU	Extensor carpi ulnaris
ESR	Erythrocyte sedimentation rate
EULAR	European League Against Rheumatism
GH	Glenohumeral
IFX	infliximab
IgM-RF	immunoglobulin M-rheumatoid factor
MCP	Metacarpophalangeal
MMP	Matrix metalloprotease
MRI	Magnetic resonance imaging

E. Naredo (✉) · M. Montoro · I. Janţă
Department of Rheumatology, Hospital General Universitario Gregorio Marañon, Madrid, Spain
e-mail: enaredo@ser.es

M. Montoro
e-mail: mariamontor1@hotmail.com

I. Janţă
e-mail: iustinajanta@yahoo.com

Y. El Miedany (ed.), *Musculoskeletal Ultrasonography in Rheumatic Diseases,*
DOI 10.1007/978-3-319-15723-8_3

MS	Musculoskeletal
MTP	Metatarsophalangeal
MTX	Methotrexate
NSAIDs	Nonsteroidal anti-inflammatory drugs
OA	Osteoarthritis
OMERACT	Outcome measures in Rheumatology
OR	Odds ratio
PD	Power Doppler
PIP	Proximal interphalangeal
PsA	Psoriatic arthritis
RA	Rheumatoid arthritis
SASD	Subacromial subdeltoid
SDAI	Simplified disease activity index
SE	Synovial fluid/effusion
SH	Synovial hypertrophy
SOLAR	Sonography of large joints in rheumatology
SpA	Spondyloarthropathy
TA	Tibialis anterior
TIMP	Metallopeptidase inhibitor
TNF	Tumor necrosis factor
TP	Tibialis posterior
US	Ultrasound
VEGF	Vascular endothelial growth factor

Introduction

Rheumatoid arthritis (RA) is a chronic, systemic, autoimmune disease affecting from 0.5 to 1% of population. It is characterized primarily by joint inflammation that affects both large and small joints. Besides intra-articular structures, inflammation can also involve periarticular structures. The natural course of the disease can lead to progressive joint destruction and physical disability [1, 2]. Over the past years, the approach to RA management has changed considerably. Early diagnosis and rapid achievement of remission became the main goals with "treat-to-target" strategy in focus. In 2010, new classification criteria which excluded radiographic evidence of joint erosion meant that the patient could be diagnosed more quickly and easily [3]. Therapies such as conventional synthetic disease-modifying antirheumatic drugs (csDMARDs) and newer therapies, such as biologic DMARDs (bDMARDs), have proven to be effective in inducing and maintaining remission in RA patients. Therefore, disease prognosis became brighter as such therapies have proven efficacy in the reduction of structural damage, physical disability as well as many other comorbidities, particularly cardiovascular [4].

Over the last decade, an increased attention has been paid to the development of standardized clinical measurements of disease activity. The development of composite indexes and the "treat-to-target" strategy represent an important step forward in a tight control of the disease [5, 6]. Despite this, disease progression has been observed in some patients who achieved clinical remission status. The development of imaging modalities such as magnetic resonance imaging (MRI) and musculoskeletal ultrasound (MSUS) have enabled more accurate assessment of RA patients compared to clinical assessment [7–16]. In particular, MSUS has proven to be a reliable, noninvasive, practical, and accessible tool for assessing all peripheral joint and periarticular structures involved in RA patients in clinical practice.

In recent years, an increasing number of publications dealing with MSUS have shown that this technique can be an accurate, reliable, and sensitive-to-change tool in clinical practice. In RA, it has been used for both diagnosis purposes—including differential diagnosis—and for monitoring disease activity. B-mode (BM) examination provides important information regarding morphological aspects of intra- and periarticular structures and Doppler mode, either color Doppler (CD) or power Doppler (PD) examination, enables detection of low-speed blood flow, as it can be found in newly formed vessels in inflamed synovial tissue. Both BM and Doppler techniques are valid tools in assessing joint inflammatory diseases.

An important issue regarding MSUS is its reliability, as ultrasound (US) is considered a highly operator-depending technique. Its accuracy depends on both acquisition and interpretation of US images. This operator-dependent nature of MSUS has promoted the need for a uniform evaluation of US-detected pathologies. Therefore, to optimize MSUS as a diagnostic and monitoring tool, universal guidelines would be needed for pathology evaluation. However, progress has been made to address this point. In 2005, the Outcome Measures in Rheumatology (OMERACT) MSUS group defined the main US findings in inflammatory arthritis, i.e., synovitis, including synovial fluid/effusion (SE) and synovial hypertrophy (SH), tenosynovitis, bone erosions, and enthesopathy [17].Various scoring systems have also been developed and validated for these pathologies. Several studies that have investigated the intra- and interobserver reliability for US RA pathologies have shown good to excellent results [7, 10, 13, 18–22].

The importance of MSUS in RA has recently been recognized by the European League against Rheumatism (EULAR), including this tool in its recommendations on the use of imaging in RA [23].

Ultrasound Findings

This section describes the most important US findings in RA patients. The evidence-based validation process of MSUS of these abnormalities is reviewed in this section.

Synovitis is one of the most important features in RA. The term "synovitis" is used to indicate the presence of SE and/or SH (Fig. 3.1a). According to the OMERACT definitions published in 2005, SE was defined as an abnormal hypoechoic or anechoic (relative to subdermal fat, but sometimes may be isoechoic or hyper-

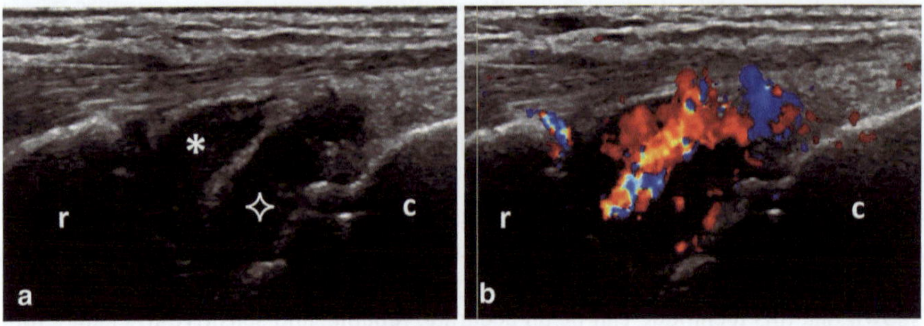

Fig. 3.1 Longitudinal view over the midline dorsal aspect of the wrist showing B-mode synovitis (**a**) with Doppler signal (**b**) in radio-carpal (*asterix*) and midcarpal (*star*) joints; *r* radius, *c* capitate

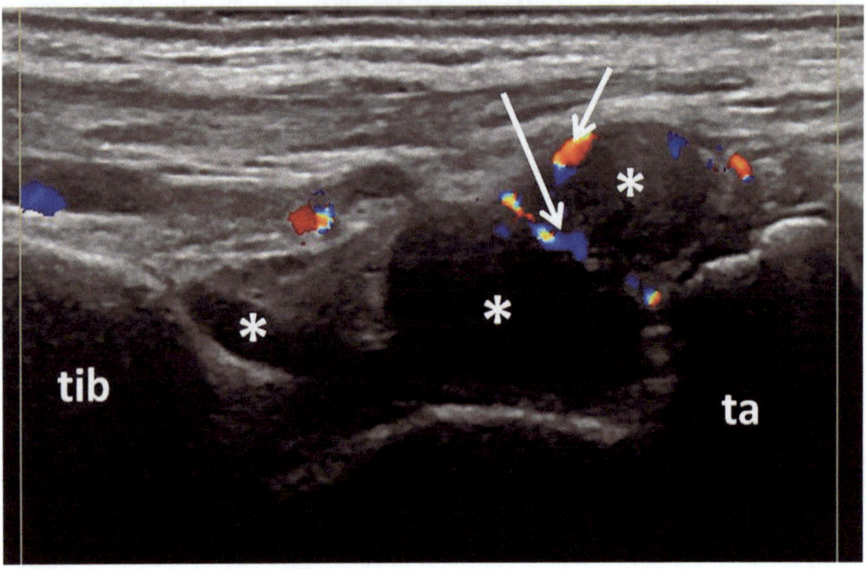

Fig. 3.2 Longitudinal view of the dorsal aspect of the tibio-talar joint showing B-mode synovitis (*asterix*) with Doppler signal (*arrows*); *tib* tibia, *ta* talus

echoic) intra-articular material that is displaceable and compressible, but does not exhibit Doppler signal, and SH as an abnormal hypoechoic (relative to subdermal fat, but sometimes may be isoechoic or hyperechoic) intra-articular tissue that is non-displaceable and poorly compressible and which may exhibit Doppler signal [17]. PD and CD are used to visualize vascularization of inflamed synovium (Figs. 3.1b and 3.2). Choosing a Doppler modality for rheumatologic practice depends on the equipment sensitivity. Differences between CD and PD are significantly reduced with newer US equipment.

While knee arthroscopy, a frequent procedure in clinical practice, has permitted a relatively easier histopathological assessment of inflamed joints, MSUS, both BM and Doppler, were reported to be accurate in detecting joint synovitis in comparison to arthroscopy and histology, respectively [7, 24]. Earlier studies reported good correlation between histological and B-mode ultrasound (BMUS) findings in knee joint synovitis. Furthermore, in patients with knee joint involvement in different diseases, BMUS has shown a high sensitivity, specificity, and accuracy (Fig. 3.3) for detecting synovitis [7, 24]. In RA inflamed joints, there was a similar good correlation between histologic and Doppler inflammatory changes in different joints [25–27]. When comparing histopathology with BMUS, power Doppler ultrasound (PDUS), and MRI, the highest correlation was found for PDUS and histopathology [28]. Although false-negative results were found for Doppler techniques when compared to histology [27, 29], the presence of a positive Doppler signal in the synovium was an indicator of active synovial inflammation.

When using others imaging techniques as comparator, MSUS, both BM and Doppler, showed considerable sensitivity and specificity. In a number of studies, moderate to good correlations were found between US-detected synovitis (either BM or Doppler) and MRI-detected synovitis in hand finger joints [12, 30]. Using MRI as reference, Szkudlarek et al. found a good to excellent sensitivity and specificity of US, both BM and PD, for the detection of synovitis at metatarsophalangeal (MTP) and metacarpophalangeal (MCP) joints [31, 32]. Similarly, Scheel et al. reported a good agreement between US and MRI in the detection of BM synovitis at MCP and proximal interphalangeal (PIP) joints [33]. In concordance with these data, Terslev et al. depicted a high significant association between Doppler US indices of inflammation and post-contrast MRI scores at wrist and hand joints [34]; whereas Fukae et al. found a good correlation between the measurements of Doppler synovitis and the enhancement rate of MRI in MCP and PIP joints [30]. Recently, Kawashiri et al. observed a moderate to good correlation between US-detected synovitis (in both BM and PD) and MRI-detected osteitis [35] (Fig. 3.4).

Good correlation was also reported in other studies investigating the link between MSUS-detected synovitis and inflammation identified through clinical examination and laboratory analysis. Naredo et al. found a moderate to good correlation between swelling joints count and MSUS-detected synovitis for both BM and PD. MSUS-detected synovitis was also found to better correlate with erythrocyte sedimentation rate (ESR) and C-reactive protein (CRP) than clinically detected synovitis [10]. In the study by Scire et al. on patients with early RA who started csDMARDs treatment, both clinical and MSUS-detected synovitis which were significantly correlated with CRP in patients with active disease, while, in patients

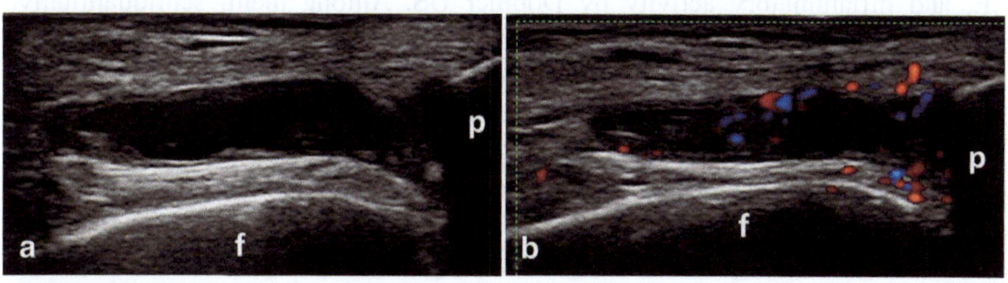

Fig. 3.3 Transverse view over the parapatellar recess of the knee showing B-mode synovitis (**a**) with Doppler signal (**b**) synovitis; *f* femur, *p* patella

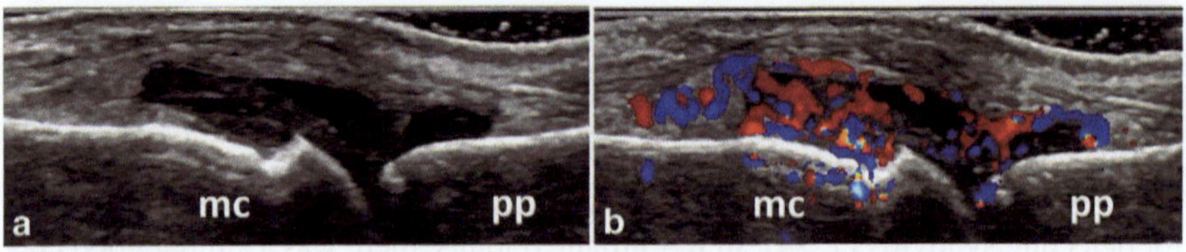

Fig. 3.4 Longitudinal view of the dorsal aspect of the metacarpophalangeal joint showing B-mode synovitis (**a**) with Doppler signal (**b**); *mc* metacarpal head; *pp* proximal phalanx

who achieved the clinical remission status, only PD correlated with CRP [14]. Kawashiri et al. found a significant moderate to good correlation between PDUS assessment of 12 joints with serum vascular endothelial growth factor (VEGF), matrix metalloprotease (MMP)-3, and metallopeptidase inhibitor (TIMP)-1 [36].

Studies that have investigated intra- and interobserver reliability in a variety of joints for MSUS-detected synovitis, for both BM and Doppler showed moderate to excellent results [7, 10, 13, 30, 37–39]. A systematic review assessing the reliability of MSUS-detected synovitis in RA showed that US, particularly PD mode, was reliable in still-image interpretation when assessed by experienced ultrasonographers, while image acquisition was less reliable. Among all joints, the knee was the most reliable joint even in terms of image acquisition [40]. Mandel et al. compared the reliability of 11 different US scoring systems for synovitis including different combinations of joint counts [42, 28, 20, 16, 12, 10, 8, 7] and found that MSUS, both BM and PD, showed a better reliability than clinical assessment in evaluating synovitis. No differences in the reliability were observed between these scoring systems [41].

The sensitivity to change of US-detected synovitis has been investigated in several published studies. Regardless of how many joints were evaluated, a decrease of BM and Doppler variables has been shown in patients treated with bDMARDs [15, 42–46] or csDMARDs [47, 48]. In a randomized control trial, Taylor et al. studied patients with early RA treated with methotrexate (MTX) and infliximab (IFX) versus MTX and placebo using US and conventional radiography (CR) for the evaluation of MCP joints. After 18 weeks of treatment, patients under IFX therapy presented significant reduction in synovial thickness and joint vascularity measured as the number of CD pixels in a defined region of interest [49]. Responsiveness of US-detected synovitis has also been shown after steroid treatment, either intra-articular or systemic [50–54].

Several scoring systems have been developed to assess synovitis, namely SH, SE, and inflammatory activity by Doppler US. Among them, semiquantitative scores have been the most used scores in clinical practice [18, 46]. In most published studies of MSUS in RA, the semiquantitative score for BM synovitis consisted of the following: grade 0—absence of synovial thickening, grade 1—mild synovial thickening, grade 2—moderate synovial thickening, grade 3—marked synovial thickening [18]. Similarly, in most published studies, the semiquantitative score for Doppler synovitis consisted of the following: grade 0—no flow in the synovium, grade 1—up to three single spots signals or up to two confluent spots or one confluent spot plus up to two single spots, grade 2—vessel signals in less than half of the area of the synovium, grade 3—vessel signals in more than half of the area of the synovium [18, 46]. Quantitative measurements of Doppler signals are obtained using a color recognition function that counts the number of total and color pixels within a region of interest. The number of color pixels is then expressed in relation to the total number of pixels as a fraction [44, 51, 55]. Terslev et al. found a good agreement between quantitative and semiquantitative scores for Doppler synovitis [56].

While small joints can be more easily assessed by clinical examination, synovitis evaluation of larger joints mostly shoulders and hips, represents a challenge in daily clinical practice. In a recent study, Sakellariou et al. found inflammatory changes at glenohumeral (GH) joint in 14 % of RA studied patients [57]. Figure 3.5 shows an example of posterior GH recess showing BM synovitis with Doppler signal in an RA patient (Fig. 3.5). Even if it is not so frequently evaluated, hip involvement is not uncommon in RA patients. In an Italian cohort of RA patients, BMUS detected hip synovitis in 24 % of patients [58]. Figure 3.6 shows an example of the anterior recess of the hip showing BM synovitis. However, Doppler activity is not detected so frequently in these joints, due to the lower sensitivity of Doppler techniques for deep areas.

Tenosynovitis is another important feature in RA patients. US-detected tenosynovitis is defined as hypoechoic or anechoic thickened tissue with or without fluid within the tendon sheath, which is seen in two perpendicular planes (Fig. 3.7) and which may exhibit Doppler signal [17]. Tenosynovitis on Doppler mode is defined as the presence of peri-tendinous Doppler signal within the synovial sheath (Fig. 3.8), seen in two perpendicular planes, excluding normal nutrient vessels in mesotenon or vinculae, only if the tendon shows peri-tendinous synovial sheath widening on BM [37].

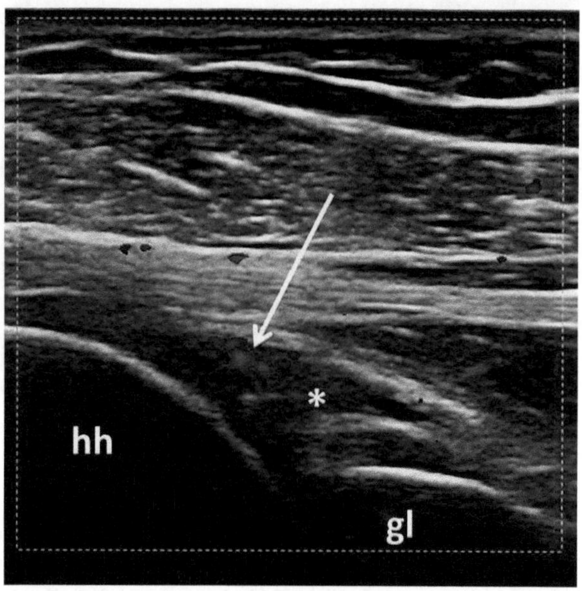

Fig. 3.5 Longitudinal view over the posterior glenohumeral recess showing B-mode synovitis (*asterix*) with Doppler signal (*arrow*); *hh* humeral head, *gl* glenoid fossa

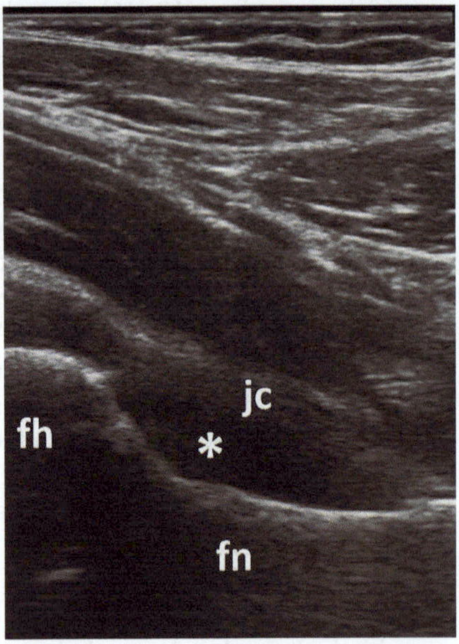

Fig. 3.6 Longitudinal view to the femoral neck showing B-mode synovitis (*asterix*) in the anterior recess of the hip; *fh* femoral head, *fn* femoral neck, *jc* joint capsule

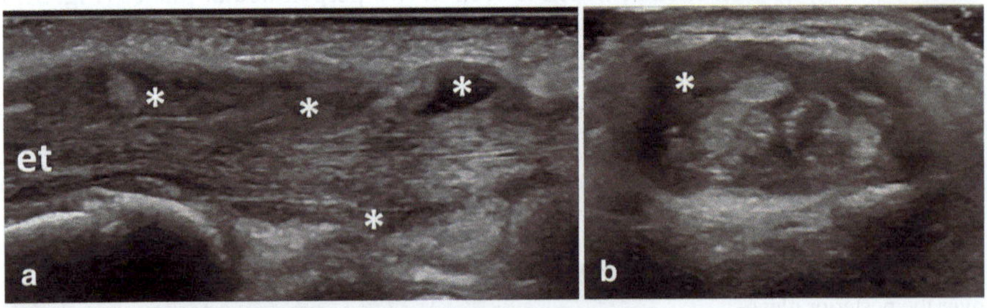

Fig. 3.7 Longitudinal (**a**) and transversal (**b**) view over the second compartment of the wrist extensor tendons showing B-mode tenosynovitis (*asterix*); *et* extensor tendons

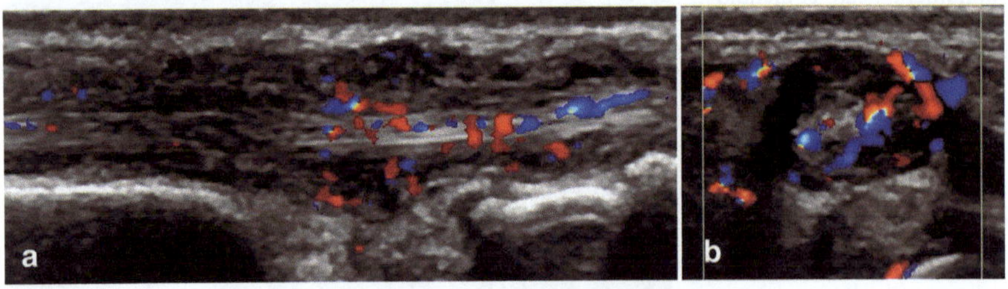

Fig. 3.8 Longitudinal (**a**) and transversal (**b**) view over the extensor carpi ulnaris tendon showing B-mode and Doppler tenosynovitis

Compared to MRI, MSUS has shown to be accurate in detection of tenosynovitis. MSUS has also shown a high specificity, but a fair to moderate sensitivity for detecting tenosynovitis [21]. A number of studies have compared US and MRI evaluation of tenosynovitis. In the study by Hoving et al., MSUS detected more tendon effusion than MRI at wrist and hand tendons [59]. Scheel et al. found a good agreement between US and MRI for tenosynovitis at ankle flexors and peroneus tendons while for ankle extensor tendons they reported a lower agreement [11].

Regarding the intra-observer consistency of US-assessed tenosynovitis, various studies have shown good to excellent reliability [19–21] and moderate to excellent interobserver reliability [19, 20, 22, 47]. Hammer et al. studied the sensitivity to change of US-detected tenosynovitis in RA patients initiating adalimumab (ADA) treatment. They assessed flexor and extensor tendons of bilateral wrist and ankle and observed a significant reduction of tenosynovitis after 12 months for all studied tendons. The MSUS assessment of a reduced number of tendons (i.e., extensor carpi ulnaris (ECU), tibialis posterior (TP), and flexor digitorum longus) was as sensitive to change as the assessment of all studied tendons [60].

For tenosynovitis, the most studied tendons have been the hand and ankle tendons. However, in RA not all tendons are affected in the same way. At hand and finger level, the most frequently involved tendons are the ECU and the flexor tendons of the second, third, and fourth fingers [61]. In an MRI study of RA patients with hindfoot pain, the most frequently involved tendons were TP and peroneal tendons and the least common involved tendons were tibialis anterior (TA) and the extensor tendons [62]. Figure 3.9 shows an example of BM tenosynovitis of TP tendon in an RA patient.

Recently, the OMERACT MSUS group developed a four-grade semiquantitative scoring system for BM and Doppler tenosynovitis which showed a good intra- and interobserver reliability [19]. This score is as follows: grade 0—normal, grade 1—minimal, grade 2—moderate, and grade 3—severe. Doppler tenosynovitis was scored as following: grade 0—no Doppler signal, grade 1—minimal, grade 2—moderate, and grade 3—severe pathological peri-tendinous Doppler signal within the synovial sheath [19].

Bone erosions are defined, according to OMERACT, as intra-articular discontinuity of the bone surface that is visible in two perpendicular planes (Fig. 3.10) [17].

Over the past decades, CR has been the primary choice in assessing bone erosions. However, early in the disease course, CR cannot always detect bone changes. MSUS was shown to be more sensitive than CR in detecting bone erosions at finger and toe joint level [12, 31, 63–65]. This was supported by various studies which

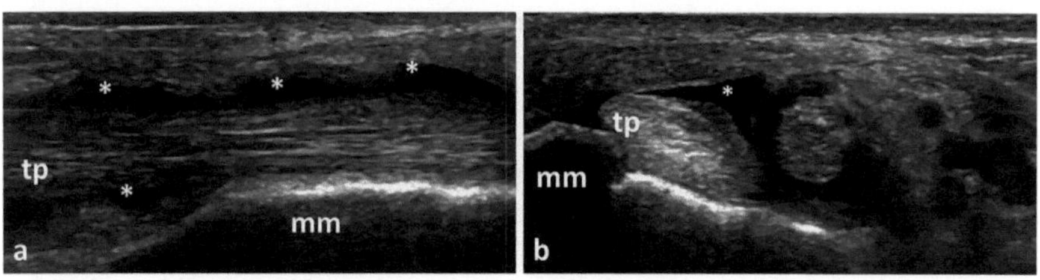

Fig. 3.9 Longitudinal (**a**) and transversal (**b**) view over tibialis posterior tendon showing B-mode tenosynovitis (*asterix*); *tp* tibialis posterior, *mm* medial malleolus

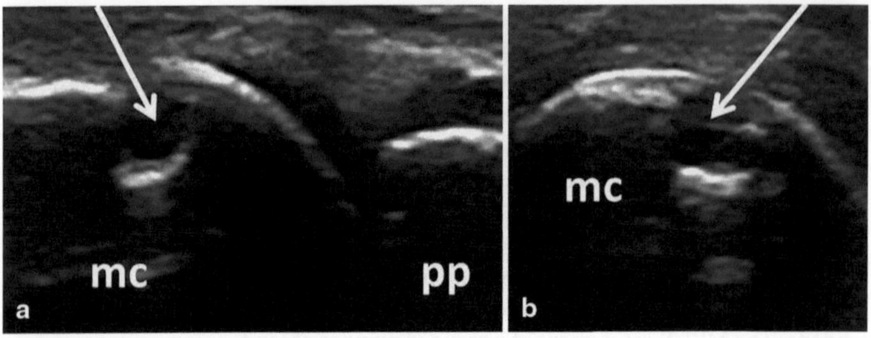

Fig. 3.10 Longitudinal (**a**) and transversal (**b**) view of the dorsal aspect of the metacarpophalangeal joint showing an erosion (*arrow*); *mc* metacarpal head, *pp* proximal phalanx

revealed a high agreement between US and MRI in detecting bone erosion at hand and foot finger joints [12, 31]. At the hand finger joint level, the agreement was higher in joints with good accessibility for US, like second and fifth MCP [12]. Finzel et al. reported a good correlation between the severity of erosions detected by US and by micro-computed tomography (micro-CT) [66]. Furthermore, using MRI and CT as reference, US has shown a high specificity with a moderate sensitivity in detecting bone erosions [12, 67]. However, sensitivity improved considerably, without losing specificity, when only US-accessible areas, i.e., radial second MCP, ulnar fifth MCP, and all dorsal/palmar aspects, were included [67]. In another study, Døhn et al. compared US- and MRI-detected erosions with CT-detected erosions at MCP joints in RA patients without CR-detected erosions. With CT as reference, US and MRI resulted in high specificity in detecting bone erosions even in normal radiographic MCP joints [68]. On the other hand, when compared to CT, false-positive results for US-detected erosions could be noted, especially in small joints. This is mostly due to the misinterpretation of normal vascular bone channels and normal grooves of the metacarpal heads that can mimic bone erosions [66].

On another front, MSUS has shown a good reliability for detecting bone erosions. Several studies revealed good to excellent intra- and interobserver reliability for both small and large joints [18, 39, 63, 69]. For small joints, the highest agreement was found at second MCP joint [18].

In published studies on RA patients, the most commonly used score for bone erosions has been semiquantitative scores. There are a number of different semiquantitative scores for bone erosion used in these studies. Szkudlarek et al. proposed a semiquantitative scoring from 0 to 3 (0—regular bone surface, 1—irregularity of the bone surface without formation of a defect seen in two planes, 2—formation of a defect in the surface of the bone seen in two planes, 3—bone defect creating extensive bone destruction) system that demonstrated a good intra-observer agreement [18]. Another semiquantitative scoring system was proposed by Wakefiel et al. based on the size of erosions (small erosion ≤ 2 mm, moderate erosion 2–4 mm, and large erosion ≥ 4 mm), that also showed good intra- and interobserver agreement [63].

The most frequent site for MSUS-detected erosions in RA patients at hand level are the second MCP and fifth MCP joints, and at foot level are the first MTP and fifth MTP joints, while the fewest erosions are detected at fourth MCP joint [63, 64, 70, 71]. An explanation for these findings could be the acoustic window for US beams at this level. The majority of erosions were detected at the metacarpal heads and on the radial or ulnar sites of the joints, while lesser erosions were detected at the phalangeal bases and on the dorsal and volar aspect of the joints [63]. In contrast, at wrist level, US evaluation of erosions is difficult due to the irregularities of the bone margins, the presence of several nutrition channels, and poor US-window for structure visualization [72]. Humeral head erosions can be seen in a significant number of healthy people [73]. Thus, clinical conclusion cannot be drawn from US-detected erosion at this level, especially for small erosions.

Tendon damage is a common finding in long-standing RA patients as repeated or persistent tendon inflammation can lead to structural damage and tendon rupture. Compared to MRI, US is seemingly more sensitive in detecting partial finger extensor tendon tear [74]. However, until recently, there was no commonly agreed definition or scoring system for tendon damage. In 2013, the OMERACT MSUS group defined tendon damage on BM as internal and/or peripheral focal tendon defect (i.e., absence of fibers) in the region enclosed by tendon sheath (Fig. 3.11), seen in two perpendicular planes [75]. For tendon damage, a three-grade semiquantitative scoring system has recently been developed (grade 0—normal, grade 1—partial, and grade 3—complete rupture). This scoring system resulted in good interobserver agreement and excellent intra-observer agreement [75]. Good to excellent intra- and interobserver agreement was found in various studies for tendon damage at hand and ankle tendons [20, 22].

Bursitis Inflammation of periarticular soft tissue, including synovial bursae, is a major cause of pain in RA patients. Accurate diagnosis of such pathologies is of utmost importance for adequate management of these patients. The most studied bursal sites are at the shoulder and foot level. Figure 3.12 shows an example of subacromial subdeltoid (SASD) bursitis in an RA patient. In a study carried out by Bruyn et al., they reported a good overall agreement between US and MRI in detecting SASD bursitis. In the same study, intra- and interobserver reliability for BM and PD ranged from poor to moderate [76]. At forefoot bursitis, the most frequently involved site was the 4/5 inter-metatarsal space [77].

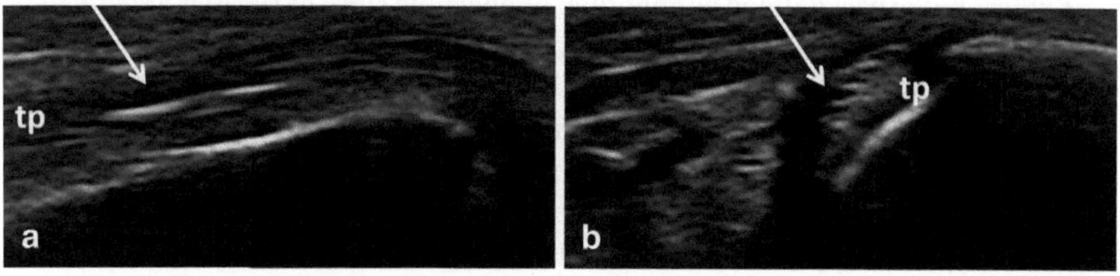

Fig. 3.11 Longitudinal (**a**) and transversal (**b**) view over tibialis posterior tendon showing tendon damage (*arrow*); *tp* tibialis posterior, *mm* medial malleolus

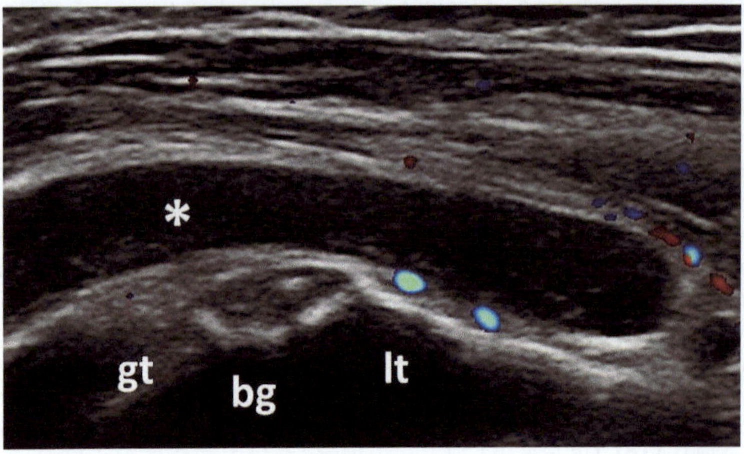

Fig. 3.12 Transversal view over the bicipital grove showing subacromial subdeltoid bursitis (*asterix*); Doppler signal is seen outside the bursa; *gt* greater tuberosity, *lt* lesser tuberosity, *bg* bicipital groove

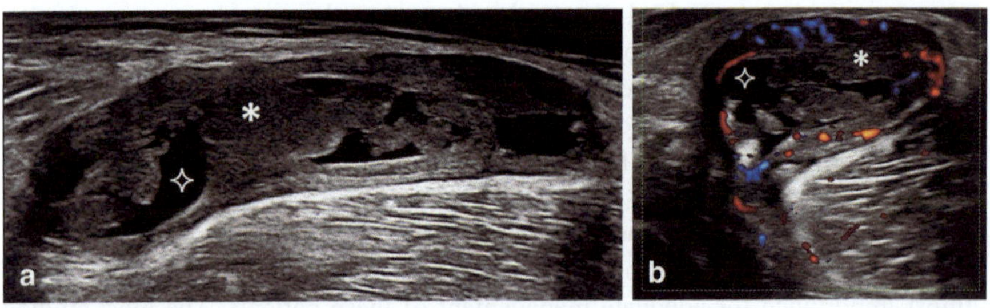

Fig. 3.13 Longitudinal extended (**a**) and transversal (**b**) view over a Baker's cyst showing synovial hypertrophy (*asterix*) with Doppler signal and synovial effusion (*star*)

An important cause of knee pain is the presence of Baker's cyst. Figure 3.13 shows an example of Baker cyst showing SH with Doppler signal and SE in an RA patient. Diagnosis is essential as treatment is different from other knee pathologies. In a number of studies, almost a half of RA patients assessed had US-detected Baker's cyst [8, 78, 79] of which less than a half were detected by clinical examination [78].

Enthesopathy was defined as abnormal hypoechoic (loss of fibrillar architecture) and/or thickened tendon at its bony attachment seen in two perpendicular planes that may exhibit Doppler signal and/or bony changes including enthesophytes, erosions, or irregularity [17]. Although entheseal abnormalities in RA patients are insufficiently studied, it seems that these are more frequent than has been previously estimated. Genc et al. compared tendon and entheseal US abnormalities of lower and upper limb in RA patients with spondyloarthropathy (SpA) and healthy controls. They found that there were no significant differences between RA and SpA patients in terms of tendon and entheseal involvement, whereas RA patients presented more tendon and entheseal pathologies than healthy controls. The most affected entheseal sites in RA and SpA patients were the distal and proximal patellar tendon and Achilles tendon. No differences from the control group were found in the involvement of plantar aponeurosis [80].

Rheumatoid nodules are more frequently found at pressure sites, usually associated with more severe disease. At US examinations, they appear oval shaped, with well-defined hypoechoic formation, generally homogenous, and in the majority of cases they are usually found close to the bone surface. They can present a central very hypoechoic, well-defined area. Compared to gout tophi, rheumatoid nodules show less frequent posterior acoustic shadowing and less erosion at adjacent bone level [81].

Clinical Applications

Besides demonstrating to be a valid, reliable, and sensitive-to-change tool in inflammatory arthritis, MSUS has also been shown to be more sensitive than clinical assessment in detecting joint inflammation in RA patients. Irrespective of the number of joints studied, disease activity, or duration, MSUS has detected inflammation in significantly more joints than clinical assessment [7–16, 39].

Currently, the main role of US assessment in RA includes diagnosis, monitoring disease activity and treatment response as well as guiding intra-articular procedures.

Diagnosis

Although several joint abnormalities can be detected by MSUS, none of them are pathognomonic for RA. However, a number of studies have shown that MSUS can be used for diagnostic purposes in addition to clinical evaluation [82, 83], especially in seronegative patients [84]. Noteworthy US findings were not interpreted out of clinical context. For the diagnostic purposes, the majority of studies have investigated the added value of US-detected abnormalities at small joint level (i.e., MCP, PIP, MTP wrist and ankle joints). Early US-detected abnormalities at this level were mostly synovitis and tenosynovitis, although erosion detection was not uncommon. Results of these studies paved the way for EULAR recommendations regarding the use of US when diagnostic doubts arise, as this would improve the certainty of an RA diagnosis above clinical criteria alone [23].

For BM synovitis, there is no consensus regarding the relevance of grade 1 of synovitis, especially in small joints. BM synovitis grade 1 can be detected in a significant percentage of healthy people, and at least for diagnosis purposes its use is debatable [85]. Figure 3.14 shows an example of grade 1 BM synovitis in healthy people. Although the presence of intra-articular Doppler signals is associated more frequently with pathology, it can be detected also in healthy people [86]. This is possible mainly due to the improvement of machines sensitivity which allows the detection of normal vessels. Thus, the sensitivity of the machine must be considered

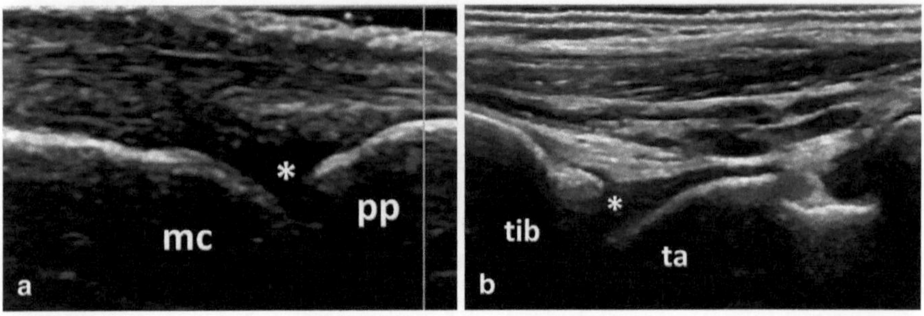

Fig. 3.14 Images showing B-mode synovitis grade 1 (*asterix*) at metacarpophalangeal joint (**a**) and tibiotalar joint (**b**) in healthy people; *mc* metacarpal head, *pp* proximal phalanx, *tib* tibia, *ta* talus

and settings adjusted accordingly. The presence of grade 1 synovitis, especially in one isolated joint, without other inflammatory changes should be carefully considered as diagnosis value. Noteworthy, all the above remarks are valid when assessing patients without any anti-inflammatory treatment, as this can mask the BM synovitis and Doppler activity. Also synovitis, erosions can also be seen in healthy people [31, 68, 87].

The minimal threshold enough to diagnose active inflammatory arthritis remains a matter of controversy. This may include the minimal degree of synovitis, number of joints with synovitis, degree of erosions, or a combination of any three that are necessary to make an RA diagnosis. Millet et al. suggested a minimum two joints showing grade 2 or 3 for BM-detected synovitis or two cases of bone erosion [87]. Other studies added tendon evaluation to US assessment of inflammatory arthritis to make an RA diagnosis. These US findings—together with clinical and laboratory findings—increase the probability of an inflammatory arthritis diagnosis. In the study by Freeston et al., US evaluation of wrists and MCP joints and flexor tendons was added to clinical examination in patients with very early inflammatory arthritis. In seronegative patients with positive CRP, swollen joints and erosion on CR, the presence of a grade 3 BM synovitis, at least a grade 1of PD synovitis, or at least one erosion increases the probability of inflammatory arthritis from 30 to 94 % [84]. In a study of early, untreated oligoarthritis, following US assessment, about one third of patients fulfilled polyarthritis classification criteria owing to the presence of subclinical synovitis [9]. According to a study by Scire et al., a PD score of two or more was highly specific for the diagnosis of RA [88]. Thus, the tendon evaluation can add valuable information about inflammatory activity. Furthermore, it is important to remember that a number of the tendons' synovial sheaths can communicate to the synovial joint (e.g., biceps tendon sheath, foot first finger flexor tendon).

In early, untreated RA patients, finger flexor tenosynovitis was observed more frequently than peri-extensor tenosynovitis (Fig. 3.15), and the most frequently involved tendons were the tendons from second and third fingers [21]. In an MRI study, hand flexor tenosynovitis was a strong predictor for RA in early, unspecified arthritis or suspected RA [89].

The detection of erosions is also useful in RA diagnosis. Although erosions can be detected in several rheumatic diseases, some areas can be considered as target

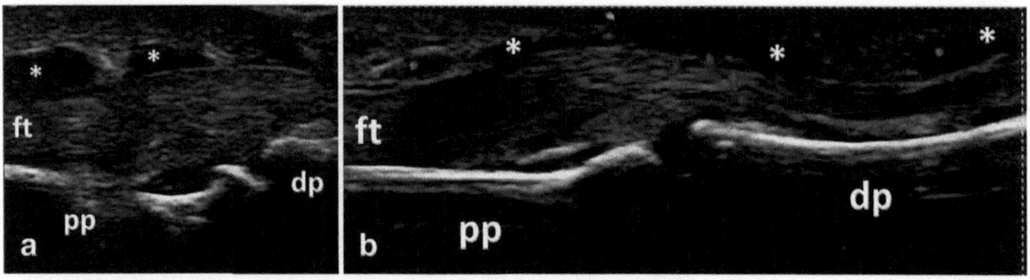

Fig. 3.15 Longitudinal view of the palmar aspect of the proximal interphalangeal joint showing B-mode flexor tenosynovitis (*asterix*) (**a**) with Doppler signal (**b**); *ft* flexor tendon, *pp* proximal phalanx, *dp* distal phalanx

for RA. US assessment of the styloid process of the ulnar, the radial part of second MCP joint, and the ulnar part of fifth MCP joint can provide important information for RA diagnosis. Zayat et al. investigated the specificity and sensitivity of US-detected erosion in RA compared to different musculoskeletal diseases (i.e., psoriatic arthritis (PsA), osteoarthritis (OA), and gout) and healthy controls. Although RA patients presented more US-detected erosions than other groups, the differences were not significant. When RA-target sites only were included, i.e., second and third MCP, fifth MTP, and distal ulna, the sensitivity improved but was still not specific for RA. However, the presence of large erosions covering between one- to two-thirds of the surface of one quadrant in any of RA-target sites was highly specific for RA. Furthermore, the presence of any erosion at the level of fifth MTP was both specific and sensitive for RA [90].

New classification criteria for RA have been developed by American College of Rheumatology (ACR) and EULAR and published in 2010. A number of recent studies have shown the value of adding US in making the diagnosis of RA [82, 91]. Furthermore, based on US finding, patients were more accurately classified as requiring MTX treatment [82].

In addition to its diagnostic role, the US-detected inflammation can also be used to predict the progression of undifferentiated inflammatory arthritis to RA. Salaffi et al. found that the strongest independent predictor factor for developing RA in early, undifferentiated arthritis was PD positivity. Moreover, the positivity of PD in more than three joints increased significantly the risk of progression to RA [92]. Furthermore, van de Stadt et al. found that the presence of both BM and PD synovitis increases the risk for the development of arthritis in patients with arthralgia, without arthritis at clinical examination and positive anti-citrullinated protein antibodies (ACPA) and/or immunoglobulin M-rheumatoid factor (IgM-RF) [93]. In another work, Navalho et al. studied the association between MRI-detected synovitis and tenosynovitis at hand level with progression to RA. They found that ECU tenosynovitis, finger flexor tenosynovitis of the second finger, and radio-carpal synovitis were significantly associated with progression to RA [94].

In other rheumatic diseases, joint inflammatory activity can also be detected. Synovitis, tenosynovitis, erosions, and Doppler signals were reported in a variety of inflammatory and noninflammatory diseases, e.g., PsA, OA. Some US findings can help differentiate RA from other inflammatory diseases. For example, peritenon

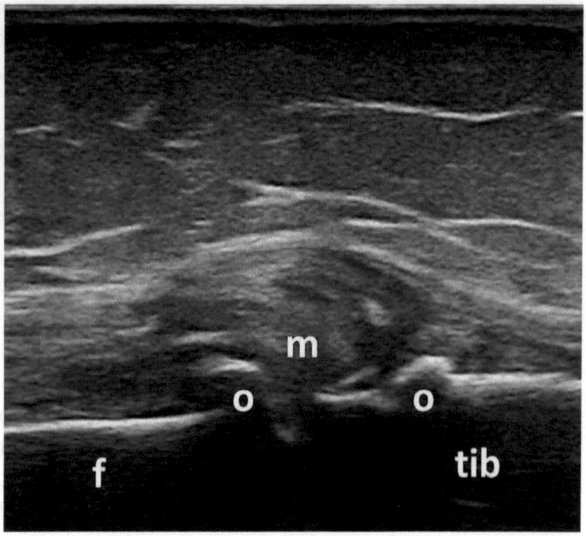

Fig. 3.16 Longitudinal view of the medial aspect of the knee joint showing degenerative changes in a patient with rheumatoid arthritis and concomitant osteoarthritis; *f* medial femoral condyle, *tib* tibia, *m* medial meniscus, *o* osteophytes

inflammation of finger extensor tendons is highly characteristic of PsA [95], while Doppler activity at the enthesis level is characteristic in SpA patients [96].

However, it should be highlighted that RA patients may also experience other rheumatic diseases that have different treatments and prognosis. Joints included in clinical scores (e.g., shoulder, knee) are often affected by degenerative processes and clinical differentiation of these pathologies can be challenging. Figure 3.16 shows an example of degenerative changes of the knee joint in an RA patient with knee pain, whereas, Fig. 3.17 shows an example of full-thickness tear of the supraspinatus tendon in an RA patient with shoulder pain. MSUS is useful in identifying pathologic changes related to degenerative musculoskeletal disorders (e.g., OA) or regional pain syndromes in RA patients, thus helping in differentiating these pathologies from active disease. Figure 3.18 shows an example of BM synovitis with Doppler signal of a PIP joint in an OA patient.

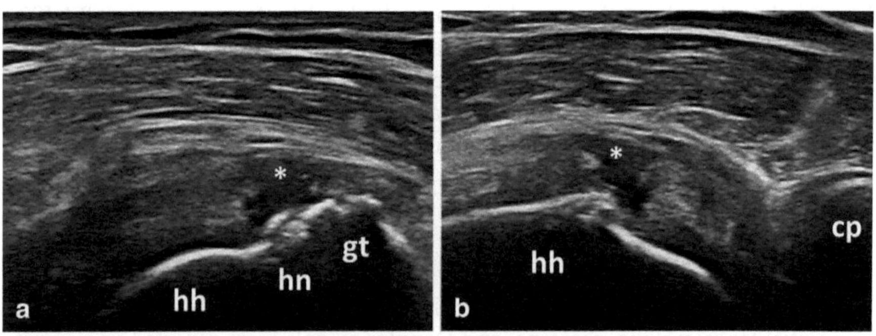

Fig. 3.17 Longitudinal (**a**) and transversal (**b**) view over the supraspinatus tendon showing full-thickness tear (*asterix*); *gt* greater tuberosity, *hn* anatomic humeral neck, *hh* humeral head, *cp* coracoid process

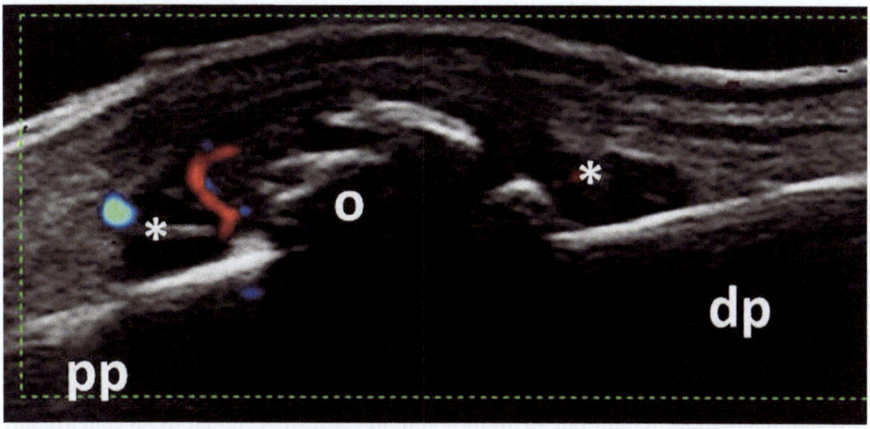

Fig. 3.18 Longitudinal view of the dorsal aspect of the proximal interphalangeal joint showing B-mode synovitis (*asterix*) with Doppler signal and osteophytes in a patient with osteoarthritis; *pp* proximal phalanx, *dp* distal phalanx, *o* osteophytes

Furthermore, in patients already diagnosed with RA, MSUS can help to differentiate active disease from chronic structural damage. Although there is no consensus in the definition of US-active synovitis, the presence of Doppler signal is considered as a sign of active inflammation. At tendon level, MSUS can help in differentiating active tendon inflammation from chronic tendon damage. Although clinical examination can identify complete tendon tear, partial tear remains undiagnosed. Thus, tendon damage although secondary to persistent tendon inflammation, does not represent active disease; thereby does not require changes in DMARDs treatment.

Monitoring Disease Activity

In several studies, MSUS has shown a good sensitivity to change [15, 42–48]. When compared to MRI, MSUS showed a high sensitivity in detecting both synovitis and tenosynovitis [21, 59, 64]. MSUS was also reported to be more sensitive than CR and as sensitive as MRI, in detecting bone erosion [69]. In addition, earlier studies revealed that MSUS was more sensitive than clinical assessment in detecting joint inflammation [31, 47, 79]. All the evidence coming from the studies points to MSUS as a useful and valuable tool in monitoring RA patients. Taking all of this into consideration, EULAR recommendations for the use of imaging in RA endorsed the use of US for more accurate assessment of inflammation [23].

As far as monitoring disease activity, a comprehensive US assessment of all accessible peripheral joints would be time-consuming and not necessarily practical in daily clinical practice. However, until now, there is no consensus on how many and which joints should be assessed. Several reduced joint counts that have been studied are discussed below. Scheel et al. suggested that US evaluation of second to fourth PIP and MCP joints with a semiquantitative score is sufficient for diagnosis and follow-up in RA patients [33]. A study carried out by Naredo et al. depicted high correlation of a reduced 12-joint count, i.e., bilateral elbow, wrist, second and third MCP and PIP, knee and ankle, with a comprehensive 44-joint count. Moreover, this reduced 12-joint count was shown to represent accurately the response to biologic therapy in RA patients [46]. In another study, Backhaus et al. developed a reduced US joint count which included assessment of only seven small joints, i.e., wrist, second and third MCP and PIP, second and fifth MTP of the clinically dominant side, combining soft tissue changes such as synovitis, tenosynovitis, paratenonitis with erosive bone lesions. This US seven-joint count showed to be a sensitive tool in monitoring patients with inflammatory arthritis (i.e., RA and PsA) in daily clinical practice [97]. This score was found to be sensitive to change in a cohort of patients with RA treated with csDMARDs or bDMARDs [98]. According to the study by Perricone et al., a reduced 6-joint count, i.e., bilateral wrist, second MCP, and knee, correlated excellently with the 12-joint count and was also shown to be sensitive to change in RA patients treated with etanercept (ETA) [99].

On the other hand, other studies focused on Doppler US as the cornerstone for monitoring the disease activity. In the study done by Ellegaard et al., Doppler US quantitative assessment of synovitis of the most symptomatic wrist showed fair to moderate correlation with disease activity index (DAS) for 28 joints (DAS28), swollen joint count, CRP, and ESR in RA patients starting anti-tumor necrosis factor (TNF) therapy. In conclusion, the authors suggested that CDUS examination of only one affected joint could be sufficient in assessing disease activity [100]. In another study, Damjanov et al. calculated the disease activity using US score (US DAS) by replacing tender joint count with PD semiquantitative score for synovitis for 22 joints, including wrists, MCP, and MTP joints, and swollen joint count with BM semiquantitative score for synovitis for 28 joints, i.e., MCP joints, PIP joints, wrists, elbows, shoulders, and knees. With MRI as reference, US DAS was more reliable than DAS28 in assessing both disease activity and further joint damage [101]. Another US score for large joints was developed by Hartung W et al. Sonography of large joints in rheumatology (SOLAR) score was used to assess the grade of inflammation in the shoulder, elbow, hip, and knee joints in RA patients. They calculated a score for each joint by summing the BM/PD scores recorded for each joint's recesses. After 12 months of treatment, all parameters showed significant improvement, except PDUS scores for shoulder and hip. They concluded that the SOLAR score is a feasible tool for evaluating large joints in patients with RA [102]. Lastly, Hammer BH and Kvien T compared previously described reduced joint counts (i.e., 7, 12, 28, and 44 joints [14]) with comprehensive 78 joints, 36 tendons, and 2 bursae count in RA patients starting bDMARDs. They observed high correlations between the comprehensive 78-joint count and all of the reduced joint count, for both BM and PD variables [103].

However, a difficult question remains unanswered, which score to be used? In an attempt to address this question, Mandel et al. compared the reliability and discriminant capacity of 11 different US scoring systems, including different combinations of 42, 28, 20, 16, 12, 10, 8, 7 joints counts. They found no significant differences between these systems, suggesting that simplified joint counts may perform at least as well as more comprehensive scores [41]. On another front, in standard daily clinical practice, the significance of BM synovitis grade 1, is another challenge which needs to be addressed especially in long-standing disease where fibrotic, irreversible changes can also be found. Witt et al. investigated the clinical relevance of grade 1 synovitis in the wrist, MCP, PIP, and MTP joints, comparing early onset and established RA patients with healthy controls. Considering the frequency of BMUS synovitis grade 1, at MCP joints there were no significant differences between the three groups; neither were significant differences found between established RA patients and healthy controls at PIP and MTP joint level. In contrast, at wrist level, they found significant more synovitis in patients with early onset and established RA disease compared to healthy controls. After 6 months of treatment, significant more joints with initial synovitis grade 1 remained unchanged compared to initial synovitis grade 2 and 3. On the other hand, as expected, the majority of joints with synovitis grade 1 were neither painful nor swollen on clinical examination nor exhibited PD activity [104]. However, in this study, the issue of predictive value in relation to structural damage was not assessed.

Attempts to define US-active synovitis have been deliberated, taking into consideration the evidence just mentioned. As structural damage has been associated mostly, in the majority of studies, with the presence of Doppler signal [105, 106], synovitis was considered as active if Doppler signal was detected [107, 108]. Other authors considered synovitis active if SH is greater than two together with the presence of Doppler activity [109]. At the patient level, a cumulative BM or PD score for defining active disease has not yet been developed. However, care must be taken with patients on anti-inflammatory medication such as nonsteroidal anti-inflammatory drugs (NSAIDs), steroidal anti-inflammatory drugs, as these may mask the BM and Doppler activity. Zayat et al. found a significant increase in BM and Doppler parameters after stopping treatment with NSAIDs for five drug half-life times [110]. Yet, pharmacokinetics of subcutaneous anti-TNF drugs showed no effect on US findings [111].

With respect to bone erosions, an important issue to be addressed was whether US could be used to monitor the erosion course. In RA patients, US-detected inflammation has shown predictive value in relation to structural damage, measured either by CR or MRI. This was documented by the results of earlier studies which revealed an increase of US-detected erosions at hand finger lever in RA patients, both at short- and long-term follow-up [47, 59, 112]. In most of the studies, the detection of vascularization in synovial proliferation by Doppler techniques has been shown to be the strongest predictor for further structural damage [48, 49, 105, 106, 113, 114].

Taylor et al. evaluated the predictive value of US-detected synovitis in a randomized controlled trial of early RA patients treated with IFX versus placebo. After 54 weeks, in the placebo group, there was a significant correlation between both baseline synovial thickness and vascularity and progression in total radiographic score, whereas in the IFX group, the negative correlation was insignificant [49]. According to a study by Naredo et al., US assessment of joint inflammation in early RA patients found that time-integrated values of Doppler variables were the strongest predictors for disease activity during subsequent visits [48]. Dougados et al. found that in active RA patients, radiographic progression was observed more frequently if BM or PD-detected baseline synovitis was present and persistent following 4 months of treatment [115]. In another study, Bøyesen et al. compared the predictive value of clinical and laboratory variables, BMUS and MRI of dominant wrist and CR of wrists and hands in early RA. After a 1-year follow-up period, only BMUS and MRI bone marrow edema (BME) were found to be independent predictors for MRI erosive progression. Moreover, BMUS inflammation, MRI synovitis, and BME performed slightly better than clinical and laboratory variables in identifying early RA patients at risk of developing MRI erosions [113]. This was in concordance with the results of another study carried out by Lillegraven et al. who investigated the predictive value of hand inflammatory US findings in patients with early RA. Baseline ECU tenosynovitis was found to be an independent factor for MRI progression at 1-year follow-up and for radiographic progression at 3-year follow-up [114].

Another interesting issue that remains relatively unknown is the predictive value of baseline US-detected inflammatory activity in relation to treatment response. In the study by Ellegaard et al., the presence of Doppler activity at baseline showed to be predictive for treatment persistence at 1 year in patients staring anti-TNF treatment, while clinical and laboratory variables showed no significant association [116]. This has been endorsed by the EULAR recommendations for the use of imaging in RA, as EULAR sanctioned the use of imaging techniques to predict response to treatment [23].

Furthermore, the role of US in identifying subclinical inflammation represents a further expansion of this tool's use in standard clinical practice. Several studies have demonstrated the presence of MSUS-detected synovitis in patients in clinical remission. This subclinical synovitis has been detected in RA patients irrespective of the treatment received, whether synthetic or biologic DMARD [14, 16, 21, 38, 105, 106, 117–121]. Figures 3.19 and 3.20 show examples of BM synovitis in an RA patient in clinical remission. Interestingly, earlier studies revealed that regardless of what remission criteria were used (DAS28 or ACR remission criteria) there were no significant

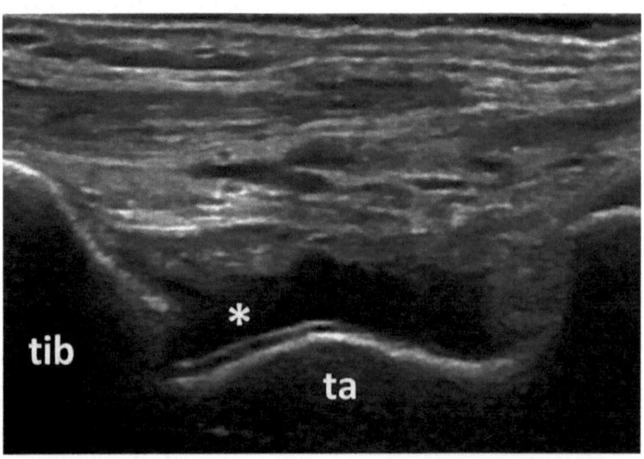

Fig. 3.19 Longitudinal view of the dorsal aspect of the tibiotalar joint showing B-mode (*asterix*) synovitis in a patient in clinical remission; *tib* tibia, *ta* talus

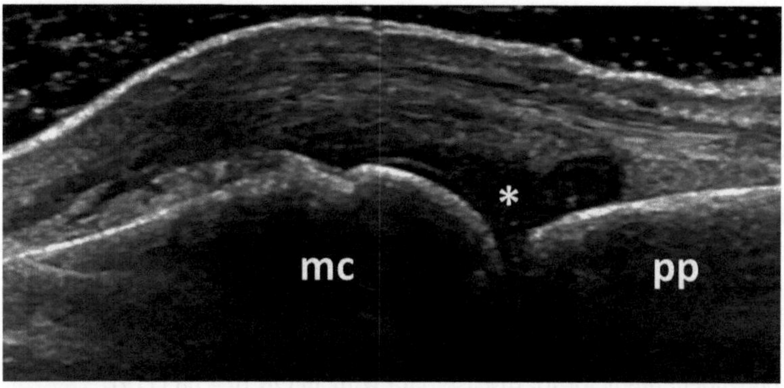

Fig. 3.20 Longitudinal view of the dorsal aspect of the metacarpophalangeal joint showing B-mode synovitis (*asterix*) in a patient in clinical remission; *mc* metacarpal head, *pp* proximal phalanx

differences in the prevalence of US-detected synovitis [38]. In the study carried out by Kawashiri et al., more than a half of RA patients in clinical remission according to simplified disease activity index (SDAI) and without any tender or swollen joint at clinical assessment, presented BM synovitis and/or Doppler activity [122]. However, patients were deemed to be in clinical remission by their attending rheumatologist. As expected, the patients who did not fulfill ACR or DAS28 remission criteria, were more likely to have a significant higher number of joints with US-detected SH. Moreover, even if clinical assessment did not detect tender and swollen joints, BMSH and PD synovitis were still detected in a high number of patients [38]. Adding the disease duration to the equation was assessed in the study performed by Peluso et al. The study revealed that the patients with early onset RA who achieved remission had lower PD synovitis score and were more likely to present no synovitis on imaging compared to patients with long-standing RA in remission [117]. Saleem et al. compared RA patients in clinical remission who received bDMARD as first-line therapy with those with delayed treatment. They found a significantly lower BM synovitis scores in the first group, but similar PD scores [118]. The findings of these studies may have an important role in providing an explanation for radiologic progressive joint damage found in patients with prolonged clinical remission according to ACR criteria [123]. An MRI study in patients with early RA suggested a direct link between synovial inflammation and structural damage as no MRI-detected new erosions were seen in any joint without synovitis. The authors of this study concluded that synovitis appears to be the primary abnormality and the likelihood of bone erosions is related to the level of synovitis [124].

Predictive value of subclinical synovitis in relation to radiographic structural damage and disease relapse/flare had been investigated in several studies. The great majority of studies found associations between Doppler variables, not BM synovitis or structural damage, and/or disease relapse/flare. An explanation of these results can be that BM SH may reflect active disease, especially in early RA, which is reversible with treatment, but also may represent a chronically thickened, fibrotic, and irreversible synovial tissue in later stages of disease. In contrast, Doppler signals reflect increased vascularity which is associated with active inflammation. The impact of the persistence of enhanced vascularity assessed by PD in RA patients who achieved clinical remission has been investigated in several other studies. Outcome of these studies revealed that PDUS predicts radiographic progression and disease relapse or flare in RA patients in clinical remission [14, 105, 106, 119]. In patients treated with either csDMARDs or bDMARDs, the baseline variables for Doppler activity were associated with radiographic progression [105] and disease relapse [106]. In the study carried out by Scire et al., the persistence of PD signal in a single joint had proved to be the main predictor for short-term relapse in RA patients in clinical remission with an odds ratio (OR) of 12.8 [14]. In concordance with these findings, the study carried out by Saleem et al. revealed that the presence of PD signal was found to be the strongest independent predictor for disease flare with an OR of 4.08 [119].

Recently, Fukae et al. studied the association between quantitative measurements of Doppler signal at hand finger level and radiographic progression at 52 weeks in patients with long-term clinical low disease activity (DAS28 < 3.2). They found that structural damage occurred more frequent in joint with synovial hypervascularity and progression of structural damage occurred irrespective to the level of cumulative synovial vascularity. These results show that joints can demonstrate radiographic progression in the presence of any Doppler activity, even if sustained low disease activity is achieved [125]. Considering that the presence of PDUS in RA patients in clinical remission may be predictive of radiographic progression and disease flare or relapse [14, 105, 118], EULAR recommended the use of US for assessing persistent inflammation even in patients in clinical remission [23].

A reduced joint count for detecting subclinical synovitis in patients in clinical remission was proposed by Naredo et al. Evaluation of bilateral wrist, second to fifth MCP joints, ankle, and second to fifth MTP joints showed a high correlation with a comprehensive 44-joint count for both BM and PD [16]. The same high correlation was found also for previous described 12-joint counts [16]. However, further studies are needed to investigate the maximal acceptable synovitis at joint level that will not produce structural damage and therefore will not require more aggressive treatment. Besides, studies are needed to established cutoff values at patient level for which a BM synovitis can be considered relevant in relation to disease activity and progression. Ongoing studies try to find if using US as a target for remission in RA patients can improve outcomes.

Guided Intra-Articular Procedures

Joint puncture for fluid aspiration purposes or intra-articular injection of different drugs represents routine procedures for rheumatologists. US guidance has a number of advantages against blinded injections. Firstly, US allows for better diagnosis and better characterization of the pattern of joint inflammation, whether it be SH or SE. Secondly, US allows direct visualization of the needle within the joint structure facilitating fluid aspiration. Therefore US-guided procedures significantly improve the accuracy of intra-articular injections [126–129]. Moreover, US-guided punctures

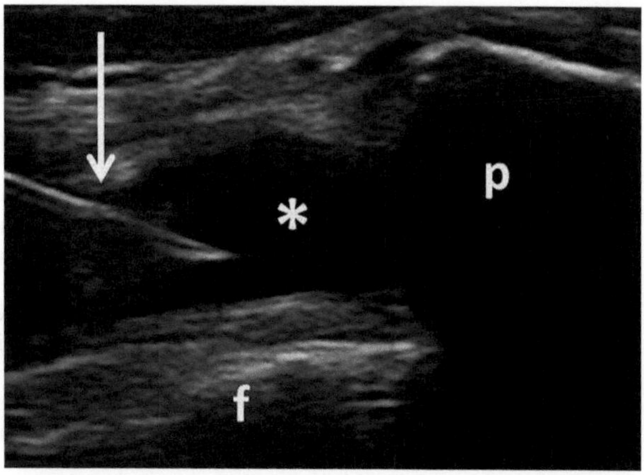

Fig. 3.21 Transverse view over the parapatellar recess of the knee showing B-mode synovitis (*asterix*); the needle (*arrow*) is used for ultrasound-guided injection; *f* femur, *p* patella

significantly reduce patient discomforts and shorten procedure (Fig. 3.21). However, in a study by Cunnington et al., no differences were found regarding improvement in clinical outcomes between US-guided and blinded corticosteroid injection [129].

Conclusions

In conclusion, MSUS is a valid, reliable, and sensitive-to-change tool in RA. Moreover, MSUS has demonstrated to be more sensitive than clinical assessment in detecting joint inflammation. The main clinical applications of MSUS in RA include diagnosis, monitoring disease activity and treatment response, and guiding intra-articular procedures. For RA diagnosis, MSUS has been successfully used in addition to clinical evaluation. For monitoring disease activity, several scoring systems including a reduced number of joints have been developed, with similar performances. US-detected joint inflammation has been observed in patients in clinical remission. Doppler-detected synovitis has showed predictive value in relation to radiographic damage and disease flare/relapse in both active and remission disease. The US guidance of procedures improves the accuracy of intra-articular injections.

References

1. Pincus T, Callahan LF, Sale WG, Brooks AL, Payne LE, Vaughn WK. Severe functional declines, work disability, and increased mortality in seventy-five rheumatoid arthritis patients studied over nine years. Arthritis Rheum. 1984;27(8):864–72.
2. Scott DL, Symmons DP, Coulton BL, Popert AJ. Long-term outcome of treating rheumatoid arthritis: results after 20 years. Lancet 1987;1:1108–11.
3. Aletaha D, Neogi T, Silman AJ, Funovits J, Felson DT, Bingham CO 3rd, et al. Rheumatoid arthritis classification criteria: an American College of Rheumatology/European League Against Rheumatism collaborative initiative. Arthritis Rheum. 2010;62(9):2569–81.

4. Barnabe C, Martin BJ, Ghali WA. Systematic review and meta-analysis: anti-tumor necrosis factor alpha therapy and cardiovascular events in rheumatoid arthritis. Arthritis Care Res. 2011;63(4):522–9.
5. Smolen JS, Aletaha D, Bijlsma JW, Breedveld FC, Boumpas D, Burmester G, et al. Treating rheumatoid arthritis to target: recommendations of an international task force. Ann Rheum Dis. 2010;69(4):631–7.
6. Nam JL, Ramiro S, Gaujoux-Viala C, Takase K, Leon-Garcia M, Emery P, et al. Efficacy of biological disease-modifying antirheumatic drugs: a systematic literature review informing the 2013 update of the EULAR recommendations for the management of rheumatoid arthritis. Ann Rheum Dis. 2014;73(3):516–28.
7. Karim Z, Wakefield RJ, Quinn M, Conaghan PG, Brown AK, Veale DJ, et al. Validation and reproducibility of ultrasonography in the detection of synovitis in the knee: a comparison with arthroscopy and clinical examination. Arthritis Rheum. 2004;50(2):387–94.
8. Kane D, Balint PV, Sturrock RD. Ultrasonography is superior to clinical examination in the detection and localization of knee joint effusion in rheumatoid arthritis. J Rheumatol. 2003;30(5):966–71.
9. Wakefield RJ, Green MJ, Marzo-Ortega H, Conaghan PG, Gibbon WW, McGonagle D, et al. Should oligoarthritis be reclassified? Ultrasound reveals a high prevalence of subclinical disease. Ann Rheum Dis. 2004;63(4):382–5.
10. Naredo E, Bonilla G, Gamero F, Uson J, Carmona L, Laffon A. Assessment of inflammatory activity in rheumatoid arthritis: a comparative study of clinical evaluation with grey scale and power Doppler ultrasonography. Ann Rheum Dis. 2005;64(3):375–81.
11. Scheel AK, Schmidt WA, Hermann KG, Bruyn GA, D'Agostino MA, Grassi W, et al. Interobserver reliability of rheumatologists performing musculoskeletal ultrasonography: results from a EULAR "Train the trainers" course. Ann Rheum Dis. 2005;64(7):1043–9.
12. Szkudlarek M, Klarlund M, Narvestad E, Court-Payen M, Strandberg C, Jensen KE, et al. Ultrasonography of the metacarpophalangeal and proximal interphalangeal joints in rheumatoid arthritis: a comparison with magnetic resonance imaging, conventional radiography and clinical examination. Arthritis Res Ther. 2006;8(2):R52.
13. Salaffi F, Filippucci E, Carotti M, Naredo E, Meenagh G, Ciapetti A, et al. Inter-observer agreement of standard joint counts in early rheumatoid arthritis: a comparison with grey scale ultrasonography–a preliminary study. Rheumatology 2008;47(1):54–8.
14. Scire CA, Montecucco C, Codullo V, Epis O, Todoerti M, Caporali R. Ultrasonographic evaluation of joint involvement in early rheumatoid arthritis in clinical remission: power Doppler signal predicts short-term relapse. Rheumatology 2009;48(9):1092–7.
15. Hammer HB, Sveinsson M, Kongtorp AK, Kvien TK. A 78-joints ultrasonographic assessment is associated with clinical assessments and is highly responsive to improvement in a longitudinal study of patients with rheumatoid arthritis starting adalimumab treatment. Ann Rheum Dis. 2010;69(7):1349–51.
16. Naredo E, Valor L, De la Torre I, Martinez-Barrio J, Hinojosa M, Aramburu F, et al. Ultrasound joint inflammation in rheumatoid arthritis in clinical remission: how many and which joints should be assessed? Arthritis Care Res. 2013;65(4):512–7.
17. Wakefield RJ, Balint PV, Szkudlarek M, Filippucci E, Backhaus M, D'Agostino MA, et al. Musculoskeletal ultrasound including definitions for ultrasonographic pathology. J Rheumatol. 2005;32(12):2485–7.
18. Szkudlarek M, Court-Payen M, Jacobsen S, Klarlund M, Thomsen HS, Ostergaard M. Interobserver agreement in ultrasonography of the finger and toe joints in rheumatoid arthritis. Arthritis Rheum. 2003;48(4):955–62.
19. Naredo E, D'Agostino MA, Wakefield RJ, Moller I, Balint PV, Filippucci E, et al. Reliability of a consensus-based ultrasound score for tenosynovitis in rheumatoid arthritis. Ann Rheum Dis. 2013;72(8):1328–34.
20. Bruyn GA, Moller I, Garrido J, Bong D, d'Agostino MA, Iagnocco A, et al. Reliability testing of tendon disease using two different scanning methods in patients with rheumatoid arthritis. Rheumatology 2012;51(9):1655–61.

21. Wakefield RJ, O'Connor PJ, Conaghan PG, McGonagle D, Hensor EM, Gibbon WW, et al. Finger tendon disease in untreated early rheumatoid arthritis: a comparison of ultrasound and magnetic resonance imaging. Arthritis Rheum. 2007;57(7):1158–64.
22. Micu MC, Serra S, Fodor D, Crespo M, Naredo E. Inter-observer reliability of ultrasound detection of tendon abnormalities at the wrist and ankle in patients with rheumatoid arthritis. Rheumatology 2011;50(6):1120–4.
23. Colebatch AN, Edwards CJ, Ostergaard M, van der Heijde D, Balint PV, D'Agostino MA, et al. EULAR recommendations for the use of imaging of the joints in the clinical management of rheumatoid arthritis. Ann Rheum Dis. 2013;72(6):804–14.
24. Fiocco U, Cozzi L, Rigon C, Chieco-Bianchi F, Baldovin M, Cassisi GA, et al. Arthroscopic synovectomy in rheumatoid and psoriatic knee joint synovitis: long-term outcome. Br J Rheumatol. 1996;35(5):463–70.
25. Walther M, Harms H, Krenn V, Radke S, Faehndrich TP, Gohlke F. Correlation of power Doppler sonography with vascularity of the synovial tissue of the knee joint in patients with osteoarthritis and rheumatoid arthritis. Arthritis Rheum. 2001;44(2):331–8.
26. Walther M, Harms H, Krenn V, Radke S, Kirschner S, Gohlke F. Synovial tissue of the hip at power Doppler US: correlation between vascularity and power Doppler US signal. Radiology 2002;225(1):225–31.
27. Andersen M, Ellegaard K, Hebsgaard JB, Christensen R, Torp-Pedersen S, Kvist PH, et al. Ultrasound colour Doppler is associated with synovial pathology in biopsies from hand joints in rheumatoid arthritis patients: a cross-sectional study. Ann Rheum Dis. 2014;73(4):678–83.
28. Takase K, Ohno S, Takeno M, Hama M, Kirino Y, Ihata A, et al. Simultaneous evaluation of long-lasting knee synovitis in patients undergoing arthroplasty by power Doppler ultrasonography and contrast-enhanced MRI in comparison with histopathology. Clin Exp Rheumatol. 2012;30(1):85–92.
29. Koski JM, Saarakkala S, Helle M, Hakulinen U, Heikkinen JO, Hermunen H. Power Doppler ultrasonography and synovitis: correlating ultrasound imaging with histopathological findings and evaluating the performance of ultrasound equipments. Ann Rheum Dis. 2006;65(12):1590–5.
30. Fukae J, Kon Y, Henmi M, Sakamoto F, Narita A, Shimizu M, et al. Change of synovial vascularity in a single finger joint assessed by power Doppler sonography correlated with radiographic change in rheumatoid arthritis: comparative study of a novel quantitative score with a semiquantitative score. Arthritis Care Res. 2010;62(5):657–63.
31. Szkudlarek M, Narvestad E, Klarlund M, Court-Payen M, Thomsen HS, Ostergaard M. Ultrasonography of the metatarsophalangeal joints in rheumatoid arthritis: comparison with magnetic resonance imaging, conventional radiography, and clinical examination. Arthritis Rheum. 2004;50(7):2103–12.
32. Szkudlarek M, Court-Payen M, Strandberg C, Klarlund M, Klausen T, Ostergaard M. Power Doppler ultrasonography for assessment of synovitis in the metacarpophalangeal joints of patients with rheumatoid arthritis: a comparison with dynamic magnetic resonance imaging. Arthritis Rheum. 2001;44(9):2018–23.
33. Scheel AK, Hermann KG, Kahler E, Pasewaldt D, Fritz J, Hamm B, et al. A novel ultrasonographic synovitis scoring system suitable for analyzing finger joint inflammation in rheumatoid arthritis. Arthritis Rheum. 2005;52(3):733–43.
34. Terslev L, Torp-Pedersen S, Savnik A, von der Recke P, Qvistgaard E, Danneskiold-Samsoe B, et al. Doppler ultrasound and magnetic resonance imaging of synovial inflammation of the hand in rheumatoid arthritis: a comparative study. Arthritis Rheum. 2003;48(9):2434–41.
35. Kawashiri SY, Suzuki T, Nakashima Y, Horai Y, Okada A, Nishino A, et al. Synovial inflammation assessed by ultrasonography correlates with MRI-proven osteitis in patients with rheumatoid arthritis. Rheumatology 2014;53(8):1452–6.
36. Kawashiri SY, Kawakami A, Iwamoto N, Fujikawa K, Satoh K, Tamai M, et al. The power Doppler ultrasonography score from 24 synovial sites or 6 simplified synovial sites, including the metacarpophalangeal joints, reflects the clinical disease activity and level of serum biomarkers in patients with rheumatoid arthritis. Rheumatology 2011;50(5):962–5.

37. Naredo E, Moller I, Moragues C, de Agustin JJ, Scheel AK, Grassi W, et al. Interobserver reliability in musculoskeletal ultrasonography: results from a †Teach the Teachers" rheumatologist course. Ann Rheum Dis. 2006;65(1):14–9.
38. Brown AK, Quinn MA, Karim Z, Conaghan PG, Peterfy CG, Hensor E, et al. Presence of significant synovitis in rheumatoid arthritis patients with disease-modifying antirheumatic drug-induced clinical remission: evidence from an imaging study may explain structural progression. Arthritis Rheum. 2006;54(12):3761–73.
39. Filer A, de Pablo P, Allen G, Nightingale P, Jordan A, Jobanputra P, et al. Utility of ultrasound joint counts in the prediction of rheumatoid arthritis in patients with very early synovitis. Ann Rheum Dis. 2011;70(3):500–7.
40. Cheung PP, Dougados M, Gossec L. Reliability of ultrasonography to detect synovitis in rheumatoid arthritis: a systematic literature review of 35 studies (1,415 patients). Arthritis Care Res. 2010;62(3):323–34.
41. Mandl P, Balint PV, Brault Y, Backhaus M, D'Agostino MA, Grassi W, et al. Metrologic properties of ultrasound versus clinical evaluation of synovitis in rheumatoid arthritis: results of a multicenter, randomized study. Arthritis Rheum. 2012;64(4):1272–82.
42. Hau M, Kneitz C, Tony HP, Keberle M, Jahns R, Jenett M. High resolution ultrasound detects a decrease in pannus vascularisation of small finger joints in patients with rheumatoid arthritis receiving treatment with soluble tumour necrosis factor alpha receptor (etanercept). Ann Rheum Dis. 2002;61(1):55–8.
43. Ribbens C, Andre B, Marcelis S, Kaye O, Mathy L, Bonnet V, et al. Rheumatoid hand joint synovitis: gray-scale and power Doppler US quantifications following anti-tumor necrosis factor-alpha treatment: pilot study. Radiology 2003;229(2):562–9.
44. Terslev L, Torp-Pedersen S, Qvistgaard E, Kristoffersen H, Rogind H, Danneskiold-Samsoe B, et al. Effects of treatment with etanercept (Enbrel, TNRF:Fc) on rheumatoid arthritis evaluated by Doppler ultrasonography. Ann Rheum Dis. 2003;62(2):178–81.
45. Filippucci E, Iagnocco A, Salaffi F, Cerioni A, Valesini G, Grassi W. Power Doppler sonography monitoring of synovial perfusion at the wrist joints in patients with rheumatoid arthritis treated with adalimumab. Ann Rheum Dis. 2006;65(11):1433–7.
46. Naredo E, Moller I, Cruz A, Carmona L, Garrido J. Power Doppler ultrasonographic monitoring of response to anti-tumor necrosis factor therapy in patients with rheumatoid arthritis. Arthritis Rheum. 2008;58(8):2248–56.
47. Scheel AK, Hermann KG, Ohrndorf S, Werner C, Schirmer C, Detert J, et al. Prospective 7 year follow up imaging study comparing radiography, ultrasonography, and magnetic resonance imaging in rheumatoid arthritis finger joints. Ann Rheum Dis. 2006;65(5):595–600.
48. Naredo E, Collado P, Cruz A, Palop MJ, Cabero F, Richi P, et al. Longitudinal power Doppler ultrasonographic assessment of joint inflammatory activity in early rheumatoid arthritis: predictive value in disease activity and radiologic progression. Arthritis Rheum. 2007;57(1):116–24.
49. Taylor PC, Steuer A, Gruber J, Cosgrove DO, Blomley MJ, Marsters PA, et al. Comparison of ultrasonographic assessment of synovitis and joint vascularity with radiographic evaluation in a randomized, placebo-controlled study of infliximab therapy in early rheumatoid arthritis. Arthritis Rheum. 2004;50(4):1107–16.
50. Newman JS, Laing TJ, McCarthy CJ, Adler RS. Power Doppler sonography of synovitis: assessment of therapeutic response–preliminary observations. Radiology 1996;198(2):582–4.
51. Terslev L, Torp-Pedersen S, Qvistgaard E, Danneskiold-Samsoe B, Bliddal H. Estimation of inflammation by Doppler ultrasound: quantitative changes after intra-articular treatment in rheumatoid arthritis. Ann Rheum Dis. 2003;62(11):1049–53.
52. Filippucci E, Farina A, Carotti M, Salaffi F, Grassi W. Grey scale and power Doppler sonographic changes induced by intra-articular steroid injection treatment. Ann Rheum Dis. 2004;63(6):740–3.
53. Stone M, Bergin D, Whelan B, Maher M, Murray J, McCarthy C. Power Doppler ultrasound assessment of rheumatoid hand synovitis. J Rheumatol. 2001;28(9):1979–82.
54. Teh J, Stevens K, Williamson L, Leung J, McNally EG. Power Doppler ultrasound of rheumatoid synovitis: quantification of therapeutic response. Br J Radiol. 2003;76(912):875–9.

55. Qvistgaard E, Rogind H, Torp-Pedersen S, Terslev L, Danneskiold-Samsoe B, Bliddal H. Quantitative ultrasonography in rheumatoid arthritis: evaluation of inflammation by Doppler technique. Ann Rheum Dis. 2001;60(7):690–3.
56. Terslev L, Ellegaard K, Christensen R, Szkudlarek M, Schmidt WA, Jensen PS, et al. Head-to-head comparison of quantitative and semi-quantitative ultrasound scoring systems for rheumatoid arthritis: reliability, agreement and construct validity. Rheumatology 2012;51(11):2034–8.
57. Sakellariou G, Iagnocco A, Filippucci E, Ceccarelli F, Di Geso L, Carli L, et al. Ultrasound imaging for the rheumatologist XLVIII. Ultrasound of the shoulders of patients with rheumatoid arthritis. Clin Exp Rheumatol. 2013;31(6):837–42.
58. Di Geso L, Filippucci E, Riente L, Sakellariou G, Delle Sedie A, Meenagh G, et al. Ultrasound imaging for the rheumatologist XL. Sonographic assessment of the hip in rheumatoid arthritis patients. Clin Exp Rheumatol. 2012;30(4):464–8.
59. Hoving JL, Buchbinder R, Hall S, Lawler G, Coombs P, McNealy S, et al. A comparison of magnetic resonance imaging, sonography, and radiography of the hand in patients with early rheumatoid arthritis. J Rheumatol. 2004;31(4):663–75.
60. Hammer HB, Kvien TK. Ultrasonography shows significant improvement in wrist and ankle tenosynovitis in rheumatoid arthritis patients treated with adalimumab. Scand J Rheumatol. 2011;40(3):178–82.
61. Filippucci E, Gabba A, Di Geso L, Girolimetti R, Salaffi F, Grassi W. H and tendon involvement in rheumatoid arthritis: an ultrasound study. Semin Arthritis Rheum. 2012;41(6):752–60.
62. Wakefield RJ, Freeston JE, O'Connor P, Reay N, Budgen A, Hensor EM, et al. The optimal assessment of the rheumatoid arthritis hindfoot: a comparative study of clinical examination, ultrasound and high field MRI. Ann Rheum Dis. 2008;67(12):1678–82.
63. Wakefield RJ, Gibbon WW, Conaghan PG, O'Connor P, McGonagle D, Pease C, et al. The value of sonography in the detection of bone erosions in patients with rheumatoid arthritis: a comparison with conventional radiography. Arthritis Rheum. 2000;43(12):2762–70.
64. Schmidt WA, Schicke B, Ostendorf B, Scherer A, Krause A, Walther M. Low-field MRI versus ultrasound: which is more sensitive in detecting inflammation and bone damage in MCP and MTP joints in mild or moderate rheumatoid arthritis? Clin Exp Rheumatol. 2013;31(1):91–6.
65. Weidekamm C, Koller M, Weber M, Kainberger F. Diagnostic value of high-resolution B-mode and doppler sonography for imaging of hand and finger joints in rheumatoid arthritis. Arthritis Rheum. 2003;48(2):325–33.
66. Finzel S, Ohrndorf S, Englbrecht M, Stach C, Messerschmidt J, Schett G, et al. A detailed comparative study of high-resolution ultrasound and micro-computed tomography for detection of arthritic bone erosions. Arthritis Rheum. 2011;63(5):1231–6.
67. Dohn UM, Terslev L, Szkudlarek M, Hansen MS, Hetland ML, Hansen A, et al. Detection, scoring and volume assessment of bone erosions by ultrasonography in rheumatoid arthritis: comparison with CT. Ann Rheum Dis. 2013;72(4):530–4.
68. Dohn UM, Ejbjerg BJ, Court-Payen M, Hasselquist M, Narvestad E, Szkudlarek M, et al. Are bone erosions detected by magnetic resonance imaging and ultrasonography true erosions? A comparison with computed tomography in rheumatoid arthritis metacarpophalangeal joints. Arthritis Res Ther. 2006;8(4):R110.
69. Baillet A, Gaujoux-Viala C, Mouterde G, Pham T, Tebib J, Saraux A, et al. Comparison of the efficacy of sonography, magnetic resonance imaging and conventional radiography for the detection of bone erosions in rheumatoid arthritis patients: a systematic review and meta-analysis. Rheumatology 2011;50(6):1137–47.
70. Riente L, Delle Sedie A, Scire CA, Filippucci E, Meenagh G, Iagnocco A, et al. Ultrasound imaging for the rheumatologist. XXXI. Sonographic assessment of the foot in patients with rheumatoid arthritis. Clin Exp Rheumatol. 2011;29(1):1–5.
71. Grassi W, Filippucci E, Farina A, Salaffi F, Cervini C. Ultrasonography in the evaluation of bone erosions. Ann Rheum Dis. 2001;60(2):98–103.
72. Salaffi F, Carotti M, Ciapetti A, Ariani A, Gasparini S, Grassi W. Validity of a computer-assisted manual segmentation software to quantify wrist erosion volume using computed tomography scans in rheumatoid arthritis. BMC Musculoskelet Disord. 2013;14:265.

73. Schmidt WA, Schmidt H, Schicke B, Gromnica-Ihle E. Standard reference values for musculoskeletal ultrasonography. Ann Rheum Dis. 2004;63(8):988–94.
74. Swen WA, Jacobs JW, Hubach PC, Klasens JH, Algra PR, Bijlsma JW. Comparison of sonography and magnetic resonance imaging for the diagnosis of partial tears of finger extensor tendons in rheumatoid arthritis. Rheumatology 2000;39(1):55–62.
75. Bruyn GA, Hanova P, Iagnocco A, d'Agostino MA, Moller I, Terslev L, et al. Ultrasound definition of tendon damage in patients with rheumatoid arthritis. Results of a OMERACT consensus-based ultrasound score focussing on the diagnostic reliability. Ann Rheum Dis. 2013;73:1929–34
76. Bruyn GA, Naredo E, Moller I, Moragues C, Garrido J, de Bock GH, et al. Reliability of ultrasonography in detecting shoulder disease in patients with rheumatoid arthritis. Ann Rheum Dis. 2009;68(3):357–61.
77. Bowen CJ, Culliford D, Dewbury K, Sampson M, Burridge J, Hooper L, et al. The clinical importance of ultrasound detectable forefoot bursae in rheumatoid arthritis. Rheumatology 2010;49(1):191–2.
78. Andonopoulos AP, Yarmenitis S, Sfountouris H, Siamplis D, Zervas C, Bounas A. Baker's cyst in rheumatoid arthritis: an ultrasonographic study with a high resolution technique. Clin Exp Rheumatol. 1995;13(5):633–6.
79. Riente L, Delle Sedie A, Filippucci E, Scire CA, Iagnocco A, Gutierrez M, et al. Ultrasound Imaging for the rheumatologist XXVII. Sonographic assessment of the knee in patients with rheumatoid arthritis. Clin Exp Rheumatol. 2010;28(3):300–3.
80. Genc H, Cakit BD, Tuncbilek I, Erdem HR. Ultrasonographic evaluation of tendons and enthesal sites in rheumatoid arthritis: comparison with ankylosing spondylitis and healthy subjects. Clin Rheumatol. 2005;24(3):272–7.
81. Nalbant S, Corominas H, Hsu B, Chen LX, Schumacher HR, Kitumnuaypong T. Ultrasonography for assessment of subcutaneous nodules. J Rheumatol. 2003;30(6):1191–5.
82. Nakagomi D, Ikeda K, Okubo A, Iwamoto T, Sanayama Y, Takahashi K, et al. Ultrasound can improve the accuracy of the 2010 American College of Rheumatology/European League against rheumatism classification criteria for rheumatoid arthritis to predict the requirement for methotrexate treatment. Arthritis Rheum. 2013;65(4):890–8.
83. Agrawal S, Bhagat SS, Dasgupta B. Improvement in diagnosis and management of musculoskeletal conditions with one-stop clinic-based ultrasonography. Mod Rheumatol. 2009;19(1):53–6.
84. Freeston JE, Wakefield RJ, Conaghan PG, Hensor EM, Stewart SP, Emery P. A diagnostic algorithm for persistence of very early inflammatory arthritis: the utility of power Doppler ultrasound when added to conventional assessment tools. Ann Rheum Dis. 2010;69(2):417–9.
85. Ten Cate DF, Luime JJ, Swen N, Gerards AH, De Jager MH, Basoski NM, et al. Role of ultrasonography in diagnosing early rheumatoid arthritis and remission of rheumatoid arthritis–a systematic review of the literature. Arthritis Res Ther. 2013;15(1):R4.
86. Terslev L, Torp-Pedersen S, Bang N, Koenig MJ, Nielsen MB, Bliddal H. Doppler ultrasound findings in healthy wrists and finger joints before and after use of two different contrast agents. Ann Rheum Dis. 2005;64(6):824–7.
87. Millot F, Clavel G, Etchepare F, Gandjbakhch F, Grados F, Saraux A, et al. Musculoskeletal ultrasonography in healthy subjects and ultrasound criteria for early arthritis (the ESPOIR cohort). J Rheumatol. 2011;38(4):613–20.
88. Scire CA, Iagnocco A, Meenagh G, Riente L, Filippucci E, Delle Sedie A, et al. Ultrasound imaging for the rheumatologist XXXIII. Sonographic assessment of the foot in early arthritis patients. Clin Exp Rheumatol. 2011;29(3):465–9.
89. Eshed I, Feist E, Althoff CE, Hamm B, Konen E, Burmester GR, et al. Tenosynovitis of the flexor tendons of the hand detected by MRI: an early indicator of rheumatoid arthritis. Rheumatology 2009;48(8):887–91.
90. Zayat AS, Ellegaard K, Conaghan PG, Terslev L, Hensor EM, Freeston JE, Emery P, Wakefield RJ. The specificity of ultrasound-detected bone erosions for rheumatoid arthritis. Ann Rheum Dis. 2014 Apr 30 [Epub ahead of print].

91. Kawashiri SY, Suzuki T, Okada A, Yamasaki S, Tamai M, Nakamura H, et al. Musculoskeletal ultrasonography assists the diagnostic performance of the 2010 classification criteria for rheumatoid arthritis. Mod Rheumatol. 2013;23(1):36–43.
92. Salaffi F, Ciapetti A, Gasparini S, Carotti M, Filippucci E, Grassi W. A clinical prediction rule combining routine assessment and power Doppler ultrasonography for predicting progression to rheumatoid arthritis from early-onset undifferentiated arthritis. Clin Exp Rheumatol. 2010;28(5):686–94.
93. van de Stadt LA, Bos WH, Meursinge Reynders M, Wieringa H, Turkstra F, van der Laken CJ, et al. The value of ultrasonography in predicting arthritis in auto-antibody positive arthralgia patients: a prospective cohort study. Arthritis Res Ther. 2010;12(3):R98.
94. Navalho M, Resende C, Rodrigues AM, Ramos F, Gaspar A, Pereira da Silva JA, et al. Bilateral MR imaging of the hand and wrist in early and very early inflammatory arthritis: tenosynovitis is associated with progression to rheumatoid arthritis. Radiology 2012;264(3):823–33.
95. Gutierrez M, Filippucci E, Salaffi F, Di Geso L, Grassi W. Differential diagnosis between rheumatoid arthritis and psoriatic arthritis: the value of ultrasound findings at metacarpophalangeal joints level. Ann Rheum Dis. 2011;70(6):1111–4.
96. D'Agostino MA, Said-Nahal R, Hacquard-Bouder C, Brasseur JL, Dougados M, Breban M. Assessment of peripheral enthesitis in the spondylarthropathies by ultrasonography combined with power Doppler: a cross-sectional study. Arthritis Rheum. 2003;48(2):523–33.
97. Backhaus M, Ohrndorf S, Kellner H, Strunk J, Backhaus TM, Hartung W, et al. Evaluation of a novel 7-joint ultrasound score in daily rheumatologic practice: a pilot project. Arthritis Rheum. 2009;61(9):1194–201.
98. Backhaus TM, Ohrndorf S, Kellner H, Strunk J, Hartung W, Sattler H, et al. The US7 score is sensitive to change in a large cohort of patients with rheumatoid arthritis over 12 months of therapy. Ann Rheum Dis. 2013;72(7):1163–9.
99. Perricone C, Ceccarelli F, Modesti M, Vavala C, Di Franco M, Valesini G, et al. The 6-joint ultrasonographic assessment: a valid, sensitive-to-change and feasible method for evaluating joint inflammation in RA. Rheumatology 2012;51(5):866–73.
100. Ellegaard K, Torp-Pedersen S, Terslev L, Danneskiold-Samsoe B, Henriksen M, Bliddal H. Ultrasound colour Doppler measurements in a single joint as measure of disease activity in patients with rheumatoid arthritis–assessment of concurrent validity. Rheumatology 2009;48(3):254–7.
101. Damjanov N, Radunovic G, Prodanovic S, Vukovic V, Milic V, Simic Pasalic K, et al. Construct validity and reliability of ultrasound disease activity score in assessing joint inflammation in RA: comparison with DAS-28. Rheumatology 2012;51(1):120–8.
102. Hartung W, Kellner H, Strunk J, Sattler H, Schmidt WA, Ehrenstein B, et al. Development and evaluation of a novel ultrasound score for large joints in rheumatoid arthritis: one year of experience in daily clinical practice. Arthritis Care Res. 2012;64(5):675–82.
103. Hammer HB, Kvien TK. Comparisons of 7- to 78-joint ultrasonography scores: all different joint combinations show equal response to adalimumab treatment in patients with rheumatoid arthritis. Arthritis Res Ther. 2011;13(3):R78.
104. Witt M, Mueller F, Nigg A, Reindl C, Leipe J, Proft F, et al. Relevance of grade 1 Gy-scale ultrasound findings in wrists and small joints to the assessment of subclinical synovitis in rheumatoid arthritis. Arthritis Rheum. 2013;65(7):1694–701.
105. Brown AK, Conaghan PG, Karim Z, Quinn MA, Ikeda K, Peterfy CG, et al. An explanation for the apparent dissociation between clinical remission and continued structural deterioration in rheumatoid arthritis. Arthritis Rheum. 2008;58(10):2958–67.
106. Foltz V, Gandjbakhch F, Etchepare F, Rosenberg C, Tanguy ML, Rozenberg S, et al. Power Doppler ultrasound, but not low-field magnetic resonance imaging, predicts relapse and radiographic disease progression in rheumatoid arthritis patients with low levels of disease activity. Arthritis Rheum. 2012;64(1):67–76.
107. Szkudlarek M, Wakefield RJ, Backhaus M, Terslev L. The discriminatory capacity of ultrasound in rheumatoid arthritis: active vs inactive, early vs advanced, and more. Rheumatology 2012;51(Suppl 7):vii6–9.

108. Cheung PP, Gossec L, Ruyssen-Witrand A, Le Bourlout C, Mezieres M, Dougados M. The relationship of patient-reported joints with active synovitis detected by power Doppler ultrasonography in rheumatoid arthritis. Clin Exp Rheumatol. 2013;31(4):490–7.
109. Ramirez J, Ruiz-Esquide V, Pomes I, Celis R, Cuervo A, Hernandez MV, et al. Patients with rheumatoid arthritis in clinical remission and ultrasound-defined active synovitis exhibit higher disease activity and increased serum levels of angiogenic biomarkers. Arthritis Res Ther. 2014;16(1):R5.
110. Zayat AS, Conaghan PG, Sharif M, Freeston JE, Wenham C, Hensor EM, et al. Do non-steroidal anti-inflammatory drugs have a significant effect on detection and grading of ultrasound-detected synovitis in patients with rheumatoid arthritis? Results from a randomised study. Ann Rheum Dis. 2011;70(10):1746–51.
111. Naredo E, Hinojosa M, Valor L, Hernandez-Florez D, Mata-Martinez C, Serrano-Benavente B, et al. Does ultrasound-scored synovitis depend on the pharmacokinetics of subcutaneous anti-TNF agents in patients with rheumatoid arthritis? Rheumatology. 2014;53:2088–94
112. Backhaus M, Burmester GR, Sandrock D, Loreck D, Hess D, Scholz A, et al. Prospective two year follow up study comparing novel and conventional imaging procedures in patients with arthritic finger joints. Ann Rheum Dis. 2002;61(10):895–904.
113. Boyesen P, Haavardsholm EA, van der Heijde D, Ostergaard M, Hammer HB, Sesseng S, et al. Prediction of MRI erosive progression: a comparison of modern imaging modalities in early rheumatoid arthritis patients. Ann Rheum Dis. 2011;70(1):176–9.
114. Lillegraven S, Boyesen P, Hammer HB, Ostergaard M, Uhlig T, Sesseng S, et al. Tenosynovitis of the extensor carpi ulnaris tendon predicts erosive progression in early rheumatoid arthritis. Ann Rheum Dis. 2011;70(11):2049–50.
115. Dougados M, Devauchelle-Pensec V, Ferlet JF, Jousse-Joulin S, D'Agostino MA, Backhaus M, et al. The ability of synovitis to predict structural damage in rheumatoid arthritis: a comparative study between clinical examination and ultrasound. Ann Rheum Dis. 2013;72(5):665–71.
116. Ellegaard K, Christensen R, Torp-Pedersen S, Terslev L, Holm CC, Konig MJ, et al. Ultrasound Doppler measurements predict success of treatment with anti-TNF- & alpha; drug in patients with rheumatoid arthritis: a prospective cohort study. Rheumatology 2011;50(3):506–12.
117. Peluso G, Michelutti A, Bosello S, Gremese E, Tolusso B, Ferraccioli G. Clinical and ultrasonographic remission determines different chances of relapse in early and long standing rheumatoid arthritis. Ann Rheum Dis. 2011;70(1):172–5.
118. Saleem B, Brown AK, Keen H, Nizam S, Freeston J, Wakefield R, et al. Should imaging be a component of rheumatoid arthritis remission criteria? A comparison between traditional and modified composite remission scores and imaging assessments. Ann Rheum Dis. 2011;70(5):792–8.
119. Saleem B, Brown AK, Quinn M, Karim Z, Hensor EM, Conaghan P, et al. Can flare be predicted in DMARD treated RA patients in remission, and is it important? A cohort study. Ann Rheum Dis. 2012;71(8):1316–21.
120. Balsa A, de Miguel E, Castillo C, Peiteado D, Martin-Mola E. Superiority of SDAI over DAS-28 in assessment of remission in rheumatoid arthritis patients using power Doppler ultrasonography as a gold standard. Rheumatology 2010;49(4):683–90.
121. Saleem B, Brown AK, Keen H, Nizam S, Freeston J, Karim Z, et al. Disease remission state in patients treated with the combination of tumor necrosis factor blockade and methotrexate or with disease-modifying antirheumatic drugs: a clinical and imaging comparative study. Arthritis Rheum. 2009;60(7):1915–22.
122. Kawashiri SY, Suzuki T, Nakashima Y, Horai Y, Okada A, Iwamoto N, et al. Ultrasonographic examination of rheumatoid arthritis patients who are free of physical synovitis: power Doppler subclinical synovitis is associated with bone erosion. Rheumatology 2014;53(3):562–9.
123. Molenaar ET, Voskuyl AE, Dinant HJ, Bezemer PD, Boers M, Dijkmans BA. Progression of radiologic damage in patients with rheumatoid arthritis in clinical remission. Arthritis Rheum. 2004;50(1):36–42.

124. Conaghan PG, O'Connor P, McGonagle D, Astin P, Wakefield RJ, Gibbon WW, et al. Elucidation of the relationship between synovitis and bone damage: a randomized magnetic resonance imaging study of individual joints in patients with early rheumatoid arthritis. Arthritis Rheum. 2003;48(1):64–71.
125. Fukae J, Isobe M, Kitano A, Henmi M, Sakamoto F, Narita A, et al. Structural deterioration of finger joints with ultrasonographic synovitis in rheumatoid arthritis patients with clinical low disease activity. Rheumatology 2014;53(9):1608–12.
126. Raza K, Lee CY, Pilling D, Heaton S, Situnayake RD, Carruthers DM, et al. Ultrasound guidance allows accurate needle placement and aspiration from small joints in patients with early inflammatory arthritis. Rheumatology 2003;42(8):976–9.
127. Naredo E, Cabero F, Beneyto P, Cruz A, Mondejar B, Uson J, et al. A randomized comparative study of short term response to blind injection versus sonographic-guided injection of local corticosteroids in patients with painful shoulder. J Rheumatol. 2004;31(2):308–14.
128. Sibbitt WL, Jr., Peisajovich A, Michael AA, Park KS, Sibbitt RR, B and PA, et al. Does sonographic needle guidance affect the clinical outcome of intraarticular injections? J Rheumatol. 2009;36(9):1892–902.
129. Cunnington J, Marshall N, Hide G, Bracewell C, Isaacs J, Platt P, et al. A randomized, double-blind, controlled study of ultrasound-guided corticosteroid injection into the joint of patients with inflammatory arthritis. Arthritis Rheum. 2010;62(7):1862–9.

Chapter 4
Ankylosing Spondylitis

Marina Backhaus MD and Yasser El Miedany MD, FRCP

Introduction

Ankylosing spondylitis (AS) is a chronic rheumatic disease known for its limited spinal mobility secondary to the inflammatory changes in the spine, entheses, joints, as well as eye (uveitis). AS is one of a wider group of inflammatory conditions known as spondyloarthritis. The European Spondyloarthropathy Study Group [1] has recently published its new classification criteria for AS in which sacroiliitis, though a very common manifestation of the disease, is no more mandatory for the diagnosis. If not diagnosed early in the disease course or ineffectively treated, a progressive loss of spinal mobility and the ability to carry out activities of daily living are the consequences which subsequently result in a reduction in the quality of life. However, in spite of all the progress made in terms of the disease diagnosis and management, a long delay between the onset of disease symptoms and the diagnosis time has been reported with an average period of 8–11 years [2].

The introduction of the new "nonradiographic spondyloarthritis" term highlighted the role of imaging for the diagnosis of AS. Though plain radiography is traditionally the initial tool for evaluating the sacroiliac joints as well as spine, its accuracy is limited by the lack of sensitivity in early inflammatory stages and by high intra- as well as inter-observer variability on interpretation [3]. While magnetic resonance imaging (MRI) has been suggested by many research studies as the best tool of detecting sacroiliitis as well as spondylitis, especially early in the course of the disease, US was reported a highly sensitive, noninvasive imaging method to evalu-

M. Backhaus (✉)
Department of Internal Medicine, Rheumatology and Clinical Immunology,
Park-Klinik Weissensee Berlin, Academic Teaching Hospital of Charité,
Berlin, Germany
e-mail: marina.backhaus@charite.de

Y. El Miedany
Department of Rheumatology, Darent Valley Hospital, Dartford, UK

Department of Rheumatology and Rehabilitation, Ain Shams University, Cairo, Egypt

Y. El Miedany (ed.), *Musculoskeletal Ultrasonography in Rheumatic Diseases*,
DOI 10.1007/978-3-319-15723-8_4

ate all the enthesis sites as well as the peripheral joints in AS. In the past decade, a large number of research studies papers regarding the role of US in AS have been published. This chapter provides an overview of the available data and discusses the possible implementations of US in standard clinical care of AS patients.

Applied Anatomy

Enthesitis represent the main pathological feature for the spondyloarthritis group. Using advanced imaging and sono-anatomy, the tendon/ligament insertion points into bone (entheses) is considered a key pathological spot in spondyloarthritis. Anatomically, this entheses form functionally integrated units with adjacent synovium, adjacent periosteal, and sesamoid fibrocartilages as well as the underlying bone in composite structures named "synovio-entheseal complexes." These fibrocartilages in addition to the fibrocartilage present at the entheses, perform as a stress shield offering the underlying bone some protection. Other studies added the bursa to this fundamental entheseal complex [4, 5]. This led to the development of a new concept of "enthesis organ" where bursa is considered part of the synovio-entheseal complex [6]. The role played by the underlying trabecular bone in normal enthesis has been studied in the work done by McGonagle et al. [7]. Bone marrow edema represents a hallmark of enthesitis, which is histologically equivalent to osteitis [8]. This osteitis response may not be perceived at all enthesitis sites, hence, its absence does not rule out enthesitis-associated pathology. Furthermore, it has been reported that some tissues that are not an enthesis perform as such due to shared anatomical, histological, biomechanical, and pathological features [24]. These structures have been termed "functional entheses," and include synovial joints lined by fibrocartilage. Examples are the sacroiliac and acromioclavicular joints [9, 10]. They also entail the wraparound tendons such as those present close to the ankle malleoli. Another observation is that the enthesis organ-associated fibrocartilages require lubrication and nutrition similar to normal articular cartilage [11]. To fulfill this physiological need, the immediately adjacent synovium provides this vascular supply. This infers that the enthesitis-associated pathology may be closely associated with synovitis without exhibiting pathological changes at the insertion proper.

In the early disease process, the enthesis as well as the adjacent tissues may show several morpho-structural variations such as entheseal thickening, fibrillar separation due to intra-tendinous edema, with or without associated bursitis, or enhanced vascularity [12]. At this stage, there are no relevant bony changes. However, during the enthesitis process, new vessels and inflammatory cells penetrate the cortical sub-entheseal bone communicating the bone marrow to the entheseal tissues, which in turn leads to thinning of cortical bone, focal loss of bone (erosions), and direct contact between fibrocartilage and underlying bone spaces (Fig. 4.1). The reparative process with reactive bone formation leads to ill-defined osteosclerosis and spur formation [13].Therefore, at later stages, bony cortex changes appear in the form of enthesophytes and/or bone erosions. In some cases, large enthesophytes may impair partially or completely the imagining of adjacent bone erosions [14].

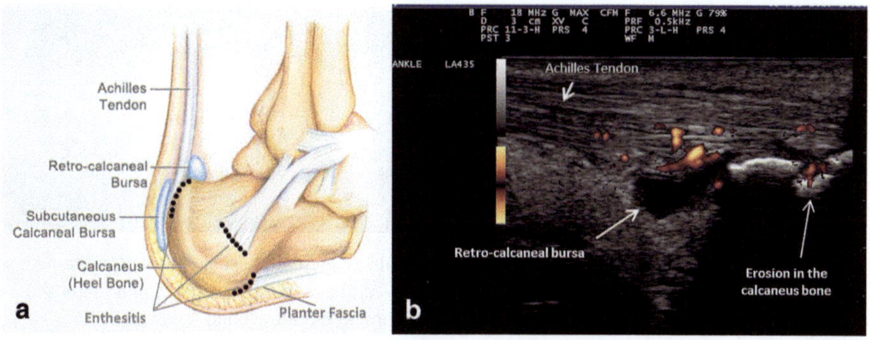

Fig. 4.1 **a** Picture of ankle depicting the distinct areas of synovio-entheseal complex. **b** Enthesis organ, composed of the ligament insertion points into bone, the adjacent synovium and bursa as well as the underlying bone, detected by ultrasound (*US*) examination in **b** mode combined with power Doppler in ankylosing spondylitis patient. Abnormal vascularization associated with marked swelling and decreased echogenicity of the enthesis at the cortical junction can be seen, as well as erosion of the calcaneus bone. Tendon thickness at the level of the enthesis is 7.2 mm and is associated with a retro-calcaneal bursa with enhanced vascularity

However, there is a difference between "enthesopathy" and "enthesitis." The Outcome Measures in Rheumatology (OMERACT) ultrasound group suggested a definition of enthesopathy: "An abnormal hypoechoic region with loss of normal fibrillar architecture and/or thickened tendon or ligament in its bony attachment, seen in 2 perpendicular planes that may exhibit Doppler signal and/or bony changes including enthesophytes, erosions or irregularity." This definition merged both acute and chronic inflammatory manifestations with structural damage findings [15]. Therefore, the association of enthesis in any pathologic process whether traumatic, inflammatory, metabolic, or degenerative is denoted "enthesopathy," whereas "enthesitis" is confined to the occurrence of tendon/ligament or capsule inflammation at its insertion site into the bone [16]. The latter is the classical feature of spondyloarthritis (Fig. 4.2).

Generally, inflammation may occur at any enthesis site. Mechanical factors, together with physiological and anatomical characteristics of the enthesis, have been proposed as contributing factors which may impact the localization of enthesitis. This might explain the finding that most of the clinically relevant enthesitis sites are present in the lower limbs; particularly heel enthesis (Achilles enthesitis) which is the main form of enthesitis included in the Assessment of SpondyloArthritis (ASAS) international society criteria [1]. Planter fasciitis is another frequently involved site [17, 18].

Sonographic Pattern of Enthesitis

A large number of research studies have applied US to the assessment of enthesitis in spondyloarthritis. The commonest entheseal sites to be ultrasonographically examined in standard clinical practice include: Achilles tendon and plantar fascia

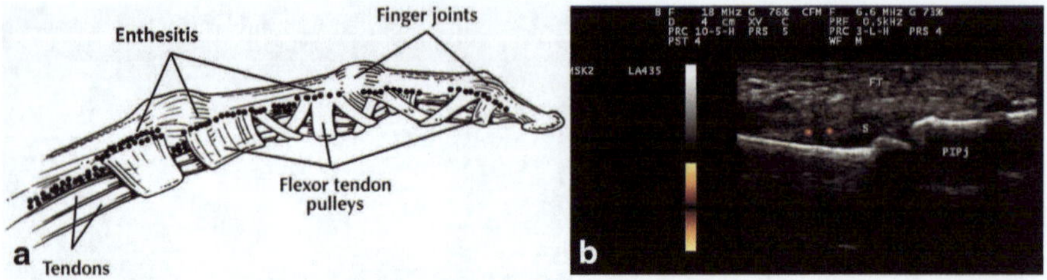

Fig. 4.2 **a** Picture of a flexor tendon clarifying the role synovio-entheseal complex in the pathogenesis of enthesitis. *Dots* show sites of inflammation—enthesis. **b** Grayscale US of a finger showing enhanced vascularity at the synovio-entheseal complex. US ultrasound

insertions to the calcaneus, tibialis anterior tendon insertion, greater trochanter, pubis, patella (at insertions of the quadriceps femoris and patellar tendons), medial, and lateral epicondyles. Sonographically, enthesitis appear as loss of normal fibrillar pattern, increased fusiform thickness (Fig. 4.3), peri-tendinous fluid, a linear intra-substance tear, calcific deposits (Fig. 4.4), or intralesional focal changes at the tendon insertion. In addition, periosteal changes such as erosions or new bone formation may also be seen at the tendon insertion. Enlarged bursa is usually considered as a sonographic manifestation of bursitis. Enhanced vascularity assessed by power Doppler (PD) at the involved entheses may also add further information about the activity status at the assessed site as reflected by its degree of vascularity (Fig. 4.5). Table 4.1 summarizes the different sonographic patterns of enthesitis.

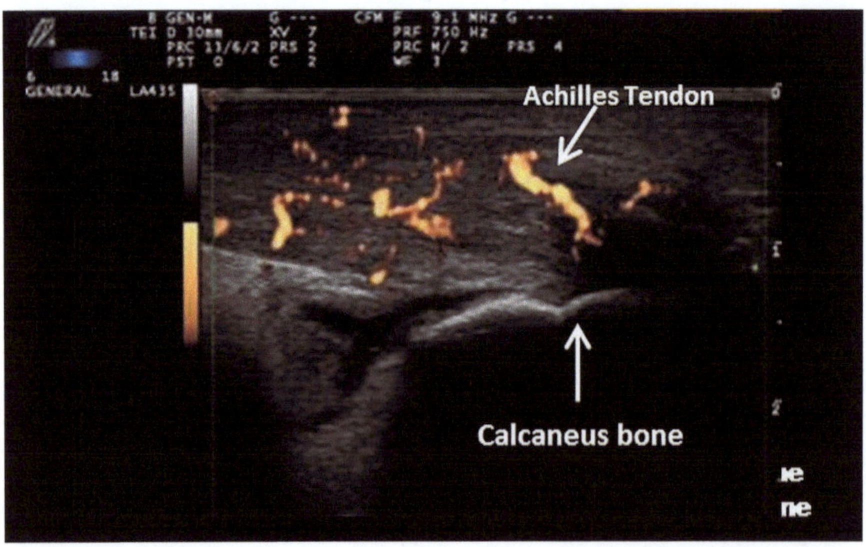

Fig. 4.3 Sonographic B-mode image of Achilles tendinitis showing increased fusiform thickness of the tendon with enhanced vascularity

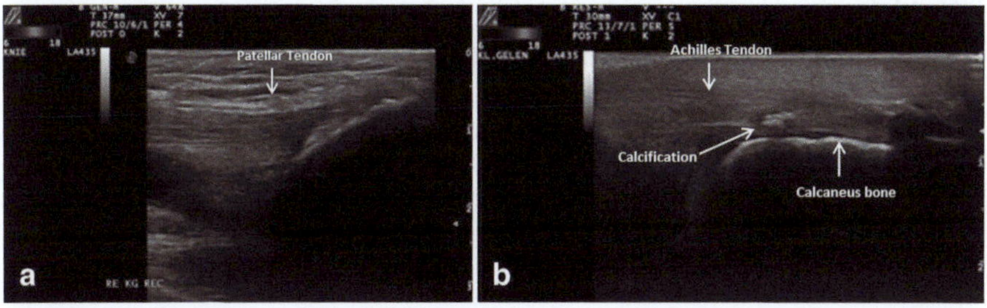

Fig. 4.4 **a** Sonographic B-mode image showing calcification of the patellar tendon. **b** Sonographic B-mode image showing calcification of the Achilles tendon

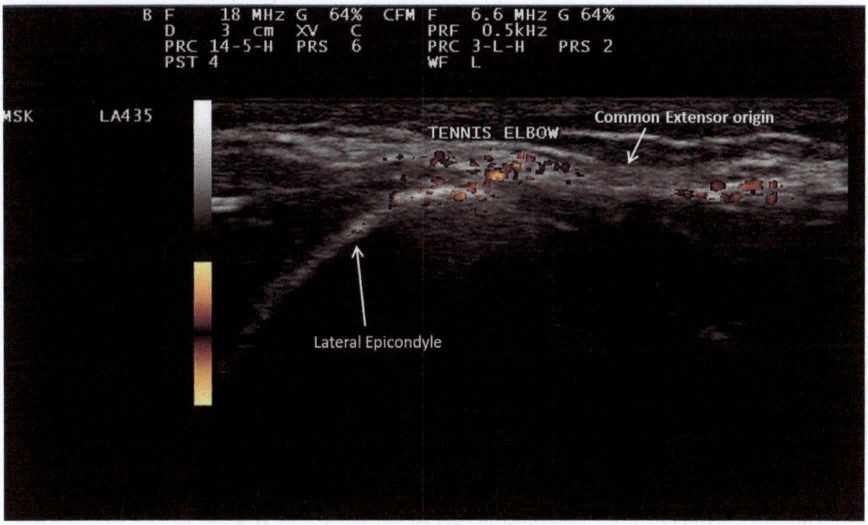

Fig. 4.5 Sonographic B-mode image with power Doppler of the elbow showing tenosynovitis of the common extensor origin with enhanced vascularity

Tips for US Assessment of Enthesitis

- Curvilinear transducer is recommended as it can cover a wider cross section of the scanned area as compared to linear transducers.
- Always assess in two perpendicular planes.
- In B mode, search for the following abnormal findings at the tendon insertion, and any one of them is considered to be a feature of enthesitis: thickening or intratendinous focal changes of the tendon, calcific deposits, or periosteal changes.
- An enlarged bursa is considered to be an evidence of bursitis
- Enhanced vascularity evaluated by PD should be assessed at these sites: cortical bone insertion, body of tendon, bursa, and the junction between tendon and enthesis. The detection of vascularization at one of these sites is considered abnormal,
- US enthesitis can be stratified according to different combinations of abnormal grayscale US and/or PD feature.

Table 4.1 Sonographic patterns of enthesitis

Lesion	Sonographic pattern
Loss of the fibrillar pattern	The tendon displays heterogeneous echogenicity
Tendon thickening	The thickness of the affected tendon is greater than that of the contralateral tendon
Intrasubstance tears	A hypoechoic band within the substance of the tendon associated with discontinuity of the tendon fibers
Intratendinous calcifications	Internal foci of high-density echoes with posterior shadowing
Peritendinous fluid	Anechoic fluid is observed along the superficial aspect of the tendon
Bursitis	Enlarged bursa

Enthesitis Scoring Tools

In a trial to standardize the quantitative assessment of enthesitis, numerous scoring tools have been developed. One of the most commonly used is the Glasgow Ultrasound Enthesitis Scoring System (GUESS) [19]. The GUESS was developed based on a study of 35 patients with spondyloarthritis (including seven psoriatic arthritis patients) who underwent independent clinical and ultrasonographic examination of both lower limbs at five entheseal sites: superior pole and inferior pole of patella, tibial tuberosity, Achilles tendon, and plantar aponeurosis. A total of 18 features, each given a score of 1, were scored on each lower limb, the maximum score being 36. The intraobserver k value for analysis of all sites was 0.9.

In another work, D'Agostino et al. [20] implemented a five-stage classification system of ultrasonography enthesitis according to B-mode and PD findings (Table 4.2) where stage 1: vascularization at the cortical junction without abnormal findings in B mode. This represents isolated vascular changes occurring during the initial phase of enthesitis; stage 2a: vascularization associated with swelling and/or decreased echogenicity at the cortical junction in B mode; stage 3a: same as stage 2a, plus erosions of cortical bone and/or calcification of enthesis, and optional surrounding bursitis. Stages 2a and 3a reflect increased vascular alterations associated with graded signs of morphologic alterations. Stage 2b: abnormal findings in B mode as in stage 2a, but without vascularization and stage 3b: abnormal findings in B mode as in stage 3a, but without vascularization. Stages 2b and 3b, which are devoid of vascularization, are suggestive of inactive lesions.

Two other scoring systems developed at the patients' level were published. The Spanish Enthesitis Index (SEI) [21] entails, after giving the patient information about variable enthesis sites, the evaluation of global patient inflammatory activity

Table 4.2 Enthesitis classification based on B-mode and power Doppler US outcomes [20]

Stage	Description
1	Vascularization at the cortical junction without abnormal findings in B mode
2a	Vascularization associated with swelling and/or decreased echogenicity at the cortical junction in B mode
3a	Same as stage 2a, plus erosions of cortical bone and/or calcification of enthesis, and optional surrounding bursitis
2b	Abnormal findings in B mode as in stage 2a, but without vascularization
3b	Abnormal findings in B mode as in stage 3a, but without vascularization

US ultrasound

or enthesis structural damage. The scoring system, however, used grayscale abnormalities only and did not differentiate between involvement of body of the tendon, enthesis, and bursa. The Madrid Sonographic Enthesitis Index (MASEI) [22] combined abnormalities detected by grayscale US and enhanced vascularity as assessed by PD (including bursal involvement). This scoring system was the first to evaluate enthesis in both lower limbs and upper limbs (the attachment of the triceps tendon to the olecranon). Also, the MASEI included assessment of bone erosions, enthesophytes, as well as PD signal.

Critical analysis of the four systems revealed that, although they all included assessment of enthesitis sites for both inflammatory and structural manifestations using grayscale US with or without PD, sensitivity to change was low making none of them a sensitive tool for monitoring the patient's enthsitis activity status. Furthermore, there is inconsistency among the four scoring tools. While the GUESS and D'Agostino scoring systems were set up to grade enthesis affection (enthesitis level). The MASEI and the SEI were established as enthesitis indices at the patients' level. Therefore, these scoring systems are uncomparable. Currently, there is an unmet need to agree on a system for use in research trials as well as standard clinical practice [23, 24].

Subclinical Versus Clinical Enthesitis

The value of US in the diagnosis of enthesitis is that it has been reported to be more sensitive than clinical examination in identify enthesitis, raising the query whether the presence of enthesitis could be missed during standard clinical assessment. A study was carried out in 2002 [25] comparing the detection of enthesitis based on clinical examination versus US. A total of 35 patients (27 AS patients, 7 psoriatic arthritis patient, and 1 patient with reactive arthritis) had their enthesitis clinically assessed in the lower limbs as well as using US and scored according to the GUESS. The study reported significantly higher enthesitis score in the US-assessed group in comparison to clinical assessment. It concluded that US is a good method to assess for enthesitis in spondyloarthritis patients. These findings were confirmed in another study carried out by by Spadaro et al. [23]. A total of 36 AS patients had both clinical and PD ultrasonography assessment of 432 enthesis sites, and 14.8 % were diagnosed to have clinical enthesitis in comparison to 44.4 % were considered abnormal on PD ultrasonography. A third study assessed enthesitis in the feet of 44 subjects with spondyloarthritis (19 AS patients, 5 with psoriatic arthritis, 8 with reactive arthritis, and 12 with undifferentiated spondyloarthritis) [26]. Pathological findings were reported in 25 patients, most of who did not have any foot problem or a complaint. Similarly, upper limb enthesitis was assessed in another study [27]. US was used to assess for enthesitis at the proximal insertion of the deltoid, which mimics clinically impingement syndrome, in 100 symptomatic consecutive spondyloarthritis outpatients, compared to 4 groups of control patients: 100 with rheumatoid arthritis, 100 with osteoarthritis, 100 with painful shoulder, and 50 with shoulders undamaged by local pathological processes. Results revealed the frequency of enthesitis in the course of spondyloarthritis was 9 %. The prognostic value of this subclinical enthesitis was assessed in a recent study [28] carried out to identify predictors of structural joint damage in patients with early spondyloarthritis. Results revealed an increased probability for structural progression in the presence of US verified enthesitis.

Differential Diagnosis of Enthesitis

US plays also an important role in the differential diagnosis of enthesitis, in particular differentiating mechanical or metabolic enthesopathy from inflammatory enthesitis. Beside spondyloarthritis patients, enthesitis have been reported among athletes as a consequence of traumatic injuries. Distinguishing fibromyalgia tender points from enthesitis is another important challenge to be considered. However, in such cases, enthesitis is not associated with intra-articular inflammation (i.e., synovitis). Furthermore, PD technology has allowed the visualization of abnormal vascularity and soft tissue hyperemia in inflammatory conditions [28]. Therefore, that abnormal enhanced vascularity, detected by PD US, at the insertion of tendons, ligaments, fascia, and capsules into the bone is considered as a primary lesion which is seen exclusively in spondyloarthritis patients and may underlie all spondyloarthritis skeletal manifestations [20]. On the other hand, enthesophytosis cannot be considered an explicit manifestation for spondylarthritis-related enthesitis, as it has also been reported in mechanical, osteoarthritic as well as in normal asymptomatic subjects [29]. Multiple and irregular enthesophytes and calcific deposits are possibly more specific for spondyloarthritis-related enthesitis, but the use of enthesophytosis alone as sign of arthritis-related enthesitis is an incorrect assumption [10]. Another diagnostic challenge is seen in spondyloarthritis patients presenting with associated arthritis which often has a similar clinical pattern to rheumatoid arthritis. US was reported to be able to differentiate between the two arthritic forms based on the entheseal findings [20]. This highlight the diagnostic utility of US for the classification of arthritis as primary synovial or entheseal based.

US as an Outcome Measure

The use of US as an objective quantitative outcome measure has expanded over the past few years and started to book its place in standard clinical practice. US allow real time image acquisition of multiple joints to assess both bony changes (erosion) and soft tissue inflammation (enthesitis). Earlier studies revealed that US can be used as an outcome measure to monitor response to therapy in inflammatory arthritis [30]. US is sensitive to inflammatory and destructive changes than plain radiographs and clinical examination. Furthermore, US demonstrated a good inter-observer agreement for bone changes (median 96% absolute agreement) as well as inflammatory changes (median 92% absolute agreement) [31]. US monitoring of enthesitis in AS patients showed significant sensitivity to change after treatment. This ability to show responsiveness to management was enhanced and appeared superior when PD evaluation was included in the examination [32]. Therefore, PD evaluation has been considered an important feature to consider when assessing for responsiveness to management. In a study carried out by Aydin et al. [33], 43 AS patients with active disease, who were treated with tumor necrosis factor alpha (TNF)-α antagonist therapy, were monitored for outcome measures in response to treatment. Grayscale and PD ultrasonography as well as physical examination were carried out to assess for Achilles enthesitis and/or retro-calcaneal bursitis before and 2 months after the initiation of therapy. While US detected subclinical Achilles enthesitis in a subset of AS patients, a significant improvement in the enthesitis score was demonstrated after 2 months of TNF-α antagonist therapy.

In addition, the use of US has been endorsed to guide positioning the needle in inflamed joints, tendon sheaths, and enthesis, for local steroid injections. Huang et al. [34] reported that AS patients diagnosed to have Achilles tendon enthesitis, and treated with US-guided local betamethasone or etanercept injection in the enthesis, experienced clinical improvement associated with a reduction in the blood flow signal.

In the research setting, US played even a role in understanding the enthesitis outcomes. US revealed that the Achilles tendon enthesitis, new bone formation or spur tend to develop distally; erosions of calcanean bone occur more proximally [7]. The underlying mechanism for this topographic variation was clarified by histology as well as sonoantomy studies. Histology of normal subjects revealed a greater density of bone trabeculae and new bone formation where there is greater tension, such as at the distal site of the todoachilles enthesitis. The inverse is also true, where bone trabeculae tend to be sparser at the proximal part of the enthesis, hence erosive changes are observed at this site [35].

Results of these studies, expands the scope for using US in AS management. In addition to its diagnostic role, US can play an important in the standard day-to-day practice to assess disease activity, differentiate variable forms of arthritis (Fig. 4.6) as well as monitor response to treatment. Being portable, with no irradiation hazards and in the meantime highly accurate as well as sensitive, these advantages favored US in contrast with conventional radiography or MRI. US can be the tool of choice in the day-to-day rheumatology practice.

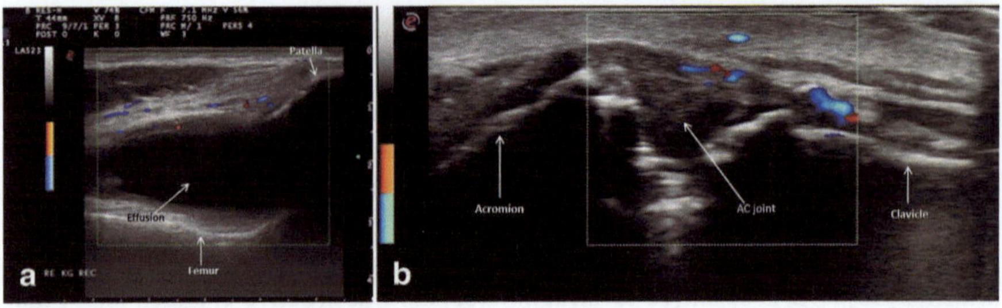

Fig. 4.6 **a** Sonographic B-mode image with power Doppler of the knee joint showing effusion in the joint with synovial hypertrophy and enhanced vascularity. **b** Sonographic B-mode image with power Doppler of the acromioclavicular joint showing synovitis with enhanced vascularity

Imaging of the Axial Skeleton

Chronic inflammatory back pain associated with progressive limitation of spinal movements secondary to syndesmophyte formation across the axial joints represent the characteristic features of AS. In addition, enthesitis and peripheral arthritis usually occur resulting in bone erosions or impairment of the tendons/ ligaments/ joints structure and function [36]. For decades, conventional radiography has been the main imaging modality in clinical practice for AS. The recent recognition of a preradiographic phase of axial skeleton inflammation and the window of opportunity for early intervention with good outcomes, and the inability of the conventional radiography to show soft tissue inflammation, highlighted the need for more sensitive tools for early diagnosis.

Both MRI and US were found able to identify sites of inflammation, however, while MRI is excellent at depicting this in both soft tissues and bones, US can only show inflammation in soft tissue structures, and only if they are superficial. As a result, to the relatively deep position of the axial skeleton in the body, US is incapable of showing these structures in full details. Therefore, MRI has been the preferred imaging modality to assess for spondylitis, in particular in its early nonradiographic stages. Ünlü et al. [37] carried out a study to investigate the role of color and duplex Doppler US in the detection of sacroiliac joint and spinal inflammation, as well as response to anti-TNF therapy in 39 AS patients. Color and duplex Doppler US were presented as "resistive index" (a measure of pulsatile blood flow that reflects the resistance to blood flow caused by microvascular bed distal to the site of measurement). Results revealed that in AS patients, the resistive index values of the sacroiliac joints, lumbar as well as dorsal vertebral paraspinal areas were lower than in controls (all $p \leq 0.01$). Furthermore, in AS patients with active disease according to Bath ankylosing spondylitis disease activity index (BASDAI), resistive index values of dorsal as well as lumbar vertebrae were significantly lower than in the inactive group. After anti-TNF therapy, there were significant increases in mean sacroiliac joint ($p=0.028$) as well as lumbar vertebrae resistive index ($p=0.039$), whereas there was a nonsignificant increase in dorsal vertebrae resistive index

($p > 0.05$). In conclusion, the study revealed that color and duplex Doppler US may be an alternative, less expensive, and easier method for detecting inflammation secondary to increased sacroiliac joint and spinal vascularization. Also it was valuable in evaluating response to anti-TNF therapy in AS patients. Further studies are still needed to assess the utility of new US modalities, such as microbubble contrast and sonoelastography in axial assessment in AS patients.

US of the Sacroiliac Joint

The anatomical features of the sacroiliac joint make it difficult for grayscale US direct imaging of synovitis of this joint. However, in 2009, Spadaro et al. [38] carried out a study to analyze the discriminative value of US-detected sacroiliac joint effusion compared to physical examination in 45 patients with spondyloarthropathy with or without inflammatory back pain. US showed joint effusion in 38.9% of sacroiliac joints of spondyloarthritis patients and in 1.7% of sacroiliac joints of controls ($p < 0.001$). The presence of inflammatory back pain was significantly associated with sacroiliac joint effusion assessed by US alone or plus at least one sacroiliac joint test. Sacroiliac joint effusion assessed by US alone or plus at least one sacroiliac test had a positive likelihood ratio of (2.67 and 4.04, respectively) for the presence of inflammatory back pain higher than likelihood ratio of single clinical tests. In another study, Bandinelli and his colleagues [39] compared US and clinical examination to traditional radiography findings of sacroiliac joins in 23 patients diagnosed with early spondyloarthritis patients. They reported that sacroiliac joint capsule distension and sacro-tuberosus ligament thickness were higher in early spondyloarthritis, but only sacro-tuberosus ligament thickness correlated significantly with clinical sacroiliac joint tenderness. As expected, that study revealed that radiographic sacroiliac joint evaluation did not correlate with US results. In conclusion, the study reported that US was more sensitive than clinical assessment as well as conventional radiography and that US is a promising technique for early sacroiliac joint assessment in standard clinical practice. In concordance with these findings, further recent studies [40, 41] revealed that color Doppler US is a valuable tool for the diagnosis, evaluation of disease activity as well as monitoring response to therapy of active sacroiliac joins. In another work, Jiang et al. [42] showed that assessment of the blood flow signals in the sacroiliac joint, assessed by PD, became weaker or disappeared and the resistive index values increased after infliximab treatment. These data endorse PD ultrasonography as a useful tool in the follow up of sacroiliac joint inflammatory status in AS patients. The value of microbubble contrast agents for color Doppler US, in comparison to MRI imaging, for the detection of active sacroiliitis was assessed in a study carried out by Klauser et al. on 103 patients (206 sacroiliac joints) [43]. Results revealed that the microbubble contrast-enhanced color Doppler US is a sensitive technique with a high negative predictive value for the detection of active sacroiliitis compared with MRI.

US-Guided Injections in Sacroiliitis

Owing to the complex anatomical structure of the sacroiliac joint, injection into this joint using blind palpation technique often results in low accuracy [44]. As a result, to perform successful sacroiliac joint injection, it is important to use image guidance. US-guided sacroiliac joint injection has been demonstrated to have a high success rate of up to 90 % [45].

Curvilinear transducer is recommended in this injection technique as it can cover a wider cross section of the scanned area as compared to a linear transducer [46]. Similar to the caudal epidural injection technique, the patient is placed in prone position to receive this injection treatment [47]. The transducer is placed in a transverse orientation, in the midline (1–2 cm above the beginning of the gluteal fold) to identify the spinous process of the sacral plate first. After identifying the spinous process, the transducer is moved in a lateral direction until the lateral edge of the sacrum is observed. With the transducer maintained in the transverse orientation, it is then moved slightly lateral until the bony contour of the ileum and posterior superior iliac spine is identified. The cleft between the bony contours of the sacrum and ileum represents the posterior aspect of the sacroiliac joint (Fig. 4.7). By tilting the transducer in a caudal direction, the lower one third of the sacroiliac joint is identified [45].

Because of its synovial component, the lower one third of the sacroiliac joint is the portion of the entire sacroiliac joint in which the injection should be performed [48].The medial to lateral approach (Fig. 4.8) is preferred for the US-guided sacroiliac joint injection [49]. It has been reported that even if the injection has not been administered accurately into the sacroiliac joint, US guidance can at least ensure periarticular deposition of the injection to the lower one third portion of the joint. Periarticular deposition of the steroid is believed to be effective as well in alleviating the pain induced by sacroiliitis [44].

In a study carried out by Migliore et al. [50], they reported that US-guided intra-articular injections of the sacroiliac joint with acetonide triamcinolone results in

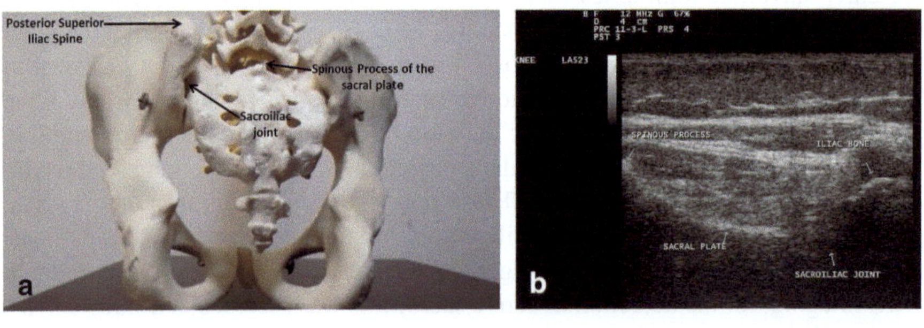

Fig. 4.7 **a** anatomy of the sacroiliac joint. **b** sonographic B-mode image of a normal sacroiliac joint. The start point is to identify the midline by identifying the spinous process of the sacral plate (1–2 cm above the gluteal fold). The probe is then moved laterally till the posterior superior iliac spine comes in view in the top of the screen. The sacroiliac joint is located between the posterior superior iliac spine/the iliac crest shadow and the sacrum

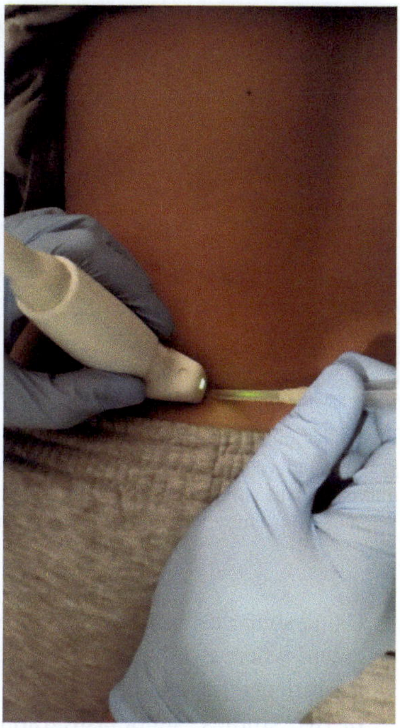

Fig. 4.8 US-guided left sacroiliac joint injection technique adopting the medial to lateral approach

articular symptoms benefits which lasted for at least 6 months. In another study, carried out by Harmon and O'Sullivan [51], they reported that intra-articular injections of etanercept into the sacroiliac joint might improve joint function and decrease the frequency of local enthesitis.

Sonoelastography and Ankylosing Spondylitis

Tendon injuries lead to alterations in the mechanical properties of the tendon. Axial-strain sonoelastography and shear-wave elastography are relatively new, real-time imaging techniques which evaluate the mechanical properties of tendons in addition to the existing morphological and vascular information that is obtained with traditional imaging tools. Axial-strain sonoelastography displays the subjective distribution of strain data on an elastogram caused by tissue compression, whereas shear-wave elastography provides a more objective, quantitative measure of the intrinsic tissue elasticity using the acoustic push-pulse. Recent studies suggest that axial-strain sonoelastography is able to distinguish between asymptomatic and diseased tendons, and is potentially more sensitive than conventional US in detecting early tendinopathy. Shear-wave elastography seems to be a feasible tool for depicting elasticity and functional recovery of tendons after surgical management [52].

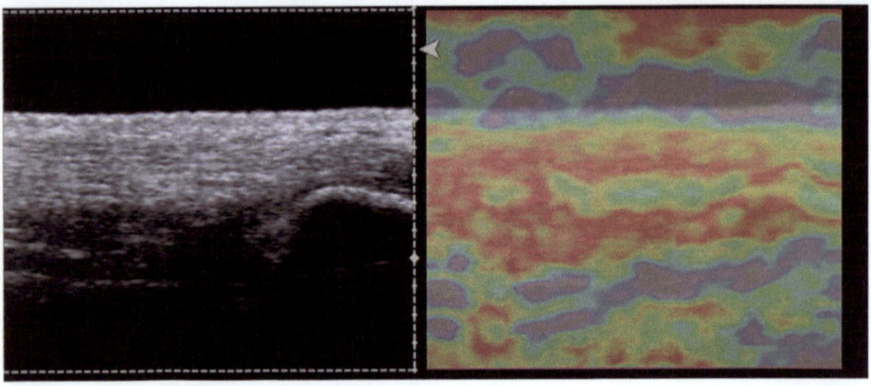

Fig. 4.9 Sonoelastography and B-mode US of a normal tendoachilles showing normal sonographic pattern on B-mode (*left*). The sonoelastogram obtained (*right*) is displayed as a color mask that was superimposed on the conventional B-mode image. (Photo copyright [Chueh-Hung Wu and Tyng-Guey Wang]. All rights reserved. Used with permission)

Using sonoelastography, the normal tendon structure (Fig. 4.9) is depicted as red or green, whereas inflamed areas (considered pathological) are seen in blue color which reflects the tendon stiffness.

In two studies, sonoelastography showed that the distal area of the Achilles tendon was the most commonly affected area in AS patients when compared to healthy subjects [53, 54]. Moreover, it was reported that pathological sonoelastography findings tended to correlate with enthesopathy findings and tendinous enlargement, thus demonstrating a moderate to good correlation with B-mode findings. In another study, Wu et al. [55], studied bilateral feet of 13 subjects with plantar fasciitis and 40 healthy controls. They reported that planter fascia softens with age and in subjects with plantar fasciitis (Fig. 4.10) Sonoelastography may be useful for the evaluation of tendon abnormalities in AS patients, but further studies are needed to standardize this technique and to assess its usefulness in clinical practice.

MRI Versus US in Ankylosing Spondylitis

Both MRI and US have shown their high sensitivity and accuracy, and therefore booked their place in AS assessment as well as management. However, while MRI is excellent at showing inflammatory changes in both soft tissues and bones, US can only show inflammation in soft tissue structures, and only if they are superficial.

In AS, and in view of the most recent guidelines [1], MRI has two major roles: as a diagnostic test and for monitoring of therapy. The latter of these, i.e., the use of MRI to monitor response to biological therapies, is well established in academic

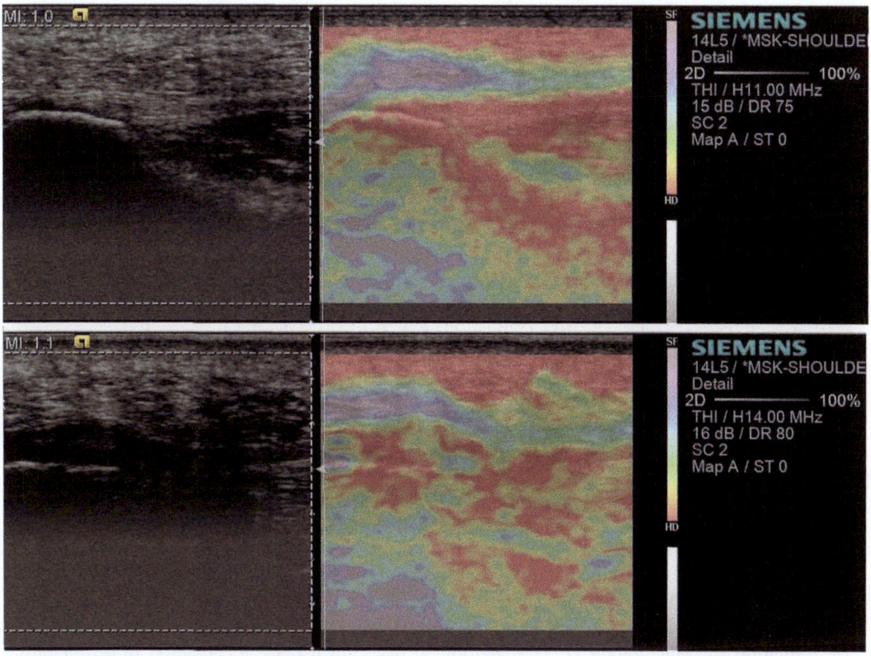

Fig. 4.10 B-mode US scan (*left*) and sonoelastogram (*right*) of a normal planter fascia (**a**), and a case of planter fasciitis (**b**). *Blue* rectangle depicts standardized area of qualitative and quantitative analysis of planter fascia. The color scheme is *red* (hard), *green* (medium stiffness) and *blue* (soft). Sonoelastography demonstrates that the plantar fascia of the healthy subjects is generally a hard structure that correspond to the red color coding. In plantar fasciitis (**b**), there is lesser red-coded area and a greater green coded area reflecting softening of the plantar fascia in plantar fasciitis patients. (Photo copyright [Chueh-Hung Wu and Tyng-Guey Wang]. All rights reserved. Used with permission)

groups for the assessment of anti-TNF therapy in the axial skeleton. In standard clinical practice, MRI is in widespread use for both the evaluation of suspected inflammatory back pain, as well as peripheral joints where the presence of diffuse osteitis/enthesitis is suspected.

On the other hand, US has various advantages in comparison to MRI. In addition to being able to identify inflammation and enthesitis, it is more comfortable for the patient, more easily accessible in clinical practice, and multiple joints can be imaged in one setting. The use of PD has also helped to increase its sensitivity in identifying enthesitis and synovitis.

Therefore, US has gained a wide variety of roles in spondyloarthritis, particularly in the area of enthesitis. The disadvantage of US is its inability to image very deep structures well, such as the spine and the sacroiliac joints commonly involved in spondyloarthritis. Standardizing US outcome measures and scoring systems still require further work.

References

1. Smolen J, Braun J, Dougados M, et al. Treating spondyloarthritis including, ankylosing spondylitis and psoriatic arthritis, to target: recommendations of an international task force. Ann Rheum Dis. 2014 Jan;73(1):6-16.
2. Feldtkeller E, Khan MA, van der Heijde D, van der Linden S, Braun J. Age at disease onset and diagnosis delay in HLA-B27 negative vs. positive patients with ankylosing spondylitis. Rheumatol Int. 2003;23:61–6.
3. van Tubergen A, Heuft-Dorenbosch L, Schulpen G, et al. Radiographic assessment of sacroiliitis by radiologists and rheumatologists: does training improve quality? Ann Rheum Dis. 2003;62:519–25.
4. Falcao S, De Miguel E, Castillo-Gallego C et al. Achilles enthesis ultrasound: the importance of the bursa in spondyloarthritis. Clin Exp Rheumatol. 2013;31:422–7.
5. Naredo E, Batlle-Gualda E, Garciavivar M, et al. Power doppler ultrasonography assessment of entheses in spondyloarthropathies: response to therapy of entheseal abnormalities. J Rheumatol. 2010;37:2110–7.
6. Benjamin M, McGonagle D. The enthesis organ concept and its relevance to the spondyloarthropathies. Adv Exp Med Biol. 2009;649:57–70.
7. McGonagle D, Wakefield R, Tan A, et al. Distinct topography of erosion and new bone formation in achilles tendon enthesitis: implications for understanding the link between inflammation and bone formation in spondyloarthritis. Arthritis Rheum. 2008;58(9):2694–9.
8. McGonagle D, Marzo-Ortega H, O'Connor P, et al. Histological assessment of the early enthesitis lesion in spondyloarthropathy. Ann Rheum Dis. 2002;61:534–7.
9. Benjamin M, McGonagle D. The anatomical basis for disease localisation in seronegative spondyloarthropathy at entheses and related sites. J Anat. 2001;199:503–26.
10. McGonagle D, Marzo-Ortega H, Benjamin M, Emery P. Report on the second International enthesitis workshop. Arthritis Rheum. 2003;48:896–905.
11. McGonagle D, Lories RJ, Tan AL, Benjamin M. The concept of a "synovio-entheseal complex" and its implications for understanding joint inflammation and damage in psoriatic arthritis and beyond. Arthritis Rheum. 2007;56:2482–91.
12. Gisondi P, Tinazzi I, El-Dalati G, et al. Lower limb enthesopathy in patients with psoriasis without clinical signs of arthropathy: a hospital based case-control study. Ann Rheum Dis. 2008;67:26–30.
13. El Miedany Y. Musculoskeletal US: taking the management of psoriatic arthritis to a new horizon. Curr Rheumatol Rev. 2012;8:12–9.
14. Falsetti P, Acciai C, Lenzi L, et al. Ultrasound of enthesopathy in rheumatic diseases. Mod Rheumatol. 2009;19:103–13.
15. Grassi W, Filippucci E, Farina A, Cervini C. Sonographic imaging of tendons. Arthritis Rheum. 2000;43:969–76.
16. De Vlam K, Mielant H, Veyes EM. Association between ankylosing spondylitis and diffuse idiopathic skeletal hyperostosis: reality or fiction? Clin Exp Rheumatol. 1996;14:5–11.
17. Wakefiled RJ, Balint PV, Szkudlarek M, et al. OMERACT 7 special interest group. Musculoskeletal ultrasound including definitions for ultrasonographic pathology. J Rheumatol. 2005;32:2485–7.
18. D' Agostino MA, Palazzi C, Olivier I. Entheseal involvement. Clin Exp Rheumatol. 2009;27(4 Suppl. 55):S50–5.
19. Balint PV, Kane D, Wilson H, McInnes IB, Sturrock RD. Ultrasonography of entheseal insertions in the lower limb in spondyloarthropathy. Ann Rheum Dis. 2002;61:905–10.
20. D'Agostino MA, Said-Nahal R, Hacquard-Bouder C, et al. Assessment of peripheral enthesitis in the spondylarthropathies by ultrasonography combined with power doppler: a cross-sectional study. Arthritis Rheum. 2003;48:523–33.
21. Alcalde M, Acebes JC, Cruz M, et al. A sonographic enthesitic index of lower limbs is a valuable tool in the assessment of ankylosing spondylitis. Ann Rheum Dis. 2007;66:1015–9.

22. De Miguel E, Cobo T, Munoz-Fernandez S, et al. Validity of enthesis ultrasound assessment in spondyloarthropathy. Ann Rheum Dis. 2009;68:169–74.
23. Spadaro A, Iagnocco A, Perrotta F, et al. Clinical and ultrasonography assessment of peripheral enthesitis in ankylosing spondylitis. Rheumatology. 2011;50:2080–6.
24. D'Agostino M. Ultrasound imaging in spondyloarthropathies. Best Pract Res Clin Rheumatol. 2010;24:693–700.
25. Balint P, Kane D, Wilson H, et al. Ultrasonography of entheseal insertions in the lower limb in spondyloarthropathy. Ann Rheum Dis. 2002;61:905–10.
26. Borman P, Koparal S, Babaoglu S, et al. Ultrasound detection of entheseal insertions in the foot of patients with spondyloarthropathy. Clin Rheumatol. 2006;25:373–7.
27. Falsetti P, Frediani B, Filippou G, et al. Enthesitis of proximal insertion of the deltoid in the course of seronegative spondyloarthritis. An atypical enthesitis that can mime impingement syndrome. Scand J Rheumatol. 2002;31:158–62.
28. El Miedany Y, El Gaafary M, Youssef S, et al. Tailored approach to early psoriatic arthritis patients: clinical and ultrasonographic predictors for structural joint damage. Clin Rheumatol. 2015 Feb;34(2):307-13.
29. Kiris A, Kaya A, Ozgocmen S, et al. Assessment of enthesitis in ankylosing spondylitis by power doppler ultrasonography. Skeletal Radiol. 2006;35:522–8.
30. Riente L, Carli L, Delle Sedie A. Ultrasound imaging in psoriatic arthritis and ankylosing spondylitis. Clin Exp Rheumatol. 2014;32(Suppl. 80):S26–33.
31. El Miedany Y, El Gaafary M, Palmer D. Imaging as an outcome measure in early inflammatory arthritis: monitoring disease activity and patients' response to therapy using ultrasonography. Arthritis Rheum. 2011;63(10):S319.
32. Wiell C, Szkudlarek M, Hasselquist M, et al. Ultrasonography, magnetic resonance imaging, radiography, and clinical assessment of inflammatory and destructive changes in fingers and toes of patients with psoriatic arthritis. Arthritis Res Ther. 2007;9–14:R119.
33. Aydin S, Omer K, Filippucci E, et al. Monitoring achilles enthesitis in ankylosing spondylitis during TNF-α antagonist therapy: an ultrasound study. Rheumatology. 2010;49:578–82.
34. Huang Z, Cao J, Lt T, et al. Efficacy and safety of ultrasound-guided local injections of etanercept into entheses of ankylosing spondylitis patients with refractory achilles enthesitis. Clin Exp Rheumatol. 2011;29:642–9.
35. Tan L, McGonagle D. Imaging of seronegative spondyloarthritis. Best Prac Res Clin Rheumatol. 2008;22(6):1045–59.
36. El Miedany Y. Spondyloarthritis with psoriasis: new insights. J Arthritis. 2012;1:106. doi:10.4172/2167-7921.1000106.
37. Unlü E, Pamuk ON, Cakir N. Color and duplex doppler sonography to detect sacroiliitis and spinal inflammation in ankylosing spondylitis. Can this method reveal response to anti-tumor necrosis factor therapy? J Rheumatol. 2007;34(1):110–6.
38. Spadaro A, Iagnocco A, Baccano G, et al. Sonographic-detected joint effusion compared with physical examination in the assessment of sacroiliac joints in spondyloarthritis. Ann Rheum Dis. 2009;68:1559–63.
39. Bandinelli F, Melchiorre D, Scazzariello F, et al. Clinical and radiological evaluation of sacroiliac joints compared with ultrasound examination in early spondyloarthritis. Ann Rheum Dis. 2011;70(Suppl. 3):374.
40. Mohammadi A, Ghasemi-Rad M, Aghdashi M, et al. Evaluation of disease activity in ankylosing spondylitis; diagnostic value of color doppler ultrasonography. Skeletal Radiol. 2013;42:219–24.
41. Hu Y, Zhu J, Xue Q, Wang N, Hu B. Scanning of the sacroiliac joint and enthuses by color Doppler Ankylosing Spondylitis ultrasonography in patients with ankylosing spondylitis. J Rheumatol. 2011;38:1651–5.
42. Jiang Y, Chen L, Zhu J, et al. Power doppler ultrasonography in the evaluation of infliximab treatment for sacroiliitis in patients with ankylosing spondylitis. Rheumatol Int. 2013;33:2025–9.

43. Klauser A, Halpern EJ, Frauscher F, et al. Inflammatory low back pain: high negative predictive value of contrast-enhanced color Doppler ultrasound in the detection of inflamed sacroiliac joints. Arthritis Rheum. 2005;53(3):440–4.
44. Hartung W, Ross CJ, Straub R, et al. Ultrasound-guided sacroiliac joint injection in patients with established sacroiliitis: precise IA injection verified by MRI scanning does not predict clinical outcome. Rheumatology. 2010;49:1479–82.
45. Klauser A, De Zordo T, Feuchtner G, et al. Feasibility of ultrasound-guided sacroiliac joint injection considering sonoanatomic landmarks at two different levels in cadavers and patients. Arthritis Rheum. 2008;59:1618–24.
46. Warner MB, Cotton AM, Stokes MJ. Comparison of curvilinear and linear ultrasound imaging probes for measuring cross-sectional area and linear dimensions. J Med Eng Technol. 2008;32:498–504.
47. Chen CP, Wong AM, Hsu CC, et al. Ultrasound as a screening tool for proceeding with caudal epidural injections. Arch Phys Med Rehabil. 2010;91:358–63.
48. Pekkafahli MZ, Kiralp MZ, Basekim CC, et al. Sacroiliac joint injections performed with sonographic guidance. J Ultrasound Med. 2003;22:553–9.
49. Chen CP, Lew HL, Tsai WC, et al. Ultrasound-guided injection techniques for the low back and hip joint. Am J Phys Med Rehabil. 2011;90:860–7.
50. Migliore A, Bizzi E, Massafra U, et al. A new technical contribution for ultrasound guided injections of sacroiliac joints. Eur Rev Med Pharmacol Sci. 2010;14:465–9.
51. Harmon D, O'Sullivan M. Ultrasound guided sacroiliac joint injection technique. Pain Physician. 2008;11:543–7.
52. Ooi CC, Malliaras P, Schneider ME, Connell DA. Soft, hard, or just right? Applications and limitations of axial-strain sonoelastography and shear-wave elastography in the assessment of tendon injuries. Skeletal Radiol. 2014;43(1):1–12.
53. Turan A, Tufan A, Mercan R, et al. Real-time sonoelastography of Achilles tendon in patients with ankylosing spondylitis. Skeletal Radiol. 2013;42:1113–8.
54. De Zordo T, Chhem R, Smekal V, et al. Real-time sonoelastography: findings in patients with symptomatic Achilles tendons and comparison to healthy volunteers. Ultraschall Med. 2010;31:394–400.
55. Wu C, Chang K, Mio S, et al. Sonoelastography of the planter fascia. Radiology. 2011;259(2):502–7.

Chapter 5
Psoriatic Arthritis

Carlos Acebes PhD and John P. Harvie MD

Psoriatic arthritis (PsA) is a clinical form of inflammatory arthritis associated with psoriasis. Worldwide prevalence of psoriasis estimated in adults varies from 0.9 to 8.5% [1] and, among these patients, up to one-third may suffer from an inflammatory arthritis. Overall, the sex ratio in PsA is close to unity with slight variations among different subsets. PsA is a very heterogeneous disease affecting musculoskeletal structures including peripheral joints, entheses, tendons and their synovial sheaths, periarticular soft tissues, in addition to the axial skeleton and sacroiliac joints. In addition, PsA can also affect extra-articular tissues such as skin, nails, and eyes. In the majority of cases (85%), psoriasis precedes joint disease but, in up to 10% of patients, arthritis can precede the skin or nail lesions. In spite of PsA being considered a benign condition, prognosis-focused studies have revealed that progression of the arthritis is associated with a moderate increase in morbidity and disability, in some cases could be permanent, though not uncommon [2, 3]. Since effective treatment, including classical synthetic disease-modifying anti-rheumatic drugs (DMARDs) as well as the more recently introduced biologic therapy, is available, the importance of an early diagnosis of this condition has recently been addressed [4]. This chapter discusses in more detail how musculoskeletal ultrasound (MSUS) can play a decisive role in its early diagnosis.

To categorize which lesions ultrasound (US) can help to identify, PsA has been stratified, classically, into five clinical patterns [5]:

1. Arthritis of the distal interphalangeal joints (DIP), almost always associated with psoriatic changes in nails.
2. Destructive arthritis (mutilans) with severely deformed joints, specially affecting hands and feet.

C. Acebes (✉) · J. P. Harvie
Department of Rheumatology, Raigmore Hospital, Inverness, Scotland
e-mail: JCAcebes@fjd.es

J. P. Harvie
e-mail: john.harvie@nhs.net

Y. El Miedany (ed.), *Musculoskeletal Ultrasonography in Rheumatic Diseases*,
DOI 10.1007/978-3-319-15723-8_5

3. Symmetric polyarthritis which can be indistinguishable from rheumatoid arthritis.
4. Asymmetric oligoarthritis consisting of a large joint, such as a knee, together with one or two interphalangeal joints and a dactylitic digit or toe.
5. Spinal involvement, frequently asymmetrical sacroiliitis which may be associated with spondylitis, sometimes clinically indistinguishable from ankylosing spondylitis (AS).

Currently, there are no specific diagnostic laboratory tests for PsA. Therefore, in general, its diagnosis is based on clinical criteria (classification criteria for psoriatic arthritis: CASPAR) [6] supported by negative serologic test, such as rheumatoid factor, and by the valuable information provided by imaging tests such as X-ray, magnetic resonance image (MRI), and US.

Radiologically, plain X-ray has been classically endorsed as a valuable tool for the diagnosis of established PsA. Characteristic radiographic features of PsA distinguishing it from rheumatoid arthritis include: asymmetric and oligoarticular distribution of joint space narrowing with erosive lesions, bony proliferation including periarticular and shaft periostitis, osteolysis including "pencil-in-cup" deformity and acro-osteolysis, ankyloses, spur formation, and spondylitis. MRI may offer some advantages over plain X-ray allowing not only tri-dimensional visualization of the tissue's structural damage, but also, direct assessment of inflammation in the peripheral and axial joints, entheses as well as lesions affecting the periarticular soft tissues. In addition, MRI is the only imaging technique able to demonstrate changes in the subchondral bone suggesting bone edema which according to some studies may represent a strong predictor of bone erosions and enthesitis [7, 8].

Over the past decade, US has booked its place as a valuable imaging tool in the evaluation of PsA, in both its early and late stages. It has been propitiated for its ability to accurately assess structural damage in peripheral joints, entheses, and soft tissues, using B mode/grayscale, along with its capacity to assess and score inflammation, through a sensitive measurement of blood flow with Doppler/color mode. Similar to its advantages in other rheumatic diseases, US in patients with PsA offers real-time and dynamic high-resolution imaging of soft tissues, enabling rapid contralateral area's assessment for comparison. In addition, it has the privilege of being portable and low expense, lack of ionizing radiation, and permitting interaction with the patient while imaging. The target anatomical structures for US examination are those frequently affected according to the clinical pattern previously described: joint synovitis and effusions, enthesitis/enthesopathy, bony erosions and proliferation, cartilage lesions, tenosynovitis, and dactylitis [9]. In addition, more recently, there has been increasing interest on the usefulness of US in axial disease including sacroiliitis [10, 11], and in the assessment of lesions affecting the nails and skin [12].

Choosing the most appropriate US equipment for PsA has become, relatively, an easy task. The continuous technological advances in the field of US have allowed the development of equipment provided with high and broadband frequency transducers (8–15 MHz for the musculoskeletal system and > 15 MHz for skin), together with a very sensitive color/Doppler mode. This ensures a detailed study of morphostructural changes and a very sensitive detection of blood flow in small vessels of the target lesions affected, in particular joints, tendons, and entheses. To ensure achieving the targeted outcomes, for most of the articular and periarticular lesions, the US examination settings entail grayscale frequency at 8–15 MHz, whereas for skin and nails the frequency is commonly above 15 MHz. Gain depth and focus should be adapted for every single exam. Power Doppler (PD) settings for all the examinations can be standardized with a pulse-repetition frequency (PRF) of around 750 Hz (1000–500 Hz) and Doppler frequency between 7 and 14 MHz. Spectral imaging can be used to confirm that PD signal obtained represents real blood flow and not an artifact.

In order to review the role of US in PsA, a detailed description of the usefulness of US in the different pathological tissues affected in this condition is presented in this chapter.

Enthesitis in PsA

Enthesitis represents inflammation at the insertion of tendons, ligaments, or articular capsule into the bone. It is an important hallmark which appears in 25–75 % of PsA patients [13] and represents a component of the classification criteria for PsA and spondyloarthritis (SpA); therefore, its identification is crucial to establish an early diagnosis [14–17]. MSUS provides high-resolution images of the entheses and has demonstrated enough content validity and reliability for the evaluation of this anatomic structure [18]. The outcome measures in rheumatology clinical trials (OMERACT) have defined enthesopathy (pathology at entheses) as an *abnormally hypoechoic and/or thickened tendon or ligament at its bone attachment seen in two perpendicular planes that may exhibit PD signal and/or bony changes including enthesophytes, erosions or irregularities* [19]. Using this definition, MSUS provides high image resolution and functional assessment of abnormalities at the entheses. Using grayscale US mode, enthese's acute inflammatory patterns include thickening and hypoechogenicity of the tendons and ligaments, as well as bursal enlargement; whereas local changes at the site of the entheses, such as enthesophytes, erosions, and calcifications represent more chronic changes (Fig. 5.1) [20–22]. Enhanced vascularity can be depicted by PD mode and appears as increased blood-flow signals, in particular when there are findings of acute enthesitis [23] (Fig. 5.2).

Clinically, enthesitis in PsA patients may be asymptomatic or can cause local pain which may vary in severity though tend to be more common and intense in the lower limbs. Common sites include enthesitis of the Achilles tendon and plantar fascia at their insertion into the calcaneus bone, which is exacerbated by weight bearing after a period of rest [13, 24] (Table 5.1). However, enthesitis may affect any of the body entheses as reported in the study carried out by Falsety et al. who detected enthesitis at the proximal insertion of the deltoid tendon in up to 17 % of PsA patients causing symptoms very similar to those of impingement syndrome [25].

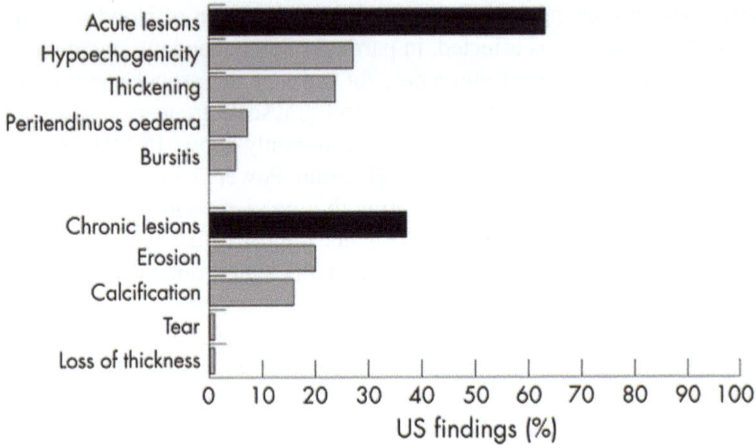

Fig. 5.1 Distribution and frequency of the entheseal lesions detected by ultrasound in patients with spondyloarthritis. Total acute lesions and total chronic lesions are represented in black bars, and each of the categories in gray; data are expressed in percentages with regard to all pathological lesions observed. (From [22])

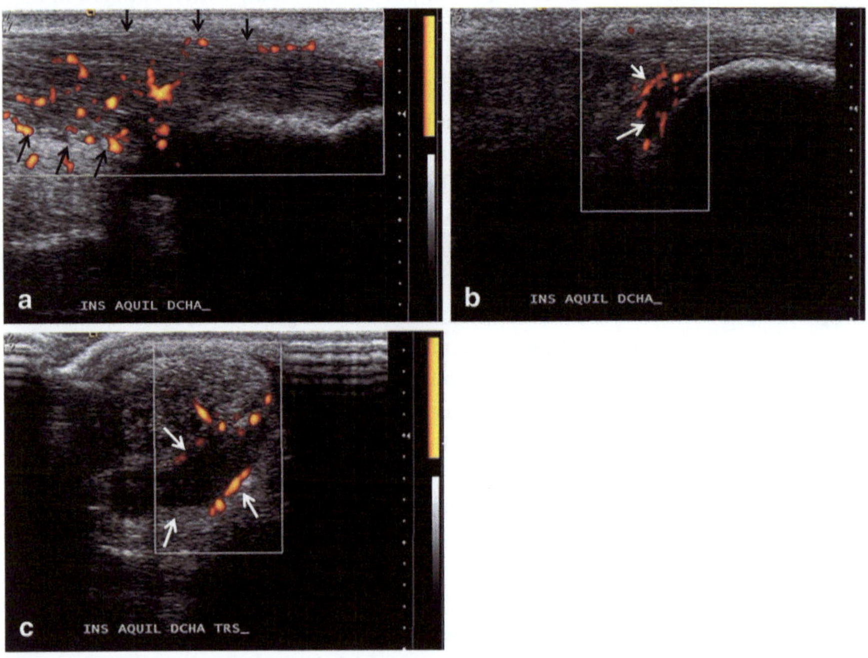

Fig. 5.2 **a** Achilles tendon enthesitis in a patient diagnosed of psoriatic arthritis. There are thickening and hypoechogenicity of the Achilles tendon (*black arrows*). In power Doppler ultrasound mode, an increased blood flow can be observed (*long axis*). **b, c** Retro-calcaneal bursitis in a patient diagnosed of psoriatic arthritis with Achilles tendon enthesitis. US image reveals hypoechoic swelling of the retro-calcaneal bursa deep to the distal Achilles tendon (*white arrows*). The peri-bursal area appears hyper-vascular at color Doppler imaging. **b** Long axis. **c** Short axis

Table 5.1 Frequency of ultrasound and Doppler-detected enthesitis at different sites in patients with psoriatic arthritis. (According to D'Agostino et al. [24])

Enthesitic sites	Frequency of US findings (grayscale and Doppler) in %
Achilles tendon	76
Plantar fascia	57
Patellar tendon	60
Quadriceps tendon	26
Great trochanter	43
Lateral epicondyle	19
Medial epicondyle	17

US ultrasound

MSUS has been demonstrated to be more sensitive in comparison to clinical examination for the detection of enthesitis. Earlier studies revealed that US was able to identify subclinical enthesitis significantly higher than clinical assessment [21, 26]. In a cohort of 162 patients diagnosed with psoriasis sine PsA compared to 60 age/sex-matched controls, another multicenter study group demonstrated a high prevalence of subclinical synovitis (3.2 vs. 1.3 %) and enthesitis (11.6 vs. 5.3 %) on US examination [27]. In another study of Gutierrez et al., enthesitis was observed in 32.9 % and 0.9 % of enthesitis sites on grayscale and PD modes, respectively, in patients with psoriasis compared to 8.4and 0 % in age-matched healthy controls [28].

Some other studies demonstrated that in patients with psoriasis, the early detection of enthesitis on US may have prognostic implications on the prediction of subsequent development of PsA. In a 3.5-year follow-up, US assessment of a psoriatic patients cohort, up to 23 % of psoriatic patients with a higher baseline enthesitis score developed PsA [29]. In addition, the detection of hypervascularity at the entheses in patients with suspected early SpA and PsA was highly predictive of a future establishment of the diagnosis [30].

Several ultrasonographic enthesitis-scoring indices for SpA, including PsA, have been published in the past decade. The Glasgow ultrasound enthesitis scoring system (GUESS) assesses lower-limb entheses including the superior and inferior pole of the patella, the distal insertion of the patellar tendon, the Achilles tendon, and the plantar fascia entheses [21]. The authors propose a scoring system of 0–36 based on the sum of the different US findings of enthesitis. Alcalde et al., based on the assessment of the same five enthesitis sites in the lower limbs, proposed a new score system of 0–36 based on US findings of acute injury and 0–40 based on findings considered to be chronic disease [22]. As these two previous studies only used grayscale US mode, de Miguel et al. added PD image information to the previous indices, with a further extension to 12 enthesitis sites, including upper-limb entheses, and published the Madrid Sonography Enthesitis Scoring Index (MASEI) [31].

Dactylitis in PsA

Dactylitis or "sausage digit" has classically been recognized as another characteristic feature of PsA [32] appearing in up to 30–48 % of patients according to different studies [33, 34]. In addition to PsA, it may be found in other SpA particularly reactive arthritis, gout, local infections of digits, sarcoidosis, and sickle cell disease. Very occasionally, it has also been reported in rheumatoid arthritis. The relatively high specificity of dactylitis in PsA leads to its inclusion in the clinical criteria for the classification of SpA suggested by Amor et al. [35]. Furthermore, due to its high sensitivity and adequate specificity, dactylitis has been included in the recent classification criteria generated by the CASPAR study group [14].

The clinical presentation of dactylitis is a uniform diffuse swelling affecting fingers in 24 %, toes in 65 % and both hands and feet in 12 % of PsA patients [34]. Dactylitis clinical manifestations may vary from significant local digital pain to an asymptomatic finding. It is accepted that dactylitis may present as an acute condition with painful inflammatory changes, or a chronic course, where the digit still remains swollen despite the disappearance of the pain [34].

Despite the incomplete agreement regarding the different tissue components involved in dactylitis [36–38], a recent meta-analysis published by Bakewell et al. [39] revealed that with the use of MRI and US for the assessment of this condition, the most common features were flexor tendon tenosynovitis and joint synovitis which appeared equally in 90 % of the patients, and extratendinous soft tissue thickening and extensor tendonitis which appeared in up to 50 % of patients. Other components, only qualitatively described in this literature review, were bone, extra and intra-articular osteoproliferation, erosions, bone edema, and plantar/palmar plate abnormalities. According to tissue-specific vascularity using Doppler-mode US or post-gadolinium MRI studies, the elementary components (not quantified) of dactylitis were joint synovitis, flexor tendon tenosynovitis, soft tissue edema, nail-bed abnormalities, bone extra and intra-articular osteoproliferation, and bone edema.

MSUS has demonstrated enough accuracy to identify the different tissue compartments involved in dactylitis, comparable to the findings obtained from MRI examination (Fig. 5.3):

- Joint synovitis, which frequently affects metacarpophalangeal, metatarsophalangeal as well as interphalangeal joints, has US pattern which is similar to and cannot be distinguished from other inflammatory conditions such as rheumatoid arthritis. It consists of a distension of the joint capsule with a hypoechoic content that can be displaced and compressed depending of the proportion of the effusion compared to synovial proliferation. Eventually, it may also exhibit tissue vascularity in Doppler mode.
- Flexor tendon tenosynovitis consists of a thickening of the diameter of the flexor tendons and their sheaths, with loss of the normal tendon echotexture and distension of the tendon sheath with hypoechoic content, consistent with effusion and synovial proliferation, which eventually may exhibit vascularity in Doppler mode.

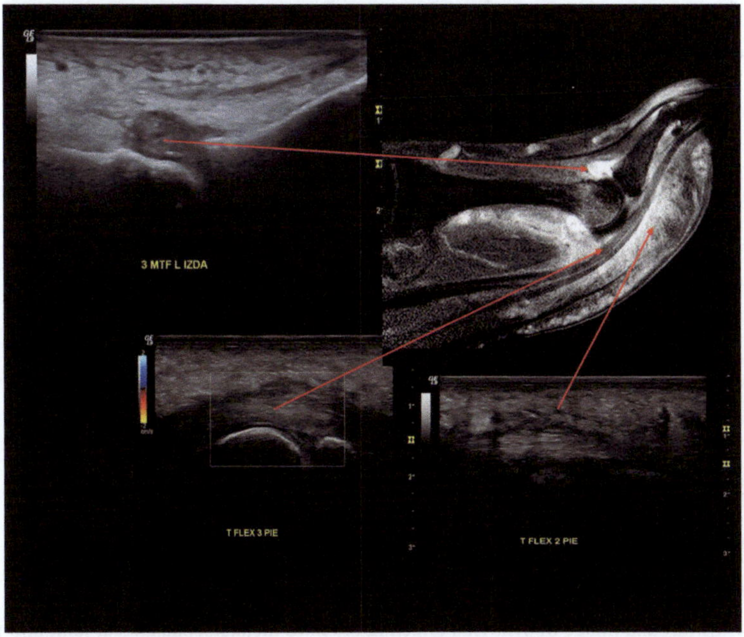

Fig. 5.3 Ultrasound image of the different tissue compartments involved in dactylitis. Comparison with MRI image from the same patient with psoriatic arthritis (*long* arrows). **a** Synovitis affecting the metatarsophalangeal joint: distension of the joint capsule with hypoechoic content. **b** Flexor tendon tenosynovitis consisting of a thickening of the diameter of the tendon, with loss of the normal tendon echotexture. **c** Extratendinous soft tissue changes which are consistent with thickening of the subcutaneous tissue, with an increase of distance from skin to tendon and changes to the echotexture pattern of the soft tissues around the tendon. *MRI* magnetic resonance imaging

- Extratendinous soft tissue changes are commonly seen in association with dactylitis and are consistent with thickening of the subcutaneous tissue, with increased distance from skin to tendon and changes in the echotexture pattern of the soft tissues around the tendon suggestive of subcutaneous tissue edema manifested clinically as local swelling.
- Other features less commonly described include extra and intra-articular osteoproliferation, erosions, and plantar/palmar plate abnormalities. These are described in more detail later in this review.

Peripheral Arthritis

Despite peripheral joint synovitis in PsA seemingly undistinguished from other chronic inflammatory arthritic conditions, such as rheumatoid arthritis, some studies have demonstrated differences in the pathology identified by MSUS. In a study published by Fournie et al., they reported extra-synovial abnormalities which seemed to

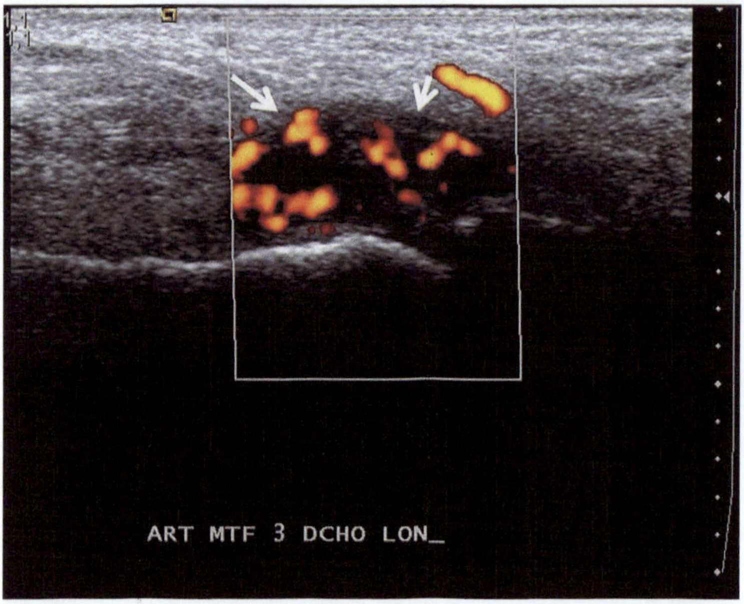

Fig. 5.4 Ultrasound image of synovitis affecting the third metatarsophalangeal joint of a patient with psoriatic arthritis. There are findings of distension of the joint capsule with an abnormal hypoechoic intra-articular content suggesting synovial hypertrophy (*white arrows*) with Doppler signal within the joint space

be specific to PsA such as enthesitis, thickening of soft tissues, and Doppler signal at the base of the nail suggestive of periungual psoriatic involvement [40].

US findings in peripheral joints with synovitis include distension of the joint capsule with an abnormal hypoechoic or anechoic intra-articular content—always relative to subdermal fat, but sometimes it may be isoechoic or hyperechoic. It potentially can be displaceable and compressible depending on the prevalence of synovial hypertrophy or effusion. When the predominant content is synovial hypertrophy, Doppler signal within the joint space can be identified (Fig. 5.4). An earlier study documented the higher sensitivity of US as a tool for the detection of synovitis in comparison to physical exam in any of the peripheral joints involved [41].

As there is not any specific scoring system for peripheral synovitis in PsA, any of the semiquantitative US scoring systems published (0–3 where 0 = normal, 1 = mild, 2 = moderate, 3 = severe) for grading synovitis in rheumatoid arthritis could be potentially used also for grading the PsA-inflamed joints [42].

Bone erosions in PsA can be found in up to 60 % of patients [43]. They appear similar to those seen in rheumatoid arthritis and do not have any specific appearance on US exam. The typical image is a discontinuation in the bone surface (cortical defect) which should be visualized in longitudinal and transversal scans (Fig. 5.5). This cortical defect may be filled with inflammatory tissue and, eventually, may exhibit vascularity in Doppler mode (Fig. 5.6). Erosions also can be scored using a binary present/absent system or a 0–3 semiquantitative score [44]

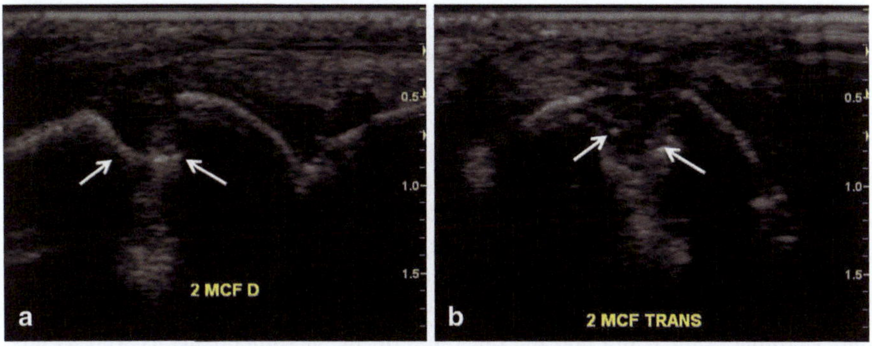

Fig. 5.5 Ultrasound image of bone erosion in psoriatic arthritis. There is a discontinuation in the bone surface (cortical defect) which is visualized in **a** Longitudinal and **b** Transversal scans (*white arrows*)

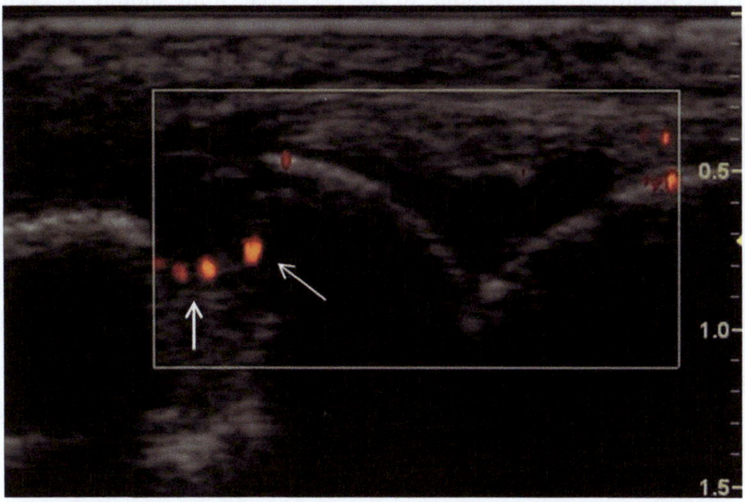

Fig. 5.6 Ultrasound image of bone erosion in psoriatic arthritis filled with inflammatory tissue exhibiting vascularity in Doppler mode (*white arrows*)

Axial Disease

Most of the information available about the use of MSUS in axial disease in SpA and PsA has been focused on the sacroiliac joints. Sacroiliitis is an early feature of SpA, including PsA, and frequently is associated with inflammatory back pain and stiffness. However, due to the deep location of the sacroiliac joint, classically there have been some technical issues in obtaining a high-quality image of its anatomy using US. Early in the disease course, when X-ray examination is negative, the inflammation of the sacroiliac joint can be detected in more than a half of patients using MRI, which allows the detection of bone edema adjacent to the sacroiliac joints, as an early finding of sacroiliitis [45]. Therefore, bearing in mind the technical limitations, only a few articles on testing the US usefulness for evaluation of sacroiliac joints have been published.

In a recent study, Spadaro et al. analyzed the ability of US to detect joint effusion in sacroiliac joints in SpA patients. The prevalence of US-detected joint effusion was 38.9 % of sacroiliac joints among the SpA patients cohort compared to 1.7 % in the control group. In addition, the presence of inflammatory back pain was significantly associated with the presence of sacroiliac joint effusion assessed by US [46].

In another study carried out by Bandinelli et al., they reported that US could be a reliable imaging method for assessing the sacroiliac joint space. They compared the sacroiliac joint width and the thickness of the sacrotuberosus as well as posterior sacroiliac ligaments in patients with early SpA to those of healthy controls, revealing that early in the disease course the sacroiliac joint space width and the sacrotuberosus ligament thickness was significantly higher in patients compared with controls [47]

The US examination technique of the sacroiliac joint should be performed with the patient in a prone or knee–chest position. The bony landmarks to identify for guiding the examination are the midline of the sacrum and the iliac wings which can be easily identified as a hyperechogenic surface laterally to the spinous processes of the sacral bone. With the transducer in a transverse position, the sacroiliac joint space is identified as a hypoechoic gap between the hyperechoic margins of the joint, sacrum, and iliac bone. In this location of the transducer, the findings of synovitis with effusion appear as the presence of a hypoechoic area within and slightly over the joint space [48, 46]. At this position, and switching to Doppler mode, eventually, an increased perfusion around or inside the sacroiliac joint can be detected in up to 48 % of patients with sacroiliitis [48] (Fig. 5.7).

Despite the lack of studies to support the real usefulness of the US in the assessment of sacroiliac joints, the few of them previously discussed, would suggest that this imaging modality could be useful in providing real-time, less expensive, comparative, and repeated images of the sacroiliac joints in SpA and PsA.

There are a very limited number of published data on the use of US in the spine in patients with SpA and PsA. Unlu [49], using color and duplex Doppler US, reported how the resistive index of blood vessels of sacroiliac joints and thoracic and lumbar vertebral paraspinal areas was significantly reduced in 39 patients diagnosed with SpA compared to controls. Interestingly, after a course of anti-TNF therapy in the same patients, the authors observed a significant increase in the resistive index at the level of sacroiliac joints and lumbar vertebral paraspinal areas [10]. The sensitivity of US to detect sacroiliac inflammation in SpA may improve with the use of contrast-enhanced color Doppler, as was demonstrated by Klauser et al. in a comparative study with MRI [50].

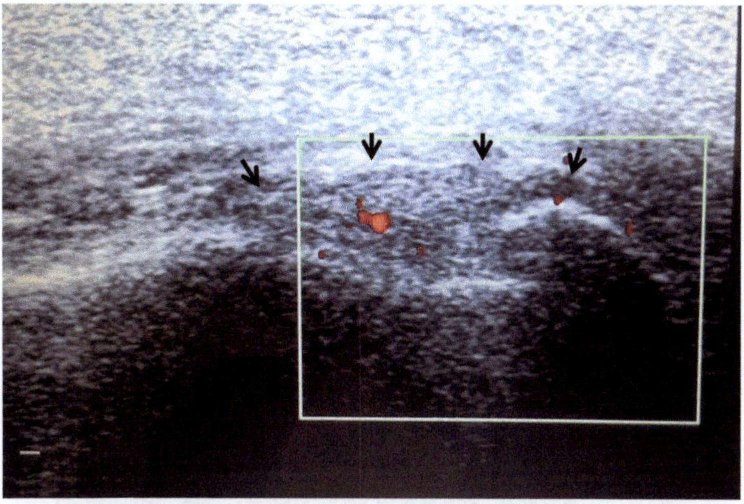

Fig. 5.7 Ultrasound (US) image of right sacroiliitis in a patient with psoriatic spondyloarthropathy (*short axis*). The US findings of synovitis with effusion appear as the presence of a hypoechoic area within and slightly over the joint space (*black arrows*). Switching to Doppler mode, there is an increased perfusion inside the sacroiliac joint

Skin and Nails

The diagnosis of skin psoriasis is routinely based on the clinical examination, and the extension of lesions is determined using clinical scores, such as the psoriasis area and severity index (PASI). In a limited number of cases, clinical suspicion needs to be confirmed with histologic examination to confirm the diagnosis. Up to 50 % of patients with psoriasis may have nail involvement.

Over the past decade, US has been increasingly used in dermatology to assess the skin and nails in patients with psoriasis. US imaging of psoriasis using high-frequency transducers (15–50 MHz) allows detailed morphologic information to be obtained about the cutaneous and ungular changes, along with the monitoring of disease activity and treatment response.

By using the enhanced detail provided by high-frequency grayscale US, it is possible to detect the typical changes of psoriasis affecting the skin consistent with thickening of the epidermis and dermis, compared to the normal surrounding skin, along with a hypoechoic band in the upper dermis. In those cases, with marked thickening of the epidermis, a posterior acoustic shadowing artifact may be observed, affecting the assessment of the underlying dermis. In psoriasis, the appearance of affected nails shows changes consistent with thickening and decreased echogenicity of the nail bed, focal hyperechoic deposits in the ventral plate, and, in more advanced lesions, thickening of both dorsal and ventral plates. The thickening of the nail bed can be measured as the distance between the ventral plate and the bone margin of the distal phalange. In the active phase of the disease, Doppler mode allows identification of an increased blood flow localized within the dermis

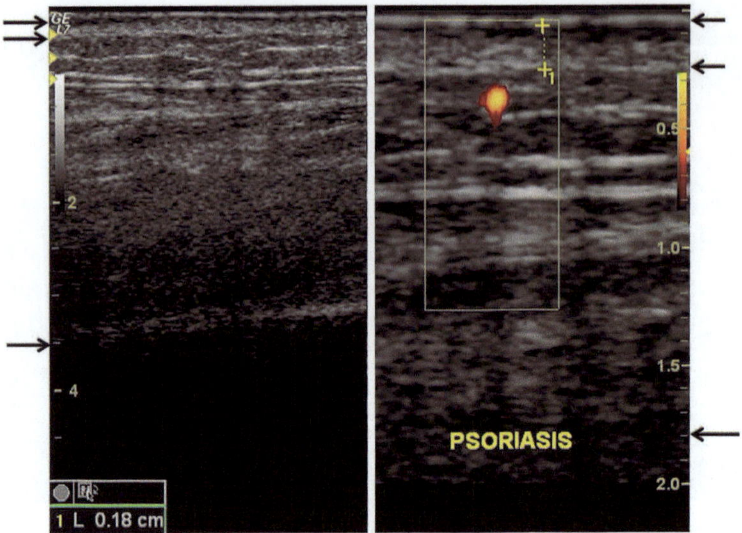

Fig. 5.8 High-frequency ultrasound of a psoriatic plaque (**b**) compared with healthy skin (**a**). The epidermis and dermis appear thicker (*black arrows*). There are hypoechoic bands in the upper dermis, and there is an increased blood-flow signal within the dermis demonstrating it is an active psoriatic plaque

of the psoriatic plaques and in the nail beds [12]. Grayscale and Doppler US findings allow monitoring of disease activity, severity of the psoriatic involvement, and assessment of treatment response, as has already been reported in the literature [51–53] (Fig. 5.8).

US-Guided Procedures in Psoriatic Arthritis

As in other rheumatic diseases, in PsA, US-guided procedures are a very reliable method to safely and accurately access the different anatomical structures involved in the pathogenic mechanisms of the lesion, such as peripheral arthritis, enthesitis, dactylitis, and, even the axial skeleton, especially the sacroiliac joints. In addition, there are advantages of the US examination, not only in establishing the diagnosis but also the ability to recognize active synovitis, providing valuable information required for decision making in terms of local therapy. Previous studies have demonstrated therapeutic effectiveness of US-guided therapy directed to musculoskeletal structures—joints and soft tissues—by choosing and ensuring the correct placement for injection as compared to blinded procedures [54–57].

In inflammatory arthritis, with peripheral joints inflammation, US-guided injection has demonstrated significantly more accuracy than those guided by the clinical examination. [57, 58]. With reference to PsA enthesitis, US may offer a very reliable anatomic image of the different entheses to differentiate pathological from normal anatomy. Despite this, there are no controlled studies assessing the effectiveness of US-guided entheseal injections specifically in PsA. Huang et al. demonstrated the efficacy and safety of US-guided local injections into entheses with steroids or Etanercept in Ankylosing Spondylitis (AS) patients [59]. In another controlled study, [60] involved over 65 patients with plantar fasciitis randomized to US-guided steroid injection/blinded steroid injection/US-guided placebo injection and demonstrated superiority of both steroids-treated groups but the difference was not significant between US guided versus blinded. Another study in a series of 24 patients with recalcitrant plantar fasciitis also did not show differences between US-guided injections and conventional injections [61]. Therefore, data supporting the superiority of US-guided procedures compared to conventional injection procedures in PsA enthesitis are still relatively limited and more controlled studies are necessary.

The Role of US in Early Diagnosis, Monitoring of Disease Activity, and Therapeutic Response in PsA

There is an undoubted potential to incorporate the US findings, especially regarding enthesitis and dactylitis, into PsA diagnostic criteria, which would allow early detection of the pathology and improve the sensitivity for the diagnosis. In addition, identification of quantitative US findings (erosions) in PsA may be helpful for the decision making of those patients who can benefit from early or more aggressive therapy.

A clear definition of the different US lesions such as enthesitis, dactylitis, peripheral arthritis, or sacroiliitis, along with an appropriate scoring system potentially would allow an objective monitoring of disease activity before and after the treatment of PsA.

References

1. Parisi R, Symmons DP, Griffiths EM, Ashcroft M. Global epidemiology of psoriasis: a systematic review of incidence and prevalence. J Investig Dermatol. 2013;133:377–85.
2. Rahman P, Nguyen E, Cheung C. Comparison of radiological severity in psoriatic arthritis and rheumatoid arthritis. J Rheumatol. 2001;28:1041–44.
3. Ruderman EM, Tambar S. Psoriatic arthritis: prevalence, diagnosis and review of therapy for the dermatologist. Dermatol Clin. 2004;22:477–86.
4. Sorensen J, Hetland ML. Decrease in diagnostic delay are supported by sensitivity analysis. Ann Rheum Dis. 2014:73:e45.
5. Moll JMH, Wright V. Psoriatic arthritis. Semin Arthritis Rheum. 1973:3:55–78.
6. Taylor W, Gladman D, Helliwell P. Classification criteria for psoriatic arthritis: development of new criteria from a large international study. Arthritis Rheum. 2006;54:2665–73.

7. McGonagle DS, Gibbon W, O'Connor P, Green M, Pease C, Emery P. Characteristic magnetic resonance imaging entheseal changes of knee synovitis in spondyloarthropathy. Arthritis Rheum. 1998;41:694–700.
8. McQueen FM, Benton N, Perry D, Crabbe J, Robinson E, Yeoman S, et al. Bone oedema scored on magnetic resonance scans of the dominant carpus at presentation predicts radiographic joint damage at the hands and feet six years later in patients with rheumatoid arthritis. Arthritis Rheum. 2003;48:1814–27.
9. Kane D. The role of ultrasound in the diagnosis and management of psoriatic arthritis. Curr Rheumatol Rep. 2005;7:319–24.
10. Unlu E, Pamuk ON, Cakir N. Colour duplex doppler sonography to detect sacroiliitis and spinal inflammation in ankylosing spondylitis. Can this method reveal response to anti-tumor necrosis factor therapy? J Rheumatol. 2007;34:110–6.
11. Spadaro A, Iagnocco A, Baccano G, Ceccarelli F, Sbatini E, Valesini G. Sonographic-detected joint effusion compared with physical examination in the assessment of sacroiliac joints in spondyloarthritis. Ann Rheum Dis. 2009;68:1559–63.
12. Wortsman X. Common applications of dermatologic sonography. J Ultrasound Med. 2012;31:97–111.
13. Sakkas LI, Alexiou I, Simopoulou T, Vlychou M. Enthesitis in psoriatic arthritis. Semin Arthritis Rheum. 2013;43:325–34.
14. Taylor W, Gladman D, Hellwell P, Marchesoni A, Mease P, Mielants H. Classification criteria for psoriatic arthritis: development a new criteria from a large international study. Arthritis Rheum. 2006;54:2665–73.
15. Coates LC, Conaghan PG, Emery P, Green MJ, Ibrahim G, McIver H, et al. Sensitivity and specificity of the classification of psoriatic arthritis criteria in early psoriatic arthritis. Arthritis Rheum. 2012;64:3150–5.
16. Dougados M, van der Linden S, Juhlin R, Huitfeldt B, Amor B, Calin A, et al. The European Spondyloarthropathy Study Group preliminary criteria for the classification of of Spondylarthropathy. Arthritis Rheum. 1991;34:1218–27.
17. Rudwaleit M, Van der Heijde D, Landewe R, Listing J, Akkok N, Brand J, et al. The development of assessment of Spondyloarthritis International Society Classification Criteria for Axial Spondyloarthropathies (part II): validation and final selection. Ann Rheum Dis. 2009;68:777–83.
18. Grandjbakhch F, Terslev L, Joshua F, Wakefield RJ, Naredo E, D'Agostino MA, et al. Ultrasound in the evaluation of enthesitis: status and perspectives. Arthritis Res Ther. 2011;13:R188.
19. Wakefield RJ, Balint PV, Szkudlarek M, Filippucci E, Backhaus M, D'Agostino MA, et al. Musculoskeletal ultrasound including definitions for ultrasonographic pathology. J Rheumatol. 2005;32:2485–7.
20. Filippucci E, de Angelis R, Salaffi F, Grassi W. Ultrasound, skin and joints in psoriatic arthritis. J Rheumatol. 2009;83:35–8.
21. Balint PV, Kane D, Wilson H, McInnes IB, Sturrock RD. Ultrasonography of entheseal insertions in the lower limb in Spondyloarthropathy. Ann Rheum Dis. 2002;61:905–10.
22. Alcalde M, Acebes C, Gonzalez-Hombrado L, Herrero-Beaumont G, Sanchez-Pernaute O. A sonographic enthesitic index of lower limbs is a valuable tool in the assessment of ankylosing spondylitis. Ann Rheum Dis. 2007;66:1015–19.
23. D'Agostino MA. Ultrasound imaging in spondyloarthropathies. Best Pract Res Clin Rheumatol. 2010;24:693–700.
24. D'Agostino MA, Said-Nahal R, Hacquard-Bourder C, Brasseur JL, Dougados M, Breban M. Assessment of peripheral enthesitis in the Spondyloarthropathies by ultrasonography combined with power doppler: a cross-sectional study. Arthritis Rheum. 2003;48:523–33.
25. Falsetti P, Frediani B, Filippou G, Acciai C, Baldi F, Storri L, et al. Enthesitis of proximal insertion of the deltoid in the course of seronegative Spondyloarthritis. An atypical enthesitis that can mime impingement syndrome. Scand J Rheumatol. 2002;31:158–62.
26. Scarpa R, Coucolo A, Peluso R, Atteno M, Gisonni P, Tervolino S, et al. Early psoriatic arthritis: the clinical spectrum. J Rheum. 2008;35:137–41.

27. Naredo E, Moller I, de Miguel E, Batlle-Gualda E, Acebes C, Brito E, et al. High prevalence of ultrasonographic synovitis and enthesopathy in patients with psoriasis without psoriatic arthritis: a prospective case-control study. Rheumatology. 2011;50:1838–48.
28. Gutierrez M, Filippucci E, de Angelis R, Salaffi F, Filosa G, Ruta S, et al. Subclinical enthesal involvement in patients with psoriasis: an ultrasound study. Semin Arthritis Rheum. 2011;40:407–12.
29. Tinazzi I, McGonagle D, Biasi D, Confente S, Caimmi C, Girolomoni G, et al. Preliminary evidence that subclinical enthesopathy may predict psoriatic arthritis in patients with psoriasis. J Rheumatol. 2011;38:2691–2.
30. D'Agostino MA, Aegerter P, Bechara K, Salliot C, Judet O, Chimenti MS, et al. How to diagnose spondyloarthritis early? Accuracy of peripheral enthesitis detection by power Doppler ultrasonography. Ann Rheum Dis. 2011;70:1433–40.
31. De Miguel E, Cobo T, Munoz-Fernandez S, Naredo E, Uson J, Acebes C, et al. Validity of enthesis ultrasound assessment in spondyloarthropathy. Ann Rheum Dis. 2009;68:196–74.
32. Moll JMH, Wright V. Psoriatic arthritis. Semin Arthritis Rheum. 1973;3:55–78.
33. Veale D, Rogers S, Fitzgerald O. Classification of clinical subsets in psoriatic arthritis. Br J Rheumatol. 1994;33:133–8.
34. Brockbank JE, Stein M, Schentag CT, Gladman DD. Dactilitis in psoriatic arthritis: a marker for disease severity? Ann Rheum Dis. 2005;64:188–90.
35. Amor B, Dougados M, Mujiyawa M. Criteria of the classification of spondyloarthropathies. Rev Rhum Osteoartic. 1990;57:85–9.
36. Kane D, Greaney T, Bresnihan B, Gibney R, Fitzgerald O. Ultrasonography in the diagnosis and management of psoriatic dactylitis. J Rheumatol. 1999;26:1746–51.
37. Wakefield RJ, Emery P, Veale D. Ultrasonography and psoriatic arthritis. J Rheumatol. 2000;27:1564–5.
38. Oliveri I, Padula A, Scarano E, Scarpa R. Dactilitis or "sausage-shaped" digit. J Rheumatol. 2007;34:1217–21.
39. Bakewell CJ, Oliveri I, Aydin SZ, Dejaco C, Ikeda K, Gutierrez M, et al. Ultrasound and magnetic resonance imaging in the evaluation of psoriatic dactilitis: status and perspectives. J Rheumatol. 2013;40:1951–7.
40. Fournie B, Margarit-Coll N, Champetier de Ribes TL, Zabraniecki L, Jouan A, Vincent V, et al. Extrasynovial ultrasound abnormalities in the psoriatic finger. Prospective comparative power-Doppler study versus rheumatoid arthritis. Joint Spine. 2006;73:527–31.
41. Grassi W. Clinical evaluation versus ultrasonography: who is the winner? J Rheumatol. 2003;30:908–9.
42. Scheel AK, Herman KG, Kahler E, Pasewaldt D, Fritz J, Hamm B, et al. A novel ultrasonographic synovitis scoring system suitable for analysing finger joint inflammation in rheumatoid arthritis. Arthritis Rheum. 2005;52:733–43.
43. Gladman DD, Farewell VT, Nadeau C. Clinical indicators of progression in psoriatic arthritis: multivariate relative risk model. J Rheumatol. 1995;22:675–9.
44. Szkudlarek M, Court-Payen M, Jacobsen S, Klarlund M, Thomsen HS, Ostergaard M. Interobserver agreement in ultrasonography of the finger and toe joints in rheumatoid arthritis. Arthritis Rheum. 2003;48:955–62.
45. Bollow M, Fischer T, Reisshauer H, Backhaus M, Sieper J, Hamm B, et al. Quantitaive analysis of sacroiliac biopsies in spondyloarthropathies: T cells and macrophages predominate in early and active sacroiliitis-cellularity correlates with the degree of enhancement detected by magnetic resonance imaging. Ann Rheum Dis. 2000;59:135–40.
46. Spadaro A, Iagnocco A, Baccano G, Ceccarelly F, Sabatini E, Valesini G. Sonographic-detected joint effusion compared with physical examination in the assessment of sacroiliac joints in spondyloarthritis. Ann Rheum Dis. 2009;68:1559–63.
47. Bandinelli F, Melchiorre D, Scazzariello F, Candelieri A, Conforty D, Matucci-Cerinic M. Clinical and radiological evaluation of sacroiliac joints compared with ultrasound examination in early spondyloarthritis. Rheumatology. 2013;52:1293–97.
48. Arslan H, Sakarya ME, Adak B. Duplex and color doppler sonographic findings in active sacroiliitis. Am J Roetgenol. 1999;173:677–80.

49. Unlu E, Pamuk ON, Cakir N. Color and duplex doppler sonography to detectsacroiliitis and spinal inflammation in ankylosing spondylitis. Can this method reveal response to anti-tumor necrosis factor therapy? J Rheumatol. 2007;34:110–6
50. Klauser A, Halpern EJ, Frauscher F, Gvozdic D, Duftner C, Springer P, et al. Inflammatory low back pain: high negative predictive value of contrast-enhanced colour doppler ultrasound in the detection of inflamed sacroiliac joints. Arthritis Rheum. 2005;53:440–4.
51. De Agustin JJ, Moragues C, De Miguel E, Moller I, Acebes C, Naredo E, et al. A multicentre study on high-frequency ultrasound evaluation of the skin and joints in patients with psoriatic arthritis treated with infliximab. Clin Exp Rheumatol. 2012;30:879–85.
52. Gutierrez M, Wortsman X, Filippucci E, De Angelis R, Filosa G, Grassi W. High-frequency sonography in the evaluation of psoriasis: nail and skin involvement. J Ultrasound Med. 2009;28:1569–74.
53. Filipucci E, De Angelis R, Salaffi F, Grassi W. Ultrasound, skin and joints in psoriatic arthritis. J Rheumatol Suppl. 2009;83:35–8.
54. Ustun N, Tok F, Yagz AE, Kizil N, Korkmaz I, Karazincir S, et al. Ultrasound-guided vs. blind steroid injections in carpal tunnel syndrome: a single-blind randomized prospective study. AmJ Phys Med Rehabil. 2013;92:999–1004.
55. Cunnington J, Marshall N, Hide G, Bracewell C, Isaacs J, Platt P. A randomized, double-blind controlled study of ultrasound-guided corticosteroid injection into the joint of patients with inflammatory arthritis. Arthritis Rheum. 2010;62:1862–69.
56. Eustace JA, Brophy DP, Gibney RP, Bresnihan B, Fitz Gerald O. Comparison of the accuracy of steroid placement with clinical outcome in patients with shoulder symptoms. Ann Rheum Dis. 1997;56:59–63.
57. Balint PV, Kane D, Hunter J, McInnes IB, Field M, Sturrock RD. Ultrasound guided versus conventional joint and soft tissue fluid aspiration in rheumatology practice: a pilot study. J Rheumatol. 2002;29:2209–13.
58. Raza K, Lee CY, Pilling D, Heaton S, Situnayake RD, Carruthers DM, et al. Ultrasound guidance allows accurate needle placement and aspiration from small joints in patients with early rheumatoid arthritis. Rheumatology. 2003;42:976–9.
59. Huang Z, Cao J, Li T, Zheng B, Wang M, Zheng R. Efficacy and safety of ultrasound-guided local injections of etanercept into entheses of ankylosing spondylitis patients with refractory Achilles enthesitis. Clin Exp Rheumatol. 2011;29:642–9.
60. Ball EM, McKeeman HM, Patterson C, Burns J, Yau WH, Moore OA, et al. Steroid injection for inferior heel pain: a randomised controlled trial. Ann Rheum Dis. 2013;72:996–1002.
61. Kane D, Greaney T, Shanahan M, Duffy G, Brenihan B, Gibney R, et al. The role of ultrasonography in the diagnosis and management of idiopathic plantar fasciitis. Rheumatology. 2001;40:1002–8.

Chapter 6
Osteoarthritis

Annamaria Iagnocco MD and Vasilia Iorgoveanu MD

Introduction

Osteoarthritis (OA) is the most common rheumatic disease affecting peripheral and axial synovial joints. It is frequently diagnosed in older adults, although it can occur relatively early in life, causing disability and work impairment. Mechanical pain, swelling, stiffness, deformity, and loss of joint motion are the most common clinical findings that cause patients' complaints and relevant public health issues [1–3].

The characteristic imaging abnormality in OA is reduced joint space, formation of osteophytes (protrusions of bone and cartilage) mostly at the joint's margins as well as sclerosis of the subchondral bone. Those changes are the result of several histopathologic phases. Biomechanical and biochemical factors, such as the production of several chemokines, cytokines, and growth factors, stimulate an aberrant chondrocyte activation that contributes to the disruption of articular homeostasis. The first recognizable change in OA is edema of the extracellular matrix, principally in the intermediate layer. The cartilage loses its smooth aspect, and microfissures start to appear. This results in a focal loss of chondrocytes, alternating with areas of chondrocyte proliferation. The imbalance between pro- and antiinflammatory molecules may influence the progression of cartilage damage. The progression of the cartilaginous fissures along the collagen fibrils produce vertical clefts that deepen into the subchondral bone cartilage. More fissures cause fragments of cartilage to detach creating osteocartilaginous loose bodies and uncovering the subchondral bone, where microcysts develop. These loose bodies also may initiate some degree of synovitis, with an episodic course that can contribute to the worsening of

A. Iagnocco (✉)
Ultrasound Unit, Department of Rheumatology, Sapienza Università di Roma, Viale del Policlinico 155, Rome, Italy
e-mail: annamaria.iagnocco1@gmail.com

V. Iorgoveanu
Department of Internal Medicine and Rheumatology, Sf. Maria Hospital, University of Medicine and Pharmacy Carol Davila, Bucharest, Romania
e-mail: iorgoveanuvasi@gmail.com

Y. El Miedany (ed.), *Musculoskeletal Ultrasonography in Rheumatic Diseases*,
DOI 10.1007/978-3-319-15723-8_6

symptoms and further cartilage deterioration. Nondestructive synovial proliferation, joint effusion, and bursitis are frequent findings in OA. Also, along with disease progression, subchondral sclerosis appears, as a result of accelerated bone turnover.

In the assessment and diagnosis of OA, conventional radiography (X-ray) remains the mainstay imaging investigation [4, 5]. X-ray is valuable for detecting structural joint changes. The radiographic features of OA include osteophytes, joint space narrowing, subchondral sclerosis, subchondral cysts, bony remodeling, and possibly joint effusions visualized as soft tissue swelling and fat pad displacement. Other arthropathies may coexist such as calcium pyrophosphate arthropathy. However, X-ray is not able to directly visualize the cartilage and employs a surrogate measure, that is joint space narrowing [6]. Moreover, as the radiographic changes occur mainly in later stages after long disease duration, it is unclear whether changes demonstrated by X-ray are sufficient enough in early OA phases to consider the diagnosis. In a subset of the patients, the disease may be asymptomatic in spite of the relative radiographic changes.

Consequently, there is unanimous consensus about the necessity to have a valid and reliable tool to study and evaluate distinct changes occurring in OA. The availability of imaging tools which would enable the treating health-care professional to directly assess cartilage in osteoarthritic joints would be of great value, especially in the early stages of disease, where there may be paucity of symptoms and few clinical findings.

Ultrasound (US) is a routinely available, dynamic, noninvasive, and relatively inexpensive bedside imaging technique which offers a multiplanar assessment of the musculoskeletal system, with high patient compliancy. The tool allows scanning of all peripheral joints as many times as required at the time of consultation. In addition, a multijoint assessment can be performed during the same scanning session. For many years, the interest of rheumatologists working in the field of musculoskeletal US has been focused mostly on the assessment of inflammatory arthritic conditions. However, lately, the interest expanded to include ultrasonographic evaluation of OA joints, endorsed by the recent advances in US technology which led to an extraordinary improvement in the quality of US equipments, and better visualization of the most superficial and deep musculoskeletal structures involved in OA. The detection of OA abnormalities demands the use of the most appropriate probe frequency for the anatomic structures examined and the correct machine settings to produce optimal visualization of the target area. This paved the way for the high-quality US machines to become the standard in day-to-day clinical practice, as broadband, multifrequency US transducers are mandatory for superior imaging of anatomic structures and for demonstrating subtle changes. By using the appropriate equipment and by optimizing the machine's settings, US can demonstrate and quantify a series of changes occurring in cartilage, in soft tissue of the joint, at the level of the bony cortex as well as in periarticular areas.

Clinical Applications

Sonography is a safe, accurate, and noninvasive imaging tool that can be used as many times as required without contraindications for evaluating the joints of osteoarthritic patients. In OA, US is able to assess a number of abnormalities at both the

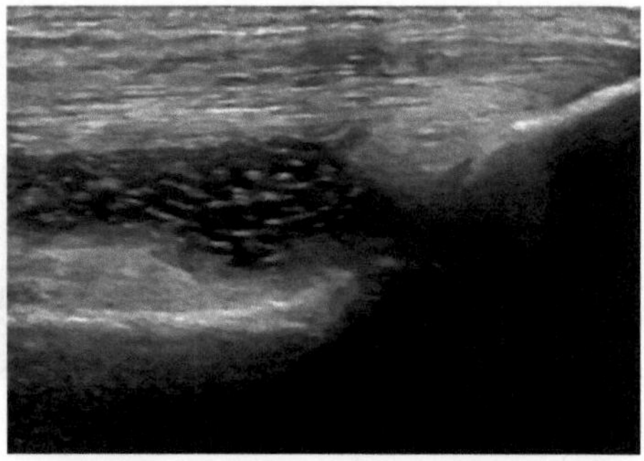

Fig. 6.1 Ultrasound of the knee in OA. Evidence of effusion at the level of the suprapatellar recess. *OA* osteoarthritis

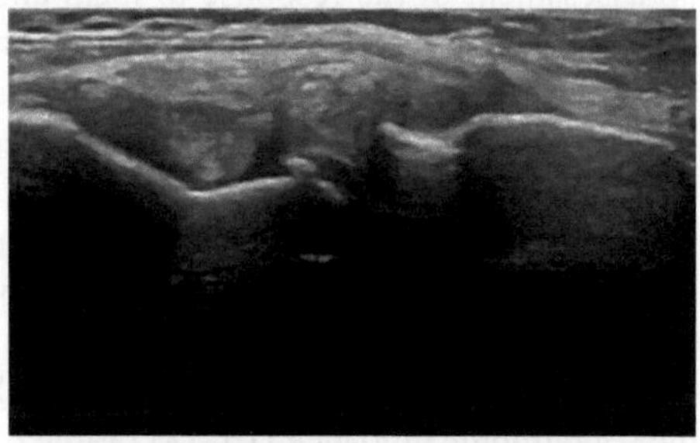

Fig. 6.2 Ultrasound of the knee in OA. Evidence of osteophytes at the level of the lateral aspect of the joint. *OA* osteoarthritis

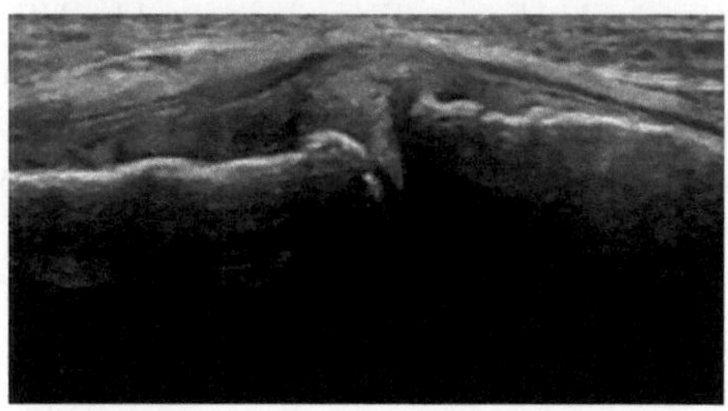

Fig. 6.3 Ultrasound of the knee in OA. Evidence of osteophytes at the level of the lateral aspect of the joint. *OA* osteoarthritis

level of joint and periarticular soft tissues as well as at the bony cortex. In addition to assessing for structural lesions, which involve the hyaline cartilage and the bone, US is able to show characteristic joint pathologies (e.g., effusion and synovitis) that may be related to the presence of pain and disability (Figs. 6.1–6.4). Moreover, in the subset of patients who may have asymptomatic osteoarthritic joint disease, US may show subclinical synovitis together with varying degree of joint affection. In OA, pain may also be due to periarticular involvement (such as trochanteric bursitis at the hip or anserine bursitis at the knee) which US is also able to depict. Psychological factors, such as anxiety and depression, as well as social problems, such as isolation and coping strategies, are all known determinants of pain in OA patients [7].

US complements the clinical evaluation of OA and can bridge the gap between the clinical and radiological findings [8–11]. It can be easily and swiftly performed in the same room used for the physical examination, reducing the patients'

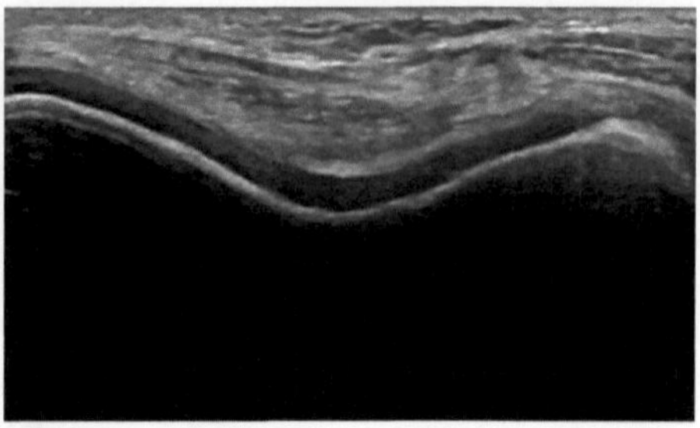

Fig. 6.4 Ultrasound of the knee in OA. Hyaline cartilage of the femoral condyles and trochlea. Abnormalities of the echotexture with loss of the anechoic structure and irregularities of the cartilage margins are visible. *OA* osteoarthritis

discomfort [12]. US facilitates monitoring of the disease progression and follow-up assessment of the response to local and systemic therapy in OA. Moreover, it can be repeated as many times as necessary [13]. Sonography consists of direct and multiplanar evaluation of different areas of interest and most of the peripheral joints involved by disease can be examined. These features allow imaging of soft tissues, such as hyaline cartilage, synovial membrane, and joint effusion; tendons, ligaments, and bursae, as well as external area of the menisci [14]. In addition, US examination can detect alterations of the cortical bone and demonstrate some of the disease typical structural changes [15]. US provides guidance for local procedures which are performed safely and are well tolerated by patients when executed under the sonographic guidance [16–18] (Table 6.1).

Importantly, US can directly image certain components of articular cartilage. Normal hyaline cartilage is imaged by US as a homogeneously anechoic layer lining the bony cortex and having a superficial as well as deep margins which characteristically appear thin, sharp, continuous, and regularly hyperechoic. In OA, a wide set of abnormalities are visualized, with evidence of loss of the anechoic texture,

Table 6.1 Indications of ultrasonographic assessment in OA

Indications
Detection of joint effusion
Detection of synovial thickening and hypertrophy
Differentiation between active and inactive synovitis
Assessment of cartilage lesions
Evaluation of osteophytes
Detection of erosions (erosive hand OA)
Evaluation of mucous cysts (hand OA)
Assessment of periarticular soft tissues abnormalities in OA (bursitis)
Execution of US-guided procedures
Monitoring of disease progression from early to late stages; follow-up of the response to local and systemic therapies

OA osteoarthritis, *US* ultrasound

irregularities of the margins, and progressive thinning (Fig. 6.4). However, for the poor contrast between hyaline cartilage and synovial fluid that are both anechoic, it may be sometimes difficult to assess small focal defects of the articular cartilage on US examination.

Sonographic Findings

For the correct execution of US joint examination and consequent interpretation of the pathologic findings, the knowledge of the normal joint and periarticular structures in US is mandatory. The most common sonographic findings in OA are reported in Table 6.2.

Table 6.2 Sonographic findings in OA

Pathology	Findings
Cartilage lesions	Blurring, loss of sharpness, and irregularities of the anterior and posterior margins
	Loss of homogeneity and loss of anechogenicity
	Focal (asymmetric) or diffuse thickening
Joint effusion	Abnormal hypoechoic or anechoic intra-articular tissue that is displaceable and compressible, but without Doppler signal
Synovial hypertrophy	Abnormal hypoechoic, intra-articular tissue that is nondisplaceable, poorly compressible, and may exhibit Doppler signal
Joint capsule hypertrophy	Joint capsule thickening
Osteophyte	A step-up bony prominence at the end of the normal bone contour, or at the margin of the joint seen in two perpendicular planes, with or without acoustic shadow
Erosion (hand OA)	A cortical breakage with the step-down contour defect seen in two perpendicular planes in the joint space
Mucoid cysts (hand OA)	Hypoechoic or anechoic cyst, similar area located over the distal interphalangeal joint
Bursitis	Abnormal hypoechoic or anechoic intrabursal material that is displaceable and compressible

OA osteoarthritis

Table 6.3 Ultrasound scan and patient position for cartilage assessment

Joint	Patient position	Scan
Metacarpophalangeal, proximal interphalangeal, and distal interphalangeal	Sitting position, joint fully flexed	Dorsal longitudinal scan Dorsal transverse scan
Elbow	Sitting position, extension of the elbow and supination of the lower arm	Anterior humeroradial longitudinal scan Anterior humeroulnar longitudinal scan Anterior transverse scan
Shoulder	Sitting position, hand in supination placed on the patient's thigh, with elbow slightly flexed	Posterior transverse scan
Hip	Supine position, leg extended and slightly externally rotated	Anterior longitudinal scan (parallel to femoral neck)
Knee	Supine position, joint fully flexed	Anterior transverse scan in the suprapatellar area Anterior longitudinal scan in the suprapatellar area (over the condyles and intercondylar notch)
Ankle (tibiotalar joint)	Supine position, flexed knee with foot on the examination table	Dorsal longitudinal scan Dorsal transverse scan
Metatarsophalangeal	Supine position, flexed knee with foot on the examination table	Dorsal longitudinal scan Dorsal transverse scan

The normal joint is visualized as a typical regularity of the bony cortex, a small amount of hypo/anechoic fluid, and a homogeneously echoic joint capsule. In order to visualize the hyaline cartilage, specific acoustic windows for dedicated scans and correct patient positioning should be implemented (Table 6.3).

Cartilage Assessment

The normal cartilage appears as an anechoic, homogenous band lining the bony profile with two sharp, continuous, and regular hypoechoic margins [19] (anterior and posterior). The anterior surface, sharper and thinner than the deeper one, represents the interface between the cartilage and the soft tissues; the correct, perpendicular penetration of the US beam allows optimal visualization [20]. The posterior margin is more echoic and thicker than the anterior one, representing the interface between the cartilage and the bony profile. The cartilage has a prevalent water content, therefore appears as a clearly defined anechoic structure lacking internal echoes [21]. Its thickness varies from 0.1 to 0.5 mm in the hand joints and 3 mm in the knee joint.

Cartilage alterations are viewed in a wide spectrum of ultrasonographic outcomes, such as those appearing in early disease showing initial blurring of the edges, which become irregular and lose their sharpness. As the disease progresses, the echogenicity varies with the typical loss of homogeneity and transparency (Fig. 6.4). At later stages, focal and asymmetric narrowing appears with the progression towards the thinning of the cartilaginous layer until disappearance and consecutive bony denudation. Care should be taken in the presence of joint effusion over the superficial cartilage edge, which may create a false thickening.

There are some limitations related to the use of US in the assessment of hyaline cartilage at different joint sites. Particularly those related to the limited width of the acoustic windows which may generate a partial visualization view of the cartilage.

Joint Effusion

The Outcome Measures in Rheumatoid Arthritis Clinical Trials (OMERACT) group's definitions for synovial effusion and synovial hypertrophy in rheumatoid arthritis can be usefully applied in OA (Table 6.2) [22]. Joint effusion appears as an anechoic or inhomogeneously hypoechoic material within the joint cavity, depending on the presence of intra-articular debris or calcified material (Fig. 6.1). The execution of multiplanar and dynamic scans facilitates the evaluation of joint effusion. To avoid misinterpretation in case of small amounts of fluid (which might be physiological) comparison to the contralateral side is recommended.

Synovial Hypertrophy

Synovial proliferation is defined as a poorly compressible hypoechoic tissue in the synovial joints. In the presence of active inflammation, Doppler signal, corresponding to the hyperemic phenomena within the synovial membrane, is demonstrated [23].

Osteophytes and Erosions

An osteophyte is defined as an abnormal step-up prominence of the bone at the end of the normal bone contour or at the margin of the joint, which can be seen in two perpendicular planes (Table 6.2). It usually has a posterior acoustic shadow. Osteophytes are clearly depicted by US at different joint sites and multiplanar assessment facilitates their detection (Figs. 6.2 and 6.3).

In erosive hand OA, erosions may be visualized when a cortical breakdown has a step-down contour defect seen in two perpendicular planes within the joint space. The lesions can be detected with various degrees of clarity related to the interposition of osteophytes, which may determine narrowing of the acoustic window.

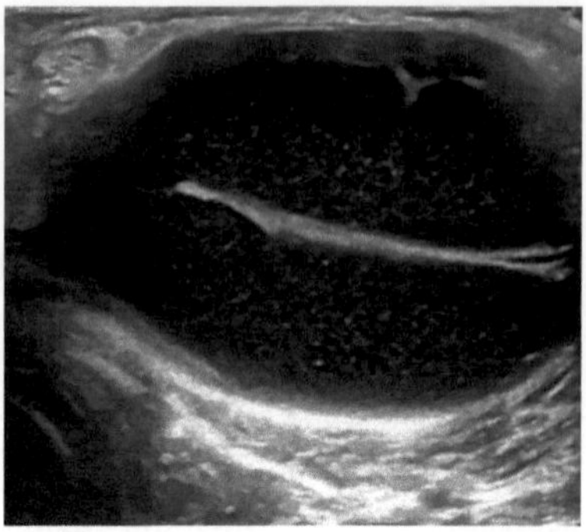

Fig. 6.5 Ultrasound of the knee in OA. Baker cyst with local effusion and synovial thickening and proliferation, causing distension of the bursal wall

US is more sensitive than conventional radiography in showing osteophytes, and the detection of osteophytes as well as evaluation of their size by US imaging has been demonstrated to be reliable at different joint sites.

Other Articular and Periarticular Findings

In hand OA, mucoid cysts are seen as hypoechoic well-defined areas located over the superolateral aspect of the distal interphalangeal joints [24]. Occasionally, joint capsule hypertrophy and fibrosis can also be seen.

At the knee joint level, periarticular involvement may also be seen. A common local sonographic abnormality is inflammation and/or swelling of local bursae such as Baker's cysts and anserine bursitis (Fig. 6.5). Similarly, when the foot is involved, bursitis over the medial aspect of the metatarsophalangeal joint can be visualized. Bursitis appears as an abnormal hypoechoic or anechoic distention of the bursa, with intrabursal material that is displaceable and compressible.

On US, menisci can be visualized as homogeneously hypoechoic, triangular structures in the joint space between the bones. The ultrasonographic view of the menisci offers indirect signs of OA as meniscal protrusion secondary to joint space narrowing in the medial or lateral compartment with the displacement of the collateral ligaments (Fig. 6.3).

Measurements/Subclinical Assessment

Currently, there is no generally accepted US grading system for cartilage changes in OA, even though the semiquantitative grading appears to be reliable and feasible in the clinical practice. Concordantly and similar to cartilage abnormalities, US assessment of other different osteoarthritic lesions such as joint effusion and synovitis, osteophytes as well as periarticular soft tissue pathologies can be graded by using semiquantitative scales [25–27]. For assessing joint effusion and synovitis, the same semiquantitative scales used in rheumatoid arthritis can be applied [28].

Osteophytes are evaluated using both dichotomous and semiquantitative scales (the latter scored at each joints as absent, mild, moderate, or severe on a scale of 0–3), whereas clinical bony enlargement was assessed as absent/present. An US atlas of osteophytes has been recently developed, and the intra as well as interreader reliability of osteophytes US scoring, on still images using the atlas as reference, has been examined [29].

Being the most common primary pathology in OA, cartilage thickness has been thoroughly investigated in OA patients. However, it was reported that the thickness of the cartilage layer varies in normal subjects and even in the same person. Thus, sonographers cannot use common cartilage thickness measures as a reference standard. The reliability of sonographic cartilage assessment has been tested in OA hands. The diagnostic sonographic pattern was loss of anechoic structure and/or thinning of the cartilage layer, as well as irregularities and/or loss of sharpness of at least one cartilage margin [30]. Furthermore, US can be reliably used for assessing cartilage abnormalities in the metacarpophalangeal (MCP) joints of patients who have MCP joint cartilage pathology. The variable results obtained by the sonographic analysis of single basic components may be improved by the application of strict definitions and the standardization of US scanning technique. These data represent a good basis for further sonographic studies on structural damage in hand pathology. Validation of cartilage degenerative pathology in OA knee joints using US was demonstrated in another study. The correlation of severity of cartilage changes (grades) between US and arthroscopy is highest at the sulcus area, also significant at the medial condyle and nonsignificant at the lateral condyle. A positive finding on US is a strong indicator of arthroscopic degenerative changes of cartilage, but a negative finding does not rule out degenerative changes. A semiquantitative scale has been recently proposed in knee arthroscopy and can be implemented in US [31]. This consists of five separate domains assessing (1) predominant morphological changes in the medial compartment, (2) predominant inflammation in the medial compartment, (3) predominant morphological changes in the lateral compartment, (4) predominant inflammation in the lateral compartment, and (5) effusion proved recently to be reliable and valid in comparison to the radiography [32].

US may also be suitable to investigate the differentiation between erosive and nonerosive OA. Even in this case, OMERACT definitions for erosions in rheumatoid arthritis can also be applied in OA. However, they can be detected with varying degrees of clarity that can be attributed to the interposition of osteophytes, which in turn may narrow the acoustic window [33]. Studies revealed high variability not only in the pathology assessed but also in clear definition and grading of the pathology. The lack of data on construct and criterion validity for US in OA as well as deficiency of demonstrations on reliability and sensitivity to change remains a matter of investigation [34]. Future development of this tool may also require revisiting the issue of erosions quantification in OA [35]. The growing number of centers implementing musculoskeletal US in their standard daily practice, would certainly have a positive impact on early diagnosis of OA patients. This earlier patient identification may offer a window of opportunity for early treatment, monitoring of response to therapy as well prevention of joint deformity.

Response to Therapy

While treatment recommendations in OA focus on a holistic approach, the pharmaceutical options are currently largely limited to analgesics. However, the spectrum of drug therapies is expanding, with a recent increase in interest in potential disease-modifying therapies in OA [36].

Local treatment is also possible and easily performed with the aid of US. Aspiration of the swollen joint may be indicated for various differential diagnostic purposes. Puncture of the joints and aspiration of fluid may be difficult without imaging guidance especially in obese persons or those with small fluid collections. Detection and aspiration of Baker's cyst is one of the traditional procedures performed under ultrasonographic guidance in knee OA. Similarly, US-guided local steroid injection has proven to be accurate and achieved best outcomes in comparison to blind injections.

Intra-articular injections of platelet-rich plasma (PRP) and hyaluronic acid (HA) represent promising medical treatments for OA, although no comparative study on long-term efficacy exists. Most of the current treatments for OA are palliative and aim at relieving the symptoms rather than influencing the physiopathology of the joint [37]. A recent study compared the clinical efficacy of PRP versus HA at 12 months of follow-up in 100 patients with hip OA and evaluated the influence of the type of injection under ultrasonographic guidance on clinical evolution. An overall improvement was detected in both groups between 1- and 3-month follow-up. Despite a slightly progressive worsening between 6- and 12-month follow-up, the final clinical scores remained higher compared with baseline, with no significant differences between PRP and HA. Intra-articular injections of PRP proved efficacious in terms of functional improvement and pain reduction but not superior to HA in patients with symptomatic hip OA at 12-month follow-up [38].

In hand OA, US assessment proved to be efficient in monitoring the potential role of high molecular weight HA in decreasing local inflammation and pain of hand OA. The joint capsule distension and local vascularization, significantly correlated with the decrease of pain [39].

Conclusion

US has demonstrated to be a valuable imaging tool for assessing osteoarthritic joints. In early and late disease stages, it depicts different changes varying from inflammation to structural damage. Sonographic findings include joint effusion and synovial hypertrophy in inflammatory arthritic joint disease, whereas cartilage lesions and osteophytes are seen in the case of structural damage. The detection of Doppler signal within synovial tissue represents the mark of local inflammatory activity. Further but inconstant findings include: bursitis, joint capsule thickening, meniscal protrusions, erosions, and mucous cysts. US permits the location and examination of the extent of tissue involvement in single and multiple joints during the same scanning session. It also enables the treating health-care professional to perform a wide and thorough evaluation of the severity of different osteoarthritic lesions. Monitoring of disease progression and response to local/systemic treatments represent significant advantages of this imaging modality, which make it particularly appropriate for use in clinical practice. The execution of US-guided procedures with safety and reliability has a relevant value in patient management.

Open questions about future applications of US in OA are mainly represented by the widespread use of new machines. Future developments in US research on OA encompass studies investigating new aspects of the disease and using novel US tools such as 3D-US, fusion imaging, and elastography. This will hopefully amplify the diagnostic abilities of sonography, analyzing early and late disease with accuracy and in minute details.

References

1. Arden N, Nevitt MC. Osteoarthritis: epidemiology. Best Pract Res Clin Rheumatol. 2006;20:3–25.
2. National Collaborating Centre for Chronic Conditions. Ostheoarthritis: national clinical guideline for care and management in adults. London: Royal College of Physicians; 2008.
3. Dieppe P: Osteoarthrosis and related disorders. Introduction and history. In: Klippel JH, Dieppe PA, editors. Rheumatology. London: Mosby; 1998. p. 8.1.1.
4. Moller I, Bong D, Naredo E, et al. Ultrasound in the study and monitoring of osteoarthritis. Osteoarthr Cartil. 2008;16:S4–7.
5. Naredo E, Acebes C, Moller I, et al. Ultrasound validity in the measurement of knee cartilage thickness. Ann Rheum Dis. 2008;67:1116–20.
6. Keen HI, Wakefield RJ, Grainger AJ, et al. An ultrasonographic study of the osteoarthritis of the hand: synovitis and its relationship to structural pathology and symptoms. Arthritis Rheum. 2008;59:1756–63.
7. Steultjens M, Dekker J, Bijlsma J. Coping, pain and disability in osteoarthritis. J Rheumatol. 2001;28:1068–72.
8. Filipucci E, Iagnocco A, Meenagh G, et al. Ultrasound imaging for the rheumatologist. Clin Exp Rheumatol. 2006;24:1–5.
9. Szkudlarek M, Klarlund M, Narvestad E, et al. Ultrasonography of the metacarpophalangeal and proximal interphalangeal joints in rheumatoid arthritis: a comparison with magnetic resonance imaging, conventional radiography and clinical examination. Arthritis Res Ther. 2006;8(2):48–52.
10. Keen HI, Wakefield RJ, Grainger A, et al. Can ultrasonography improve on radiographic assessment in osteoarthritis of the hands? A comparison between radiographic and ultrasonographic detected pathology. Ann Rheum Dis. 2007;67:1116–20.
11. Iagnocco A, Filippucci E, Ossandon A, et al. High resolution ultrasonography in detection of bone erosions in patients with hand osteoarthritis. J Rheumatol. 2005;32:2381–3.
12. Manger B, Kalden JR. Joint and connective tissue ultrasonography—a rheumatologic bedside procedure? A German experience. Arthritis Rheum.1995;38:736–42.
13. Iagnocco A, Filipucci E, Meenagh G, et al. Ultrasound imaging for the rheumatologist. Ultrasonography of the shoulder. Clin Exp Rheumatol. 2006;24:6–11.
14. Grassi W, Lamanna G, Farina A, et al. Sonographic imaging of normal and ostheoarthritic cartilage. Semin Arthritis Rheum. 2000;28:398–403.

15. Iagnocco A, Filipucci E, Ossandon A, et al. High resolution ultrasonography in detection of bone erosions in patients with hand osteoarthritis. J Rheumatol. 2005;32:2381–3.
16. Balint PV, Kane D, Hunter J, et al Ultrasound guided versus conventional joint and soft tissue fluid aspiration in rheumatology practice: a pilot study. J Rheumatol. 2002;29:2209–13.
17. Epis O, Iagnocco A, Meenagh G, et al. Ultrasound for the rheumatologist. XVI. Ultrasound guided procedures. Clin Exp Rheumatol. 2008;26:515–8.
18. Qvistgaard E, Christensen R, Torp-Pedersen S, et al. Intra-articular treatment of hip osteoarthritis: a randomised trial of hyaluronic acid, corticosteroid and isotonic saline. Osteoarthr Cartil. 2006;14:163–70.
19. Meenagh G, Iagnocco A, Filipucci E, et al. Ultrasound for the rheumatologists IV. Ultrasonography of the knee. Clin Exp Rheumatol. 2006;24:357–60.
20. Filipucci E, Iagnocco A, Meenagh G, et al. Ultrasound for the rheumatologists IV. Ultrasonography of the hand and wrist. Clin Exp Rheumatol. 2006;24:118–22.
21. Naredo E, Acebes C, Moller I, et al. Ultrasound validity in the measurement of knee cartilage thickness. Ann Rheum Dis. 2009;68:1322–7.
22. Wakefield R, Balint PV, Szkudlarek M, et al. Musculoskeletal ultrasound including definitions for ultrasonographic pathology. J Rheumatol. 2005;32:2485–7.
23. Iagnocco A, Epis O, Delle Sedie A, et al. Ultrasound for the rheumatlogist.XVII. Role of colour doppler and power doppler. Clin Exp Rheumatol. 2008;26:759–62.
24. Grassi W, Filipucci E, Farina A. Ultrasonography in osteoarthritis. Semin Arthritis Rheum. 2005;34:19–23.
25. Iagnocco A, Coari G, Zoppini A. Sonographic evaluation of femoral condylar cartilage in osteoarthritis and rheumatoid arthritis. Scand J Rheumatol. 1992;21:201–3.
26. Moller B, Bonel H, Rotzetter M, et al. Measuring finger joint cartilage by ultrasound as a promising alternative to conventional radiograph imaging. Arthritis Rheum. 2009;61:435–41.
27. D'Agostino MA, Conaghan P, Le Bars M, et al. EULAR report on the use of ultrasonography in painful knee osteoarthritis. Part 1: prevalence of inflammation in osteoarthritis. Ann Rheum Dis. 2005;64:1703–9.
28. Wakefield RJ, Balint PV, Szkudlarek M, et al. OMERACT 7 Special Interest Group. Musculoskeletal ultrasound including definitions for ultrasonographic pathology. J Rheumatol. 2005;32:2485–7.
29. Mathiessen A, Haugen IK, Slatkowsky-Christensen B, et al. Ultrasonographic assessment of osteophytes in 127 patients with hand osteoarthritis: exploring reliability and associations with MRI, radiographs and clinical joint findings. Ann Rheum Dis. 2013;72:51–6.
30. Iagnocco A, Conaghan PG, Aegerter P, Möller I, et al. The reliability of musculoskeletal ultrasound in the detection of cartilage abnormalities at the metacarpo-phalangeal joints. Osteoarthr Cartil. 2012; 20:1142–6.
31. Saarakkala S, Waris P, Waris V, et al. Diagnostic performance of knee ultrasonography for detecting degenerative changes of articular cartilage. Osteoarthr Cartil. 2012;20:376–81.
32. Riecke BF, Christensen R, Torp-Pedersen S. An ultrasound score for knee osteoarthritis: a cross-sectional validation study. Osteoarthr Cartil. 2014;22:1675–91.
33. Naredo E, Acebes C, Moller I, et al. Ultrasound validity in the measurement of knee cartilage thickness. Ann Rheum Dis. 2009;68:1322–7.
34. Keen HI, Wakefield RJ, Conaghan PG. A systematic review of ultrasonography in osteoarthritis. Ann Rheum Dis. 2009;68:611–9.
35. Keen HI, Lavie F, Wakefield RJ, D'Agostino MA. The development of a preliminary ultrasonographic scoring system for features of hand osteoarthritis. Ann Rheum Dis. 2008;67:651–5.
36. Tonge DP, Pearson MJ, Jones SW. The hallmarks of osteoarthritis and the potential to develop personalised disease-modifying pharmacological therapeutics. Osteoarthr Cartil. 2014;22:609–21.
37. Sampson S, Reed M, Silvers H, et al. Injection of platelet-rich plasma in patients with primary and secondary knee osteoarthritis: a pilot study. Am J Phys Med Rehabil. 2010;89:961–9.

38. Battaglia M, Guaraldi F, Vannini F, et al. Efficacy of ultrasound-guided intra-articular injections of platelet-rich plasma versus hyaluronic acid for hip osteoarthritis. Orthopedics. 2013;36:1501–8.
39. Klauser AS, Faschingbauer R, Kupferthaler K, et al. Sonographic criteria for therapy follow-up in the course of ultrasound-guided intra-articular injections of hyaluronic acid in hand osteoarthritis. Eur J Radiol. 2012;81:1607–11.

Chapter 7
Crystal-Induced Arthritis

Rodina Nestorova MD and Daniela Fodor MD, PhD

Introduction

One of the most frequent situations that a rheumatologist confronts within the standard daily practice is the crystal-deposition-induced pathology. The types of crystals associated with joint disease include monosodium urate (MSU), calcium pyrophosphate (CPP), basic calcium phosphate (BCP), including hydroxyapatite (HA) and calcium oxalate. These microcrystals are found not only in the articular structures but also in the periarticular tissues.

The crystal-induced inflammatory musculoskeletal (MS) changes are frequently encountered in general population. The prevalence of gout, for example, was found to be 1.4 % in the UK and Germany [1]. The prevalence of chondrocalcinosis of the lower limbs in the elderly Italian population was reported in the range of 10 % of the studied population [2]. The prevalence of the crystal-induced arthritis may even be higher if the medical inpatients whose intercurrent illness and dehydration may trigger acute arthritic attacks were included in the statistical analysis [3].

The attacks of acute arthritis are the characteristic clinical manifestation. These attacks of articular inflammation are considered one of the most painful conditions in rheumatology [4]. Over time, repetitive episodes may produce chronic arthropathy with subsequent damage in the articular or juxta-articular spaces [3, 5].

In addition to clinical examination and synovial fluid analysis which forms the cornerstone for the crystal-related arthropathies diagnosis, recently, imaging techniques booked its place for both the diagnosis and follow-up of the these patients.

R. Nestorova (✉)
Department of Musculoskeletal Ultrasound, Rheumatology Centre "St. Irina", Sofia, Bulgaria
e-mail: rodina_usd@yahoo.com

D. Fodor
Department of Internal Medicine, "Iuliu Hatieganu" University of Medicine and Pharmacy, Cluj-Napoca, Romania
e-mail: dfodor@umfcluj.ro

Y. El Miedany (ed.), *Musculoskeletal Ultrasonography in Rheumatic Diseases*,
DOI 10.1007/978-3-319-15723-8_7

Table 7.1 Main indications for using US in crystal induced arthritis

Detection of the joint effusion (quantity, aspect)
Detection of the synovitis (with clinically manifestation or asymptomatic)
Differentiation between active or inactive synovitis
Hyaline cartilage and fibrocartilage study (deposits)
Bone contour (erosions, osteophytes)
Tendons evaluations (deposits, tears, inflammation)
Crystal depositions evaluation (articular and/or juxta-articular spaces)
Execution of US-guided procedures (for diagnosis and/or therapeutic purposes)
Monitoring the disease evolution and progression
Differential diagnosis with other inflammatory arthritis

US ultrasonography

Ultrasonography (US) is now considered to be one of the most useful tools for diagnosis, monitoring, as well as guided interventional procedures, with results comparable to magnetic resonance imaging (MRI) or *computed tomography (CT)* [6, 7] and better than conventional X-rays [8]. The recent progress in US technology (high-resolution machines, high-frequency transducers, new techniques), led to a surge of its use by the treating clinicians to confirm the diagnosis or manage the acute as well as chronic forms of the disease (Table 7.1).

Gout

Gout, one of the oldest recognized diseases, is a complex form of inflammatory arthritis, first identified by the Egyptians in 2640 BC and later recognized by Hippocrates in the fifth century BC. Six centuries later, Galen was the first to describe tophi, whereas Antonie van Leeuwenhoek was the first to describe the appearance of the crystals from a gouty tophus in 1679, although their chemical composition was unknown at that time [9, 10].

Gout is the consequence of deposition of MSU crystals in joints and other tissues, affecting up to 1–2 % of men and occurs as a result of persistent hyperuricemia [11]. This is associated with excess morbidity, disability, and poorer quality of life [9]. More than 80 % of gout patients have a positive family history of gout or hyperuricemia. Four phases are described in the disease evolution: asymptomatic hyperuricemia, acute, intercritical, and chronic gout [12].

Diagnosis of suspected gout is based on typical clinical and laboratory (i.e., hyperuricemia) findings, while definitive diagnosis requires identification of MSU crystals in aspirated synovial fluid or tophi [11].

Review of the literature revealed that the management of gout is often suboptimal. At the early stage of gout, radiography may not be useful for the disease

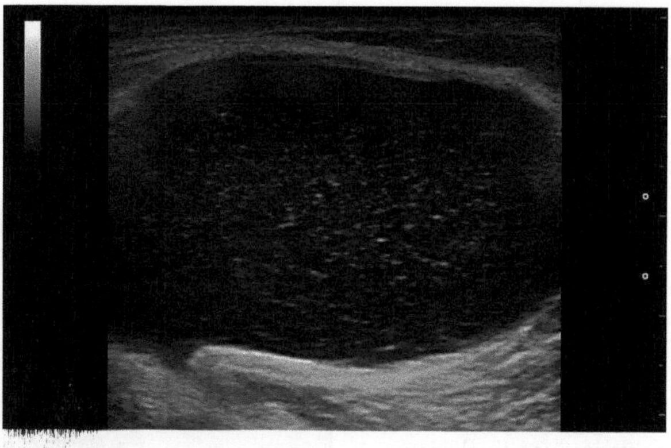

Fig. 7.1 Gout. The "snowstorm" aspect of the synovial fluid in a popliteal cyst

diagnosis, as abnormal findings may take several years to develop a radiologic pattern (apart from nonspecific soft tissue swelling overlying the inflamed joint) [13].

Nonspecific Features in Gout

Synovial Fluid

A joint effusion is an early but nonspecific finding in gout patients [12, 13, 15, 16]. During acute gout, synovial fluid may vary from complete anechogenicity (usually in an early phase of the disease) to aggregates of variable echogenicity (more often after multiple acute attacks) [15, 17–19].

Multiple aggregates of MSU microcrystals are visualized as hyperechoic spots or bright stippled foci and gentle pressure on the skin surface makes them float in the joint cavity, creating a "snowstorm" appearance in some cases [12, 13, 15, 18–22] (Fig. 7.1).

Decrease of the gain could improve the ability to detect hyperechoic structures [13, 14, 18].

Gout has a predilection for the first metatarsophalangeal (MTP I) joint, with as many as 50–70 % of first gout attacks occurring there [16, 18]. Ultrasound detects more joint effusion than clinical examination especially on dorsal view of the MTP joints [23, 24].

In a recent study, the MTP I as well as the radiocarpal joints were the most frequently involved joints for hyperechoic aggregates in gouty arthritis patients (57.1 % and 38.5 % of patients, respectively). The midcarpal and the knee joints were also frequently involved in patients with gout (28.6 % and 25.3 %, respectively) [16].

Synovial Proliferation and Hypervascularization

Similar to joint effusion, synovial hypertrophy is not specific to gout, but their association with hyperechoic spots (bright dotted foci) in the synovium is strongly evocative of gout [13, 23].

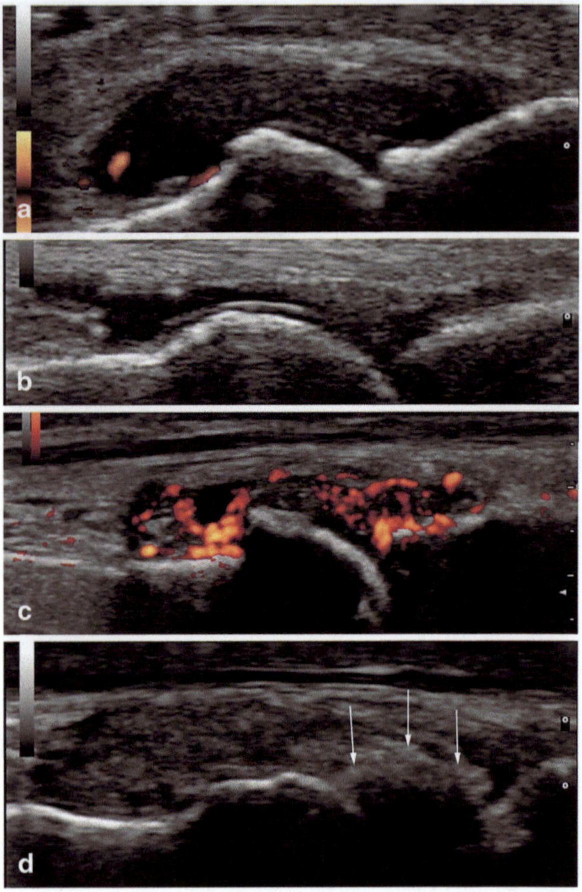

Fig. 7.2 Dorsal longitudinal scans of the first metatarsophalangeal joint in patients with gout: **a** acute gout—large hypoechoic content of the joint cavity, with discrete vascularization; **b** intercritical gout—small quantity of transonic fluid, double contour of the hyaline cartilage; **c** acute flare of chronic gout; **d** tophaceous gout—a large hard tophus (*arrows*) situated intra-articular. The posterior acoustic shadow makes the interosseous space non-examinable

Doppler mode (power Doppler, PD; color Doppler, CD) is able to differentiate active, inflamed, from inactive, noninflamed/fibrotic synovial tissue, by estimating the vascular pattern (Fig. 7.2).

While US features of synovial effusion, synovial proliferation, or synovial Doppler signal are not specific for gout, as these US findings are frequently found in any type of inflammatory arthritis, all of them are crucial for the diagnosis of gout and in the monitoring of the patient and therapy. In addition, PD could be a way to monitor acute gouty attacks [25, 26].

Recent reports have shown that the PD signal disappears with treatment, which suggests that PD could be a useful method to monitor gouty synovitis response to therapy.

Interestingly, other studies reported that the PD signal was even present in asymptomatic joints from gout patients and in asymptomatic hyperuricemia [15, 25, 26], suggesting that subclinical chronic inflammation may be at work [25]. With colchicine therapy, PD signal has decreased and finally disappeared [15].

Bone Erosions

Bone erosions on US have been defined as breaks in the hyperechoic outline of the bony cortex (step-down lesions), seen in two perpendicular planes, following the Outcome Measure in Rheumatology (OMERACT) criteria [27].

The OMERACT MSUS group has produced a definition of erosion specific for gout as follows: intra- and/or extra-articular discontinuity of the bone surface (visible in two perpendicular planes) [28].

In gout, bone erosions are due to tophus [12–14, 18, 20, 21, 25, 29]. The tophus eroding the underlying bone is pivotal for the development of bone erosions in gouty arthritis [29], therefore the specificity of an US diagnosis of erosions is increased when there are adjacent tophi [12, 29]. Erosions have been found more commonly in patients with frequent acute gouty attacks, long disease duration, and tophi proven by US [23, 30].

The MTP I joint (medial surface), and metacarpophalangeal (MCP) joints are the main targets for erosions in gout [14, 23, 30, 31], but they were also found commonly in MTP joints that have never been clinically affected by gout [23, 26]. Unlike erosions in rheumatoid arthritis, gouty erosions are usually deeper and more destructive [18, 30].

US has proven to be three times more sensitive than plain films in the detection of bone erosions <2 mm ($P<0.001$) [26, 30]. So far, there is no standardized US scoring system for erosions in gout. In contrast, in rheumatoid arthritis erosions have been classified according to the bones involved (metatarsal or phalangeal), the position of the erosions (dorsal or medial) and whether they were unifocal or multifocal. They have been categorized into a semiquantitative scale as small erosion <2 mm, moderate erosion 2–4 mm, and large erosion >4 mm) [21, 30] (Fig. 7.3).

Specific US Features in Gout

Articular Cartilage—"Double Contour Sign"

The double contour sign (DCS) is a highly specific US feature for gout. This DCS is a result of uric acid (UA) predilection to crystallize forming deposits of MSU microcrystals on the hyaline cartilage surface, which increases the cartilage surface interface giving it a thickness similar to that of subchondral bone [18, 20, 26]. OMERACT MSUS group has given the following definition of the DCS: abnormal hyperechoic band over the superficial margin of the articular hyaline cartilage, independent of the angle of insonation and which may be either irregular or regular, continuous or intermittent and can be distinguished from the cartilage interface sign [28].

Thus, DCS is seen as an echogenic band with an irregular surface detected paralleling the bony contour with an anechoic region between. This anechoic region represents hyaline cartilage [12–14, 16, 23, 32] (Fig. 7.4).

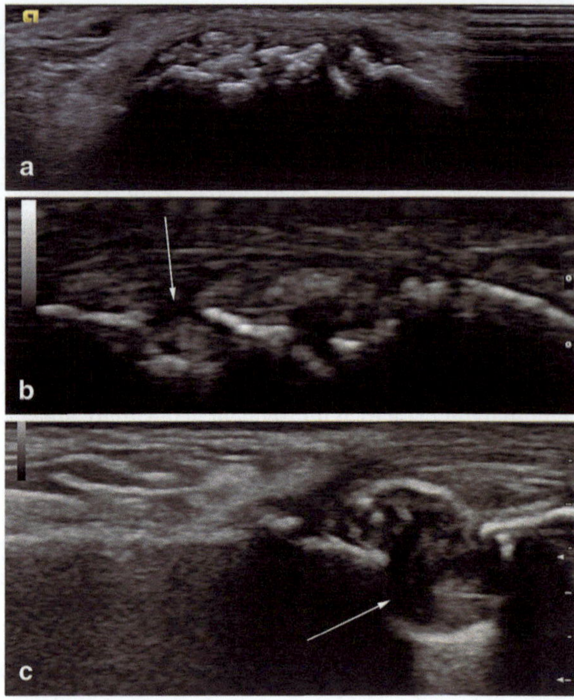

Fig. 7.3 Chronic gout. Different aspects of the erosions at the first metatarsal head: **a** small erosions covered with a hyperechoic and inhomogeneous material (*medial scan*); **b** large erosion (*arrow*) filled with hyperechoic material (tophus; *dorsal scan*); **c** very large and deep erosion (*arrow*) realized by an intraosseous tophus, quite similar with the radiological aspect of erosions with overhanging edges (*dorsal scan*). Note the posterior "dirty shadow" artifact

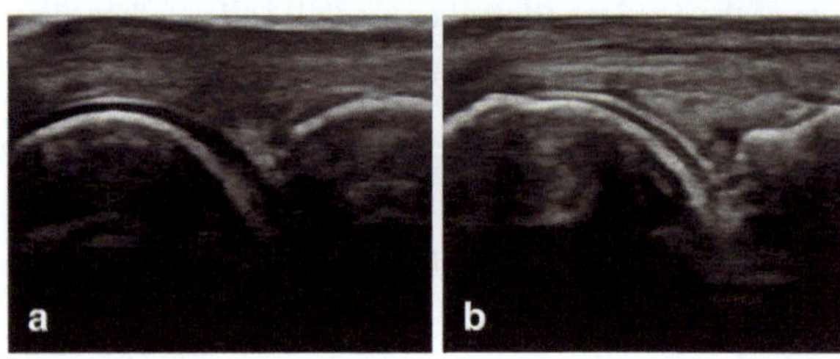

Fig. 7.4 Dorsal longitudinal scan of the metacarpophalangeal joint: **a** normal aspect of the hyaline cartilage; **b** gout— the double contour sign. Slight irregularities of the superficial margin of the hyaline cartilage can be observed

DCS adheres to the cartilage, moving together with bone and cartilage layer in dynamic moves [13–15].

There is a correlation between the presence of DCS, serum UA level, and the occurrence of UA arthropathy, which suggests that DCS could be a marker of severity [13].

DCS is most frequently detected in the dorsal aspect of the first MT head (61.5 %) and the femoral articular cartilage (41.8 %) and is seen in the majority of joints with intra-articular tophi [16, 23].

DCS should be searched in the trochleal cartilage of knees (suprapatellar plane in maximal flexion [12, 13, 16, 23, 31] (Fig. 7.5).

It has been reported that the frequency of the "double contour" sign in the knee is higher in clinically affected versus unaffected knees of gout patients, although

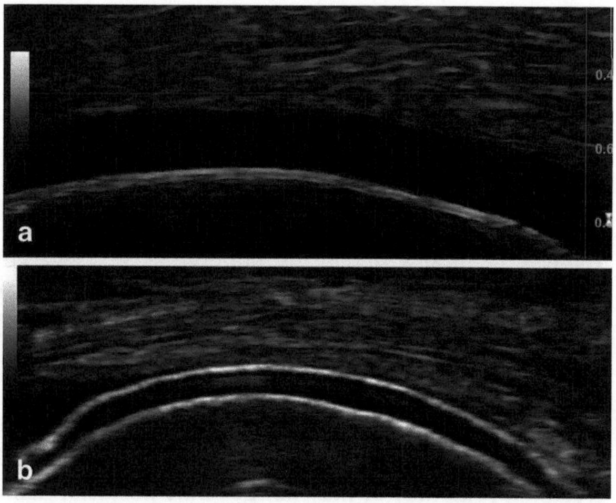

Fig. 7.5 Longitudinal scan of the femoral condyle: **a** normal aspect; **b** gout—the double contour sign

the difference does not achieve statistical significance [21]. DCS has been noted in patients with an acute gout flare-up, with a history of prior gout attacks, in clinically unaffected joints, and also in patients with asymptomatic hyperuricemia. The sensitivity of this finding ranges from 25 to 95 % in patients with gout [12–18, 20, 21, 23, 25, 26, 30, 33]. False-positive results for DCS are possible in the following circumstances [13, 15, 19, 20, 23, 30, 31]:

1. The hyperechoic aspect of the normal synovium is a regular hyperechoic band, as if written with a pen. DCS is irregular and adheres to the cartilage in dynamic moves.
2. The joint effusion induces a posterior-wall echo reinforcement (improvement of propagation of ultrasound) and can accentuate the normal hyperechoic aspect of the synovium.
3. A thin cartilage in small joints.
4. Damaged cartilage associated to osteoarthritis (OA; the presence of synovial effusion in the MTP I joint, which is a quite common finding even in OA, can enhance the visualization of the superficial margin of the metatarsal hyaline cartilage).
5. Hyper-echoic spots located in the middle layer of the cartilage and the presence of meniscal calcification are associated with calcium pyrophosphate deposition (CPPD) disease. MSU deposition in gout occurs on the surface of the hyaline cartilage producing the DCS, whereas CPPD deposition appears as a central hyperechoic focus or line within the cartilage.

Responsiveness of US to cartilage changes in gout has been demonstrated by the disappearance of the DCS in response to urate-lowering therapy (ULT) in gouty patients whose serum UA levels were lowered to 6 mg/dl for 7 months or more, but not in patients whose levels remained above 6 mg/dl [34].

A wide range of sensitivity (e.g., 21–92 %) with very high specificity (98–100 %) for US-detected DCS in gouty arthritis patients have been reported in the literature

[12–14, 16, 23, 30–32]. The DCS was found to be present in subjects with gout significantly more often than in healthy subjects [14, 20, 23]. Inter-reader reliability assessing DCS was found to be excellent [13, 14, 21, 23, 31].

MSU Deposits (Tophi and Aggregates)

Tophi have unique diagnostic importance to gout. They represent deposits composed of MSU crystals core, encased by dense connective tissue [35]. Earlier studies comparing US to MRI in the detection of gouty tophi revealed that MRI confirmed 81 % of the tophi reported by US and US detected 90 % of the reported tophi on MRI [7, 36].

The OMERACT MSUS group defined tophus and aggregations as follows [28]:

Tophus is a circumscribed, inhomogeneous, hyperechoic, and/or hypoechoic aggregation (that may or may not generate posterior acoustic shadow) which may be surrounded by a small anechoic rim. Aggregates are heterogeneous hyperechoic foci that maintain their high degree of reflectivity even when the gain setting is minimized or the insonation angle is changed and which occasionally may generate posterior acoustic shadow. Most commonly, tophi are hyperechoic and always heterogeneous [12–14, 16, 19, 23].

Tophi have been also described by US as “wet sugar clumps” with an oval or irregular shape [14].

Intra-articular and intrabursal tophi have been defined as heterogeneous hyperechoic (relative to subdermal fat) aggregates with poorly defined margins with or without areas with acoustic shadowing within the synovial recesses or bursae, respectively [16].

Tophi are walled off and their contour is poorly defined [14, 37]. They could be surrounded by a thin anechoic rim. The histopathological equivalent of the anechoic rim is formed by macrophages, lymphocytes, and large foreign body giant cells [37]. This rim allows distinction of tophaceous material from proliferative synovial tissue and hyperechoic joint capsule. At the same time, this halo would be in close contact with subchondral bone in the case of pannus invasion of an erosion [14].

According to the degree of compaction of the deposits, tophi could be divided into soft/inhomogeneous echogenicity and soft to palpation/ hard/ hyperechoic aggregate generating a posterior acoustic shadow and hard to palpation/and mixed/ features of both soft and hard tophus. Tophi that are sonolucent have been termed as “soft tophi.” Hard tophi occur in longstanding tophaceous gout, and generation of posterior acoustic shadow is due to calcification within the tophus/calcified tophi [15, 17, 18, 20] (Fig. 7.6).

The Doppler US can make the difference between active/”hot” tophi and inactive/”cold” tophi by the presence or absence of the Doppler signal.

Nevertheless, there is a frequent presence of persistent low-grade inflammation in asymptomatic chronic tophaceous gouty arthritis on US [29]. Main sonographic features of tophi are presented in Table 7.2.

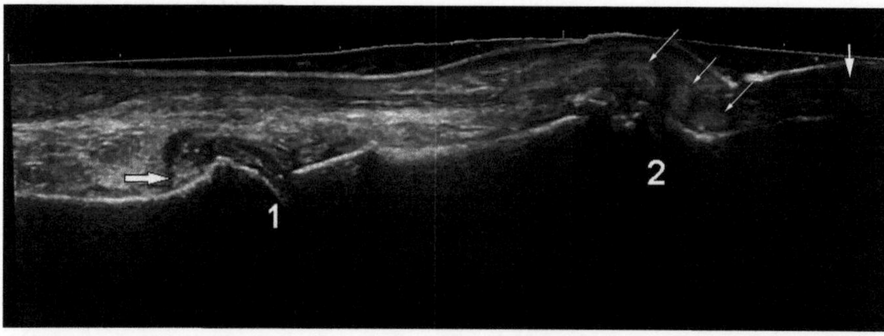

Fig. 7.6 SieScape imagine of the first toe in patient with chronic gout. Metatarsophalangeal (1) and interphalangeal (2) joints are affected. Note the soft tophi in and around the interphalangeal joint (*small arrows*), the hard tophus in the nail bed (*large arrow*), and a hyperechoic cloudy areas (*horizontal arrow*). Irregularities of the bone contour of the phalangeal head represent erosions

Table 7.2 US features of Tophi

Echogenicity	Hyperechoic, hypoechoic, anechoic, and heterogeneous
	Mostly hyperechoic and heterogeneous
Contour	Poorly defined, could be surrounded by a small anechoic rim
Shape	Oval (spherical or ellipsoid) or irregular
Types according to the degree of compaction	Soft
	Hard
	Mixed
Types according to the degree of inflammation	Hot
	Cold
Dynamic assessment	Tophi do not move with bone and cartilage

US is able to detect tophi both inside and around the erosions [19]. Naredo et al. found that the assessment of one joint (i.e., radiocarpal joint), two tendons (i.e., patellar tendon and triceps tendon), and three articular cartilages (i.e., first metatarsal dorsal and plantar cartilage, talar cartilage, and either second metacarpal cartilage (dorsal aspect) or femoral condyle cartilage) showed the best balance between sensitivity and specificity (84.6 and 83.3 %, respectively), positive predictive value and negative predictive value (91.7 and 71.4 %, respectively) [16].

Intra-articular Tophi

Preferred locations for the disposal of tophi are symptomatic joints and their periarticular tissue. Tophi may occur at any site but their frequent locations are toes, especially MTP I, wrist, fingers, knee, and ankle [16].

In small joints, intra-articular tophi are often associated with bone erosions.

Nearly half of the patients with gout have US-proved tophi in at least one of the examined joints, predominantly in subjects with a higher level of UA and subjects who do not receive ULT [13, 21, 23].

Main US features of tophi in the MTP I are presented in Table 7.3 [14, 25].

Tophi in Periarticular Tissue

Bursae

The prepatellar and olecranon bursae are the most commonly involved sites [37].

The bursitis has chronic features of inflammation from the beginning. The walls are thick, poorly delimited from the surrounding tissues, and the cavity is always filled with proliferated synovia. There is not a good correlation with the adjacent joint involvement. Peritendinous bursae (infrapatellar, preachillian) are inflamed only in cases of concomitant tendon pathology [17] (Fig. 7.7).

The most frequent characteristics of tophi in the olecranon bursa were: hyperechogenicity (91.7%), poorly defined contours (88.6%), multiple grouped nodules (85.6%), and heterogeneity (68.6%) [38].

Tendons and Ligaments

Tophi in tendon and ligaments were defined as heterogeneous hyperechoic (relative to tendon/ligament fibers) aggregates with poorly defined margins with or without areas with acoustic shadowing within and/or around the tendon or ligament, respectively. MSU crystal deposition in tendons was also defined as hyperechoic (relative to tendon fibers) linear bands within the tendon substance [13–20, 25, 31] (Fig. 7.8).

Table 7.3 The main US features of tophi in the MTP I

Place of the tophi or aggregates	In the dorsal proximal recess of the synovial membrane
	Medial to the metatarsal head
	In the central area of the joint space
Echogenicity	In inflamed joints, intra-articular tophi are more echogenic on the background of the hypoechoic or anechoic fluid
Tophi and erosions	Formed, oval-shaped tophi riding on the hyaline cartilage of the metatarsal head, impinging on the proximal phalanx, and eroding into the bone

MTP metatarsophalangeal, *US* ultrasound

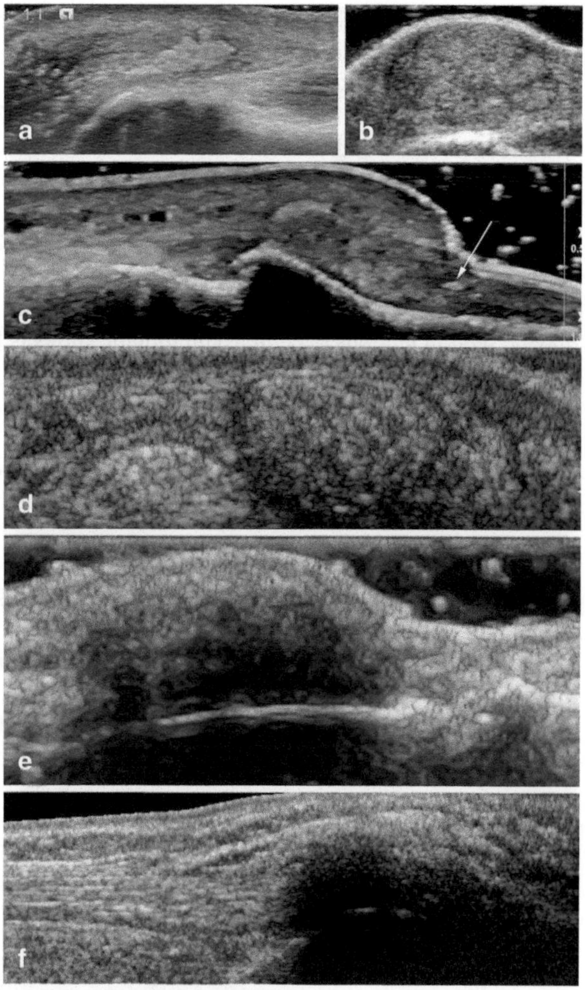

Fig. 7.7 Different aspects of the tophi gout: **a** soft tophus in the olecranon bursa; **b** pseudotumoral aspect of a soft tophus in the olecranian region; **c** soft tophi in the dorsal aspect of the distal interphalageal region. Note the extension of the deposits in the nail bed (*arrow*); **d** mixed tophi in the olecranian region; **e** mixed tophus over the calcaneus. The calcaneus bone contour is still visible; **f** hard tophus in the Achilles tendon. The posterior acoustic shadow covers the calcaneus bone contour

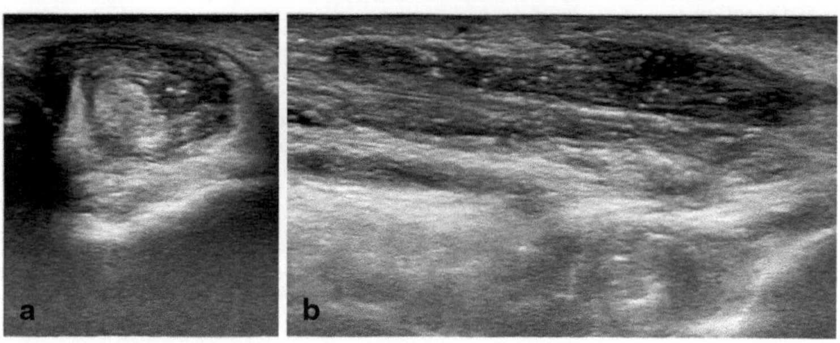

Fig. 7.8 Gout. Transversal (**a**) and longitudinal (**b**) scan of the peroneal tendons. The tendon sheath is filled with hypoechoic inhomogeneous material with hyperechoic spot

The literature reveals that the most frequently affected tendons and ligaments by tophi are the patellar ligament, the triceps tendon, the quadricipital tendon, under the lateral collateral ligament, the Achilles tendon and the anterior tibialis tendon, and plantar fascia [13, 15, 17, 20, 29, 30, 39, 40] (Fig. 7.9).

Naredo et al. found tendon abnormalities most frequently in the patellar tendon (60.4 %) and the triceps tendon (47.3 %), followed by the quadriceps tendon (38.5 %) and the Achilles tendon (34.1 %) [16].

Aggregation of MSU crystals at the insertion of the tendons or ligaments (enthesis) leading to gouty enthesopathy [17, 20].

MSU microdeposits, with a small hypoechoic halo due to local inflammation, may be seen in tendons even in asymptomatic subjects [15–18, 20, 33].

Subcutaneous Aggregates

In contrast to rheumatoid nodules, 80 % of the gouty nodules appeared as heterogeneous masses with generally more of the nodule being hyperechoic [22, 41].

Tophus Measurement and Assessment of Treatment Response by US

Eight methods of tophus measurement have been described in the literature: counting the total number of tophi, physical measurement using tape measure, physical measurement using Vernier calipers, digital photography, US, MRI, CT, and dual energy CT (DECT). US has an important role in monitoring tophus measurement

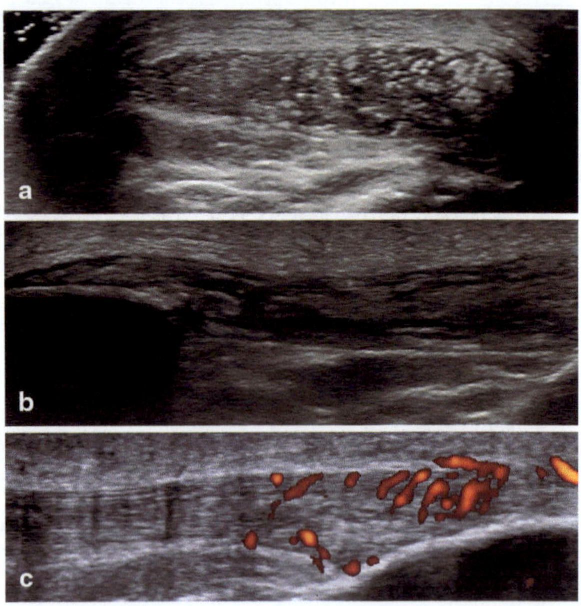

Fig 7.9 Microdeposits of monosodium urate crystals and intra-tendinous tophi in the patellar tendon, associated with hypervascularization: **a** transversal scan; **b** longitudinal scan; **c** power Doppler examination

in patients receiving ULT. The literature reports good sensitivity to change, associated with ULT [40] and good intra-observer correlation at follow-up [22], so tophi measurement could become an outcome measure for ULT in chronic gout [15, 25].

Tophus size can be directly measured using US calipers and can be reported as longest diameter and total volume. Diameters can range from less than 1 mm (microtophi) to centimeters. Decrease of tophus volume objectifies a treatment response. Ultrasound probes with three-dimensional capabilities may be used for an operator-independent measurement of volume [13, 14].

The measurements of tophus size by MRI have correlated well with US measurements [7, 36].

In a study by Perez et al., 68% of tophi detected and measured with US showed significant reduction in tophus size in patients with an average serum urate <6 mg/dl after 1 year of ULT (allopurinol and benzbromarone), compared with tophi where the urate remained >6 mg/dl [7].

After treatment with pegloticase for 3 months, 22% of subjects experienced complete resolution of the target tophus. After 6 months, 45% experienced complete resolution [36].

Responsiveness of US cartilage changes in gout has been demonstrated by disappearance of the DCS and/or complete dissolution of tophi in response to ULT [7, 13, 14, 23].

Asymptomatic Hyperuricemia

The US examination should not be limited to the painful joints. The search for deposits of MSU crystals in patients suspected of gout should also be extended to asymptomatic joints previously involved in acute attacks [19, 30].

There is a wide spectrum of subclinical structural damage in both intra- and extra-articular structures in asymptomatic individuals induced by hyperuricemia [33].

Urate deposits in tendons and the synovium and the prevalence of patellar and Achilles enthesopathy occur more frequently in subjects with asymptomatic hyperuricemia than in asymptomatic, normouricaemic individuals [29, 33, 40].

Some authors have observed that the PD signal was even present in asymptomatic joints from gout patients and in asymptomatic hyperuricemia [26, 37].

Tophi were found in tendons (distal patellar tendon), synovium (knees and ankles), and other soft tissues in 34% of patients with asymptomatic hyperuricemia [37].

Another published US study found significant number of patients with asymptomatic hyperuricemia and MSU crystals (joints with US-proved effusions were aspirated). DCS or "hyperechoic cloudy areas" were seen in 11/26 with 100% sensitivity and 88% specificity. [42].

Some author have suggested the need for use of the terms "subclinical gout" or "asymptomatic gout" instead of the old term "asymptomatic hyperuricemia" [21, 36, 39].

A study found DCS in 25% of the MTP I and 17% of the knees, erosions in 12% of the MTP I's and tophi/intra- and extra-articularly/in 18% of hyperuricemic patient but in none of the normouricemic controls [21]. Another study found small tophaceous deposits in 34% of patients, and an increased power Doppler signal in 23% of them [26].

Urate deposition is present in more than half of gout or hyperuricemic patients who do not require ULT according to the European League Against Rheumatism (EULAR) recommendations [13].

These US findings open a new battlefront in the current debate about the need for treatment necessity of urate-lowering agents in asymptomatic patients with persistent hyperuricemia and indisputable US features of MSU microcrystal tissue deposition such as the DCS or the presence of tophi [33].

US-Guided Interventions

US is also the primary imaging modality used for needle guidance during diagnostic and therapeutic interventions.

US can help the clinician to localize synovial fluid in symptomatic or asymptomatic gout joints for aspiration [13, 25].

US can assist in the diagnosis of gout, allowing needle guidance for fluid aspiration for microscopic assessment of microcrystals [12].

In addition, US can guide aspiration of accessible extra-articular tophi for microscopic confirmation of the presence of microcrystals [16].

In conclusion, MSUS is a method of choice for diagnosis and management of gout and asymptomatic hyperuricemic patients.

Calcium Pyrophosphate Deposition Disease

CPPD is associated with significant morbidity due to the pain in acute attack of arthritis or to symptoms of chronic arthropathy. The risk of CPPD greatly increases with age, but it can occur in young people too. The prevalence of the disease is rather uncertain. In the Framingham Knee Osteoarthritis Study, the prevalence of radiological knee cartilage calcification was 8.1% in a sample of 1425 subjects aged more than 63 years [43]. In the general adult population, the CPP-associated arthritis is the third most common inflammatory arthritis (0.42%) after rheumatoid arthritis and gout [44]. There are some risk factors recognized for this pathology: previous joint injury, age (the major risk), familial predisposition, or metabolic diseases (e.g., hemochromatosis, primary hyperparathyroidism, hypophosphatasia, hypomagnesemia) [8, 45]. The best evidence for previous joint injury that may predispose to CPPD comes from studies of postmeniscectomy knees in which the risk of cartilage calcification was five times greater comparing with the contralateral unoperated knees [46].

Due to a large variety of terms concerning the CPP-crystals-related diseases, the terminology and classification of CPPD used today was established by expert consensus in 2011 [8]. Clinical presentations of the patients with CPPD can be asymptomatic CPPD (incidentally identified be imaging techniques), OA with CPPD, acute CPP crystal arthritis (severe pain and swelling in one or more joints with self-limiting evolution, replacing the term "pseudogout"), and chronic CPP crystal inflammatory arthritis (the chronic oligo/polyarthritis is associated with CPPD and presented sometimes as destructive arthropathy).

CPPD is confined to hyaline cartilage, fibrocartilage, and areas of chondroid metaplasia (e.g., degenerated areas of tendons, ligaments, and joint capsule) within the joint. The formation of calcium crystals occurs within cartilage and other connective tissue and the shedding of these crystals into the joint space may trigger an acute inflammatory arthritis [47].

The knee is the most common site followed by wrist, shoulder, ankle, hip, and elbow [4, 5, 48]. The deposition of CPP crystals frequently involves at least two sites with a mean of four sites involved [49] and it was reported that hip, wrist, and symphysis pubis cartilage calcification and synovial/capsular calcification in MCP joints commonly occur without knee involvement [50]. In conclusion, when a patient with CPPD is investigate scanning multiple sites for searching CPP aggregates is necessary, even in asymptomatic patients.

Hyaline Cartilage

Typically, the CPP crystals are deposited within the hyaline cartilage. The high spatial resolution of the high-frequency transducers (> 13 MHz) and the sparkling reflectivity of CPP crystals allow a clear depiction of even minimal aggregates within cartilage (< 1 mm). The CPP aggregates will create hyperechoic images inside the hypoechoic cartilage, in the middle layer (frequently observed in the femoral cartilage of the knee or in the metacarpal heads) [51]. The aspect of CPP deposits is related to the evolution phase of the disease (Fig. 7.10). Crystal depositions can be focal (isolated spots or extended deposits) or as a thin hyperechoic band, parallel to the surface of the hyaline cartilage, which is created by the permeability of the crystal layer, allowing the ultrasounds to penetrate and depict the bone profile. No posterior shadowing is observed [51–53]. A posterior shadowing can be observed only when the calcifications have dimensions more than 10 mm [15, 54]. The deposits create a "double contour" different from gout in which the crystals are situated in the superficial margin of the cartilage [15, 19, 20, 51]. The presence of CPPD in the hyaline cartilage may partially explain the damage of the cartilage in chronic CPP crystal inflammatory arthritis [20].

Due to the high sonographic reflectivity (similar to the bone) the CPP deposits (and also the MSU deposits) are poorly influenced by the ultrasound beam incidence. That is why reducing the gain level or changing the angle of ultrasound

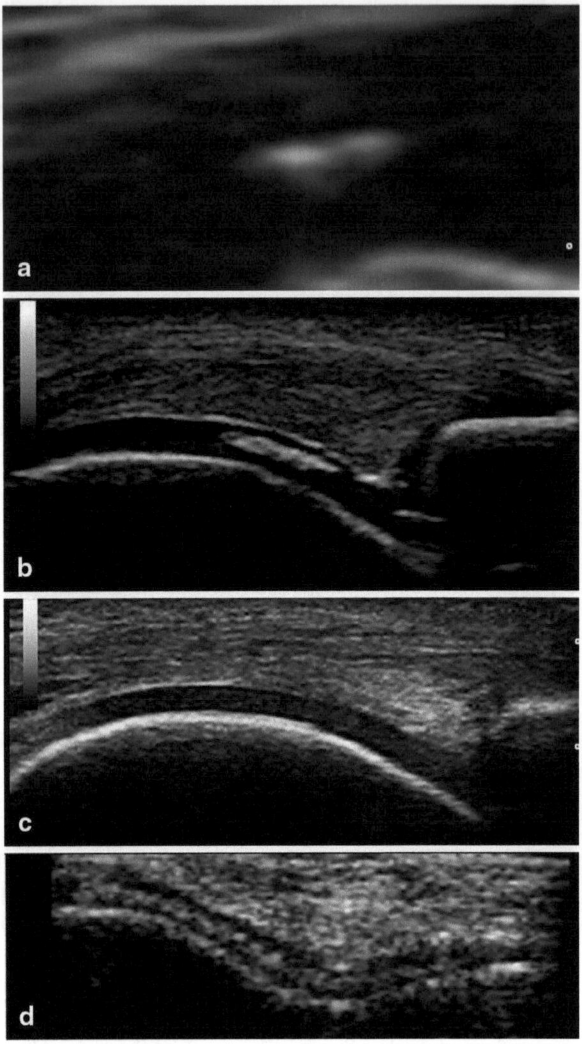

Fig. 7.10 Longitudinal or transversal scans of the femoral condyles—progressive CPPD within the hyaline cartilage **a** hyperechoic isolated spot; **b** focal linear deposit; **c** diffuse but slightly deposit; **d** hyperechoic band, parallel to the surface of the hyaline cartilage. *CPPD* calcium pyrophosphate deposition

incidence are useful methods to differentiate between crystal deposits and other hyperechoic structures [19, 51] (Fig. 7.11).

Fibrocartilage

Calcific deposits in fibrocartilage have a punctuated pattern, composed of several thin hyperechoic spots [19, 54] but sometimes compact deposits can be seen. The aggregates create a "punctuate" pattern due to the hyperechoic rounded or amorphous-shaped areas [15, 20, 51]. The dynamic evaluation of the joint (real-time examination) help to confirm their location [20, 51]. Other US findings of

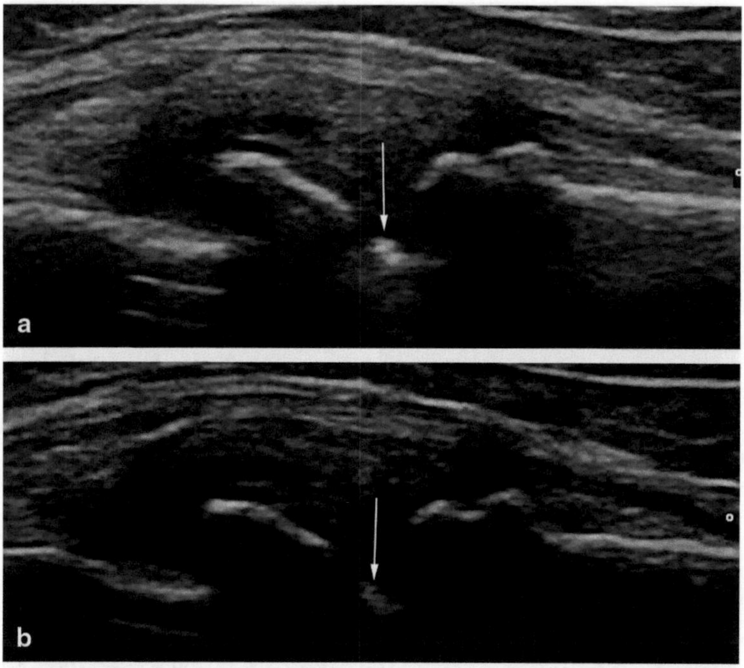

Fig. 7.11 Longitudinal scan of the medial knee. The CPPD (*arrows*) in the medial meniscus (**a**) are well seen after the gain level is reduced (**b**). *CPPD* calcium pyrophosphate deposition

fibrocartilage degeneration (mucoid cysts, fibrous tissue, or tears) can be concomitantly found in patients with OA.

In the menisci, the CPP crystals may deposit on the surface, not only inside. In this situation, a hyperechoic line (sometimes very thin line) on the meniscus surface can be identified (to be distinguished by the surrounding structure, especially hyaline cartilage or synovial fluid during the dynamic examination) [49]. Generally, any lesion detected by US must be confirmed by examination in a perpendicular section in order to avoid the false positive and negative results. For the menisci, the longitudinal scan is very difficult to be obtained so the CPP deposits may be misinterpreted.

In 1995, Brandes et al. [55] described three types of meniscal calcification: type 1A disseminated calcification, which affected equally all four menisci and represented primary chondrocalcinosis; type 1B calcification occurring in limited areas, associated with meniscal degeneration and corresponding to secondary chondrocalcinosis; type 2 cloud-like diffuse calcification, with fine granular amorphous materials, interpreted as dystrophic and postnecrotic. The real composition of the meniscal calcification (CPP and/or BCP) is difficult to be established. Majority are probably mixed, depending probably of the etiology and the stage of the underlying disease [56–58]. There are no US studies concerning this problem, but if the posterior acoustic shadow of a meniscal calcification is well defined, probably the BCP is the majority component of the calcification (Fig. 7.12).

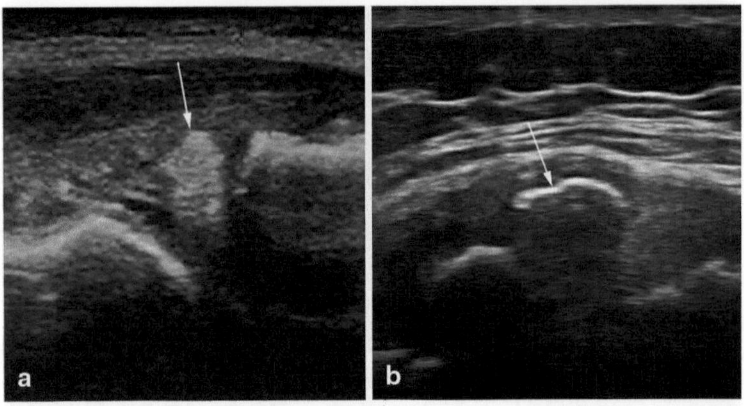

Fig. 7.12 The different aspect of calcifications (*arrows*) in medial meniscus: **a** CPPD; **b** basic calcium phosphate. *CPPD* calcium pyrophosphate deposition

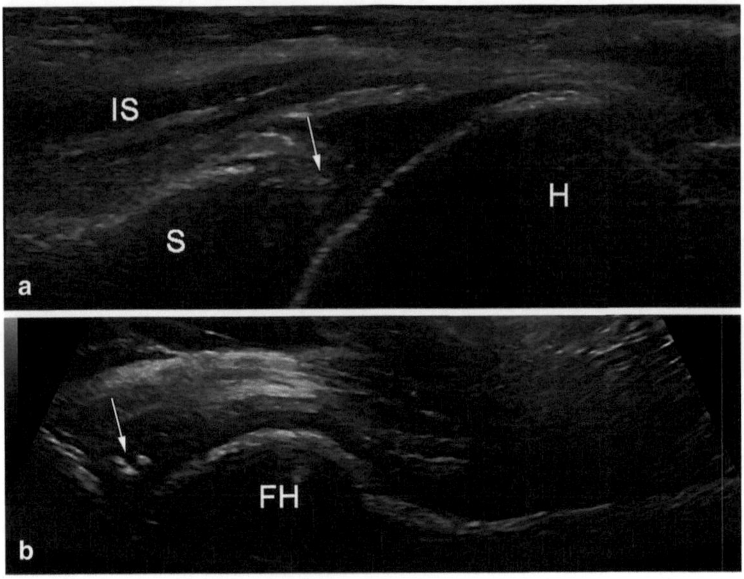

Fig. 7.13 CPPD in **a** glenoid labrum (*arrow*), posterior transversal scan and **b** acetabular labrum (*arrow*), anterior longitudinal scan. *IS* infraspinatus muscle, *S* scapula, *H* humerus, *FH* femoral head, *CPPD* calcium pyrophosphate deposition

The deposits are identified more frequently in menisci or in the triangular fibrocartilage of the wrist but the intra-articular disks from acromioclavicular, symphysis pubis, temporomandibular joints, the labrum from shoulder or hip, and even intervertebral disks can be places for CPPD [48, 50] (Figs. 7.13 and 7.14).

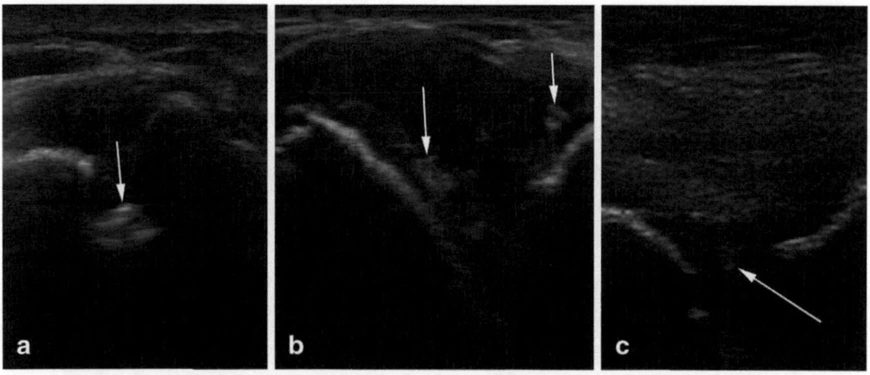

Fig. 7.14 CPPD (*arrows*) in **a** acromioclavicular joint, **b** sternoclavicular joint, and **c** symphysis pubis. *CPPD* calcium pyrophosphate deposition

Synovial Fluid

The synovial fluid from acute arthritis may have an anechoic aspect or may contain debris and proteinaceous material. Sometimes typically aggregates of CPP can be detected. These are small hyperechoic structures, uniformly rounded, with defined margins, floating in the synovial fluid (by pressing the joint with the transducer the movement of the aggregates can be observed). Due to their reflectivity, by adjusting the gain settings, the differentiation from other hyperechoic and floating structure can be done. In bursa or in articular recesses, homogeneous hyperechoic nodular or oval-shaped deposits can be identified [15].

Soft Tissues

There are rare reports of extra-articular CPPD deposition or tophus-like deposits [59] and periarticular inflammations, like tendinitis, bursitis, or tophi, are uncommon presentations.

The CPPD in tendons is rarer comparing with gout. The calcifications are variable in aspect: hyperechoic spots or subtle hyperechoic linear deposits parallel to the tendon fascicles, with or without posterior acoustic shadow and not in continuity with the bone [15, 19, 20, 51, 52] (Figs. 7.15 and 7.16). Pathological vascularization of the tendon is rarely detected, only in case of significant inflammatory response induced by CPP [19].

Validity of US for detecting CPPD crystals in the hyaline cartilage is not well known as there are only few studies concerning this aspect. Filipou et al. [60] found a sensitivity of 86.7% and a high specificity (96%) of US in identifying CPPD calcifications in 47 patients with CPPD comparing with 29 patents in the control group. Filippucci et al. [31] found a specificity of 97.6% in a blind evaluation of

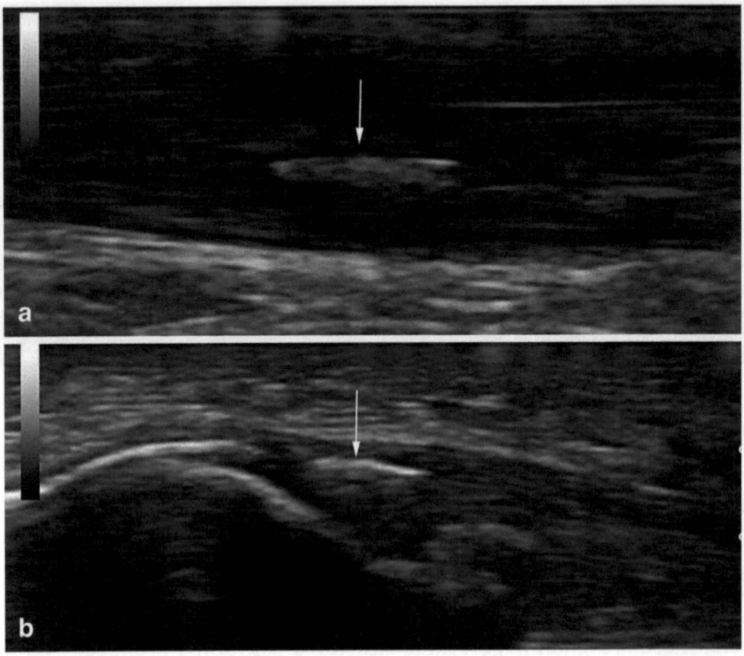

Fig. 7.15 Longitudinal scan of the Achilles tendon (**a**) and plantar fascia (**b**). Hyperechoic linear deposits parallel to the tendon fascicles, without acoustic shadowing, not in continuity with the bone, indicative for CPPD. *CPPD* calcium pyrophosphate deposition

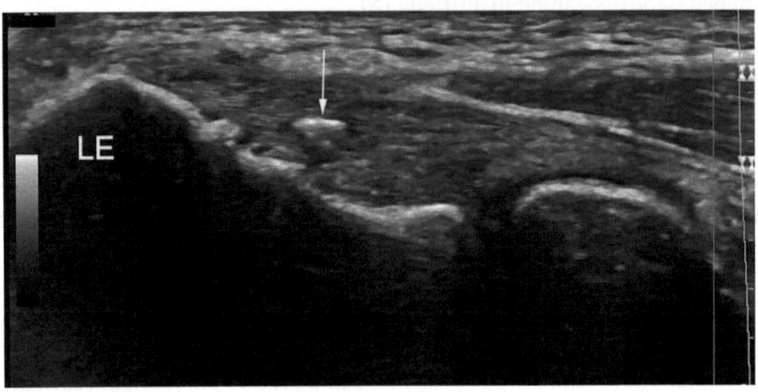

Fig. 7.16 Longitudinal scan at the lateral epicondyle (*LE*) of the elbow—lateral epicondylitis with calcification (*arrow*)

264 knees in 132 patients comparing 48 CPPD subjects with 32 gouty patients and other 52 disease controls. In this study, the presence of hyperechoic spots in the hyaline cartilage showed a relatively low sensitivity, 68.7%.

As concerning the fibrocartilage CPPD, Filipou et al. [61] demonstrated that there are important differences between in vivo and *ex vivo* examination of 11 menisci in 6 patients undergone knee replacement surgery. Higher sensitivity and specificity values were obtained with the *ex vivo* US examination comparing to the in vivo US (67 and 100 % vs. 44 and 50 %, respectively). The technical limitations (obese patients, late stage of knee OA with marked osteophytosis), the presence of small vessels generating posterior enhancement, and the difficulty of examining menisci using perpendicular scans are the explanation for these differences.

The CPPD in tendons was studied by Falsetti et al. [62]. They found in patients with CPPD the presence of the Achilles tendon calcifications in 57.9 % patient, plantar fascia calcifications in 15.8 % patients, deep retrocalcaneal bursitis in 10.5 % patients, and plantar fasciitis in 40.3 % patients. For the diagnosis of CPPD, the presence of Achilles tendon calcifications has 39 % sensitivity and 100 % specificity and the plantar fascia calcifications 15 % sensitivity, 98 % specificity. An excellent agreement was found between US and radiography in detecting Achilles tendon and plantar fascia calcifications. Quite similar results were found recently by Ellabban et al. They found a sensitivity of US for detection Achilles tendon and plantar fascia calcifications of 57.9 % and 15.8 %, respectively, and the specificity of 100 % for both. The authors concluded that US Achilles tendon and plantar fascia calcifications are in fact frequent findings in patients with CPPD [63].

The high specificity of US indicative for crystal deposits at hyaline cartilage level was found also in a recent study concerning the shoulder in patients with crystal-related arthropathies [64]. The authors demonstrated that the most frequently affected structures of the shoulders in these patients are the supraspinatus tendon and the fibrocartilage of the acromioclavicular joint.

US demonstrated a sensitivity and specificity at least equal to that of radiography in identifying CPP deposits [54]. The explanation for the calcifications not seen on the radiological examination could be the presence of low-density calcifications. The small size of CPP deposits and the significant cartilage loss due to OA may be other explanations for radiological non-visualization [65]. Furthermore, the CPP deposits are not stable and according to the crystal-shedding theory these deposits decreasing after an acute attack of arthritis making their radiological detection more difficult [66].

Until now there is only one study in which US was compared with conventional radiology and CT [67]. In this study on 25 patients with CPP crystals detected in the synovial fluid from knees, the conventional radiology found CPPD in 52 %, CT in 72 % patients, and US in 100 % of patients. The study showed the advantages of US in revealing CPPD, especially for the patients without evidence of diseases at classic radiological examination.

No comparative studies between US and MRI in CPPD were published.

Basic Calcium Phosphate Crystals Deposition Disease

BCP crystals is a family of crystals in which the HA is the representative component. These crystals have been found in and around the joints and have been associated with several forms of arthritis and periarthritis [68]. In the last years, there are more and more discussions about the role of these crystals in the OA pathogenesis [69, 70]. The presence of HA was correlated with the degree of articular degeneration (elicit the synovial inflammation and cartilage degradation) suggesting that HA have a direct pathogenic role in OA [57]. The technical difficulties for identifying these crystals is the most important problem of the diagnosis limitation. The factors responsible for HA accumulation in the intra-articular space are not fully clarified but the crystals were found in higher quantity in the femoral cartilage of osteoarthritic patients [71]. Local trauma, surgical procedures, injections, acute illness, or hemodialysis favor the HA aggregation in the conjunctive tissue.

Virtually, any joint can be involved, but the most common involved areas include the shoulder, great trochanter of the hip, lateral epicondyle of the elbow, or wrist tendons; the shoulder is by far most commonly involved [72].

Calcific Tendinopathy of the Rotator Cuff

One of the most frequent localization of HA deposits is the rotator cuff. There are three phases in the evolution of a calcification: the formative phase (in which calcium crystals are deposited into matrix vesicles that coalesce and form foci of calcification), the resting phase (fibrocollagenous tissue encircles and isolates the foci of calcification), and the resorptive phase (associated with acute inflammatory response). The first two are chronic, lasting for years, usually asymptomatic or with pain related to movements [73, 74].

The US morphology of calcifications had been divided into four types: *arc shaped* with clear shadowing; *fragmented or punctuated* with or without shadowing; *nodular,* with no shadow; and *cystic,* a bold echogenic wall with an anechoic area, weak internal echoes or layering content [75] (Fig. 7.17). The first type of calcification is in formative phase and the last ones are in the resorptive stage. The resorptive phase is the painful phase of the calcific tendonitis, the pathology being a dynamic process [76]. In 10 % of adults on plain radiographs, the calcifications of the rotator cuff can be observed; only in half of these cases, the patients are symptomatic. Large and fragmented calcifications were associated with symptomatic shoulder (especially with pain). The presence of the PD signal within the calcification and the widening of subacromial–subdeltoidian (SASD) bursa are the more powerful US findings associated with pain [77]. Sometimes the calcium deposits may drain within the SASD bursa and calcium-related bursitis will appear (Fig. 7.18). To mention that the HA deposits are less sparkling and the posterior acoustic shadowing can be detected even at small dimensions (<2–3 mm) [15].

US is important also for the conservative treatment of calcifications. By US guidance, the SASD bursa can be injected with corticosteroid or this maneuver can be associated with US-guided needling and lavage (barbotage) of the calcification (Fig. 7.19). There are two methods for needling and barbotage, using one or two needle [78], but the purpose is the same: to dissolve the calcium and to aspirate it. Clinical and radiological improvement is significant better when the two procedures are combined [79].

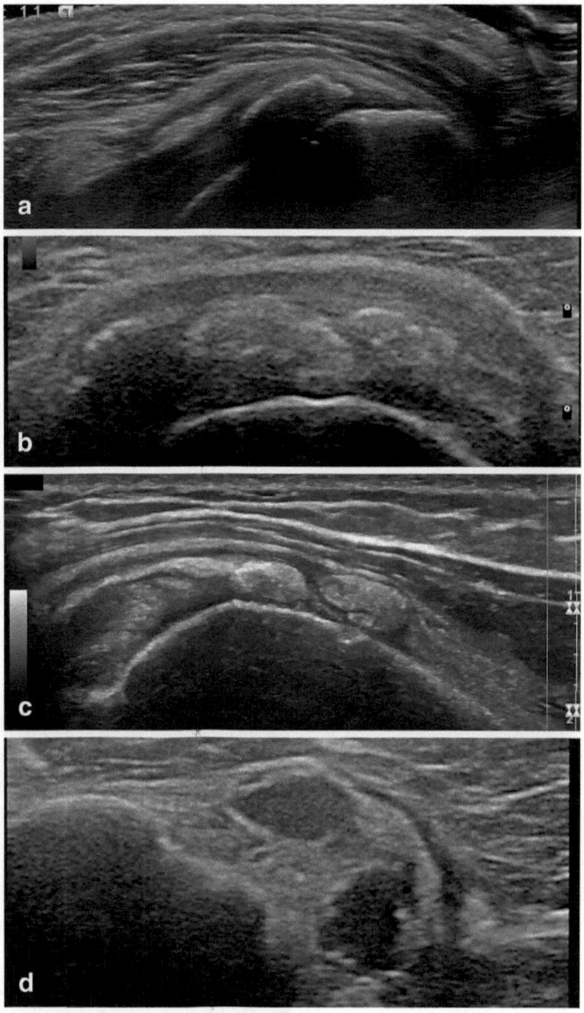

Fig. 7.17 Rotator cuff calcifications: **a** arc-shaped, with clear posterior acoustic shadow; **b** fragmented, without shadowing; **c** nodular, with no shadow; **d** cystic, with echogenic wall and anechoic content

Milwaukee Shoulder

Milwaukee shoulder or apatite-associated destructive arthropathy is an association between rotator cuff arthropathy and intra-articular presence of HA crystals. It affects mainly the elderly women and is characterized by intra-articular or periarticular deposition of HA crystals and rapid destruction of the rotator cuff and the glenohumeral joint [80, 81].

Great Trochanteric Pain Syndrome

The pathologic changes of the gluteus medius and minimus tendons are the main causes of the pain in the great trochanteric region. Apart from tendinopathy

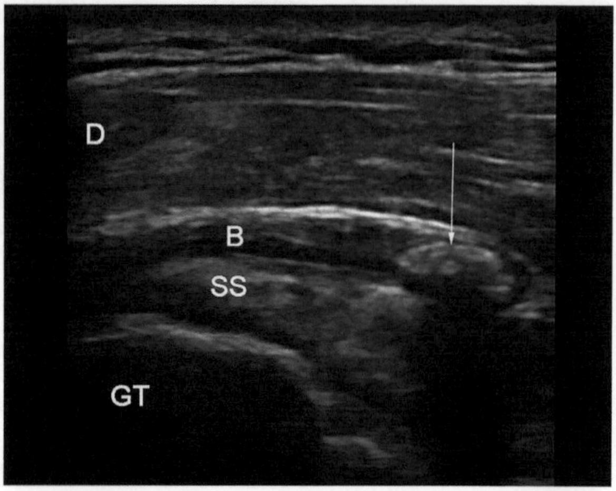

Fig. 7.18 Calcification (*arrow*) inside the subdeltoid bursa (*B*). *D* deltoid muscle, *SS* supraspinatus muscle, *GT* great tuberosity

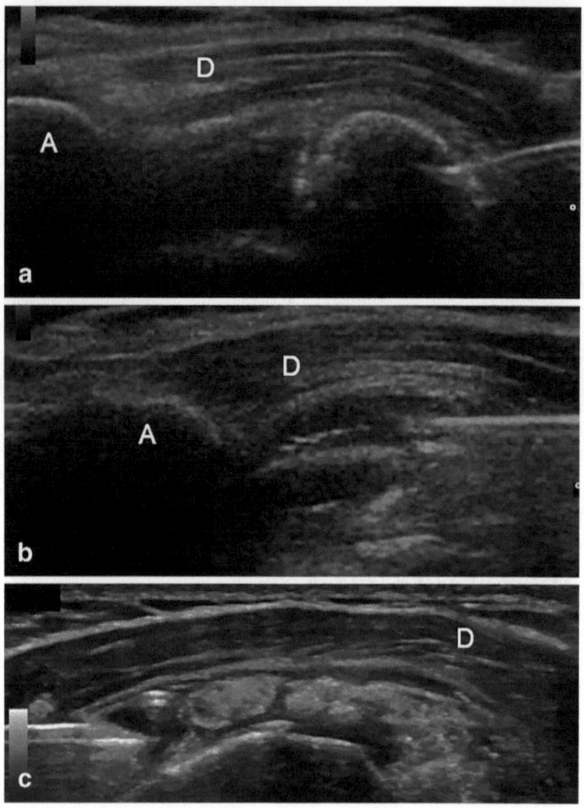

Fig. 7.19 Supraspinatus tendon calcifications: **a** US-guided needling followed by **b** corticosteroid injection of the subdeltoid bursa; **c** aspiration of a cystic calcification. *US* ultrasonography

(generally degenerative) or tears, calcification at the tendon insertion (hyperechoic spots) or intratendinous (one or more hyperechoic foci within the body of the tendon with a posterior acoustic shadow) may be the explanation of this syndrome

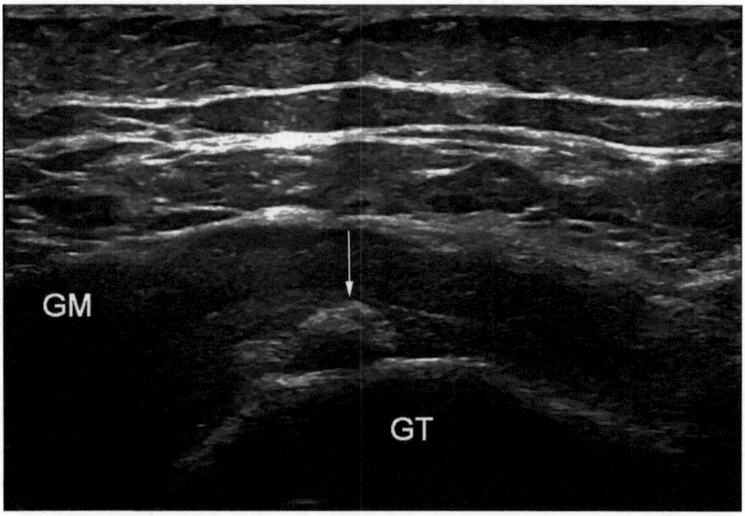

Fig. 7.20 Calcification (*arrow*) in the gluteus medius (*GM*) tendon, longitudinal scan. *GT* great trochanter

(Fig. 7.20). Sometimes fluid in the trochanteric bursa may accompany this situation [82, 83].

Oxalate Crystal Deposition Disease

Oxalate crystal deposition disease is rarely seen: in patients with primary hyperoxaluria or in patients with end-stage renal disease managed with long-term dialysis. Oxalate crystal deposits are found mainly in the kidneys, bone, skin, and vessels, and less often inside the joints. Musculoskeletal and systemic manifestations of oxalate crystal deposition disease are not specific and may be confused to other crystal deposition diseases [84].

In Table 7.4 are summarized the main US findings in crystal-related arthropathies.

Conclusions

In conclusion, in the assessment of the patients with crystal-related arthropathies, US is an important and valuable tool. Solid US knowledge about normal and pathological aspects of the musculoskeletal structures is needed for a good examination and a correct interpretation of the scans. Sonographers should be fully trained before applying the technique to clinical use, and should work with a qualified sonographer in the beginning so as to avoid misinterpretations. Combining the clinical

Table 7.4 The main ultrasonographic findings in crystal-related arthropathies

–	Gout	Calcium pyrophosphate deposition disease	Hydroxyapatite deposition disease
Hyaline cartilage	Double contour: enhancement (focal or diffuse) of the superficial articular cartilage layer	Double contour: thin hyperechoic band (focal or diffuse) within the cartilage	–
	Slight irregularities of the superficial margin of the hyaline cartilage;		
	Independent of the insonation angle		
Fibrocartilage	–	"Punctuate" pattern: hyperechoic rounded or amorphous-shaped areas	Hyperechoic deposits with well-defined posterior acoustic shadow
		Hyperechoic thin line on the meniscus surface	
Synovial fluid	With hyperechoic spots (< 1 mm) floating within the joint cavity	With small hyperechoic aggregates, uniformly rounded and defined margins	
	"Snowstorm" appearance		
Bone	Intra-articular erosions, profound, and destructive		
	Extra-articular breaks of the bony cortex due to intraosseous tophaceous deposits		
Tendon	Microdeposits: ovoid-shaped hyperechoic densities Intra-tendinous tophi Small hypoechoic halo in inflammatory response	Hyperechoic spots or subtle hyperechoic linear deposits parallel to the tendon fascicles, ± acoustic shadowing; not in continuity with the bone	Arc shaped with clear shadowing Fragmented or punctuated ± shadowing Nodular, with no shadow Cystic, a bold echogenic wall with an anechoic area

Table 7.4 (continued)

–	Gout	Calcium pyrophosphate deposition disease	Hydroxyapatite deposition disease
Soft tissue deposits	Soft tophi: nodular, small, hypoechoic, homogeneous structures, soft on palpation	Rare extra-articular tophus-like deposits	Hyperechoic, with posterior acoustic shadow
	Hard tophi: hyperechoic band with posterior acoustic shadow, harder in consistency on palpation		
	Mixed tophi		

examination with US scan in patients with suspicion of microcrystalline arthritis will offer a more complete picture about underlying pathology and will permit to take proper and rapid decisions for the patient management.

References

1. Annemans L, Spaepen E, Gaskin M, Bonnemaire M, Malier V, Gilbert T, Nuki G. Gout in the UK and Germany: prevalence, comorbidities and management in general practice 2000–2005. Ann Rheum Dis. 2008;67(7):960–6.
2. Ramonda R, Musacchio E, Perissinotto E, Sartori L, Punzi L, Corti MC, et al. Prevalence of chondrocalcinosis in Italian subjects from northeastern Italy. The Pro. V. A. (PROgetto Veneto Anziani) study. Clin Exp Rheumatol. 2009;27:981–4.
3. Lillicrap M. Crystal arthritis: contemporary approaches to diseases of antiquity. Clin Med. 2007;7:60–4.
4. Ramonda R, Frallonardo P, Oliviero F, Lorenzin MG, Ortolan A, Scanu A, Punzi L. Pain and microcrystalline arthritis. Reumatismo. 2014;66(1):48–56.
5. Olivieri F, Punzi L. L'inflammazione articolare da microcristalli. Reumatismo. 2003;55:16–27.
6. Gerster JC, Landry M, Dufresne L, Meuwly JY. Imaging of tophaceous gout: computed tomography provides specific images compared with magnetic resonance imaging and ultrasonography. Ann Rheum Dis. 2002;61:52–54.
7. Perez-Ruiz F, Martin I, Canteli B. Ultrasonographic measurement of tophi as an outcome measure for chronic gout. J Rheumatol. 2007;34:1888–93.
8. Zhang W, Doherty M, Bardin T, Barskova V, Guerne PA, Jansen TL, et al. European league against rheumatism recommendations for calcium pyrophosphate deposition. Part I: terminology and diagnosis. Ann Rheum Dis. 2011;70:563–70.
9. Dieppe P, Doherty M, Macfarlane D. Introduction. Crystal-related arthropathies. Ann Rheum Dis. 1983;42(Suppl 1):1–3.
10. Nuki G, Simkin PA. A concise history of gout and hyperuricemia and their treatment. Arthritis Res Ther. 2006;8(Suppl 1):S1.
11. Sivera F, Andrés M, Carmona L, et al. Multinational evidence-based recommendations for the diagnosis and management of gout. Ann Rheum Dis. 2014;73(2):328–35.

12. Girish G, Glazebrook K, Jacobson J. Advanced imaging in gout. AJR Am J Roentgenol. 2013;20193:515–25.
13. Ottaviani S, Bardin T, Richette P. Usefulness of ultrasonography for gout. Joint Bone Spine. 2012;79(5):441–5.
14. Thiele RG, Schlesinger N. Diagnosis of gout by ultrasound. Rheumatology. 2007;46(7):1116–21.
15. Delle Sedie A, Riente L, Iagnocco A, et al. Ultrasound imaging for the rheumatologist X. Ultrasound imaging in crystal-related arthropathies. Clin Exp Rheumatol. 2007;25(4):513–7.
16. Naredo E, Uson J, Jiménez-Palop M, et al. Ultrasound-detected musculoskeletal urate crystal deposition: which joints and what findings should be assessed for diagnosing Gout? Ann Rheum Dis. 2014;73(8):1522–8.
17. Fodor D, Albu A, Gherman C. Crystal associated synovitis—ultrasonographic feature and clinical correlation. OTR. 2008;2(6):99–110.
18. Grassi W, Gutierrez M, Filippucci E. Crystal associated synovitis. In: Wakefield RJ, D'Agostino MA, editors. Essential applications of musculoskeletal ultrasound in rheumatology. Philadelphia: Saunders Elsevier; 2011. p. 187–97.
19. Filippucci E, Di Geso L, Grassi W. Tips and tricks to recognize microcrystalline arthritis. Rheumatology (Oxford). 2012;51(Suppl 7):vii18–21.
20. Grassi W, Meenagh G, Pascual E, et al. 'Crystal clear'-sonographic assessment of gout and calcium pyrophosphate deposition disease. Semin Arthritis Rheum. 2006;36(3):197–202.
21. Howard RG, Pillinger M, Gyftopoulos S, Thiele R, Swearingen C, Samuel S. Reproducibility of musculoskeletal ultrasound for determining monosodium urate deposition: concordance between readers. Arthritis Care Res. 2011;63(10);1456–62.
22. Rettenbacher T, Ennemoser S, Weirich H, et al. Diagnostic imaging of gout. Comparison of high-resolution US versus conventional X-ray. Eur Radiol. 2008;18(3):621–30.
23. Chowalloor PV, Keen HI. A systematic review of ultrasonography in gout and asymptomatic hyperuricaemia. Ann Rheum Dis. 2013;72(5):638–45.
24. Filippucci E, Meenagh G, Sedie AD, et al. Ultrasound imaging for the rheumatologist XXXVI. Sonographic assessment of the foot in gout patients. Clin Exp Rheumatol. 2011;29(6):901–5.
25. Perez-Ruiz F, Naredo E. Imaging modalities and monitoring measures of gout. Curr Opin Rheumatol. 2007;19(2):128–33.
26. Perez-Ruiz F, Urresola A, Miguel E, Schlesinger N. Gout. Imaging of gout: findings and utility. Arthritis Res Ther. 2009;11(3):232.
27. Thiele R. Ultrasound in the diagnosis of crystal deposition disease. In: Terkeltaub R, editor. Gout and other crystal arthropathies. Philadelphia: Saunders Elsevier; 2012. p. 331–42.
28. Gutierrez M, Smith W, Thiele R, et al. Defining elementary ultrasound lesions in gout. Preliminary results of delphi consensus and web-exercise reliability. Ann Rheum Dis. 2014;73(Supp l2):302.
29. Schlesinger N, Thiele RG. The pathogenesis of bone erosions in gouty arthritis. Ann Rheum Dis. 2010;69(11):1907–12.
30. Wright SA, Filippucci E, McVeigh C, et al. High-resolution ultrasonography of the first metatarsal phalangeal joint in gout: a controlled study. Ann Rheum Dis. 2007;66(7):859–64.
31. Filippucci E, Gutierrez M, Georgescu D, et al. Hyaline cartilage involvement in patients with gout and calcium pyrophosphate deposition disease: an ultrasound study. Osteoarthr Cartil. 2009;17(2):178–81.
32. McQueen FM, Reeves Q, Dalbeth N. Advanced imaging in the diagnosis and management of gout. Postgrad Med J. 2013;89(1048):87–93.
33. Pineda C, Amezcua-Guerra LM, Solano C, et al. Joint and tendon subclinical involvement suggestive of gouty arthritis in asymptomatic hyperuricemia: an ultrasound controlled study. Arthritis Res Ther. 2011;13(1):R4.
34. Thiele RG, Schlesinger N. Ultrasonography shows disappearance of monosodium urate crystal deposition on hyaline cartilage after sustained normouricemia is achieved. Rheumatol Int. 2010;30(4):495–503.

35. Dalbeth N, Pool B, Gamble GD, et al. Cellular characterization of the gouty tophus: a quantitative analysis. Arthritis Rheum. 2010;62(5):1549–56.
36. Dalbeth N, Schauer C, MacDonald P, et al. Methods of tophus assessment in clinical trials of chronic gout: a systematic literature review and pictorial reference guide. Ann Rheum Dis. 2011;70:597–604.
37. Puig JG, de Miguel E, Castillo MC, et al. Asymptomatic hyperuricemia: impact of ultrasonography. Nucleosides Nucleotides Nucleic Acids. 2008;27(6):592–5.
38. Fernandes EA, Lopes MG, Mitraud SA, Ferrari AJ, Fernandes AR. Ultrasound characteristics of gouty tophi in the olecranon bursa and evaluation of their reproducibility. Eur J Radiol. 2012;81(2):317–23.
39. Rodas G, Pedret C, Catala J, Soler R, Orozco L, Cusi M. Intratendinous gouty tophus mimics patellar tendonitis in an athlete. J Clin Ultrasound. 2013;41(3):178–82.
40. Abate M, Schiavone C, Salini V, Andia I. Occurrence of tendon pathologies in metabolic disorders. Rheumatology. 2013;52(4):599–608.
41. Nalbant S, Corominas H, Hsu B, et al. Ultrasonography for assessment of subcutaneous nodules. J Rheumatol. 2003;30(6):1191–5.
42. De Miguel E, Puig JG, Castillo C, et al. Diagnosis of gout in patients with asymptomatic hyperuricaemia: a pilot ultrasound study. Ann Rheum Dis. 2012;71(1):157–8.
43. Felson DT, Anderson JJ, Naimark A, et al. The prevalence of chondrocalcinosis in the elderly and its association with knee osteoarthritis: the Framingham study. J Rheumatol. 1989;16:1241–5.
44. Salaffi F, De Angelis R, Grassi W, MArche Pain Prevalence, INvestigation Group (MAPPING) study. Prevalence of musculoskeletal conditions in an Italian population sample: results of a regional community-based study. I. The MAPPING study. Clin Exp Rheumatol. 2005;23(6):819–28.
45. Ciancio G, Bortoluzzi A, Govoni M. Epidemiology of gout and chondrocalcinosis. Reumatismo. 2012;63(4):207–20.
46. Doherty M, Watt I, Dieppe PA. Localised chondrocalcinosis in post-meniscectomy knees. Lancet. 1982;1(8283):1207–10.
47. Oliviero F, Scanu A, Punzi L. Metabolism of crystals within the joint. Reumatismo. 2012;63(4):221–9.
48. Ferrone C, Andracco R, Cimmino MA. Calcium pyrophosphate deposition disease: clinical manifestations. Reumatismo. 2012;63(4):246–52.
49. Filippou G, Filippucci E, Tardella M, Bertoldi I, Di Carlo M, Adinolfi A, et al. Extent and distribution of CPP deposits in patients affected by calcium pyrophosphate dihydrate deposition disease: an ultrasonographic study. Ann Rheum Dis. 2013;72(11):1836–9.
50. Abhishek A, Doherty S, Maciewicz R, Muir K, Zhang W, Doherty M. Chondrocalcinosis is common in the absence of knee involvement. Arthritis Res Ther. 2012;14:R205.
51. Ciapetti A, Filippucci E, Gutierrez M, Grassi W. Calcium pyrophosphate dihydrate crystal deposition disease: sonographic findings. Clin Rheumatol. 2009;28:271–6.
52. Filippucci E, Di Geso L, Girolimetti R, Grassi W. Ultrasound in crystal-related arthritis. Clin Exp Rheumatol. 2014; 32(1 Suppl 80):S42–7.
53. Gutierrez M, Filippucci E, Salaffi F, Grassi W. The current role of ultrasound in the assessment of crystal-related arthropathies. Reumatismo. 2009;61(3):216–21.
54. Frediani B, Filippou G, Falsetti P, et al. Diagnosis of calcium pyrophosphate dihydrate crystal deposition disease: ultrasonographic criteria proposed. Ann Rheum Dis. 2005;64:638–40.
55. Brandes A, Muller KM. Calcinosis of the meniscus. Morphologic and roentgenographic findings for zonal classification. Pathologe. 1995;16:269–77.
56. MacMullan PA, McCarthy GM. The meniscus, calcification and osteoarthritis: a pathologic team. Arthritis Res Ther. 2010;12(3):116.
57. Ea HK, Chobaz V, Nguyen C, Nasi S, van Lent P, Daudon M, et al. Pathogenic role of basic calcium phosphate crystals in destructive arthropathies. PLoS ONE. 2013;8(2):e57352.
58. Sun Y, Mauerhan DR, Honeycutt PR, Kneisl JS, Norton HJ, Zinchenko N, et al. Calcium deposition in osteoarthritic meniscus and meniscal cell culture. Arthritis Res Ther. 2010;12(2):R56.

59. Fam AG. Calcium pyrophosphate crystal deposition disease and other crystal deposition diseases. Curr Opin Rheumatol. 1992;4:574–82.
60. Filippou G, Frediani B, Lorenzini S, Galeazzi M, Marcolongo R. A "new" technique for the diagnosis of chondrocalcinosis of the knee: sensitivity and specificity of high-frequency ultrasonography. Ann Rheum Dis. 2007;66:1126–8.
61. Filippou G, Adinolfi A, Bozios P, Lorenzini S, Picerno V, Di Sabatino V, et al. Do not hallow until you are out of the wood! Ultrasonographic detection of CPP crystal deposits in menisci: facts and pitfalls. Sci World J. 2013:181826.
62. Falsetti P, Frediani B, Acciai C, et al. Ultrasonographic study of Achilles tendon and plantar fascia in chondrocalcinosis. J Rheumatol. 2004;31(11):2242–50.
63. Ellabban AS, Kamel SR, Abo Omar HA, El-Sherif AM, Abdel-Magied RA. Ultrasonographic findings of Achilles tendon and plantar fascia in patients with calcium pyrophosphate deposition disease. Clin Rheumatol. 2012;31(4):697–704.
64. Filippucci E, Delle Sedie A, Riente L, Di Geso L, Carli L, Ceccarelli F, Sakellariou G, Iagnocco A, Grassi W. Ultrasound imaging for the rheumatologist. XLVII. Ultrasound of the shoulder in patients with gout and calcium pyrophosphate deposition disease. Clin Exp Rheumatol. 2013;31(5):659–64.
65. Genant HK. Roentgenographic aspects of calcium pyrophosphate dihydrate crystal deposition disease (pseudogout). Arthritis Rheum. 1976;19:307–28.
66. Ryan LM, McCarty DJ. Calcium pyrophosphate crystal deposition disease, pseudogout and articular chondrocalcinosis. In: McCarty DJ, Koopman WJ, editors. Arthritis and allied conditions. 13th edn. Philadelphia: Lea & Febiger; 1997. p. 2103–2125.
67. Barskova VG, Kudaeva FM, Bozhieva LA, Smirnov AV, Volkov AV, Nasonov EL. Comparison of three imaging techniques in diagnosis of chondrocalcinosis of the knees in calcium pyrophosphate deposition disease. Rheumatology (Oxford). 2013;52(6):1090–4.
68. Bardin T, Richette P. Basic calcium phosphate crystal deposition disease. Presse Med 2011;40:850–5.
69. McCarthy GM, Cheung HS. Point: hydroxyapatite crystal deposition is intimately involved in the pathogenesis and progression of human osteoarthritis. Curr Rheumatol Rep. 2009;11:141–7.
70. Pritzker KP. Counterpoint: hydroxyapatite crystal deposition is not intimately involved in the pathogenesis and progression of human osteoarthritis. Curr Rheumatol Rep. 2009;11:148–53.
71. Garcia GM, McCord GC, Kumar R. Hydroxyapatite crystal deposition disease. Semin Musculoskelet Radiol. 2003;7(3):187–93.
72. Ohira T, Ishikawa K. Hydroxyapatite deposition in osteoarthritic articular cartilage of the proximal femoral head. Arthritis Rheum. 1987;30:651–60.
73. Chiou HJ, Chou YH, Wu JJ, et al. Evaluation of calcific tendonitis of the rotator cuff. Role of color Doppler ultrasonography. J Ultrasound Med. 2002;21:289–95.
74. Halverson PB. Crystal deposition disease of the shoulder (including calcific tendonitis and Milwaukee shoulder syndrome). Curr Rheumatol Rep. 2003;5:244–7.
75. Le Goff B, Berthelot JM, Guillot P, Glémarec J, Maugars Y. Assessment of calcific tendonitis of rotator cuff by ultrasonography: comparison between symptomatic and asymptomatic shoulders. Joint Bone Spine. 2010;77(3):258–63.
76. Bureau NJ. Calcific tendinopathy of the shoulder. Semin Musculoskelet Radiol. 2013;17(1):80–4.
77. Carcia CR, Scibek JS. Causation and management of calcific tendonitis and periarthritis. Curr Opin Rheumatol. 2013;25(2):204–9.
78. Sconfienza LM, Viganò S, Martini C, Aliprandi A, Randelli P, Serafini G, Sardanelli F. Double-needle ultrasound-guided percutaneous treatment of rotator cuff calcific tendinitis: tips & tricks. Skeletal Radiol. 2013;42(1):19–24.
79. de Witte PB, Selten JW, Navas A, Nagels J, Visser CP, Nelissen RG, Reijnierse M. Calcific tendinitis of the rotator cuff: a randomized controlled trial of ultrasound-guided needling and lavage versus subacromial corticosteroids. Am J Sports Med. 2013;41(7):1665–73.

80. Jensen KL, Williams GR, Russell IJ, Rockood CA. Rotator cuff tear arthropathy. J Bone Joint Surg. 1999;9:1312–22.
81. Popov HI, Gherman C, Rogojan L, Botar-Jid C, Barna C, Fodor D. Milwaukee shoulder syndrome associated with pigmented villonodular synovitis. Case report. Med Ultrason. 2012;14:67–70.
82. Iagnocco A, Filippucci E, Meenagh G, et al. Ultrasound imaging for the rheumatologist III. Ultrasonography of the hip. Clin Exp Rheumatol. 2006;24:229–32.
83. Nestorova R, Vlad V, Petranova T, Porta F, Radunovic G, Micu MC, Iagnocco A. Ultrasonography of the hip. Med Ultrason. 2012;14(3):217–24.
84. Maldonado I, Prasad V, Reginato AJ. Oxalate crystal deposition disease. Curr Rheumatol Rep. 2002;4:257–64.

Chapter 8
Skin, Nail, and Hair in Rheumatology

Ximena Wortsman MD

Introduction

Ultrasound has been present in the medical arsenal for more than 70 years. However, in the past 30 years, US took a giant leap as the technique passed from a phase of purely experimental use to a widely available imaging procedure used in standard daily clinical practice. The usage of ultrasound for studying skin started in 1951 with the work of Meyer et al. [1] but has continued its development through the years. In 1979, Alexander and Miller studied the skin using pulsed ultrasound [2]. Later on, during the 1980s and 1990s, several groups started to use compact high-frequency ultrasound machines with fixed-frequency probes that currently can vary from 20 to 100 MHz [3–7]. Due to their high-frequency probes and configuration, these machines present high resolution but low penetration. Though this enables the examiner to detect the dermis and upper hypodermis, they lack color Doppler capabilities which means that they cannot detect blood flow. Nevertheless, they provide valuable information in a bitmap-type image about local skin changes. In the year 2004, the first experience in the use of high- and variable-frequency probes with multichanneled color Doppler machines in dermatologic lesions was reported [8]. Since then, dermatologic applications of ultrasound have been growing rapidly, covering a wide range of common lesions such as benign and malignant skin tumors, vascular anomalies, inflammatory diseases, nail lesions, scalp and hair pathologies, and cosmetic applications, among others [9–11]. However, such imaging technique requires both an appropriate ultrasound machine and an operator trained in dermatologic pathologies. This combination is essential to attain the desired outcomes, as it is the integration of the information generated by the visual inspection of the lesion as well as the anatomical data provided by sonography which would

X. Wortsman (✉)
Departments of Radiology and Dermatology, IDIEP-Institute for Diagnostic and Reasearch of the Skin and Soft Tissue, Clinica Servet, Faculty of Medicine, University of Chile, Santiago, Chile
e-mail: xworts@yahoo.com

Y. El Miedany (ed.), *Musculoskeletal Ultrasonography in Rheumatic Diseases*,
DOI 10.1007/978-3-319-15723-8_8

enable achieving the assessment targets [12]. The limitations for this type of imaging study are lesions that measure ≤ 0.1 mm, epidermal only lesions and pigments such as melanin [10, 12, 13]. However, having all these settings, ultrasound presents an accuracy of diagnosis up to 97 % in common dermatologic entities [13].

Normal Ultrasound Anatomy of the Skin, Nail, and Hair

Skin

The epidermis is the most superficial layer of the skin and shows on ultrasound as a hyperechoic layer on non-glabrous skin (i.e., that not from the palms and soles) and as a bilaminar hyperechoic layer in glabrous skin (i.e., that from palms and soles). Its echogenicity is due to the keratin content. The dermis presents as a hyperechoic band, less bright than the epidermis, and its echogenicity is mainly due to the presence of collagen. With age, a hypoechoic band can be observed in the upper dermis. This is called subepidermal low-echogenicity band (SLEB) and corresponds to the deposit of glycosaminoglycans in the skin due to photoaging. The hypodermis appears as a hypoechoic layer due to its fatty lobules that are separated by hyperechoic septa. Slow flow arterial and venous vessels are commonly detected in the hypodermis and occasionally in the dermis ([10, 12–15]; Fig. 8.1).

Nail

This is an enthesis organ that makes up a complex with the interphalangeal joint and the distal insertion of the extensor tendon. The nail unit can be divided in the

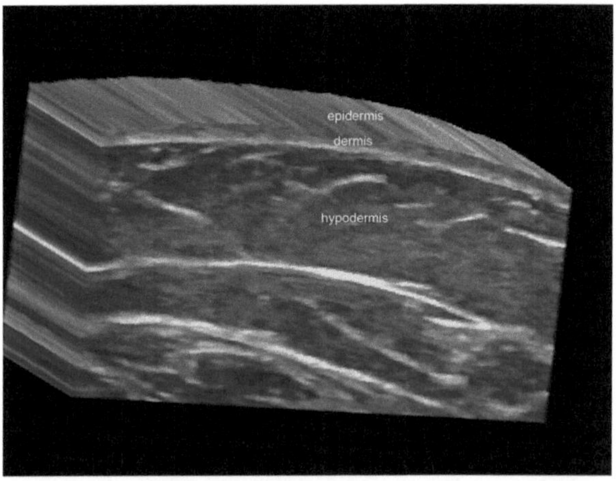

Fig. 8.1 Sonographic anatomy of the normal skin (3D reconstruction dorsal forearm, *transverse* view)

ungual and periungual regions. The ungual region is composed of a hypoechoic nail bed mainly made up of collagen that may turn to slightly hyperechoic in the proximal region where the matrix is located. Above the nail bed, there is a hyperechoic bilaminar structure that is called the nail plate which is composed of hard keratin and has two layers: the dorsal and ventral plates separated by a hypoechoic interplate space. This virtual space turns to hyperechoic on higher frequencies (≥20 MHz). Underlying the nail bed, there is a hyperechoic line that corresponds to the bony margin of the distal phalanx. The periungual region is mostly composed of skin without fatty lobules and is divided into the lateral and proximal nail folds. However, hypodermal fatty tissue is prominent in the pulp of the fingers or toes ([16–20]; Fig. 8.2).

Hair

Scalp hair can be divided into two parts: the deep part which is the hair follicle located in the dermis, and the superficial (visible) part that makes up the hair tract. The hair follicle appears on ultrasound as an oblique hypoechoic dermal band and the scalp hair tract mostly shows as a trilaminar hyperechoic structure and ≤20 % may show a bilaminar hyperechoic pattern. The vellus hair present in the rest of the body appears as a hyperechoic monolaminar structure. The hair follicles can show different depth according to the cycle of the hair, also called the "hair clock hair." Thus, immature hair follicles (telogen phase) are located in the upper dermis and mature hair follicles (anagen phase) are located in the deep dermis close to the upper hypodermis. In between these phases, there is a catagen phase. Also, the orientation of the hair follicles may present some differences according to ethnic parameters. Hence, hair follicles tend to be more oblique in Caucasians, almost parallel in individuals with African ancestry and more vertical in persons with Asian ancestry. The blood flow of the scalp shows a centripetal pattern and mainly originates from branches of the temporal and occipital arteries. ([21–24]; Figs. 8.3 and 8.4).

Technical Considerations

Ultrasound examinations start with the visual inspection of the lesion and the regulation of the lighting in the examination room in order to assess the presence of single or multiple lesions. A copious amount of gel is applied in the lesional and perilesional region. Comparisons with the contralateral side are also performed. The sonographic test begins with a gray scale sweep that is followed by a color or power Doppler sequence and then the distribution, thickness, type, and speed of the blood vessels is determined by spectral curve analysis. The ultrasound examinations

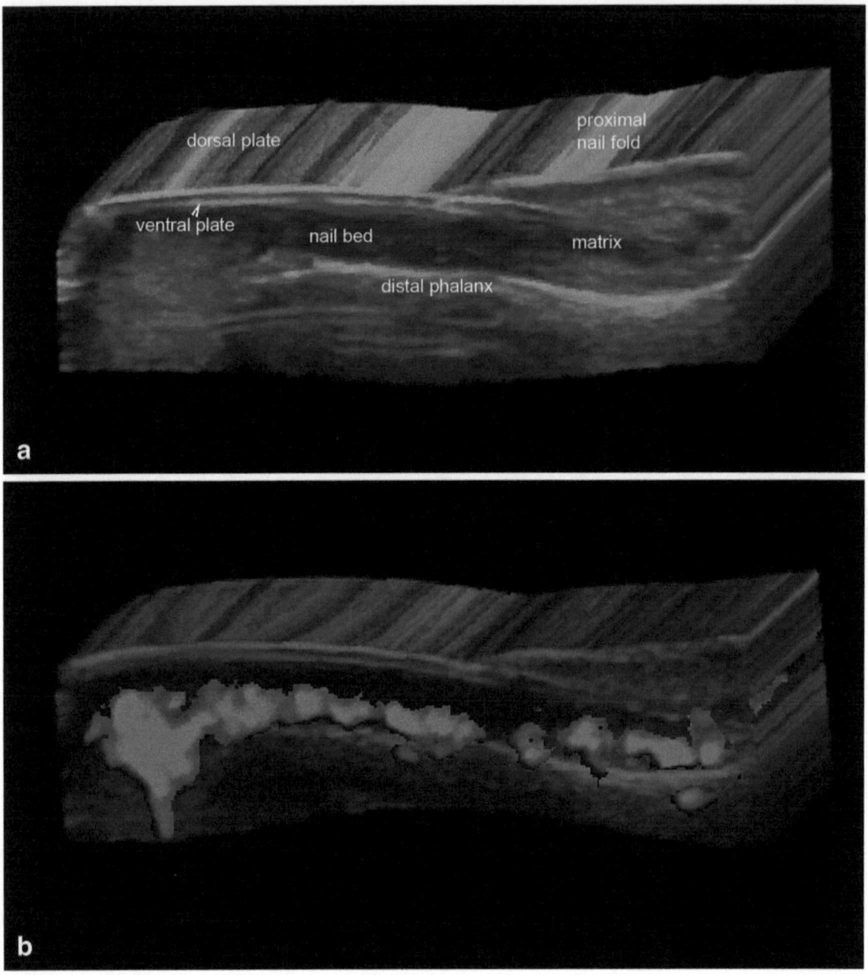

Fig. 8.2 Sonographic anatomy of the normal nail. **a**, **b** 3D reconstructions (**a**: *gray* scale and **b**: power Doppler; *longitudinal* views, index finger; *color* filter) show the parts of the nail unit (**a**) and the ungual vessels (**b**), in *color*)

should include at least two sweeps in perpendicular axes (transverse and longitudinal). Frequently, 3D reconstructions are performed in order to clearly demonstrate the pathology to the clinicians. In children ≤ 4 years old, sedation is commonly used in order to avoid movements or crying of the child that can produce artifacts on the screen. Normally, chloral hydrate or melatonin is administered 30 min before the ultrasound examination. The chloral hydrate dosage used is 50 mg/kg and the melatonin dosage varies according to age. Chloral hydrate allows deeper sedation than melatonin; therefore, melatonin is normally not used in periorificial lesions such as around the eyes or mouth because the child can easily wake up. In these

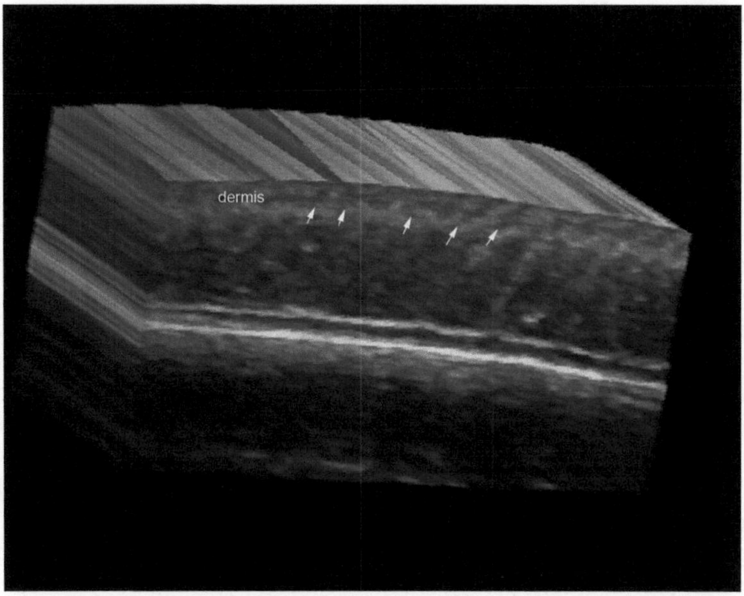

Fig. 8.3 Sonographic anatomy of the normal hair follicles (*arrows*) of the scalp

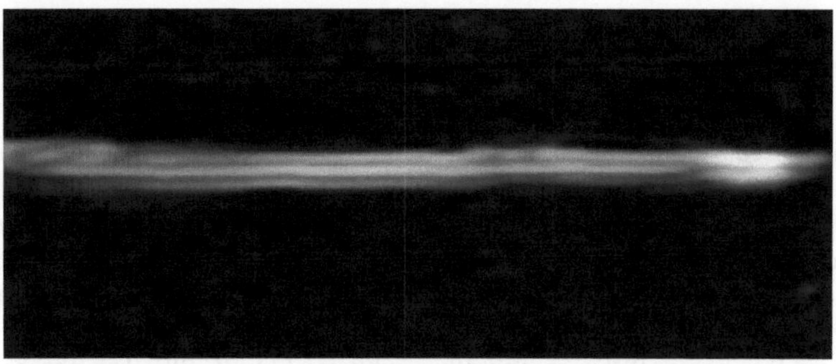

Fig. 8.4 Sonographic anatomy of a normal trilaminar hair tract of the scalp

last cases, the preferred sedative is chloral hydrate. The child is monitored using the modified Aldrete scoring and only discharged when the score is ≥ 9 points [9, 12, 13, 25–27].

Pathologies

There are several dermatologic pathologies that can be scanned on ultrasound. However for academic purposes, we have divided the rheumatologic topics into the following parts:

Inflammatory Diseases

Psoriasis: Skin and Nail Involvement

This is an autoimmune disease that affects the skin, nails, enthesis, and joints. The role of ultrasound is to support an early diagnosis and to assess activity and monitoring of the disease [28, 29]. Cutaneous psoriatic plaques appear on ultrasound as thickening and/or undulation of the epidermis with decreased echogenicity and thickening of the upper dermis. Increased blood flow is usually detected in the dermal region of the plaque during active phases. Psoriatic onychopathy presents from early to late phases with thickening (i.e., increased distance between the ventral plate and the bony margin of the distance phalanx) and decreased echogenicity of the nail bed, loss of definition of the ventral plate, hyperechoic deposits in the ventral plate, loss of definition of both plates, wavy and thickened dorsal and ventral plates. Importantly, early changes in psoriatic onychopathy may be subclinical [20]. Commonly, hypervascularity can be detected in the proximal part of the nail bed during the active stages of the disease. Decreased or heterogeneous echogenicity of the tendons (due to enthesopathy), anechoic fluid or hypertrophic synovium in the joints as well as periarticular bony erosions can also be present ([29–35]; Figs. 8.5 and 8.6).

Morphea

This is the cutaneous form of scleroderma, a connective tissue disease, and can show dermal, hypodermal, and fascial involvement. This latter fascial type is also called eosinophilic fasciitis. Morphea can go from inflammatory to atrophic stages and generate multiple plaque-type lesions; affect one or both sides of the body or produce a linear type of involvement which is frequently seen in children and can present as "en coup de sabre" (ECDS), progressive hemifacial atrophy (Parry–Romberg Syndrome, PRS), among other subtypes [34, 36–38]. The phases show different patterns on ultrasound which can be used for assessing the activity of the disease [38]. Usually during the inflammatory phase, there is thickening and decreased echogenicity of the dermis and increased echogenicity of the underlying hypodermis. In the atrophy phase, there is thinning of the dermis and hypodermis, sometimes with absence of fatty lobules, which can generate direct contact between the muscle and the dermis. In between these phases, there is an inactive phase. Frequently, in patients with multiple morphea plaques, there are asynchronic activity between different plaques or parts of the same plaque. This subclinical data could be critical for the early management of these cases [34, 38].

Increased regional blood flow and hyperechogenicity of the hypodermis have been reported as the most sensitive sonographic signs for assessing activity in

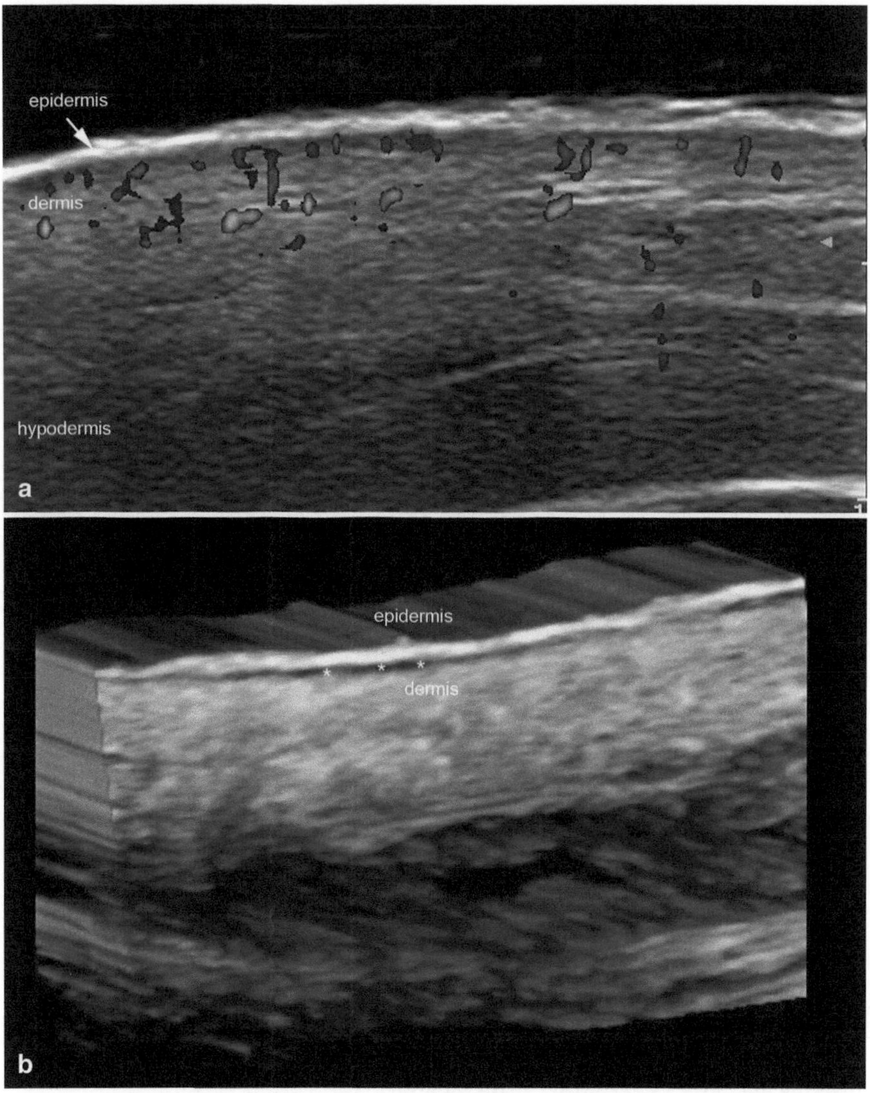

Fig. 8.5 Cutaneous psoriasis plaque. **a** Power Doppler and **b** 3D reconstruction *gray* scale. Notice the thickening of the epidermis and the hypoechoic band in the upper dermis (*). Increased vascularity (**a**, in *color*) is also shown in the dermal part of the plaque (**a**)

morphea. Also, inflammatory signs of the ipsilateral parotid gland (i.e., decreased echogenicity and/or hypervascularity) have been reported in concomitance with Parry–Romberg Syndrome [34, 38]. Upward displacement of the nail plate, increased thickness, and decreased echogenicity of the nail bed due to edema, have also been reported in scleroderma ([20]; Fig. 8.7).

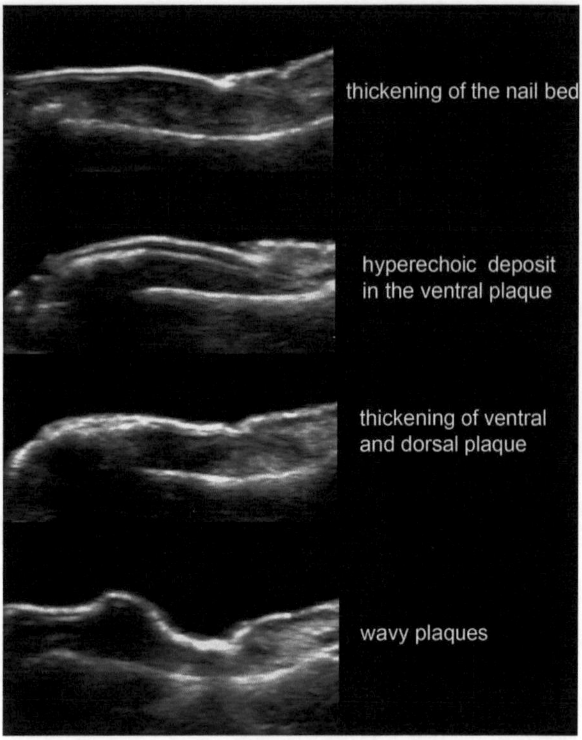

Fig. 8.6 Common sonographic alterations in psoriatic onychopathy

Dermatomyositis

This is an autoimmune disease that affects skeletal muscle, skin, and lungs. On sonography, increased echogenicity of the muscle due to inflammation has been reported. Calcinosis cutis is a common finding in this condition which can be detected on ultrasound as hyperechoic dermal or hypodermal deposits that show a posterior acoustic artifact. These calcinosis deposits may also be seen in the periungual regions at the tip of the fingers. Increased echogenicity of the hypodermis due to panniculitis may also be seen ([20, 34]; Fig. 8.8).

Cutaneous Lupus

This is the cutaneous manifestation of lupus erythematosus which may precede its systemic involvement. According to activity, cutaneous lupus has three phases: acute, subacute, and chronic, also called discoid lupus. On sonography, the acute phase shows as thickening and decreased echogenicity of the dermis with increased echogenicity of the underlying hypodermis and regional hypervascularity (Fig. 8.9). Occasionally, the dermal involvement may take a fusiform shape. In the discoid or chonic lupus, there is atrophy and hypoechogenicity of the dermis with hypovascularity. Isolated increased echogenicity of the hypodermis in the acute phase can also be observed

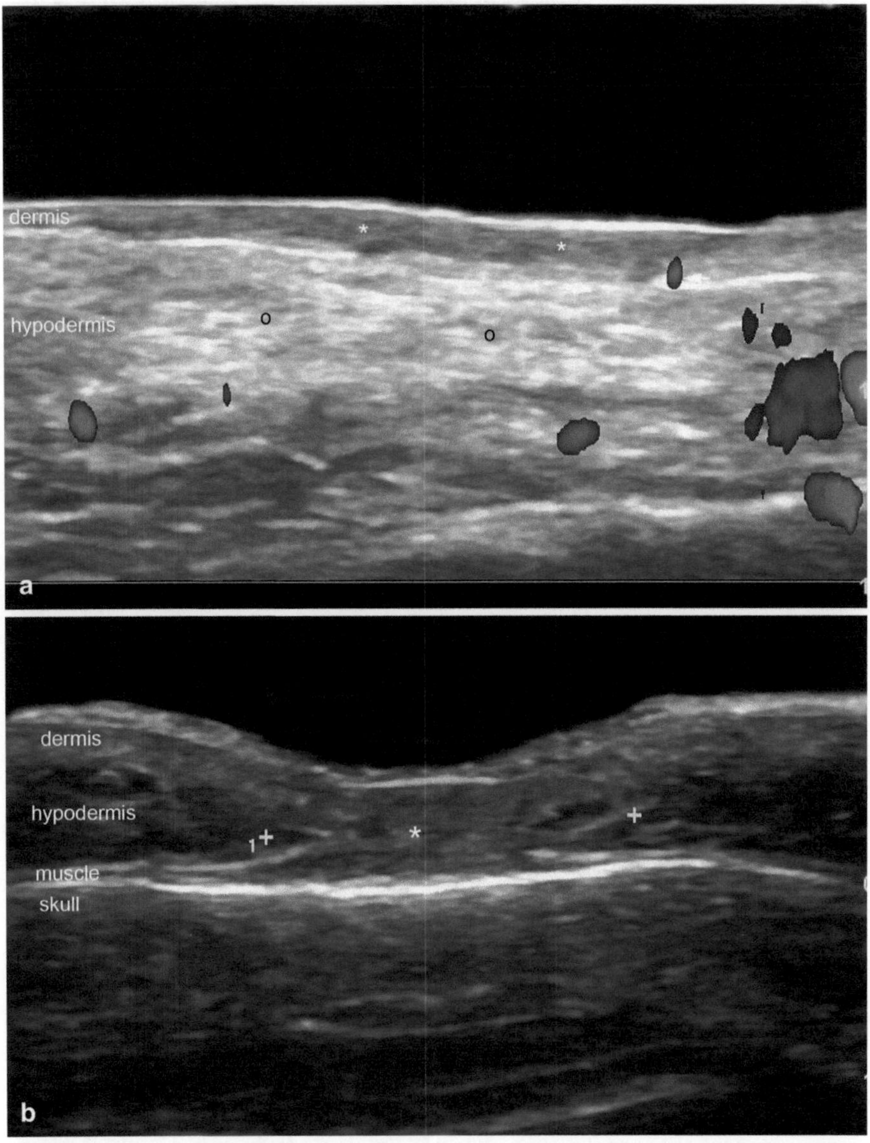

Fig. 8.7 Morphea. **a** Active phase (*color* Doppler; *transverse* view; *right* cheek) shows thickening and decreased echogenicity of the dermis (*) with increased echogenicity of the underlying hypodermis (*o*). A vessel is observed in the dermal part of the lesion (*arrow*). **b** Atrophy phase (*gray* scale; *frontal* region; *transverse* view) demonstrates lack of fatty tissue in the hypodermis (*, *between markers*) in the lesional area

which is called "lupus panniculitis" or "lupus profundus." Additionally, thrombosis and vasculitis may complicate the disease. The thrombosis appears as hypoechoic material within the vessels that show lack of blood flow. The disease may also affect the nail matrix and bed and cause inflammatory and/or atrophic changes in the ungual region with dystrophy (i.e., irregularities) of the nail plate [20, 34].

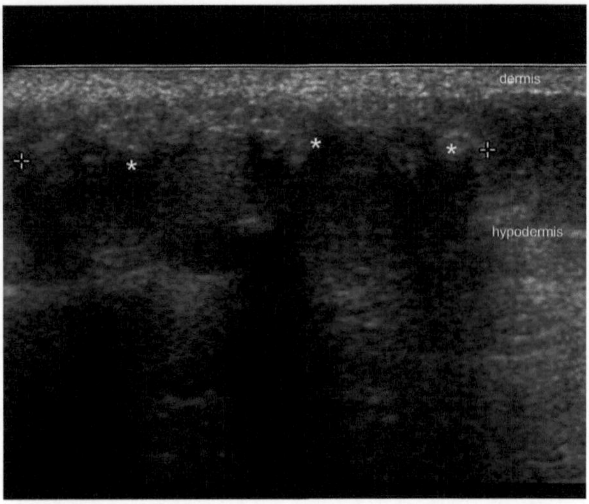

Fig. 8.8 Dermatomyositis, calcinosis. Ultrasound (*gray* scale, *transverse* view, right gluteal region) shows hyperechoic calcified deposits (*, *between markers*) in the hypodermis with posterior acoustic shadowing artifact

Rheumatoid Arthritis

This is a systemic autoimmune inflammatory disease that affects the joints in concomitance with diverse organs which include the nails. Ultrasound may support the early assessment of anatomic changes within the joints such as synovitis, periarticular bony erosions, narrowing of the joint space, tendinopathy, tear or atrophy of the tendons, among other signs. In the nail unit, rheumatoid arthritis produces thickening and decreased echogenicity of the nail bed with regional hypervascularity during the active phases of the disease. Upward displacement of the ungual plaques can also be detected due to edema ([20]; Fig. 8.10). In contrast with psoriatic onychopathy, these patients do not show prominent involvement of the nail plate [35].

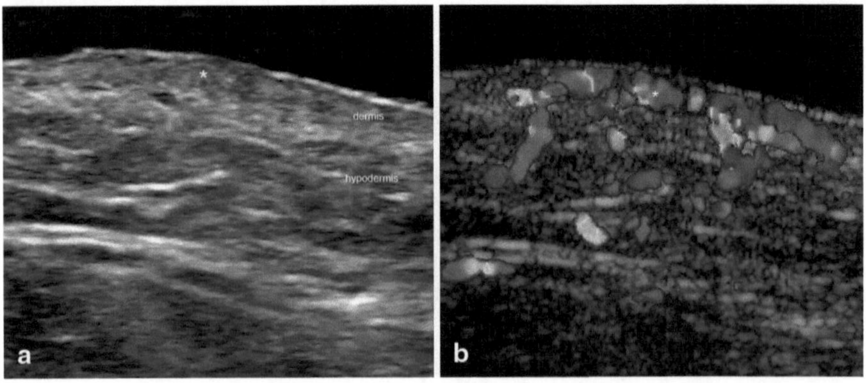

Fig. 8.9 Acute Lupus. **a** Ultrasound (*gray* scale, *transverse* view, *right* mandible region) demonstrates thickening and decreased echogenicity of the dermis in the lesional area (*). **b** Color Doppler ultrasound shows increased vascularity in the region (*)

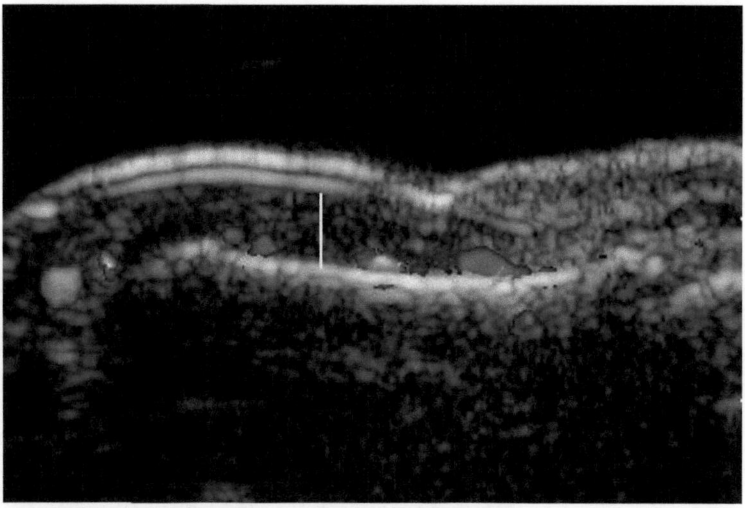

Fig. 8.10 Rheumatoid arthritis. Color Doppler ultrasound of the nail demonstrated upward displacement of the ungual plaques due to thickening of the nail bed (*vertical white* line)

Immunosuppression Complications

Rheumatologic patients are commonly immunosuppressed, which could be related to their primary diseases and/or long-term immunosuppression treatment. Thus, common secondary dermatologic problems may be detected or monitored under sonography [39, 40]. Among these entities are:

Cellulitis or Panniculitis

These are inflammations of the hypodermal layer and could have several origins such as autoimmune issues, trauma, or infection. This inflammatory process usually involves both the lobules and septa; however, it can be divided in lobular or septal panniculitis according to the principal involved part. On sonography, there is increased echogenicity of the hypodermis, commonly with hypervascularity shown on color or power Doppler. However, in the mainly septal forms of panniculitis such as erythema nodosum, there is prominent thickening and hypoechogenicity of the septa. Anechoic fluid or hypoechoic inflammatory interlobular tissue may show a “cobblestone” appearance of the hypodermis ([34]; Fig. 8.11). In the presence of fat necrosis, anechoic pseudocystic structures may be seen due to liquefaction of the fatty tissue [41, 42].

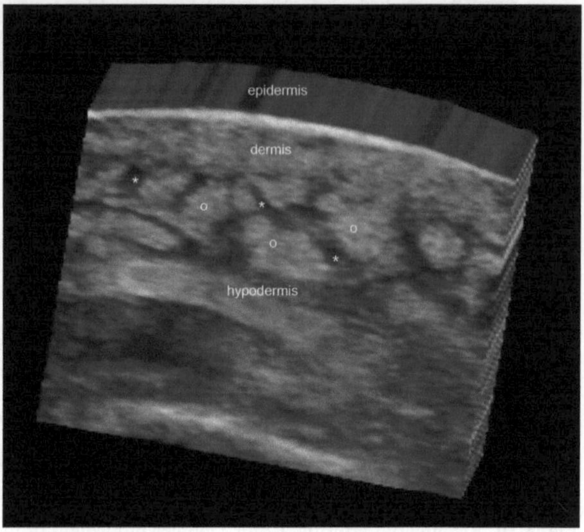

Fig. 8.11 Septal panniculitis with "cobblestone appearance." 3D reconstruction ultrasound (*gray* scale, *transverse* view, anterior aspect of the *right* leg) shows increased echogenicity and thickening of the hypodermal fatty lobules (*). Thickening and decreased echogenicity of the hypodermal septa (*) is also noted

Abscesses

These are infectious anechoic or heterogeneous fluid collections normally with echoes due to detritus commonly found in the hypodermis (Fig. 8.12). Sometimes hyperechoic septa can be found within the collections, and with color or power Doppler increased blood flow may be detected in the periphery [34].

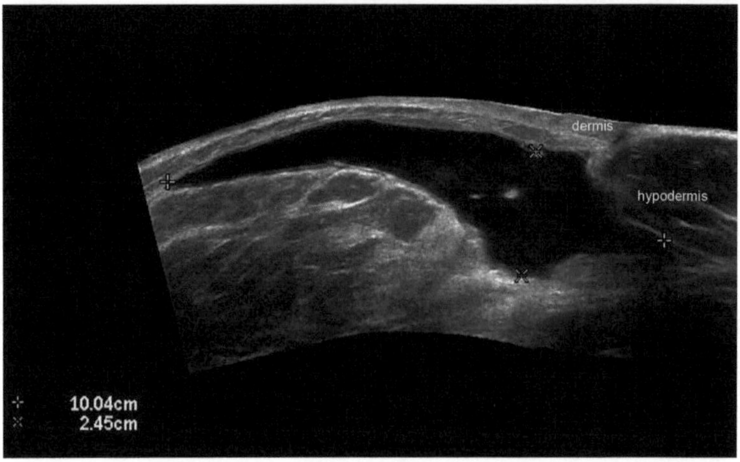

Fig. 8.12 Abscess. Ultrasound (*gray* scale, *transverse* view, inframammary region) shows 10.4 cm transverse × 2.5 cm depth, anechoic fluid hypodermal collection (*between markers*) with some hyperechoic echoes due to debris

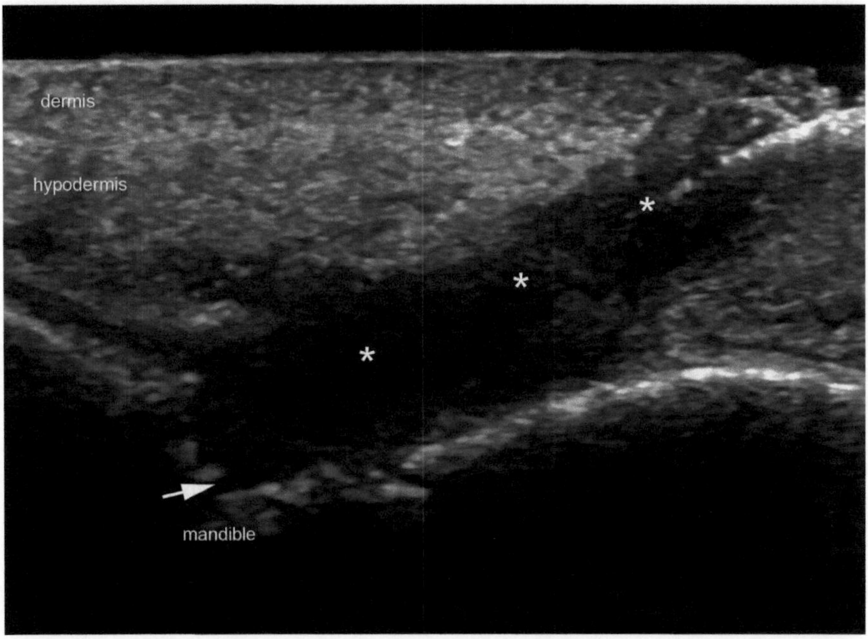

Fig. 8.13 Odontogenic fistula. Ultrasound (*gray* scale, *longitudinal* view, chin region) demonstrates hypoechoic fistulous tract (*) that connects the subepidermal region with the bony margin of the mandible. Notice the erosion of the bone (*arrow*) at the end of the fistula

Odontogenic Fistula

This is a dental inflammatory and infectious process that produces a fistulous connecting tract that goes from the bony margin of the maxilla or mandible to the skin. Clinically, these lesions can mimic other dermatologic diseases including malignant tumors such as basal cell carcinoma. On sonography, there is a hypoechoic or heterogeneous band connecting the bony margin of the upper maxilla or mandible with the skin. Usually an erosion of the bony margin is detected (Fig. 8.13). Hypervascularity may also be seen in the periphery of the tract [40, 43].

Warts

These are caused by infection with the human papillomavirus and commonly affect the plantar regions, although they can also be seen in the hands or other corporal locations. Clinically, they may mimic Morton's neuromas or foreign bodies. Sonographically, they appear as hypoechoic oval fusiform-shaped structures in the dermis with thickening of the epidermis. Frequently, there is hypervascularity with slow flow vessels in the dermal part of the wart which usually correlates well with

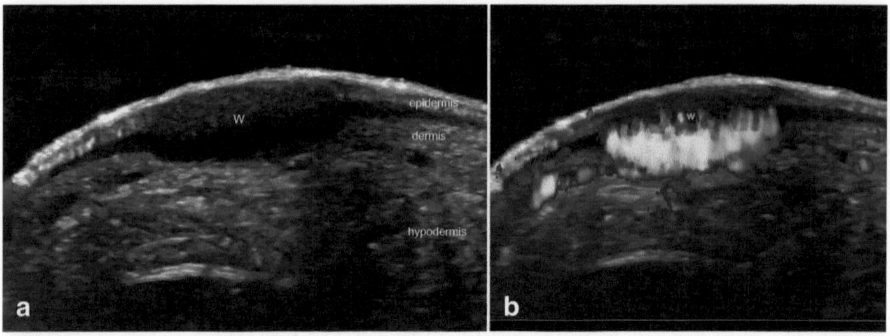

Fig. 8.14 Plantar wart (*W*). **a** Ultrasound (*gray* scale, *transverse* view) shows hypoechoic fusiform structure in the epidermal and dermal region. **b** Power Doppler demonstrates increased blood flow to the lesion (in *color*) in the dermis and the underlying hypodermis

the level of pain (Fig. 8.14). Thus, hypervascular warts tend to be very painful. Additionally, when the wart is located in the plantar region, commonly there is inflammation with anechoic fluid in the underlying bursa which may contribute to the patient's pain [44, 45].

Skin Cancer

This is the most frequent cancer in human beings and can be divided into nonmelanoma skin cancer (NMSC) and melanoma. NMSC can be separated into basal cell (BCC) and squamous cell carcinoma (SCC). BCC is the most frequent type and accounts for 75–80 % of NMSC [46]. The use of immunosuppressive therapies for example in the recipients of renal transplants seems to be related to the increase in the prevalence of skin cancer, usually nonmelanoma type [47]. This type of cancer affects areas of the body highly exposed to the sun such as the face. Even though skin cancer shows low mortality, it can cause significant disfigurement which can be an important cosmetic problem and may decrease the quality of life of patients. Ultrasound allows the detection and measurement of the tumor in all axes. It commonly shows as a hypoechoic oval shaped lesion with hyperechoic spots. Slow flow vessels may be seen within the lesion. In tumors with a facial location, the involvement of muscle or cartilage may be detected [48, 49]. Ultrasound can also support the detection of histologic subtypes of BCC with a high risk of recurrence, such as the morpheiform, sclerosing, or micronodular forms. The latter types of BCC show a higher number of hyperechoic spots within the lesions. Thus, a cut-off point of $\geq$ 7 hyperechoic spots within the BBC tumor is suggestive of a high risk of recurrence histologic subtype showing a sensitivity of 79 % and specificity of 53 % ([50]; Fig. 8.15).

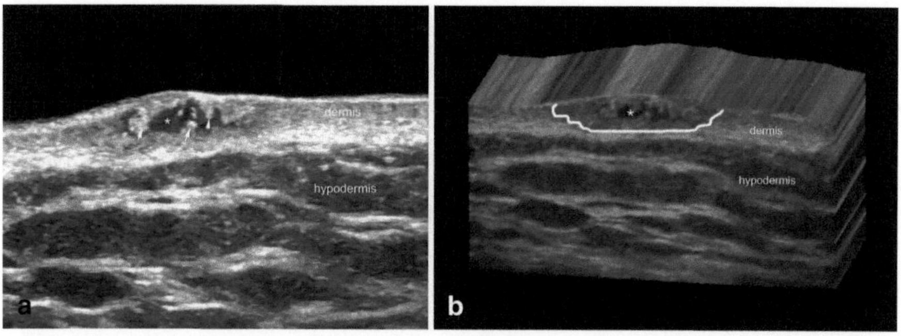

Fig. 8.15 Basal cell carcinoma of the skin (*transverse* views, posterior aspect of the *left* shoulder). **a** Ultrasound (*gray* scale) and **b** 3D reconstruction show hypoechoic dermal lesion (*, *outlined*) with irregular borders and hyperechoic spots (*arrows*)

Conclusion

Ultrasound is a powerful and first-line imaging tool for studying common dermatologic manifestations of rheumatologic disorders. It can provide subclinical detailed anatomical information that can be relevant for management.

References

1. Meyer J, Sans G, Rodallec C. Ultrasonics in dermatology. Bull Soc Fr Dermatol Syphiligr. 1951;58:266–7.
2. Alexander H, Miller DL. Determining skin thickness with pulsed ultrasound. J Invest Dermatol. 1979;72:17–9.
3. Murakami S, Miki Y. Human skin histology using high-resolution echography. J Clin Ultrasound. 1989;17:77–82.
4. El Gammal S, El Gammal C, Kaspar K, Pieck C, Altmeyer P, Vogt M, Ermert H. Sonography of the skin at 100 MHz enables in vivo visualization of stratum corneum and viable epidermis in palmar skin and psoriatic plaques. J Invest Dermatol. 1999;113:821–9.
5. Gropper CA, Stiller MJ, Shupack JL, Driller J, Rorke M, Lizzi F. Diagnostic high-resolution ultrasound in dermatology. Int J Dermatol. 1993;32:243–50.
6. Harland CC, Bamber JC, Gusterson BA, Mortimer PS. High frequency, high resolution B-scan ultrasound in the assessment of skin tumours. Br J Dermatol. 1993;128:525–32.
7. Seidenari S, Pagnoni A, Di Nardo A, Giannetti A. Echographic evaluation with image analysis of normal skin: variations according to age and sex. Skin Pharmacol. 1994;7:201–9.
8. Wortsman XC, Holm EA, Wulf HC, Jemec GB. Real-time spatial compound ultrasound imaging of skin. Skin Res Technol. 2004;10:23–31.
9. Wortsman X. Common applications of dermatologic sonography. J Ultrasound Med. 2012;31:97–111.
10. Echeverría-García B, Borbujo J, Alfageme F. The use of ultrasound imaging in dermatology. Actas Dermosifiliogr. 2014;pii:S0001–7310(14)00187–2. doi:10.1016/j.ad.2014.03.007.
11. Alfageme Roldán F. Ultrasound skin imaging. Actas Dermosifiliogr. 2014;pii:S0001–7310(14)00204-X. doi:10.1016/j.ad.2013.11.015.
12. Wortsman X. Ultrasound in dermatology: why, how, and when? Semin Ultrasound CT MR. 2013;34:177–95.

13. Wortsman X, Wortsman J. Clinical usefulness of variable-frequency ultrasound in localized lesions of the skin. J Am Acad Dermatol. 2010;62:247–56.
14. Gniadecka M. Effects of ageing on dermal echogenicity. Skin ResTechnol. 2001;7:204–7.
15. Wortsman X, Wortsman J, Carreño L, Morales C, Sazunic I, Jemec GBE. Sonographic Anatomy of the skin, appendages and adjacent structures. In: Wortsman X, Jemec GBE, editors. Dermatologic ultrasound with clinical and histologic correlations. 1st edn. New York: Springer; 2013. p. 15–35.
16. Jemec GB, Serup J. Ultrasound structure of the human nail plate. Arch Dermatol. 1989 May;125:643–6.
17. Wortsman X, Jemec GBE. Ultrasound imaging of nails. Dermatol Clin. 2006;24:323–8.
18. Singh R, Bryson D, Singh HP, Jeyapalan K, Dias JJ. High-resolution ultrasonography in assessment of nail-related disorders. Skelet Radiol. 2012;41:1251–61.
19. Thomas L, Vaudaine M, Wortsman X, Jemec GBE, Drapé JL. Imaging the nail unit. In: Baran R, de Berker D, Holzberg M, Thomas L, editors. Baran & Dawber's diseases of the nails and their management. 4th edn. West Sussex: Wiley; 2012. p. 132–53.
20. Wortsman X. Sonography of the nail. In: Wortsman X, Jemec GBE, editors. Dermatologic ultrasound with clinical and histologic correlations. 1st edn. New York: Springer; 2013. p. 419–76.
21. Wortsman X, Wortsman J, Matsuoka L, Saavedra T, Mardones F, Saavedra D, Guerrero R, Corredoira Y. Sonography in pathologies of scalp and hair. Br J Radiol. 2012;85:647–55.
22. Yagyu K, Hayashi K, Chang SC. Orientation of multi-hair follicles in non baldmen: perpendicular versus parallel. Dermatol Surg. 2006;32:651–60.
23. Wortsman X, Wortsman J. Sonography of the scalp and hair. In: Wortsman X, Jemec GBE, editors. Dermatologic ultrasound with clinical and histologic correlations. 1st edn. New York: Springer; 2013. p. 477–503.
24. Seery GE. Surgical anatomy of the scalp. Dermatol Surg. 2002;28:581–7.
25. Wortsman X. Sonography of cutaneous and ungual lumps and bumps. Ultrasound Clin. 2012. doi:10.1016/j.cult.2012.08.006.
26. Wortsman X. How to start on skin, nail and hair ultrasound: guidance and protocols. In: Wortsman X, Jemec GBE, editors. Dermatologic ultrasound with clinical and histologic correlations. 1st edn. New York: Springer; 2013. p. 597–607.
27. Aldrete JA. Modifications to the post anesthesia score for use in ambulatory surgery. J Perianesth Nurs. 1998;13:148–55.
28. Gutierrez M, Wortsman X, Filippucci E, De Angelis R, Filosa G, Grassi W. High-frequency sonography in the evaluation of psoriasis: nail and skin involvement. J Ultrasound Med. 2009;28:1569–74.
29. Gutierrez M, Filippucci E, De Angelis R, Filosa G, Kane D, Grassi W. A sonographic spectrum of psoriatic arthritis: the five targets. Clin Rheumatol. 2010;29:133–42.
30. Aydin SZ, Castillo-Gallego C, Ash ZR, Marzo-Ortega H, Emery P, Wakefield RJ, Wittmann M, McGonagle D. Ultrasonographic assessment of nail in psoriatic disease shows a link between onychopathy and distal interphalangeal joint extensor tendon enthesopathy. Dermatology 2012;225:231–5.
31. Cucoş M, Crişan M, Lenghel M, Dudea M, Croitoru R, Dudea SM. Med ultrason conventional ultrasonography and sonoelastographyin the assessment of plaque psoriasis under topical corticosteroid treatment—work in progress. Med Ultrason. 2014;16:107–13.
32. De Agustín JJ, Moragues C, De Miguel E, Möller I, Acebes C, Naredo E, Uson J, Rejon E, Mayordomo L, Garrido J. A multicentre study on high-frequency ultrasound evaluation of the skin and joints in patients with psoriatic arthritis treated with infliximab. Clin Exp Rheumatol. 2012;30:879–85.
33. Gutierrez M, De Angelis R, Bernardini ML, Filippucci E, Goteri G, Brandozzi G, Lemme G, Campanati A, Grassi W, Offidani A. Clinical, power Doppler sonography and histological assessment of the psoriatic plaque: short-term monitoring in patients treated with etanercept. Br J Dermatol. 2011;164:33–7.

34. Wortsman X, Carreño L, Morales C. Inflammatory diseases of the skin. In: Wortsman X, Jemec GBE, editors. Dermatologic ultrasound with clinical and histologic correlations. 1st edn. New York: Springer; 2013. p. 73–117.
35. Sandobal C, Carbó E, Iribas J, Roverano S, Paira S. Ultrasound nail imaging on patients with psoriasis and psoriatic arthritis compared with rheumatoid arthritis and control subjects. J Clin Rheumatol. 2014;20:21–4.
36. Li SC, Liebling MS, Haines KA. Ultrasonography is a sensitive tool for monitoring localized scleroderma. Rheumatology (Oxford) 2007;46:1316–9.
37. Li SC, Liebling MS. The use of Doppler ultrasound to evaluate lesions of localized scleroderma. Curr Rheumatol Rep. 2009;11:205–11.
38. Wortsman X, WortsmanJ, Sazunic I, Carreño L. Activity assessment in morphea using color Doppler ultrasound. J Am Acad Dermatol. 2011;65:942–8.
39. Wortsman X. The traces of sound: taking the road to skin. Curr Rheumatol Rev. 2011;7:231–8.
40. Wortsman X, Gutierrez M, Saavedra T, Honeyman J. The role of ultrasound in rheumatic skin and nail lesions: a multi-specialist approach. Clin Rheumatol. 2011;30:739–48.
41. Walsh M, Jacobson JA, Kim SM, Lucas DR, Morag Y, Fessell DP. Sonography of fat necrosis involving the extremity and torso with magnetic resonance imaging and histologic correlation. J Ultrasound Med. 2008;27:1751–7.
42. Avayú E, Rodríguez C, Wortsman X, et al. Newborn fat necrosis: case-report. Revista Chilena Pediatria. 2009; 80:60–4 (Spanish).
43. Cohen PR, Eliezri YD. Cutaneous odontogenic sinus simulating a basal cell carcinoma: case report and literature review. Plast Reconstr Surg. 1990;86:123–7.
44. Wortsman X, Sazunic I, Jemec GBE. Sonography of plantar warts: role in diagnosis and treatment. J Ultrasound Med. 2009;28:787–93.
45. Wortsman X, Jemec GBE, Sazunic I. Anatomical detection of inflammatory changes associated with plantar warts by ultrasound. Dermatology 2010;220:213–7.
46. Gniadecki R, Dam TN. Basal cell carcinoma-clinical guidelines, Danish Dermatological Society. Forum Nord Derm Ven. 2009;14:4–6.
47. Specchio F, Saraceno R, Chimenti S, Nistico S. Management of non-melanoma skin cancer in solid organ transplant recipients. Int J Immunopathol Pharmacol. 2014;27:21–4.
48. Bobadilla F, Wortsman X, Muñoz C, Segovia L, Espinoza M, Jemec GBE. Pre-surgical high resolution ultrasound of facial basal cell carcinoma: correlation with histology. Cancer Imaging. 2008;8:163–72.
49. Uhara H, Hayashi K, Koga H, Saida T. Multiple hypersonographic spots in basal cell carcinoma. Dermatol Surg. 2007;33:1215–9.
50. Wortsman X, Vergara P, Castro A, Saavedra D, Bobadilla F, Sazunic I, Zemelman V, Wortsman J. Ultrasound as predictor of histologic subtypes linked to recurrence in basal cell carcinoma of the skin. J Eur Acad Dermatol Venereol. 2014. doi:10.1111/jdv.12660.

34. Wortsman X, Carreño L, Morales C. Inflammatory diseases of the skin. In: Wortsman X, Jemec GBE, editors. Dermatologic ultrasound with clinical and histologic correlations. 1st ed. New York: Springer; 2013. p. 73–117.
35. Sandobal C, Carbó E, Iribas J, Roverano S, Paira S. Ultrasound nail imaging on patients with psoriasis and psoriatic arthritis compared with rheumatoid arthritis and control subjects. J Clin Rheumatol. 2014;20:21–4.
36. [illegible]
37. [illegible]

Chapter 9
Pediatric Musculoskeletal Disorders

Paz Collado Ramos MD, PhD and Eva Álvarez Andrés MD

Introduction

The use of musculoskeletal ultrasound (US) in pediatric rheumatology is quickly evolving, having gained significant interest long after its application in adult rheumatology [1]. US is well suited for the examination of the pediatric musculoskeletal system due to the high amount of cartilaginous skeleton which let evaluate deeper structures. Moreover, US is a friendly technique, easily acceptable bedside tool without exposure to radiation or the need of sedation [2, 3]. Although US has been widely used in the diagnosis of the infant hip disorders, the current aim for using it is to reach an early diagnosis and to improve routine care for children [4]. US helps to distinguish intra-articular from periarticular disorders. Nevertheless, due to the unique features of the growing skeleton, it should pay special attention in the interpretation of pediatric US findings. Some of the challenges in such interpretation arise from lack of available data about ultrasonographic normality for child's anatomy and standardization of US scanning. In addition, US enables the treating doctor to examine several joints at the same session lending to early and more accurate classification of patients with arthritis [5]. However, the role of US for monitoring disease and therapeutic response has not been fully defined. This chapter evaluates the spectrum of joint diseases and other unique disorders in childhood concentrating on the aspects that are unique to the immature skeleton (Table 9.1).

P. Collado Ramos (✉) · E. Álvarez Andrés
Department of Rheumatology, Paediatric Rheumatology Unit,
Hospital Universitario Severo Ochoa, Madrid, Spain
e-mail: paxcollado@yahoo.es

E. Álvarez Andrés
e-mail: evalvareza@hotmail.com

Y. El Miedany (ed.), *Musculoskeletal Ultrasonography in Rheumatic Diseases*,
DOI 10.1007/978-3-319-15723-8_9

Table 9.1 Main applications of pediatric musculoskeletal ultrasound in routine practice

Juvenile idiopathic arthritis
Septic arthritis
Other unique disorders in childhood
Transient synovitis
Legg–Perthes–Calvé disease
Slipped upper femoral epiphysis
Acute and chronic recurrent multifocal osteomyelitis
Enthesopathy and apophysitis in sport activities

Musculoskeletal Ultrasound Imaging in Children: Indications and Techniques

Screening for developmental hip dysplasia [6] was the first indication of musculoskeletal US in children. Gradually, the role of musculoskeletal US in standard clinical practice has expanded aiming at improving the care of children with musculoskeletal disorders, particularly juvenile idiopathic arthritis (JIA) [1, 7]. Both US and magnetic resonance (MR) imaging are the two most suitable tools used for the examination of pediatric patients' immature joints. However, it is important to keep in mind that these methods are complementary rather than exclusive, as they should not replace or precede clinical evaluation. From a pediatric rheumatology perspective, the principal indication of US is direct visualization of joint and tendon sheath inflammatory changes which can be assessed in few minutes, even in the absence of abnormalities on clinical examination [8, 9]. The particular advantage of US over MRI is its ability to scan multiple joints in a single session, which is of paramount importance for the current JIA classification criteria [10]. However, in contrast to its ability to assess peripheral joints, it is not applicable for axial involvement, including the temporomandibular joint [11–13]. In addition to its diagnostic role, US seems to have a promising role in monitoring disease activity and therapeutic response [14], though prognostic evaluation has not been defined yet.

Regarding technical considerations, high-resolution equipment is essential for pediatric musculoskeletal work. As in adults, the choice of transducer (high/low-frequency transducer) will depend on the type of examinations likely to be undertaken. Nevertheless, for pediatric musculoskeletal examination, high-frequency (7.5–20 MHz) linear transducers are generally equally suited in demonstrating superficial structures such as tendons, ligaments, and small joints, as well as larger or deeper joints (Fig. 9.1). Color and Power Doppler techniques provide information on the vascularization of the synovial and soft tissues reflecting vascular abnormalities and active inflammation. However, Doppler setting should be carefully considered due to the finding of some Doppler signals that can be detected at cartilaginous epiphyses as a physiological finding [15]. The size of the footprint (i.e., the surface area of the transducer in contact with the skin) is also an important

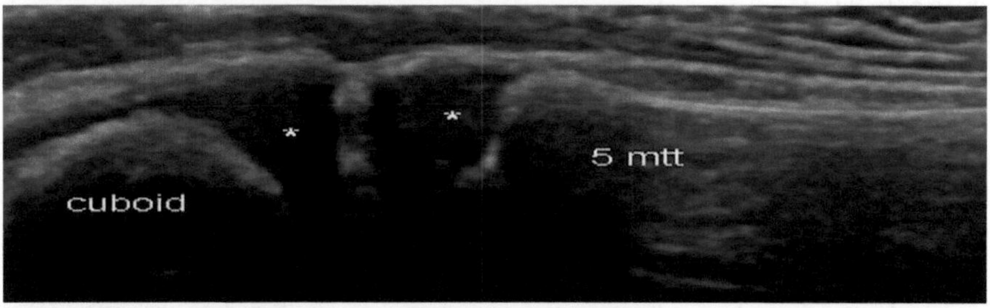

Fig. 9.1 Healthy feet in a 3-year-old child. Longitudinal US scan over the cuboid fifth metatarsal joint showing cuboid and metatarsal bones partially ossified. US depicts this small joint well-defined, epiphyseal hyaline cartilage of those bones appears as anechoic structure (*star*)

factor in examination technique. In children, transducers with a large footprint are often inadequate for child's joint as they cannot be maneuvered adequately. A transducer with a smaller footprint, such as the hockey stick transducer, allows better angulations between the small joints of the hand and the foot, reducing the risk of artifacts.

Despite US standardization has been shown in adults being essential for good inter-reader reliability [16], there is a paucity of data on the standardized use of US in pediatric rheumatology [14].

US Imaging: Appearance of the Growing Skeleton

Children differ significantly from adults with respect to skeletal anatomy and physiology. In children, the end of each long bone is composed of the metaphysis, physis (growth plate), and epiphysis. Whereas the skeleton of an adult person is completely ossified except for the articular cartilage, a varying degree of epiphyseal hyaline cartilage is present in addition to the articular cartilage in children. Since the growth cartilage contains a high amount of water; it usually appears as anechoic structure (Fig. 9.2). The US appearance of the joint cartilage in children seems thicker than in adults because the individual components of the joint cartilage (articular and growth cartilages) are initially continuous with each other. They are usually displayed as the only anechoic structure, particularly in early ages. Depending on the age of the child, the sonographer will encounter a variable view of the joint with regard to hyperechoic (cortical bone) and hypo or anechoic portions (epiphyseal cartilage) [17]. However, this also depends on skeletal maturity. Sometimes, the outline of the epiphyseal cartilage can be visualized as hyperechoic line depending on the angle of insonation. A particular feature of the growing skeleton that can also be seen on US is the presence of one or more secondary ossification centers in the epiphyses of the long bones and within the short bones. At birth, cortical bone is present in most long bone diaphyses representing the primary ossification center while many ossification centers in short bones, for example in the

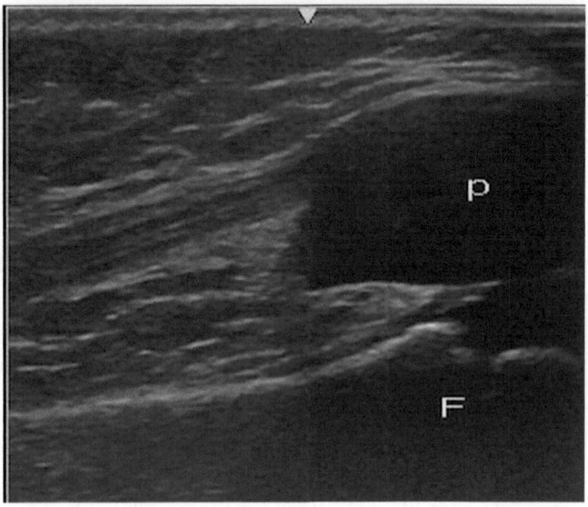

Fig. 9.2 Healthy knee in an 8-month-old child. Longitudinal US scan over the anterior surface of the patella shows the patella (*p*) as an unossified structure located anterior to the distal epiphysis of the femoral condyle (*F*). The immature skeleton usually appears as anechoic structure because the growth cartilage contains a high amount of water

wrist, are not present and hardly any epiphyseal bone is present on imaging with radiographs. Since imaging appearance of multiple secondary ossification centers may cause confusion with fragmentation observed in traction apophysitis, it is important to have a deep knowledge of the childhood anatomy. However, there is a paucity of data on the joints-specific imaging features present during growth and development in healthy children [18].

Currently, the pediatric subgroup of the Outcome Measures in Rheumatology Clinical Trials (OMERACT) Special Interest Group for musculoskeletal US has published definitions for the individual components of the healthy joint on B-mode US (or gray-scale US) that will support the interpretation of US findings in children having arthritis. The definition for the growing hyaline cartilage on B-mode describes it as a well-defined hypo/anechoic structure that is noncompressible and in which surface can (but does not have to) be detected as a hyperechoic line with evenly spread brighter echoes reflecting vascular channels.

Juvenile Idiopathic Arthritis

JIA, the most common chronic rheumatologic disease of childhood, includes a group of clinically heterogeneous arthritis of unknown etiology presenting before the age of 16 years and lasting for at least 6 weeks [10]. It relies heavily on physical examination to determine the disease activity per current validated definitions [19], although the detection of joint swelling and limited range of motion in pediatric patients varies significantly among experienced clinicians, especially in the small joints of the hand [20, 21]. US has demonstrated higher sensitivity for detecting joint involvement compared to clinical examination

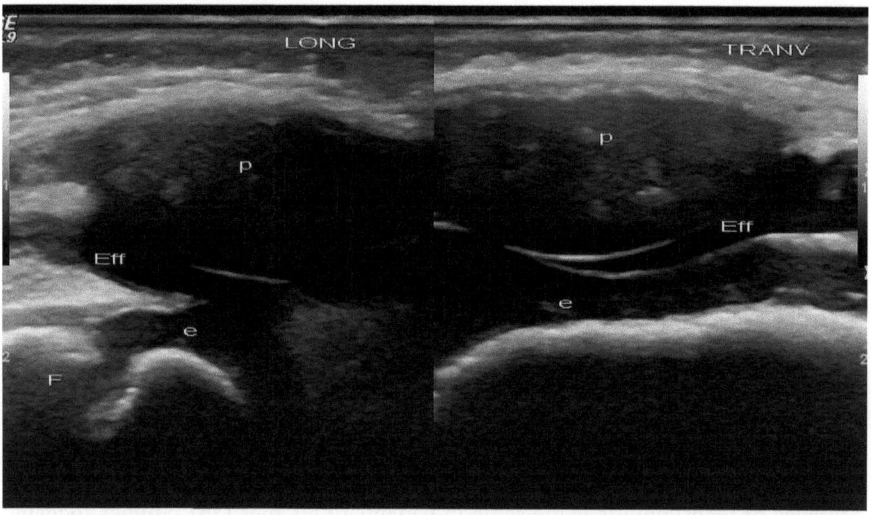

Fig. 9.3 Knee joint in an 8-month-old child. Longitudinal (*left side*) and transversal (*right side*) US images of the knee joint. Besides effusion (*Eff*) within suprapatellar bursa, effusion is visible under the unossified patella (*p*). Surfaces of the growth cartilage of patella and femur are detected as hyperechoic lines (*interface sign*) allowing to detect fluid. Dynamic examination would prove that effusion is compressible and displaceable tissue, whereas the epiphyseal cartilage (*e*) is not

[8, 22–25]. The presence of joint involvement in JIA may be expressed by ultrasonographic findings on B-mode such as synovial hypertrophy, effusion, bone erosions, and enthesitis.

Definitions of JIA pathology have varied in different US studies as a recent systematic review showed [14]. The current tendency is to apply the definitions of pathological US findings provided for adult rheumatoid arthritis (RA) by the OMERACT US Group [26]. Effusion and synovial hypertrophy can easily be imaged in peripheral joints. In children, the real-time (or dynamic) examination is the most reliable method of scanning for distinguishing synovitis from epiphyseal cartilage. Effusion is defined as compressible and displaceable, synovial hypertrophy as poorly compressible whereas the epiphyseal cartilage is not (Fig. 9.3). Although joint effusion is an early indicator of active joint disease [27], it is nonspecific. Additionally, effusion, as small as 1 mL, can be detected with US and may represent only a physiological finding in some joints [28]. It is important to keep in mind that there is no effusion-specific imaging features for different diseases or supporting a specific diagnosis. US cannot accurately differentiate whether a fluid collection is inflammatory, infectious, or hemophilic arthropaty. Aspiration of fluid, which is more successful with US guidance, remains the key for the diagnosis of joint disorder. The thin synovial membrane is undetectable under normal circumstances but it can be detected as a hypoechoic structure (relative to adjacent hypoechoic tissues) in case of synovial hypertrophy. US can be useful in monitoring disease activity and evaluating any response to therapy based on the assessment of volume and distribution of the pannus as well as synovial vasculature [29–31].

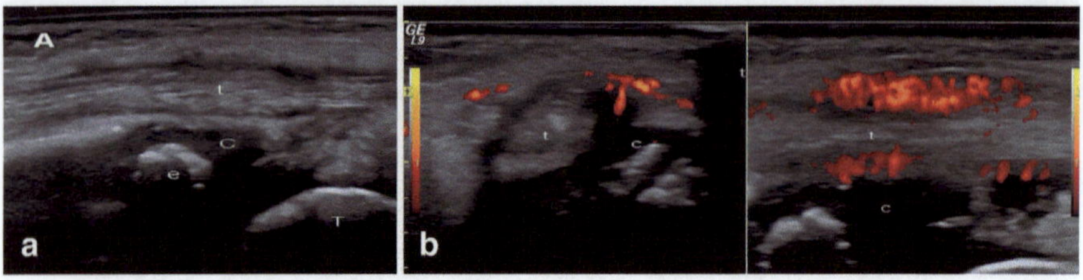

Fig. 9.4 Tenosynovitis of the posterior tibialis tendon. **a** B-mode US image: Longitudinal US image shows tenosynovitis as the presence of anechoic synovial fluid surrounding the tendon and loss of the homogeneous fibrillar pattern of the tendon (*t*). *C*: cartilage of the distal end of tibia. *e*: secondary ossification center in tibial epiphysis. **b** Power Doppler US images: Longitudinal and transversal US images showing active inflammation

Tendon sheath widening is the hallmark of tenosynovitis that can be detected by US in JIA patients, particularly in the ankle and wrist joints [32–35]. The US appearance of tenosynovitis is the same in patients of all ages [26]. Studies focusing on healthy children depicted the presence of small amount of fluid within tendon sheath as a physiological finding [28]. Color and Power Doppler techniques provide information on the vascularization of the synovial and soft tissues reflecting active inflammation and vascular abnormalities (Fig. 9.4;[31]).

Entheses inflammation is a distinctive feature of enthesitis-related arthritis (ERA), which is associated with human leukocyte antigen (HLA) B-27 and represents nearly 20 % of JIA categories [36]. Typically, involvement of the lower limb enthesitis (i.e., the plantar aponeurosis, calcaneal entheses, and distal and proximal insertions of the patellar tendon), the tarsal joints, and the small toe joints is common in ERA. Definition of enthesitis includes abnormal hypoechoic (loss of normal fibrillar architecture) and/or thickened tendon or ligament as its bone attachment (may contain hyperechoic foci consistent with calcification) seen in two perpendicular planes that may exhibit a Doppler signal and/or bony changes such as enthesophytes, erosions, or irregularities [26]. Although enthesitis may be associated with bursitis, it must be noted that a minimal amount of bursal fluid can be detected with US in healthy children (Fig. 9.5). The examiner should be aware that enthesitis ultrasonographic image in children differs from adults mainly in the appearance of bony structure on which the tendon attaches. Depending on the child's age, tendons and ligaments attach to layers of fibrocartilage and hypoechoic cartilage covering the secondary ossification center of the bone to which they insert. It is important to recognize the real tendon insertion site. For example, the distal portion of the patellar tendon lies in close apposition with the tibial epiphysis before inserting on the tuberosity. Additionally, when a growing child is examined by Doppler US, some vessels within the nonossified apophysis and epiphyseal cartilage can be detected indicating physiologic blood flow (Fig. 9.6; [37]).

US is able to show the ossification nuclei in the epiphyseal cartilage much earlier than when it becomes visible radiographically. However, US assessment of structural damage in children with JIA remains a challenge because of the peculiarities of the growing skeleton. Physis (or growth plate) showed as a discontinuity of

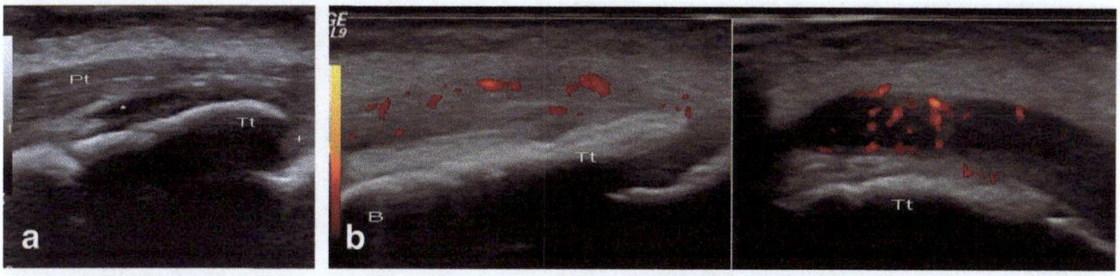

Fig. 9.5 Distal patellar tendinopathy. **a** Longitudinal B-mode US image shows distal insertion of the patellar tendon (*Pt*). Transducer should be positioned parallel to tendon insertion in order to assume a homogeneously hyperechoic appearance (avoiding anisotropy). Deep to the patellar tendon, a minimal amount of bursal anechoic fluid can be observed (*star*). Note that under the deep infrapatellar bursa, a layer of hypoechoic fibrocartilage is covering the secondary ossification center of the tibial tuberosity (*Tt*) to which the patellar tendon inserts. Apophyseal cartilage (*cross*) of the tibial tuberosity remains between tibial diaphysis and secondary ossification center. **b** Longitudinal (*left side*) and transversal (*right side*) US images reveal a focal thickened and heterogeneous hypoechoic patellar tendon insertion with Power-Doppler signal reflecting active tendinopathy

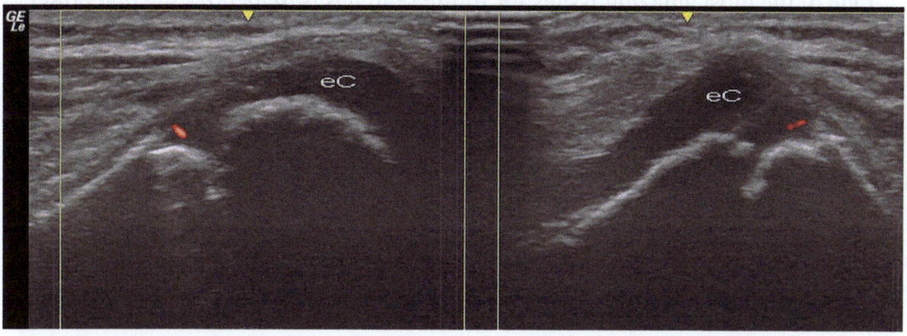

Fig. 9.6 Healthy knee in a 6-year-old child. Longitudinal (*left side*) and transversal (*right side*) US images of the knee joint. Power Doppler US scan over the anterior-lateral surface of distal femur shows normal vascularization in epiphyseal cartilage (*eC*) of the femoral condyle. As the ossification process of the distal end of the femur has not been ended, the physis appears as cortical discontinuity in both images

cortical bone in the oldest children and physiological irregularities visualized by US in some ossification centers may be misinterpreted as erosions. Nevertheless, bony erosions can be detected by US with even higher sensitivity than that of conventional radiography [38]. As joint space narrowing is secondary to erosion and loss of articular cartilage, early US assessment showing decrease in joint cartilage thickness seems crucial in JIA [39]. On the other hand, epiphyseal vascularization can become conspicuous in inflammatory states of chronic disease (Fig. 9.7; [40, 41, 34]). Nevertheless, Doppler sensitivity to inflammatory flow (low-velocity flow) could depend partly on the settings and partly on the equipment.

US is the best available imaging technique for guiding needle position within inflamed joints and tendon sheaths in local procedures [34, 35]. Fluid aspiration for analyzing is important to exclude septic arthritis [42] or hemarthrosis in routine practice [43].

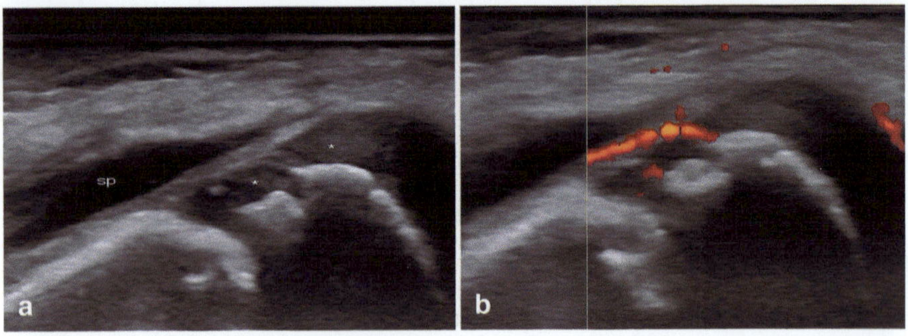

Fig. 9.7 Juvenile idiopathic arthritis in the knee joint. **a** Longitudinal B-mode US image shows the presence of suprapatellar bursitis (*sp*), erosion and decreased homogeneous hypoechogenicity of the femoral epiphyseal cartilage (*stars*). **b** Longitudinal Power Doppler US image pathologic vascularization involving the synovial tissue and epiphyseal cartilage close to the erosion. *e* secondary ossification center in epiphysis

One of the most amazing issues for future searches is to identify the potential US role in the detection of subclinical disease in the pediatric population. US has recently been used to confirm the disease activity status and detect patients in disease remission. Earlier studies demonstrated that a significant number of JIA patients in clinical remission, have abnormal US findings, particularly in the wrist and ankle joints [44, 7]. A recent paper showed the superiority of US for detecting subclinical activity through the presence of positive PD signal in patients having inactive disease on and off medication [45]. The clinical relevance of those observations detected by US remains a critical issue that need to be addressed.

Septic Arthritis

Septic arthritis is infection of a joint which can occur either by direct spread of contiguous osteomyelitis, hematogenous dissemination from a remote site, or direct inoculation due to penetrating trauma. The most common presentation pattern of septic arthritis is monoarticular. In children younger than 3 years old, the hip is commonly affected, though the knee and ankle joints are also often involved. On the other hand, in neonates, due to the metaphyseal vascularization the association of osteomyelitis and septic arthritis is particularly more frequent. Clinical findings along with erythrocyte sedimentation rate greater than 40 mm/h, and white blood cell (WBC) count greater than 12,000/mm^3 are strong predictors of an infected joint. Irritability and refusing to be breast fed are more common symptoms in neonates.

The main role of US in septic arthritis is the detection of joint effusion in a child with suspicion of infection. US is a handy tool for guiding pus joint aspiration [42]; It relieves intracapsular pressure, decreases damage to the articular surfaces, and helps to guide antibiotic treatment [46]. Although septic effusion often contains

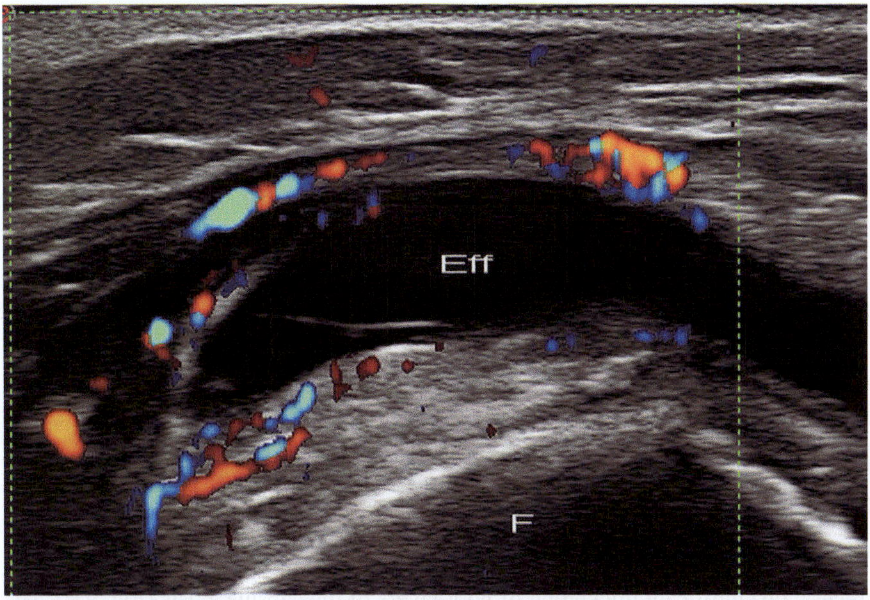

Fig. 9.8 Septic arthritis. Longitudinal color Doppler US image of the knee shows suprapatellar bursitis with synovial thickness and hyperemia reflecting acute arthritis

bright echoes and debris [47], US cannot be reliably used to differentiate septic arthritis from noninfectious arthritis based on the size and echogenicity of the effusion. On the other hand, an anechoic collection resulting from infected joint fluid is quite rare. Usually Doppler signal may be detected surrounding the joint at early disease (Fig. 9.8), although the absence of US features does not exclude it [48, 49].

Transient Synovitis

In the field of pediatric rheumatology, US has been used in the initial workup of the hip disease in children having limited or no weight-bearing capability [50]. Transient synovitis (TS or toxic synovitis) represents the most common cause of acute nontraumatic painful hip in children between 5 and 10 years of age [51]. US allows early detection of effusion as the plausible cause. Managed with only supportive measures, such as avoidance of weight bearing and anti-inflammatory medication, it usually resolves in 3–10 days with no long-term consequences. US is also used for monitoring and follow-up of children with longer or repeated episodes of TS in order to rule out early Perthes disease.

The differential diagnosis of TS includes infection (septic arthritis, osteomyelitis [52], Legg–Calvé–Perthes disease, slipped capital femoral epiphysis (SCFE), JIA, or tumors [53]).

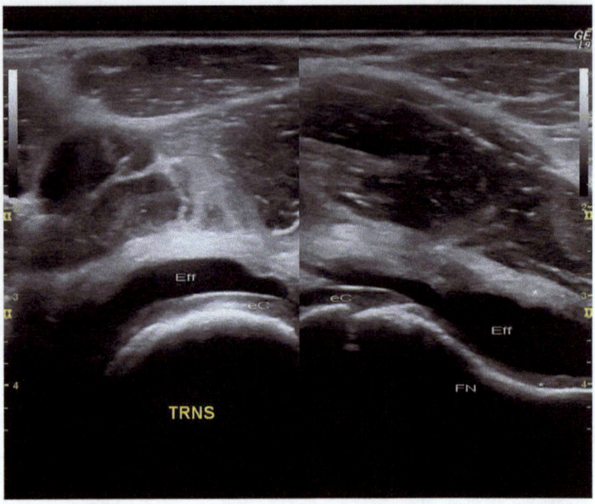

Fig. 9.9 Transient synovitis. Transverse and longitudinal B-mode US images of the hip joint show distended anterior recess due to the presence of anechoic effusion (*Eff*) without thickening of the hip capsule layers associated (*stars*). The surface of the hypoechoic cartilage of the femoral head is clearly visible as a strong reflection (*interference sign*) between the effusion and the epiphyseal cartilage (*eC*)

In TS, effusion is identified as the only cause of distention of the anterior joint recess; no significant thickening of both layers of the capsule exists (Fig. 9.9). The anterior-sagittal approach clearly demonstrates joint effusion within the anterior recess of the hip capsule. The joint capsule is normally concave anteriorly and close to the femoral neck. A minimum of about 1 mm^3 of fluid may be detected intracapsular, allowing differentiation of both layers of the capsule [54]. On B-mode ultrasonographic image, an effusion is diagnosed when the joint capsule bows anteriorly in addition to: (1) distension more than 5.2 mm with fluid measured from the middle of the femoral neck to the capsular outer margin [28]; or (2) when the distension of the capsule is 2 mm or more than the contralateral asymptomatic side. Often the asymptomatic contralateral joint offers the best standard for comparison. Hip aspiration should be recommended to patients who pose a diagnostic problem, such as a high level of suspicion for septic arthritis or a chronic rheumatologic disorder [42, 55]. Unlike JIA and infectious diseases, Doppler signal is not visualized in the hip capsule.

Legg–Perthes–Calvé Disease

Legg–Perthes–Calvé disease (LPCD) is an idiopathic osteonecrosis of the immature capital femoral epiphysis. It occurs in children between 2 and 14 years of age, with a peak incidence at 5–6 years old, male predominance, and mostly unilateral involvement. Although the causes of LPCD disease are unknown, there is considerable evidence that the disease might have an underlying vascular mechanism [56, 57]. LPCD may also occur as a result of recurrent TS in patients who have long-lasting effusion and symptoms [58]. Clinical symptoms include limp and pain in the hip, thigh, or knee.

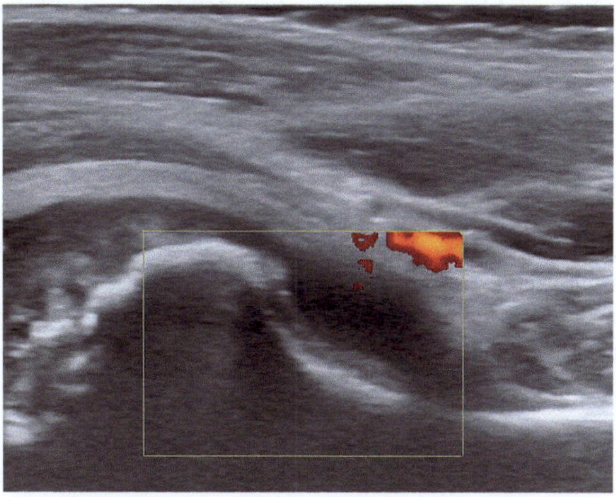

Fig. 9.10 Legg–Perthes–Calvé disease. Longitudinal US scan of the hip joint shows hypoechoic effusion (*Eff*). The hyperechoic surface of the epiphyseal cartilage of the femoral head is not visible due to the disease. Note irregularities in femoral head contour because of fragmentation of epiphysis and deformity. Doppler was applied towards the anterior and posterior layers of capsule in order to identify active synovial membrane but only a vessel is detected outside of capsule

In early LPCD, the radiography may be normal or show collapse of the femoral head [59]. Though there are no specific US features, hip effusion can be visualized in early disease. The permanence of effusion or irregular echogenicity may help to suspect the disease. After the ischemic phase, US can show irregularities in femoral head contour because of fragmentation of epiphysis, and deformity ([60]; Fig. 9.10). Using Doppler technique, Doria [57] has described some changes in the intraosseous/deep transphyseal vascularity of the proximal femur in children with LPCD.

Slipped Capital Femoral Epiphysis

SCFE is a well-described cause of adolescent hip pathology. SCFE is an idiopathic disorder—even though a trauma was reported in 50 % of cases—characterized by a displacement of the capital femoral epiphysis from the femoral neck before fusion of the physis. It should be sought in obese adolescents [61], mainly boys, presenting with unilateral limp and pain. Contralateral hip may be affected later, turning it to bilateral involvement. Overall, the prognosis with conservative treatment is very good and surgery procedures are indicated for high-risk patients to avoid permanent deformity and avascular necrosis.

Though the initial concept was that US may not be the tool of choice for the diagnosis of SCFE, this was challenged by the results of several small studies [61–63] published earlier. These data were supported by the results of a recent study [64] reporting that US may be more sensitive than radiography for the diagnosis of

SCFE. Joint effusion, which often accompanies SCFE, was readily recognized and quantified by US. US also permitted recognition of the initial SCFE and assessment of the slip degree by measuring the height of the physeal step [65]. In addition to its role in diagnosis, US proved to play an important role in the follow-up of SCFE patients. In a prospective study of patients with SCFE, serial US evaluation seems to be more sensitive than radiography in showing epiphyseal displacement and reduction [66]; while joint effusion represented physeal instability, recent progression suggests that SCFEs should be operatively fixed. On the other hand, remodeling was reported as a sign of chronicity. In contrast, MRI easily allows the detection of early widening of the physis, which represents a high risk of displacement of the epiphysis.

Acute and Chronic Recurrent Multifocal Osteomyelitis

Osteomyelitis is defined as an infection of the bone, bone marrow, and surrounding soft tissue. In childhood, the most common route of infection is hematogenous spread of a microorganism. In growing bone, the diaphysis and metaphysis share the same nutrient arteries and veins that lead to the formation of so-called sinusoidal lakes in the metaphysis [67]. These sinusoidal lakes make favorable the implantation of microorganism, thus leading to a focus of osteomyelitis.

In contrast to neonates, who are more prone to acute osteomyelitis and where acute hematogenous osteomyelitis primarily arises in the metaphysis, in children younger than 18 months of age, transphyseal vessels allow vertical spread of infection from the metaphysis through the physis into the epiphysis. On the other hand, after 18 months of age, these transphyseal vessels disappear and the physis acts as a natural border preventing the spread of osteomyelitis towards the epiphysis. Secondary spread by contiguity and direct spread of infection or after a trauma are seen less often in children. The most common site of infection is the long bones, especially the femur and tibia. Most infections are mono-ostotic, but polyostotic involvement is reported in up to 6.8 % infants, whereas its prevalence was 22 % in neonates [68]. Clinical presentation can be diverse and therefore confusing (pain, fever, swelling, tenderness, and reluctance to move limbs). Acute hematogenous osteomyelitis (AHOM) is defined as the presence of complaints for fewer than 14 days, whereas the subacute form persists longer than 14 days. The most common organism causing AHOM is *Staphylococcus aureus* but other pathogens were identified [69]. Ideally, antibiotic treatment is based on the isolation of the pathogen from the focus of infection or blood.

In osteomyelitis, conventional radiography is the initial modality of choice to evaluate osseous changes and is valuable to rule out other disease conditions [70]. US can provide useful information for an early diagnosis and it allows comparing the affected side with the opposite side even though it has limitations in evaluating bone structures (Fig. 9.11; [71]). In acute osteomyelitis, US has a high sensitivity for the detection of early abnormalities in the soft tissues overlying the bone and fluid collections just 1–2 days after the onset of symptoms and before any radiographic

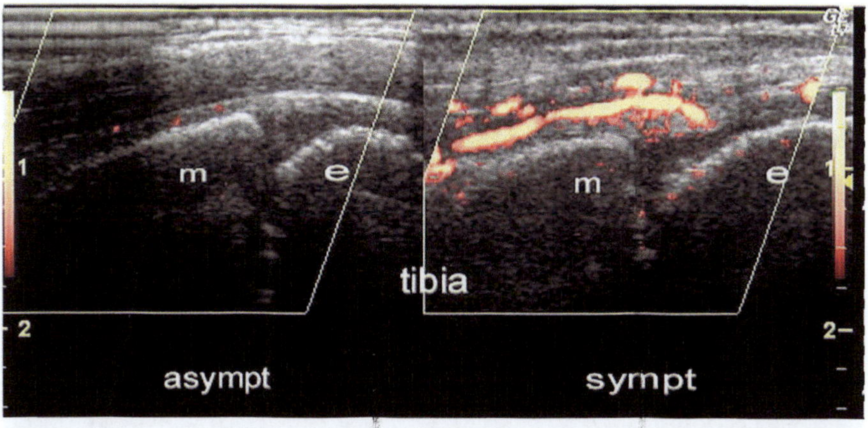

Fig. 9.11 Acute osteomyelitis at the distal end of tibia. Longitudinal power Doppler US of the symptomatic and asymptomatic ankles. The power Doppler signal was asymmetrically increased over the metaphysis and epiphysis of the distal left tibia (*symptomatic*). In children younger than 18 months of age, transphyseal vessels allow vertical spread of infection from the metaphysis, through the physis into the epiphysis and joint space. *m* metaphysis, *e* epiphysis, *asympt* asymptomatic, *sympt* symptomatic

sign is manifest. [72]. As infection progresses, US can also identify subperiostal abscesses (Fig. 9.12), fistulas, and sinus track formation [72, 73]. Furthermore, it is possible to perform US-guided direct intervention, such as aspiration or drainage of an abscess, or joint effusion. During the course of the disease, color and power Doppler imaging can reveal a hypervascular pattern within and around the infected periosteum: this finding can be used as a parameter for monitoring the progression or regression of inflammation and to assess the response to antibiotic therapy [73]. Although MR imaging is the most sensitive and specific imaging modality for evaluating bone and extraosseous changes in osteomyelitis, scanning more than once to improve the detection rate is restricted by the need for sedation.

Chronic recurrent multifocal osteomyelitis, a disease that primarily occurs in childhood, is associated with the so-called SAPHO syndrome (Synovitis, Acne, Pustolosis, Hyperostosis, Osteitis). It is characterized by a variety of subacute and chronic foci of osteomyelitis with the absence of causative microorganism detectable from specimens, septic features, and abscesses [74]. Besides guiding aspiration, US can be used to identify polyarticular involvement.

Enthesopathy and Apophysitis in Sport Activities

The fibrillar appearance of tendons and ligaments is similar in all ages, but differs in its attachment to bone. Apophyses are secondary growth centers that serve as tendinous attachment sites. Pediatric athletes involved in sport activities suffer from increasing forces acting on the "muscle–tendon–bone" unit. Unlike adults, the

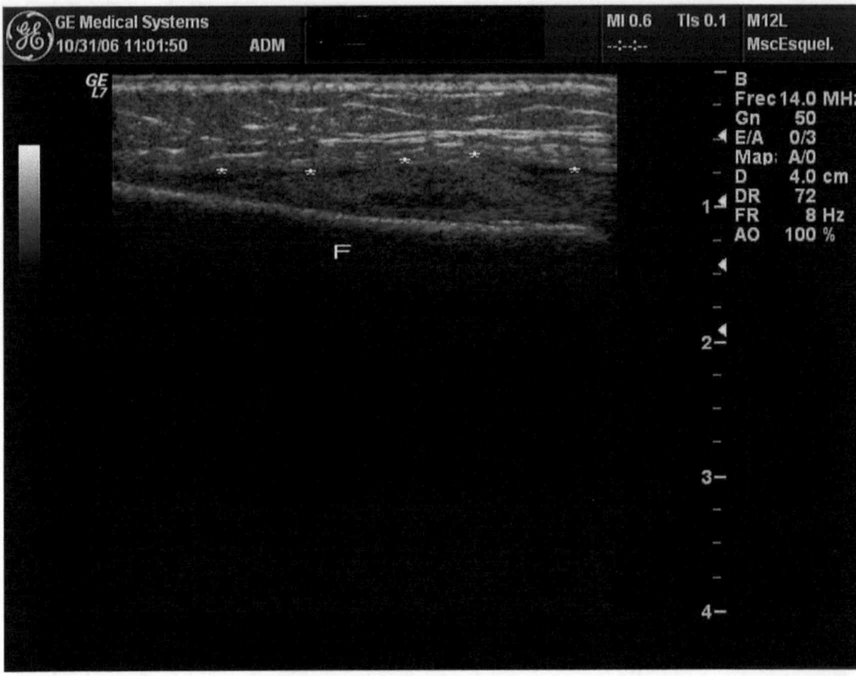

Fig. 9.12 Abscess in osteomyelitis. Longitudinal B-mode US image of distal leg over the anterior surface of the metaphysis of femur (*F*) shows swelling of the soft tissues with nonhomogeneous echogenicity overlying the affected bone (between *stars*). Note small fluid collections into the abscess and a thin layer of subperiosteal fluid causing elevation of the periosteum (*arrow*). US allows to perform guided aspiration or drainage of abscess lending to accuracy in antibiotic therapy

weakest point of this unit in children is the attachment of the tendon to the immature bone (i.e., fibrocartilage and hypoechoic cartilage covering the secondary ossification center of bone). Unossified apophyses are prone to both chronic traction injury (apophysitis) and acute avulsion fractures.

Chronic repetitive microtrauma secondary to overload could lead to osseous or cartilage fragmentation, as well as local inflammation (e.g., Osgood–Schlatter disease that results from repetitive traumatic traction on the tibial apophysis by the patellar tendon). Not only US is able to detect the disease pathologic feature but it also enables the treating doctor to monitor the Osgood–Schlatter disease course. This varies from focal hypoechoic swelling of the physeal cartilage, fragmentation of the ossification center (i.e., small calcified fragments and irregularities in the bony outlines), patellar tendon lesions (i.e., hypoechoic appearance), and reactive bursitis of the deep or superficial tibial patellar bursae (Fig. 9.13; [75, 76]). In the acute phases, local hyperemia can be demonstrated with color and power Doppler imaging [76]. In Sinding-Larsen–Johansson syndrome, US findings are the same as those in Osgood–Schlatter disease but located at proximal insertion of the patellar tendon.

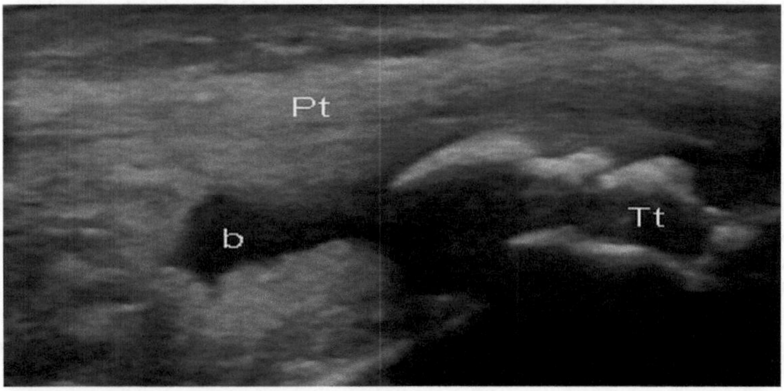

Fig. 9.13 Osgood–Schalatter disease. Longitudinal B-mode US image of the knee shows distal patellar (*Pt*) tendinosis, fragmentation of the ossification center (small calcified fragments and irregularities in the bony outlines) of the tibial tuberosity (*Tt*), and deep infrapatellar bursitis (*b*)

In avulsion injuries (apophyseal separation), the insertion of the tendon is torn away from its origin. It usually results from violent muscle contraction. US image shows a thickened tendon associated with elevation and fragmentation of the involved bone cortex forming two hyperechoic layers instead of one. This appearance (the "double cortical sign") indicates minimally displaced cortical avulsion. On radiographs, acute apophyseal avulsion is diagnosed when there is displacement of the apophysis from its expected position. US comparison with the opposite side will be useful at early disease if no abnormalities are observed. MRI has the advantage to show marrow edema within the apophysis [77].

References

1. Magni-Manzoni S, Collado P, Jousse-Joulin S, Naredo E, D'Agostino MA Muratore V, Merli P, Roth J, (on behalf of the) Paediatric ultrasound group of the OMERACT Ultrasound Task Force Current state of musculoskeletal ultrasound in paediatric rheumatology: results of an international survey. Rheumatology 2014;53:491–6.
2. Buchmann RF, Jaramillo D. Imaging of articular disorders in children. Radiol Clin North Am. 2004;42:151–68.
3. Azouz EM. Juvenile idiopathic arthritis: how can the radiologist help the clinician? Pediatr Radiol. 2008;38(Suppl 3):S403–8.
4. Collado P, Naredo E. Ecografía osteoarticular en reumatología pediátrica. Ann Pediatr Contin. 2007a;5:366–8.
5. Collado P, Naredo E, Calvo C, Gamir ML, Calvo I, García ML, Merino R, Graña J, Bustabab S, Garrido J; ECO-JIA Study Group. Reduced joint assessment vs comprehensive assessment for ultrasound detection of synovitis in juvenile idiopathic arthritis. Rheumatology. 2013;52:1477–84.
6. Graf R. Fundamentals of sonographic diagnosis of infant hip dysplasia. J Pediatr Orthop. 1984;4:735–40.
7. Magni-Manzoni S, Scirè CA, Ravelli A, Klersy C, Rossi S, Muratore V, Visconti C, Lanni S, Merli P, Montecucco C. Ultrasound-detected synovial abnormalities are frequent in clinically inactive juvenile idiopathic arthritis, but do not predict a flare of synovitis. Ann Rheum Dis. 2013;72:223–8.

8. Magni-Manzoni S, Epis O, Ravelli A, Klersy C, Veisconti C, Lanni S, Muratore V, Sciré CA, Rossi S, Montecucco C. Comparison of clinical versus ultrasound-determined synovitis in juvenile idiopathic arthritis. Arthritis Rheum. 2009;61:1497–504.
9. Haslam KE, McCann LJ, Wyatt S, Wakefield RJ. The detection of subclinical synovitis by ultrasound in oligoarticular juvenile idiopathic arthritis: a pilot study. Rheumatology. 2010;49:123–7.
10. Petty RE, Southwood TR, Manners P, Baum J, Glass DN, Goldenberg J, He X, Maldonado-Cocco J, Orozco-Alcala J, Prieur AM, Suarez-Almazor ME, Woo P. International league of associations for rheumatology classification of juvenile idiopathic arthritis: second revision, Edmonton 2001. J Rheumatol. 2004;31:390–2.
11. Müller L, Kellenberger CJ, Cannizzaro E, Ettlin D, Schraner T, Bolt IB, Peltomäki T, Saurenmann RK. Early diagnosis of temporomandibular joint involvement in juvenile idiopathic arthritis: a pilot study comparing clinical examination and ultrasound to magnetic resonance imaging. Rheumatology. 2009;48:680–5.
12. Melchiorre D, Falcini F, Kaloudi O, Bandinelli F, Nacci F, Matucci Cerinic M. Sonographic evaluation of the temporomandibular joints in juvenile idiopathic arthritis. J Ultrasound. 2010;13:34–7.
13. Assaf AT, Kahl-Nieke B, Feddersen J, Habermann C. Is high-resolution ultrasonography suitable for the detection of temporomandibular joint involvement in children with juvenile idiopathic arthritis? Dentomaxillofac Radiol. 2013;42(3):20110379.
14. Collado P, Jousse-Joulin S, Alcalde M, Naredo E, D'Agostino MA. Is ultrasound a validated imaging tool for the diagnosis and management of synovitis in juvenile idiopathic arthritis? a systematic literature review. Arthritis Care Res. 2012;64:1011–9.
15. Bearcroft PW, Berman LH, Robinson AH, Butler GJ. Vascularity of the neonatal femoral head: in vivo demonstration with power Doppler US. Radiology. 1996;200:209–11.
16. Naredo E, Wakefield RJ, Iagnocco A, Terslev L, Filippucci E, Gandjbakhch F, Aegerter P, Aydin S, Backhaus M, Balint PV, Bruyn GA, Collado P, Finzel S, Freeston JE, Gutierrez M, Joshua F, Jousse-Joulin S, Kane D, Keen HI, Moller I, Mandl P, Ohrndorf S, Pineda C, Schmidt WA, Szkudlarek M, Conaghan PG, D'Agostino MA. The OMERACT ultrasound task force-status and perspectives. J Rheumatol. 2011;38:2063–7.
17. Collado P, Naredo E. Sonographic image of children's joints. Badalona: EUROMEDICE, Ediciones Médicas; 2007b.
18. Khanna PC, Thapa MM. The growing skeleton: MR imaging appearances of developing cartilage. Radiol Clin Am. 2009;47:899–909.
19. Wallace CA, Ruperto N, Giannini E. Preliminary criteria for cinical remission for select categories of juvenile idiopathic arthritis. J Rheumatol. 2004;31:2290–4.
20. Guzman J, Burgos-Vargas R, Duarte-Salazar C, Gomez-Mora P. Reliability of the articular examination in children with juvenile rheumatoid arthritis: interobserver agreement and sources of disagreement. J Rheumatol. 1995;22:2331–6.
21. Goff I, Rowan A, Bateman BJ, Foster HE. Poor sensitivity of musculoskeletal history in children. Arch Dis Child. 2012;97:644–6.
22. Filippou G, Cantarini L, Bertoldi I, Picerno V, Frediani B, Galeazzi M. Ultrasonography vs clinical examination in children with suspected arthritis. Does it make sense to use poliarticular ultrasonographic screening? Clin Exp Rheumatol. 2011;29: 345–50.
23. Breton S, Jousse-Joulin S, Cangemi C, de Parscau L, Colin D, Bressolette L, Saraux A, Devauchelle-Pensec V. Comparison of clinical and ultrasonographic evaluations for peripheral synovitis in juvenile idiopathic arthritis. Semin Arthritis Rheum. 2011;41:272–8.
24. Jousse-Joulin S, Breton S, Cangemi C, Fenoll B, Bressolette L, de Parscau L, Saraux A, Devauchelle-Pensec V. Ultrasonography for detecting enthesitis in juvenile idiopathic arthritis. Arthritis Care Res. 2011;63:849–55.
25. Janow GL, Panghaal V, Trinh A, Badger D, Levin T, Ilowite N. Detection of active disease in juvenile idiopathic arthritis: sensitivity and specificity of the physical examination vs ultrasound. J Rheumatol. 2011;38:2671–4.

26. Wakefield RJ, Balint PV, Szkudlarek M. on behalf of the OMERACT Special Interest Group. The OMERACT 7 Special Interest Group: diagnosed musculoskeletal ultrasound including definitions for ultrasonographic pathology. J Rheumatol. 2005;32:2485–7.
27. Frosch M, Foell D, Ganser G, Roth J. Arthrosonography of hip and knees joints in the follow up of juvenile rheumatoid arthritis. Ann Rheum Dis. 2003;62:242–4.
28. Collado P, Naredo E, Calvo C, Crespo M. Assessment of the joint recesses and tendon sheaths in healthy children by high-resolution B-mode and power Doppler sonography. Clin Exp Rheumatol. 2007;25:915–21.
29. Lamer S, Sebag G. MRI and ultrasound in children with juvenile chronic arthritis. Eur J Radiol. 2000;33:85–93.
30. El-Miedany YM, Housny IH, Mansour HM, Mourad HG, Mehanna AM, Megeed MA. Ultrasound versus MRI in the evaluation of juvenile idiopathic arthritis of the knee. Joint Bone Spine. 2001;68:222–30.
31. Doria AS, Kiss MHB, Lotito AP, Molnar LJ, de Castro CC, Medeiros CC, Cerri GG. Juvenile rheumatoid arthritis of the knee: evaluation with contrast-enhanced color Doppler ultrasound. Pediatr Radiol. 2001;31:524–31.
32. Rooney M, Mcallister C, Burns JF. Ankle disease in juvenile idiopathic arthritis: ultrasound findings in clinically swollen ankles. J Rheumatol. 2009;36:1725–9.
33. Pascoli L, Wright S, Mcallister C, Rooney M. Prospective evaluation of clinical and ultrasound findings in ankle disease in juvenile idiopathic arthritis: importance of ankle ultrasound. J Rheumatol. 2010;37:2409–14.
34. Laurell L, Court-Payen M, Nielsen S, Zak M, Boesen M, Fasth A. Ultrasonography and color Doppler in juvenile idiopathic arthritis: diagnosis and follow-up of ultrasound- guided steroid injection in the ankle region. A descriptive interventional study. Pediatr Rheumatol Online J. 2011;9:4.
35. Laurell L, Court-Payen M, Nielsen S, Zak M, Fasth A. Ultrasonography and color Doppler in juvenile idiopathic arthritis: diagnosis and follow-up of ultrasound-guided steroid injection in the wrist region. A descriptive interventional study. Pediatr Rheumatol Online J. 2012;10:11.
36. Fisher C, Ioannou Y, Hall-Craggs M, Sen D. Enthesitis related arthritis; a new era of understanding. Ann Paediatr Rheum. 2012;1:8–16.
37. Grechenig W, Mayr J, Peicha G, Hammerl R, Schatz B, Grechenig S. Sonoanatomy of the Achilles tendon insertion in children. J Clin Ultrasound. 2004;32:338–43.
38. Malattia C, Damasio MB, Magnaguagno F, Pistorio A, Valle M, Martinoli C, Viola S, Buoncompagni A, Loy A, Ravelli A, TomÃ P, Martini A. Magnetic resonance imaging, ultrasonography, and conventional radiography in the assessment of bone erosions in juvenile idiopathic arthritis. Arthritis Rheum. 2008;59:1764–7.
39. Pradsgaard DØ, Spannow AH, Heuck C, Herlin T. Decreased cartilage thickness in juvenile idiopathic arthritis assessed by ultrasonography. J Reumatol. 2013;40:1596–603.
40. Karmazyn B, Bowyer SL, Schidt KM, Ballinger SH Buckwalter K, Beam T. US findings of metacarpophalangeal joints in children with idiopathic juvenile arthritis, Pediatr Radiol. 2007;37:475–82.
41. Daldrup-Link HE, Steinbach L. MR imaging of pediatric arthritis. Magn Reson Imaging Clin N Am. 2009;17:451–67.
42. Zawin JA, Hoffer FA, R and FF, Teele RL. Joint effusion in children with an irritable hip: US diagnosis and aspiration. Radiology. 1993;187:459–63.
43. Zukotynski K, Jarrin J, Babyn S, Carcao M, Pazmino-Canizares J, Stain AM, Doria AS. Ultrasonography of joints and correlation with function in hemophilic arthropathy in children and young adults: interim results of a German clinical pilot trial. Haemophilia. 2007;13:293–304.
44. Rebollo-Polo M, Koujok K, Weisser C, Jurencak R, Bruns A, Roth J. Ultrasound findings on patients with juvenile idiopathic arthritis in clinical remission. Arthritis Care Res. 2011;63:1013–9.
45. Collado P, Gamir ML, López-Robledillo JC, Merino R, Modesto C, Monteagudo I. Detection of synovitis by ultrasonography in clinically inactive juvenile idiopathic arthritis on and off medication. Clin Exp Rheumatol. 2014;32:597–603.

46. Jaramillo D, Treves SD, Kasser JR, Harper M, Sundel R, Laor T. Osteomyelitis and septic arthritis in children: appropriate use of imaging to guide treatment. AJR Am J Roentgenol 1995;165:399–403.
47. Chau CLF, Griffith JF. Musculoskeletal infections: ultrasound appearances. Clin Radiol. 2005;60:149–59.
48. Strouse PJ, DiPietro MA, Adler RS. Pediatric hip effusions: evaluation with power Doppler sonography. Radiology. 1998;206:731–5.
49. Gordon JE, Huang M, Dobbs M, Luhmann SJ, Szymanski DA, Schoenecker PL. Causes of false-negative ultrasound scans in the diagnosis of septic arthritis of the hip in children. J Pediatr Orthop. 2002;22:312–6.
50. Crow A, Cheung A, Lam A, Ho E. Sonography for the investigation of a child with a limp. AJUM. 2010;13:23–30.
51. Do TT. Transient synovitis as a cause of painful limps in children. Curr Opin Pediatr. 2000;12:48–51.
52. Luhmann SJ, Jones A, Schootman M, Gordon JE, Schoenecker PL, Luhmann JD. Differentiation between septic arthritis and transient synovitis of the hip in children with clinical prediction algorithms. J Bone Joint Surg Am. 2004;86:956–62.
53. Vlychou M, Athanasou NA. Radiological and pathological diagnosis of paediatric bone tumours and tumour-like lesions. Pathology. 2008;40:196–216.
54. Robben SG, Lequin MH, Diepstraten AF, den Hollander JC, Entius CA, Meradji M. Anterior joint capsule of the normal hip and in children with transient synovitis: US study with anatomic and histologic correlation. Radiology. 1999;210:499–507.
55. Liberman B, Herman A, Schindler A, Sherr-Lurie N, Ganel A, Givon U. The value of hip aspiration in pediatric transient synovitis. J Pediatr Orthop. 2013;33:124–7.
56. Aksoy MC, Aksoy DY, Haznedaroglu IC, Sayinalp N, Kirazli S, Alpaslan M. Thrombomodulin and GFC levels in Legg-Calve-Perthes disease. Hematology. 2008;13:324–8.
57. Doria AS, Cunha FG, Modena M, Maciel R, Molnar LJ, Luzo C, Moineddin R, Guarniero R. Legg-Calvé-Perthes disease: multipositional power Doppler sonography of the proximal femoral vascularity. Pediatr Radiol. 2008;38:392–402.
58. Bhargava S, Bhargava SK. Infective arthritis of hip: role of sonography. JIMSA. 2013;26:15–6.
59. Dimeglio A, Canavese F. Imaging in Legg-Calvé-Perthes disease. Orthop Clin North Am. 2011;42:297–302.
60. Murray AW, Wilson NIL. Changing incidence of slipped capital femoral epiphysis a relationship with obesity? J Bone Joint Surg Br. 2008;90-B(1):92–4.
61. Kallio PE, Lequesne GW, Paterson DC, Foster BK. Ultrasonography in slipped capital femoral epiphysis. Diagnosis and assessment of severity. J Bone Joint Surg Br. 1991;73:884–9.
62. Terjesen T. Ultrasonography for diagnosis of slipped capital femoral epiphysis. Comparison with radiography in 9 cases. Acta Orthop Scand. 1992;63:653–7.
63. Magnano GM, Lucigrai G, De Filippi C, et al. Diagnostic imaging of the early slipped capital femoral epiphysis. Radiol Med. 1998;95:16–20.
64. McCabe A. Is ultrasound or plain film radiography a more sensitive diagnostic modality for diagnosing slipped capital femoral epiphysis? Emerg Med J. 2014;31:80.
65. Castriota-Scanderbeg A, Orsi E. Slipped capital femoral epiphysis: ultrasonographic findings. Skeletal Radiol. 1993;22:191–3.
66. Kallio PE, Paterson DC, Foster BK, Lequesne GW. Classification in slipped capital femoral epiphysis. Sonographic assessment of stability and remodeling. Clin Orthop Relat Res. 1993;294:196–203.
67. Jaramillo D, Villegas-Medina O, Doty D, Rivas R, Strife K, Dwek JR. Age-related vascular changes in epiphysis, physis and metaphysis: normal findings on gadolinium-enhanced MRI of piglets. AJR. 2004;182:353–60.
68. Riise OR, Kirkhus E, Handeland KS, Flato B, Reiseter T, Cvancarova M, Nakstad B, Wathne KO. Childhood osteomyelitis-incidence and differentiation from other acute onset musculoskeletal features in a population-based study. BMC Pediatr. 2008;8:45.
69. Sreenivas T, Nataraj AR, Menon J, Patro DK. Acute multifocal haematogenous osteomyelitis in children. J Child Orthop. 2011;5:231–5.

70. Van Schuppen J, van Doorn MAC, van Rijn RR. Childhood osteomyelitis: imaging characteristics. Insights Imaging. 2012;3:519–33.
71. Riebel E, Nash R, Nazarenko O. The value of sonography in the detection of osteomyelitis. Pediatric Radiol. 1996;26:291–7.
72. Zamzam MM. The role of ultrasound in differentiating septic arthritis from transient synovitis of the hip in children. J Pediatr Orthop B. 2006;15:418–22.
73. Collado P, Naredo E, Calvo C, Crespo M. The role of power Doppler sonography in the early diagnosis of osteomyelitis in children. J Clin Ultrasound. 2008;36:251–3.
74. Tlougan BE, Podjasek JO, O'Haver J, Cordova KB, Nguyen XH, Tee R, Pinckard-Hansen KC, Hansen RC. Chronic recurrent multifocal osteomyelitis (CRMO) and synovitis, acne, pustulosis, hyperostosis, and osteitis (SAPHO) syndrome with associated neutrophilic dermatoses: a report of seven cases and review of the literature. Pediatr Dermatol. 2009;26:497–505.
75. Blankstein A, Cohen I, Heim M, Diamant L, Salai M, Chechick A, Ganel A. Ultrasonography as diagnostic modality in Osgood-Schlatter disease. A clinical study and review of the literature. Arch Orthop Trauma Surg. 2001;121:536–9.
76. Carr JC, Hanly S, Griffin J, Gibney R. Sonography of the patellar tendon and adjacent structures in pediatric and adult patients. AJR. 2001;176:1535–9.
77. Pisacano RM, Miller TT. Comparing sonography with MR imaging of apophyseal injuries of the pelvis in four boys. AJR. 2003;181:223–30.

Chapter 10
Carpal Tunnel Syndrome

Yasser El Miedany MD, FRCP

Introduction

The use of ultrasonography (US) has changed the traditional concept of idiopathic carpal tunnel syndrome (CTS), caused by median nerve entrapment in the carpal tunnel (CT), to a miniature closed compartment syndrome with its own causes and pathophysiology. The ability of US to depict normal and pathologic median nerve morphology, abnormalities of the tunnel wall, as well as its contents has had a positive impact not only on using US as a diagnostic tool but also as a management guide and outcome predictor. US has been compared to nerve conduction study testing, and was recommended as a new alternative diagnostic modality for the evaluation of CTS [1, 2]. Furthermore, the development of high-resolution US transducers gave US several advantages over MRI, including being relatively fast, inexpensive, dynamic, and allowing blood flow imaging with relatively little additional time [3, 4]. To explain the findings seen on US scanning of the CT, it is important to start with a brief review of the applied anatomy and pathophysiology. This chapter then discusses US of the CT, tips to enhance it reproducibility, its use for diagnosis as well as management including an algorithm for patient diagnosis and the use of US as a biomarker for tailoring a treatment approach for CTS patients.

Y. El Miedany (✉)
Department of Rheumatology, Darent Valley Hospital, Dartford, UK
e-mail: drelmiedany@rheumatology4u.com

Rheumatology and Rehabilitation Department, Ain Shams University, Cairo, Egypt

Y. El Miedany (ed.), *Musculoskeletal Ultrasonography in Rheumatic Diseases*,
DOI 10.1007/978-3-319-15723-8_10

Applied Anatomy of the Carpal Tunnel

The Carpal Tunnel

Although the CT appears as an open passageway communicating the flexor compartment of the forearm proximally and the mid-palmar space of the hand distally, it functions as a closed compartment and maintains its own tissue fluid pressure levels [3]. Anatomically, the CT is an inelastic fibro-osseous tunnel on the palmar aspect of wrist extending from distal volar wrist crease to the mid-palm (Fig. 10.1). The osseous component (carpal bones) together with the palmar radio-carpal ligament and palmar ligament complex between the carpal bones forms its floor and sides. The arch formed by the carpal bones is defined by four bony prominences: the tubercle of scaphoid and trapezium radially as well as the pisiform and hook of hamate medially. The tunnel roof is formed broadly by the flexor retinaculum or transverse carpal ligament (1–3 mm thick and 3–4 cm wide). However, the flexor retinaculum is formed by three contiguous distinct components. The proximal portion is a direct continuation of the deep investing fascia of the forearm (antebrachial fascia). Distally, the transverse carpal ligament represents the central portion of the flexor retinaculum. Most distally is an aponeurosis between the thenar and hypothenar muscles [4]. The anatomic zone of the flexor retinaculum extends from the distal radius to the proximal metaphysis of the third metacarpal. The distal volar flexion crease crosses proximal end of scaphoid and pisiform and marks proximal edge of transverse carpal ligament. Longitudinally, the CT is widest at its proximal and distal margins (25 mm), whereas it gets narrower midway approximately at the hook of the hamate (about 2–2.5 cm distal to its most proximal margin) with a mean width of 20 mm. The transverse carpal ligament is thickest at the level of its attachment to the hook of the hamate and the tubercle of the trapezium, which is also the narrowest region of the CT.

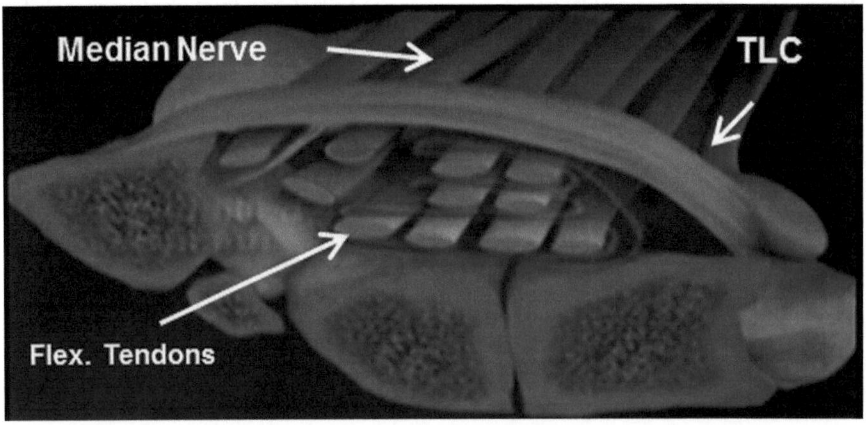

Fig. 10.1 Anatomy of the carpal tunnel: the proximal level of the carpal tunnel, delineated by the pisiform and the scaphoid carpal bones. The transverse carpal ligament forms the carpal tunnel roof. The median nerve and flexor tendons (surrounded by their tendon sheaths) pass through the tunnel, with the median nerve lying palmar and radial, *TLC* Transverse carpal ligament

The principle contents of the CT include: (1) the median nerve, typically accompanied by the highly variable arteria mediana and (2) nine extrinsic flexor tendons (four tendons of flexor digitorum superficialis and four tendons of flexor digitorum profundus in common flexor sheath as well as one tendon and sheath of flexor pollicis longus). The muscles of the nine extrinsic flexor tendons which traverse the CT originate from the medial epicondyle of the humerus and the anterior aspect of the radius, ulna, and interosseous membrane. The musculotendinous junctions are found proximal to the proximal edge of the CT. In the tunnel flexor digitorum superficialis, tendons of middle and ring finger are superficial than that of index and little fingers, whereas the median nerve lies most superficially, just beneath the flexor retinaculum passing between the tendons of flexor digitorum superficialis ulnarly and flexor carpi radialis radially, dorsal or dorso-radial to the palmaris longus tendon. The tendons are surrounded by mesodermal tissue, which provides vincular blood supply to the tendons as well as extratendinous lubrication and nutrition. The mesodermal tissue is composed of a continuous layer of mesoderm, forming invaginated loops around the individual tendons. The source of the blood supply to the tendinous vincula is the anterior interosseous artery [5]. Just above this layer of mesodermal tissue is the fibrous layer of the anterior wrist joint capsule which is continuous with the periosteum of the carpal bones and the transverse carpal ligament.

Recently, lumbrical muscle incursion into the CT has also been implicated as a possible cause for CTS. Besides the margins and the anterior surface, the origin of the lumbricals also extends on the posterior surface of flexor digitorum profundus tendons. In a study carried out by Joshi et al. [6], they found that the proximal attachment of the second lumbrical was extending into the CT compared to the rest of the lumbrical muscles whereas in some cases the first lumbrical was found in continuation with the fleshy fibers of flexor digitorum profundus. Siegel et al. [7] have described the origin of lumbricals being significantly proximal in patients with CTS, whereas, Cobb et al. [8] have described that lumbrical muscle incursion into the CT during finger flexion is a normal occurrence. The incursion of the lubmrical muscles into the CT varies according to the hand position. All the lumbricals extend proximally into the CT when passive finger flexion was performed whereas they extend only up to the intermediate/distal parts of the CT in extended position of fingers. Siegel et al. [7] have stated that in those performing repetitive hand motions, lumbricals had significantly larger and proximal origins in the CT which can be a contributory factor in the causation of CTS. Cobb et al. [9] described "first test" to ascertain the CTS due to lumbricals muscle incursion in the CT. According to this test, if a person is asked to keep the hand in a sustained fist position for 45 s it would result in numbness in the area of distribution of median nerve.

Median Nerve

The median nerve is composed of branches from the C5 through T1 spinal cord nerve roots. It enters the anterior compartment of the forearm via the cubital fossa. It travels between the flexor digitorum superficialis and profundus muscle bellies, often within the deep epimysium of the flexor digitorum superficialis. Further distal in the forearm, the median nerve becomes more superficial, passing between the tendons of flexor digitorum superficialis ulnarly and flexor carpi radialis radially, dorsal or dorsoradial to the palmaris longus tendon. The palmar cutaneous branch of the median nerve arises proximal (approximately 5 cm proximal to the wrist crease) to, and does not pass through, the CT; thus, sensation in the central palm remains unaffected in CTS patients. It parallels the median nerve for 1.6–2.5 cm and then courses separately under the antebrachial fascia between the tendons of palmaris longus and flexor carpi radialis [10]. After exiting the CT, the median nerve divides into six terminal branches. The recurrent motor branch supplies innervation to the thenar muscles (flexor pollicis brevis, abductor pollicis brevis, and opponens pollicis, in addition to the first and second lumbrical muscles); three proper digital nerves, including the radial and ulnar proper digital nerves to the thumb and the radial proper digital nerve to the index finger; as well as two common digital nerves emerge from the median nerve. The point of departure of the recurrent motor branch from the median nerve may vary in its relation with the distal edge of the transverse carpal ligament. Frequently (46 %), the recurrent motor branch passes in a retrograde fashion into the thenar musculature, whereas in 31 %, the recurrent motor branch diverges from the median nerve deep to the transverse carpal ligament, within the confines of the CT, and passes around the distal edge of the transverse carpal ligament to enter the thenar musculature. Less frequently (23 %), the recurrent motor branch diverges from the median nerve within the limits of the CT and appears to perforate the ligament in its course to the thenar musculatures [11]. The fascicles ultimately destined for the recurrent motor branch have been found in 60 % of dissections to arise from the extreme radial aspect of the median nerve, in 22 % from the central–anterior aspect, and in the remaining 18 % between the extreme radial-anterior and the central aspect of the median nerve. The variation in the location of the motor branch of the median nerve with respect to the other fascicles may predispose to thenar muscle wasting in CTSs with unequal distribution of demyelination and degeneration between fascicles.

Vascular Anatomy

The vascular anatomy of the median nerve shows that the median artery of the forearm is an accessory artery arising from the ulnar artery at the proximal forearm, which courses alongside the median nerve to the CT. Along with the interosseous artery, the median artery is the main route of blood supply to the hand in embryos. After the eighth week of gestation, it regresses and is replaced by the ulnar and the radial arteries. Pecket et al. [12] described three different variations of median nerve vascularization. The first one (70 %) entails radial and ulnar arteries converging into the superficial and deep palmar arches, with the median nerve supplied by the superficial palmar arch and an anastomosis between the two arteries as well as by the forearm muscular branches. The second type (10 %) originates from the radial and ulnar arteries and a median artery that branches out directly from the brachial artery or from the bifurcation of the latter into radial and ulnar arteries. The median artery runs superficial to the median nerve, and at the palm it separates into the finger branches which supply the second, third, and fourth fingers; with no evidence of either the superficial or deep palmar arches. The third type (20 %) shows the median artery associated with the superficial palmar arch. In case the median artery persisted into adult life, it runs on its ulnar side in normal median nerve, whereas, when associated with a bifid median nerve, the artery courses between the two nerve trunks.

Studying the intrinsic vascular anatomical pattern of the median nerve [13] revealed that the rich surface epineurial capillary plexus is distinct from the peri-fascicular capillary plexus, and is fed and drained, both by the arteries and veins of the surface epineurium and by the underlying interfascicular vessels. This arrangement is not peculiar to the median nerve, for it has been demonstrated also in the antebrachial ulnar nerve. The deep venous drainage of the surface "epineurium" may serve to prevent the venous engorgement resulting from a relatively minor degree of external compression of a segment of the nerve trunk. The characteristic finding of relatively large arteries in the interfascicular connective tissue at the distal end of the median nerve appears of interest since Denny-Brown and Brenner [14] have shown that the "compartmented structure" of a peripheral nerve trunk may serve to protect intraneural vessels from the effects of external compression. The architecture of the median nerve in approximately its distal 3 in. is extremely fascicular, and expectedly these large interfascicular vessels would be highly resistant to the effects of external compression. This would explain the US power Doppler (PD) findings which are discussed later in this chapter.

High division of the median nerve proximal to the CT (known as a bifid median nerve) is a median nerve anomaly with an incidence rate of 2.8 % [15]. This rare anomaly is often associated with various clinically relevant abnormalities, such as vascular malformations (persistent median artery), [16] aberrant muscles [17], and CTS. The two parts of the nerve may be equal in size [18], or predominance may exist of either the radial or ulnar part [19]. Larger radial trunks approximately equal in size to a normal median nerve [16], however, and larger ulnar trunks were also reported [19]. When pathology of the median vessels exists, the bifidity may be missed if the small ulnar trunk is surrounded by the abnormal median vessels. Accessory lumbrical muscles passing through the bifid nerve, and bifid median nerves in which the radial part passes through a separate compartment of the CT [20, 21] were also reported.

Pathophysiology

In many cases, the onset of CTS is insidious and progressive. In these instances, the etiology is usually illusive and consequently labeled as "idiopathic." A growing body of evidence indicates that the common pathway for the CTS development is increased pressure within the carpal canal. Experimental studies suggested that the changes in the CTS are linked to the amount and duration of the interstitial fluid pressure and may be reversible up to a point with management [22]. Therefore,

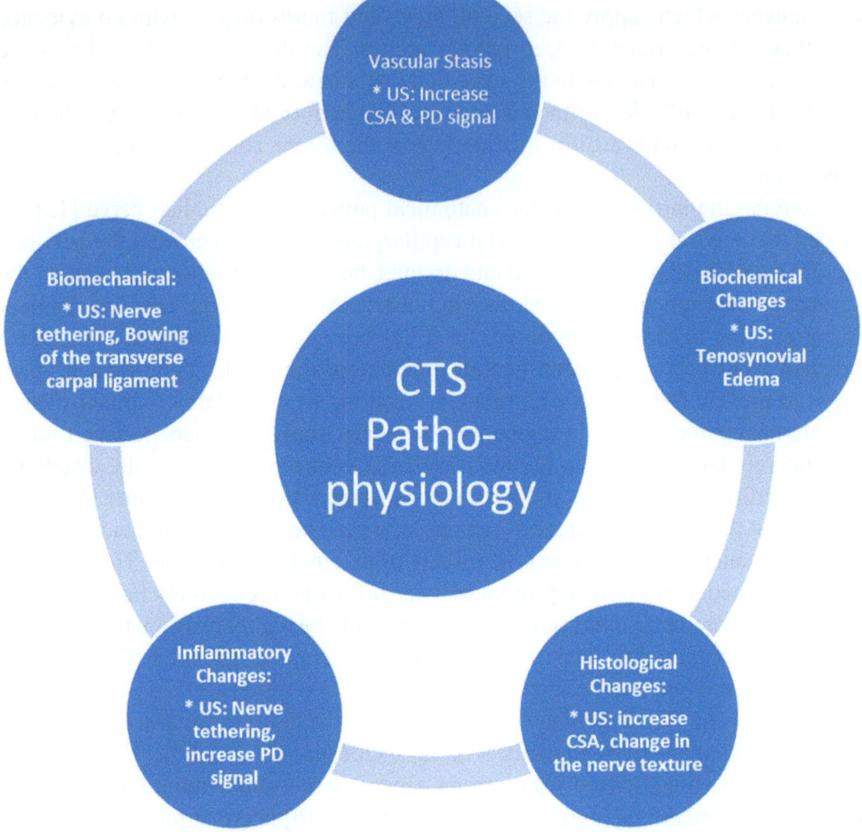

Fig. 10.2 Pathophysiologic changes in carpal tunnel syndrome and their impact on the median nerve sonography

understanding the pathophysiological changes in the CT and median nerve would have a positive impact not only on the diagnosis but also on the management of the condition (Fig. 10.2). Earlier reports revealed that CTS pathophysiology is multifactorial including anatomic, physiologic, biochemical, mechanical, histologic as well as pathological components which integrate to explain and characterize this syndrome.

Anatomic (Mechanical Pressure)

While the CT cross-sectional area (CSA) diminishes with progressive wrist flexion or extension, its interstitial pressure increases. Normal pressure in the tunnel has been recorded in the range of 2–10 mm Hg [23]. Extension increases the pressure tenfold whereas wrist flexion increases it eightfold [24]. Therefore, repetitive hand movements have been implicated as one of the many risk factors for CTS. Experimental studies have suggested a dose-response curve—the greater the duration and amount of pressure, the more significant is the neural dysfunction [22].

Physiologic (Stasis-Ischemic Injury)

Earlier reports revealed that exposure of the nerve to repetitive mechanical forces leads to a consequent demyelination [25]. Nerve demyelination starts at the compression site, and can then spread to the entire internodal segment, leaving the axons intact. This leads to a block of the nervous transmission (neuroapraxia). Persistence of the nerve compression leads to interruption of the blood flow to the endoneural capillary system which in turn would lead to venous stasis followed by ischemia at the capillary level. This in turn leads to alterations in the blood–nerve barrier and development of endoneural edema. This starts a vicious cycle consisting of venous congestion, ischemia, and local metabolic alterations [25, 26], resulting in axonal degeneration, macrophage attraction and activation, release of inflammatory cytokines, nitric oxide, and development of "chemical neuritis" if it continues for a substantial amount of time. This ends in further decrease in the intraneural blood flow followed by fibroblastic activity and scar formation, resulting in destruction and replacement of the normal epineurium with scar tissue and later intraneural damage [26]. Morphological changes in the form of nerve deformity (flattening and hourglass deformity) may also occur at this stage.

Biochemical

Intermittent local ischemia and reperfusion during periods of recovery lead to local biochemical changes in the loose connective tissue in the CT. Under normal conditions, the tendency of loose connective tissue to expand is usually countered by collagen and microfibrillar networks. During periods of ischemia and reperfusion, there is a negative interstitial fluid pressure, suggesting that the tensile forces exerted on the tenosynovium are reduced due to altered gylocosamine and hyaluronan content, augmenting the diffusion of fluid into the tenosynovium. In a study carried out by Sud et al. [27], the authors reported that fluid diffusion occurs at a faster rate and in higher volume in tenosynovium harvested from CTS patients than those from normal tenosynovium. Progressive tenosynovial edema tends to compress the structures in the CT. Furthermore, the altered proteoglycan ratios may reduce the ability of the synovium to bear the compression forces, thus increasing the force incident on the median nerve inside the carpal canal [28]. In contrast with the tendons which are firm and resistant to ischemia, the median nerve is pliable and vulnerable to pressure.

Biomechanical

Nerve fibers have layers of connective tissue: the mesoneurium, epineurium, perineurium, and endoneurium; which is the most intimate layer. The extensibility of these layers is critical to nerve gliding, which is necessary to accommodate joint motion; otherwise nerves are stretched and become injured [29]. The median nerve is expected to move up to 9.6 mm with wrist flexion and slightly less with extension [30]. Chronic compression resulting in fibrosis would inhibit nerve gliding, leading to injury and therefore scarring of the epineurium as well as mesoneurium. This causes the nerve to adhere to the surrounding tissue, or at least cause variable degrees of excursion between the flexor tendons and the median nerve, resulting in traction of the nerve during movement as the nerve attempts to glide from this fixed position [31]. Consequently, this would cause strain and microdamage to the median nerve.

Histological

The inner cells of the perineurium and the endothelial cells of endoneurial capillaries form the blood–nerve barrier which accompanies the median nerve through the CT. These microvessels are formed from nutrient branches that arise from the radial and ulnar arteries, proximal to the flexor retinaculum. An increase in pressure within the tunnel can cause a breakdown of vasculature within this barrier, leading to accumulation of proteins and inflammatory cells [31]. This may induce a miniature closed compartment syndrome by increasing the permeability, contributing to increased endoneurial fluid pressure and development of an intrafascicular edema [32], hence the nerve appears swollen. Similarly, repetitive exposure of the tendons to compression or tensile strength can lead to micro-damage of the synovial tissue lining the tendons within the CT. This leads to thickening of the synovial tissue, and hypertrophy of the tendon, increasing its CSA which in turn increases the pressure within the CT [33].

Pathological

Tenosynovitis, inflammation of the synovial tissue of the flexor tendons, is one of the most common pathological predisposing factors to CTS [34]. This was supported by the presence of increased levels of prostaglandin E2 and vascular endothelial growth factor in synovial biopsy tissue from patients with symptomatic CTS [35]. This leads to an increase in fibroblast density, collagen fiber size, vascular proliferation, and type III collagen in the synovial connective tissue [36]. The outcome is constrictive scar tissue formation around the median nerve [37], which in turn can result in the nerve tethering.

Table 10.1 Causes of carpal tunnel syndrome

Osseous causes of tunnel narrowing	Increased volume causes of tunnel narrowing
Malalignment of carpal bones	Tendon sheath enlargement (tenosynovitis) (Fig. 10.3)
Displaced fractures	Synovial proliferation (rheumatoid arthritis)
Callus formation	Hypertrophied muscles (occupational)
Hypertrophic bone changes	Increased fat (obesity)
	Focal swelling, e.g., ganglion (Fig. 10.4)

Causes of Carpal Tunnel Syndrome

Though median nerve entrapment symptoms are the primary manifestation of CTS, the encroachment on the median nerve is usually secondary to another focal pathology. This can be either a decrease in the tunnel size (mainly osseous causes) or increase in the volume within the tunnel-confined space. Table 10.1 shows the most common causes in each category. However, the most recently found common cause of CTS is repetitive stress injury, such as in computer keyboard operators, tablets, or mobile phones.

US Scanning of the Carpal Tunnel

Anatomic Landmarks and Imaging Technique

One of the most important factors which are essential for the routine median nerve US is patient positioning and identifying landmarks with high diagnostic accuracy and reliability for median nerve assessment. It is advisable that the patient sits with the forearm resting comfortably on a flat surface, with the elbow in mid-flexion and the wrist in supination while the fingers are semi-extended. In most studies, US assessment of median nerve was performed at the proximal CT (at the level of the pisiform bone or distal forearm crease) [38–40]. Furthermore, the US outcome measures recorded at this level demonstrated the highest sensitivity and specificity [41, 42]. In addition, the measurement at this level was found easier to perform and achieved high reproducibility factors.

Imaging should be carried out with the transducer in a plane perpendicular to the tendon surface to eliminate the anisotropic effect. The weight of the probe should be applied without additional pressure. The tunnel contains the flexor digitorum tendons which are hyperechoic as well as the median nerve (anterior to the tendons). The nerve appears as hypoechoic with multiple bright reflectors and a hyperechoic border (Fig. 10.5). At the proximal wrist, the normal median nerve appears elliptical (Fig. 10.6) and flattens progressively as it courses distally. However, while its shape may get quite variable, its size remains constant and within the tunnel, the nerve remains in intimate contact with the flexor retinaculum. In the longitudinal imaging

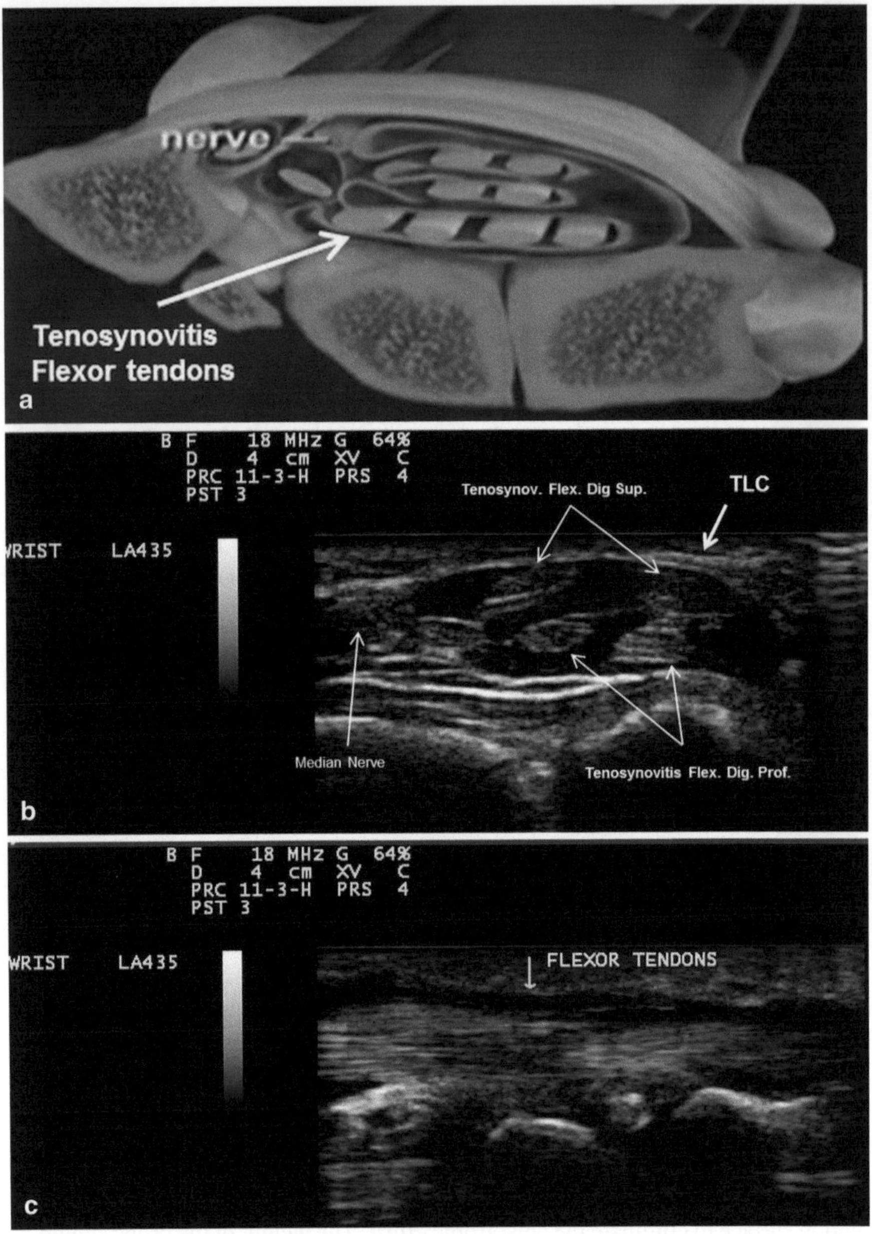

Fig. 10.3 **a** A diagram showing the compression of the median nerve by inflamed tendons' sheath. **b** Gray scale US using Esaote Mylab 25 system: transverse. **c** longitudinal view showing tenosynovitis flexors of the hand. *TLC* transverse carpal ligament. *Flex. Dig. Sup.* flexor digitorum superficialis, *Flex. Dig. Prof.* flexor digitorum profundus

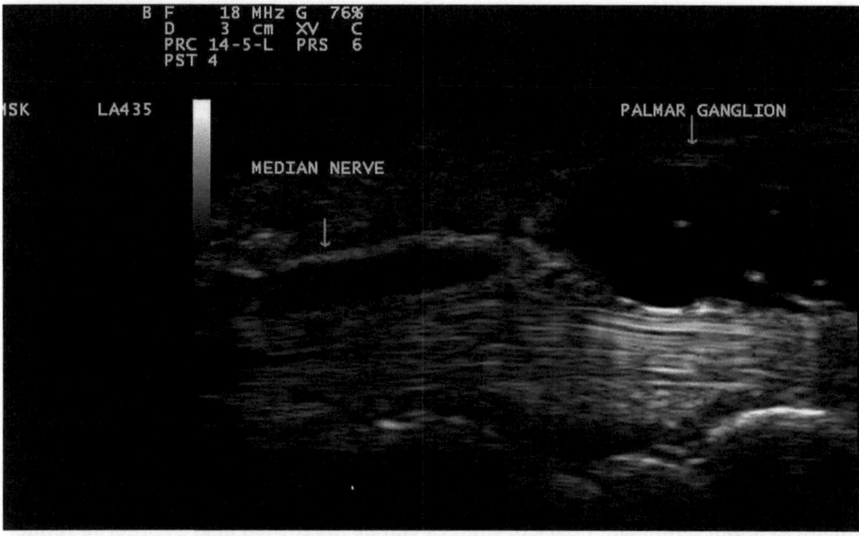

Fig. 10.4 Gray scale US using Esaote Mylab 25 system: palmar longitudinal view showing a space occupying lesion (ganglion) compressing the median nerve at the carpal tunnel proximal entrance

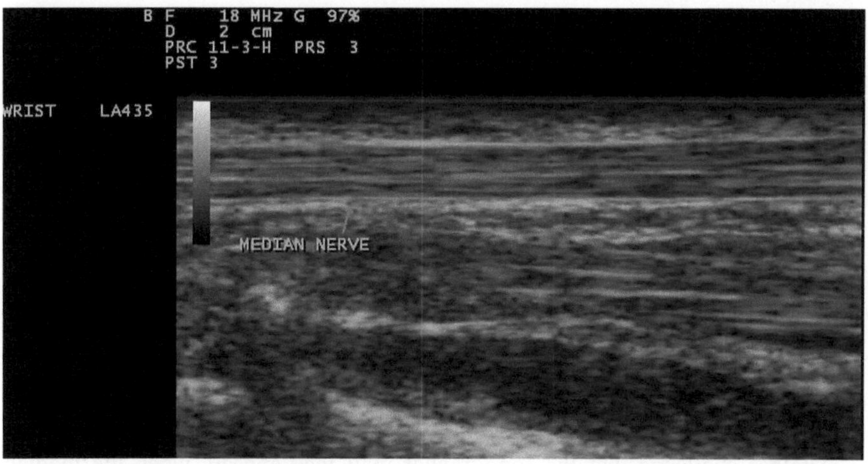

Fig. 10.5 Gray scale US using Esaote Mylab 25 system: Palmar longitudinal view at the proximal carpal tunnel inlet showing the normal sonographic median nerve fascicular pattern which appears as multiple bright reflectors with a hyperechoic border

plane, the nerve enters the tunnel parallel and superficial to the flexor digitorum tendons. Similar to the transverse view, the nerve in this plane demonstrates hyperechoic continuous anterior and posterior borders (the nerve sheath) and is easily distinguished from the distinctive fibrillar tendons sonographic pattern which lies posteriorly. To minimize errors due to differential loads, a gel pad (Sonar Aid) of

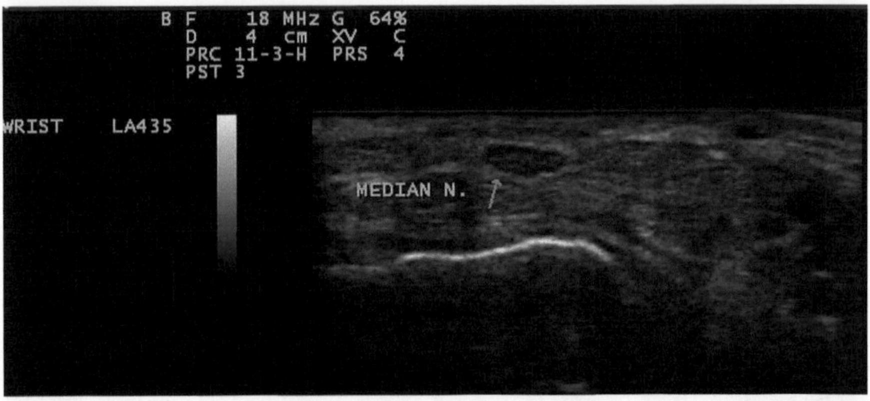

Fig. 10.6 Gray scale US using Esaote Mylab 25 system: Palmar transverse view at the proximal inlet of the carpal tunnel showing normal elliptical median nerve

thickness about 3 mm can be used. Images can be magnified also in order to reduce measurement error.

Scanning Tips

As the median nerve changes in both shape and size as the probe is moved along the nerve from the forearm to the hand, the nerve can be assessed at different levels. To ensure the reliability of the scanning, most studies have chosen to standardize their "level" by reference to anatomical landmarks such as the pisiform bone or the hook of the hamate bone, or the distal wrist crease (which closely approximates to the pisiform bone). This is important not only to a new ultrasonographer but is also confounded by the fact that the median nerve is not actually longitudinally fixed in place. With the movement of the wrist and fingers, the normal median nerve slides up and down through the CT and also across the tunnel. On transverse scanning of the median nerve, at the wrist crease, the nerve may change in appearance as the wrist and fingers move or the tendons may push the nerve around and deform its outline. Therefore, standardizing the hand and finger position is essential for accurate reproducible measures.

Another challenge that the sonographer may face appears when the median nerve starts to divide early. While in most subjects the median nerve at the CT appears as a single structure, in some patients, it starts to divide into branches above the CT and a transverse image may show two or even three nerves, providing yet a challenge regarding how to measure them. Options include: either to measure them all individually and add them together, draw a boundary which goes round them all (which may be what the nerve sheath does in some cases), or just conclude that these cannot be measured and compared with subjects who only have a single nerve trunk at the CT.

US Findings in Carpal Tunnel Syndrome

An early diagnosis based on clinical and electrodiagnostic findings has been the standard for assessment of CTS patients aiming at preventing permanent nerve damage and functional sequelae [43]. The development of high-resolution US transducers (7–15 MHz) has allowed noninvasive evaluation of normal and abnormal appearances of the median nerve as well as the other contents of the CT. Over the past years, several studies [44–48] were published showing the value of US as a tool for the diagnosis of the condition. US enables the treating physician to detect changes in nerve shape and exclude anatomic variants and space-occupying alterations such as ganglion cysts and tenosynovitis [49, 50]. It has several advantages over magnetic resonance imaging (MRI), including being relatively fast and inexpensive and allowing additional dynamic as well as blood flow imaging with relatively little additional time. Table 10.2 depicts US findings seen on assessment of the CT which can be stratified into subjective and objective outcomes. The next section focuses on the objective US outcomes and methods of their assessment/measurement.

US Measures of the Median Nerve

Median nerve compression is revealed on US by the classic triad of nerve swelling, nerve flattening, and palmar bowing of the flexor retinaculum.

Cross-Sectional Area

Considering the underlying pathophysiological changes, compression of median nerve leads to congestion of epineural and endoneural veins, subsequently causing nerve edema which manifests as increase in the median nerve CSA (Fig. 10.7). Usu-

Table 10.2 Sonographic carpal tunnel changes stratified into subjective and objective outcomes

Subjective criteria	Objective criteria
Flattening of the nerve, especially at the level of the hamate bone	The mean cross-sectional area of the median nerve is greater than 10 mm^2 at the pisiform bone level
Volar bulging of the flexor retinaculum	The flattening ratio of the nerve (transverse diameter divided by Antero-posterior (AP) diameter) is greater than 4:1 at the level of the hamate bone
Enlargement of the median nerve as it enters the carpal tunnel	Volar bulging of the flexor retinaculum is greater than 3.1 mm
Large fluid or fat layer surrounding the tendons	Power doppler assessment of the median nerve vasculature
Decreased mobility of the median nerve on flexion and extension of the fingers, hand, or wrist	

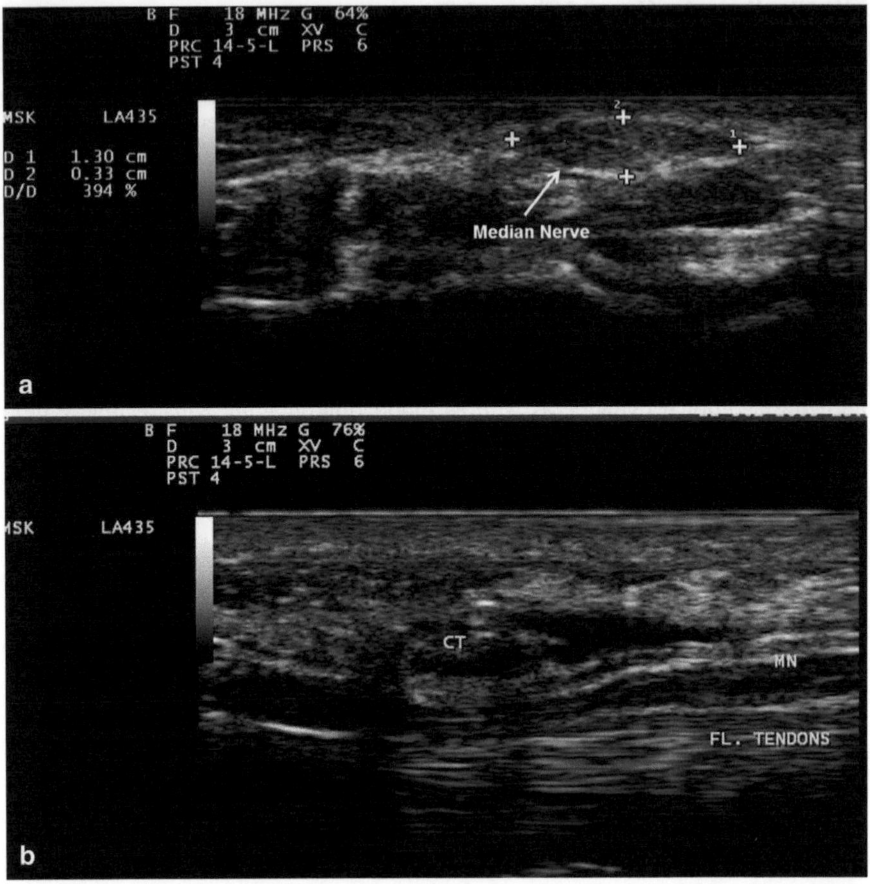

Fig. 10.7 Gray scale US using Esaote Mylab 25 system: Palmar view at the proximal inlet of the carpal tunnel showing: **a** Transverse view of the median nerve swelling manifested by increased median nerve cross-section area in a subject with carpal tunnel syndrome. **b** Longitudinal view showing compression of the median nerve at the proximal inlet of the carpal tunnel whereas there is increased nerve width at both tunnel inlet and outlet. *MN* median nerve, *Fl. Tendons* flexor tendons, *CT* Carpal Tunnel

ally, the CSA of the median nerve (recorded in mm^2) is measured at the proximal CT (at the level of the pisiform bone). Duncan et al. [46] endorsed a direct method for measurement, which is calculated automatically by tracing the inner margin of the median nerve epineurium.

The diagnostic accuracy of this method has been reported high [51–57] although the concluded cutoff values vary because of differences in the gold standards or devices used for diagnosis, as well as the severity of the disease or other factors. The diagnostic sensitivity and specificity of this approach (CSA measurement at the tunnel inlet/the distal wrist crease) was found to be 67 and 97 %, respectively [56]. When this criterion was combined with the results of nerve conduction studies (NCSs), the sensitivity and specificity were 84 and 94 %, respectively [57]. CSAs larger than 9 mm^2 [58, 59] and 10 mm^2 [60, 61] recorded at the tunnel inlet in patients with CTS were reported the most appropriate median nerve cutoff values to identify abnormal median nerve CSA in comparison to control patients.

Median Nerve Flattening

Flattening of the median nerve (Fig. 10.8) reflects the morpho-pathologic changes in response to persistent pressure on the nerve. To calibrate the degree of flattening, flattening ratio (FR) was suggested to quantify the change of the median nerve shape by measuring the ratio of the major axis to the minor axis at the tunnel proximal inlet (at the level of pisiform bone) [62–64]. This is the ratio of the mediolateral diameter to the anteroposterior diameter on cross sections. A ratio of one would indicate a round structure. Earlier studies [36, 56–61] showed a trend of increase in the FR measure with increase in the severity of CTS (as evident from other US measures and electromyography findings). A recent study [65] discussed the value of median nerve flattening for the prediction of management outcome, which is discussed later in this chapter.

Bowing of the Transverse Carpal Ligament

The palmar bowing of the flexor retinaculum is a measure used to quantify the internal pressure exerted on the retinaculum from the contents within the CT. Exaggerated palmar bowing of the flexor retinaculum has been reported to predict the CTS diagnosis [66]. The palmar bowing of the flexor retinaculum is displacement

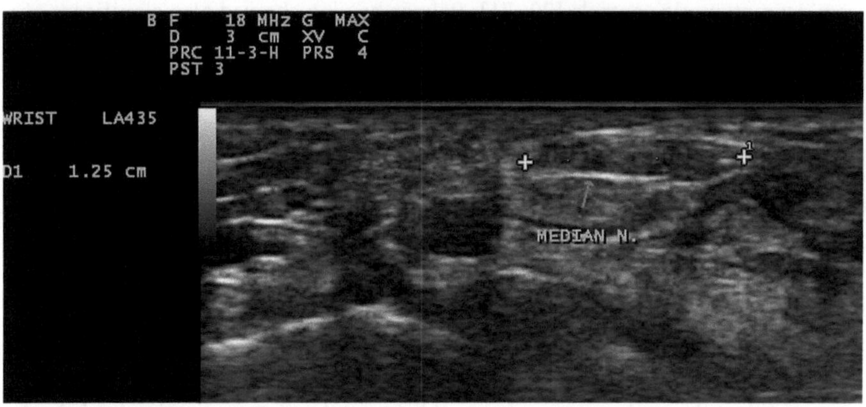

Fig. 10.8 Gray scale US using Esaote Mylab 25 system: Palmar transverse view at the proximal inlet of the carpal tunnel showing flattening of the median nerve (The flattening ratio (*FR*) is defined as the ratio of the nerve's transverse axis to the anteroposterior axis)

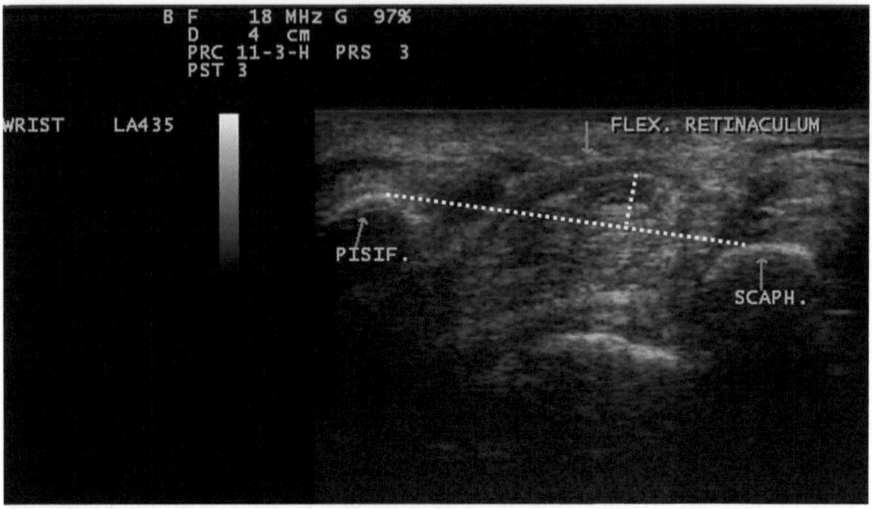

Fig. 10.9 Gray scale US using Esaote Mylab 25 system-Palmar transverse view at the proximal inlet of the carpal tunnel showing the palmar bowing of the transverse carpal ligament (displacement of the flexor retinaculum from its attachments to both pisiform and scaphoid bones). Volar bulging of the flexor retinaculum > 3.1 mm is suggestive of increase intra-tunnel pressure. *Pisif.* Pisiform bone, *Scaph.* Scaphoid bone, *Flex. Retincaculum* Flexor retinaculum

(measured in mm) of the retinaculum of a line connecting the pisiform bone with the scaphoid bone (Fig. 10.9). Earlier studies [67–70] showed that, in addition to CSA and FR, palmar bowing was significantly increased in the CTS group than the control group. Among these US outcome measures, CSA and palmar bowing were reported to have a relatively higher accuracy than FR according to the receiver operating characteristic (ROC) curve. Therefore, measurement of CSA and/or palmar bowing can be considered as an alternative modality to distinguish CTS patients from asymptomatic controls [70, 71]. Palmar bowing showed also significant correlation with both patients symptoms and nerve conduction testing [71].

Other Less Common Measures

Multiple Median Nerve Measures

The median nerve CSA can be measured at variable sites in the forearm, CT, and hand. Studied sites include: (a) CSA at the proximal border of the pronator quadratus muscle, (b) area of the proximal third of the pronator quadratus muscle, (c) level of largest CSA of the median nerve observed between the area proximal to the CT inlet and the tunnel outlet, (d) CT inlet defined as the proximal margin of the flexor retinaculum, and (e) in the carpal canal, level of the scaphoid tubercle and pisiform bone [72]. One study has calculated that a 2 mm^2 difference in nerve cross section between

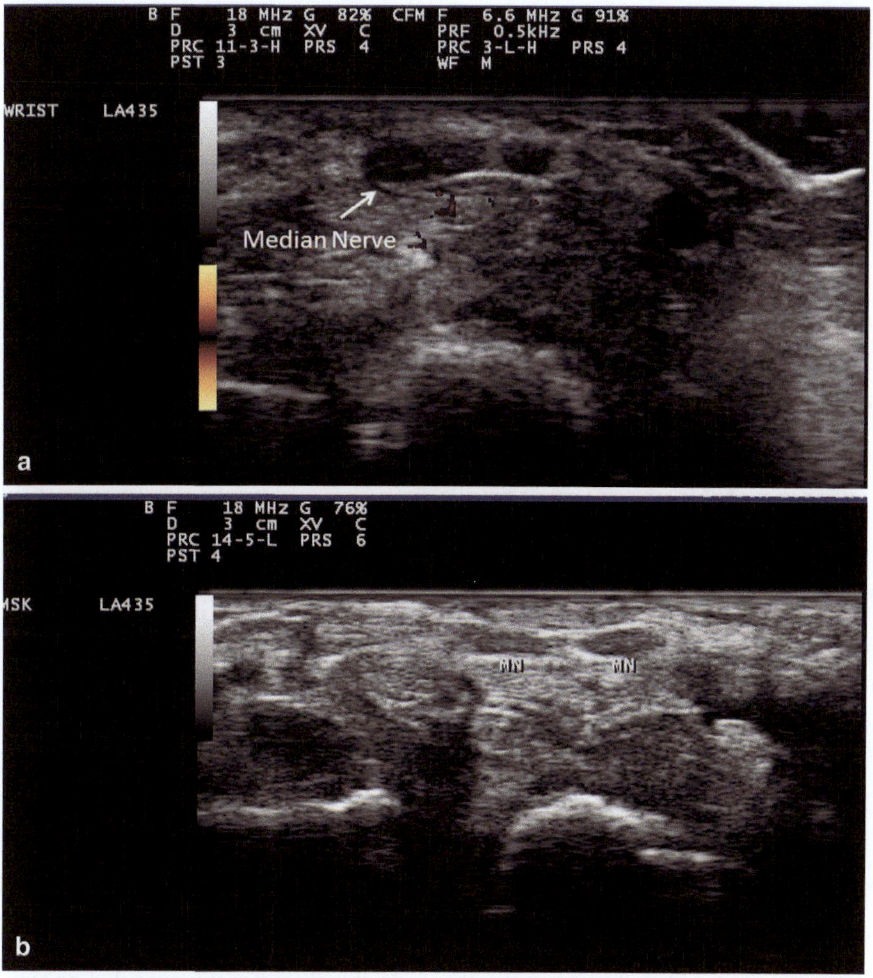

Fig. 10.10 Gray scale US using Esaote Mylab 25 system: Palmar transverse view at the proximal inlet of the carpal tunnel showing bifid median nerve in two separate sheaths (**a**), and contained in one sheath (**b**). (*MN* median nerve)

the level of the pronator quadratus and the CT has 99% sensitivity and 100% specificity for CTS [73]. Another challenge is when the median nerve is bifid or split into branches (Fig. 10.10) above the CT (transverse scanning may show two nerves). In such case, it is advisable to measure them all individually and add them together, or to draw a boundary round them all (which the nerve sheath may do in some cases).

Median Nerve Indices

As the shape of the nerve varies while passing through the tunnel, indices have been introduced to better quantify abnormal findings. Relations between measures of the median nerve at different sites have been suggested. Calculating the ratio between the median nerve CSA at the proximal forearm and the CT was reported to improve diagnostic accuracy [74–76]. However, in those studies reliability testing and adequate control groups were lacking. The swelling ratio is another index obtained by dividing the CSA of the median nerve proximal to the flexor retinaculum by the CSA of the same nerve at the distal end of the flexor retinaculum [77, 78]. A third index, the nerve/tunnel index, was suggested by calculating the ratio between the proximal or distalCSA of the median nerve to that of the CT and multiplying the value by 100 (proximal nerve/tunnel index = (proximal median nerve CSA)/proximal CT CSA) × 100 [79]. However, there is no definite consensus on the normal values and ranges of these sonographic indices [80].

Nerve Mobility

The median nerve slides up and down through the tunnel with wrist and finger movement. Also it moves in the transverse plane in relation to the flexor tendons. Chronic compression results in fibrosis, which prohibits nerve gliding and leads to adherence to the surrounding tissues resulting in traction of the nerve during movement as the nerve attempts to glide from this fixed position [81, 82]. This is the basis of the tethered median nerve stress test which can be used to diagnose chronic low-grade CTS [83].

Nerve Texture

Nerve ischemia and subsequent damage render the affected nerve more hypoechoic on ultrasound images in both the transverse and longitudinal views. This is in contrast with the normal proximal median nerve which shows much more internal structure seen within the nerve sheath giving the characteristic speckled appearance in transverse section and linear streaks in longitudinal images [80]. However, quantifying this as a number for measurement purposes is difficult. Furthermore, the absolute brightness of a pixel in an ultrasound image is not only determined by the imaged tissue but also by the ultrasound beam angle on the tissue as well as the machine setting (the nerve may appear brighter or darker simply by tilting the transducer or twiddling the knobs on the scanner).

Subsynovial Connective Tissue Thickness

A major pathologic finding in patients with idiopathic CTS is noninflammatory fibrosis and thickening of the subsynovial connective tissue. Earlier, a study [84] was carried out to determine the ability of sonography to depict this thickening and compare the subsynovial connective tissue thickness in patients with CTS and

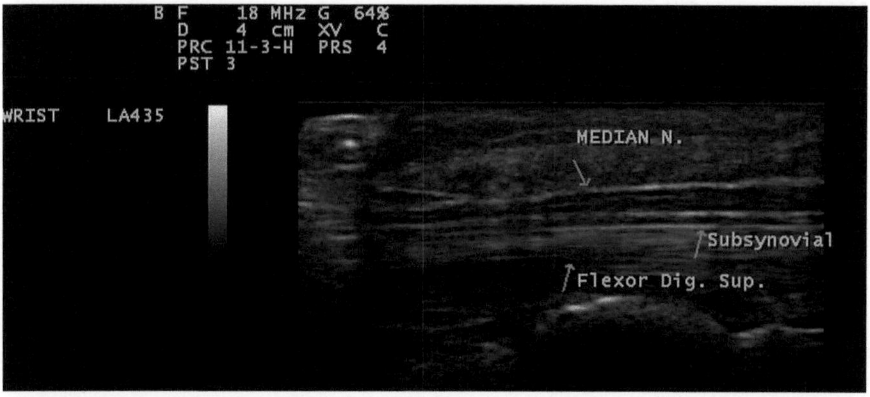

Fig. 10.11 Gray scale US using Esaote Mylab 25 system: Palmar longitudinal view at the proximal inlet of the carpal tunnel showing subsynovial connective tissue thickening. The subsynovial connective tissue is defined as the thin echogenic layer at the border of the tendon, between the median nerve and the flexor digitorum superficialis tendon. *Flexor Dig. Sup.* flexor digitorum superficialis

healthy control participants. The subsynovial connective tissue is defined as the thin echogenic layer at the border of the tendon, between the median nerve and the flexor digitorum superficialis tendon (Fig. 10.11). To compensate for differences in hand size, especially between men and women, the subsynovial connective tissue thickness was normalized to the flexor digitorum superficialis tendon diameter at the same level, using the thickness ratio, calculated as subsynovial connective tissue thickness/tendon thickness = thickness ratio. Results revealed that the subsynovial tissue was significantly thicker in patients with CTS than in healthy controls [85].

Power Doppler of the Median Nerve

Although compression of the median nerve beneath the flexor retinaculum is the primary basis of CTS, many of the early symptoms are thought to be due to altered intraneural hemodynamics. The pathophysiological findings of prominent median nerve intraneural microvasculature due to inflammation and compression led to the hypothesis that the evaluation of intraneural vasculature may be useful for detecting CTS. In turn, this would also augment the diagnostic efficiency of US in CTS patients. Earlier studies revealed that, comparing the sonography findings and electrodiagnostic test results, nerve hypervascularization and nerve swelling yielded the best detectability of CTS with nerve hypervascularization showing higher accuracy of 91–94% [86, 87] and positive predictive value of 96% [88]. Comparing the state of median nerve hypervasularity to the severity of CTS based on nerve conduction results depicted an inverse relationship between intraneural vascular flow in the median nerve and an increasing severity of the median nerve compression [89].

US Versus NCS

The definition of CTS in study populations has varied; some studies used positive NCS as an inclusion criteria [90–95], while others enrolled patients diagnosed based on clinical criteria [96–103]. Though in most cases, where the complaints and findings are typical, the diagnosis of CTS is relatively easy, this is not always the case. CTS is often accompanied by other entrapment neuropathies or tenosynovitis of the flexor tendons. The patient may complain of pain rather than numbness. Moreover, the presence of a cervical lesion may further complicate the symptoms. Clinical studies revealed that the traditional provocative tests are not even valid tests for the CTS diagnosis [104, 105]. Therefore, in such cases, CTS cannot be confirmed merely by physical examination; and objective examinations are necessary.

NCSs are the most common tests used to assess the function of the median nerve. A plethora of research has been published providing sensitivity and specificity data for the testing and supporting the utility of electrodiagnostic studies for the diagnosis of CTS [106–108]. However, abnormal electrodiagnostic test results do not necessarily equate to CTS and are not necessarily predictive of future disorders [106]. Furthermore, in some cases presenting with typical CTS picture, their NCS may be normal. This discordance between NCS and clinical presentation was attributed to variability in protocols, diagnostic threshold values, and overall diagnostic accuracy of neurophysiologic testing [109–113]. In addition to these limitations, electrodiagnostic testing evaluates nerve dysfunction status that only appears in more chronic stages of the disorder. The limitations of electrodiagnostic testing, with relatively low diagnostic accuracy (sensitivity and specificity reported to be around 70 and 82%, respectively) failing to diagnose approximately 20–30% of CTS cases [114], highlighted the lack of a reliable step in the diagnostic guidelines and provided a ground for continued investigation for complementary or alternative diagnostic methods.

The use of US for the diagnosis has been receiving increasing attention. In contrast with electrodiagnostic tests which are based on physiologic malfunctions of the median nerve, US enables morphologic assessment including nerve swelling, edema, bowing of the flexor retinaculum, focal swelling in the tunnel as well as evaluation of functional disturbance manifested by nerve hypervascularization. The role of US is augmented, in particular, in those patients presenting with CTS manifestations, yet, have normal nerve conduction studies. Paresthesia has been shown to occur before conduction failure in myelinated sensory fibers, as measured with nerve conduction tests. A study by Koyuncuoglu et al. [115] reported positive US findings in patients who had CTS positive clinical results, with negative electrodiagnostic findings in 30.5% of these patients. Such data provide an advantage to using US, especially during the early stages of CTS, when the median nerve shows no functional impairment on electrodiagnostic examination. Therefore, in view of its relatively painless nature, high specificity, some have suggested that US be used as a screening test for CTS before performing NCS [116].

US As a Diagnostic Tool for CTS

Despite some limitations, high-resolution sonography may be an appropriate alternative diagnostic tool for CTS. Its ability to visualize the median nerve morphologic changes as well as the intraneural vascularity throughout various stages of CTS gives the treating doctor the privilege of not only assessing the median nerve compression status but also the cause of this entrapment. Attempts to calculate the diagnostic accuracy of measurements with sonography have been challenging because of the variability in methods and equipment parameters [117]. A systematic review [118] revealed that when a combination of sonographic measures of the median nerve CSA, FR, hypoechogenicity, and hypervascularity, resulted in 90 % agreement with the clinical presentation of CTS [119–121] which could provide the necessary evidence for combining complementary methods to improve accuracy.

An algorithm for diagnosing CTS (Fig. 10.12) was suggested by El Miedany et al. [39] which relies mainly on US measures. It begins with a baseline of history and clinical examination followed by gray scale evaluation of the wrist to screen patients with abnormalities from those without. Based on combined measurement approach and identifying the cutoff points, it is possible to stratify the patients according to their CTS severity. The next tier of evaluation should be the use of electrodiagnostic testing for assessing the physiologic dysfunction of the nerve and grading the severity of the median nerve compression. In concordance with this combined approach, Doppler examinations could provide diagnostic information about the severity of the nerve compression and perhaps the stage of the entrapment as well as the location of the injury [122].

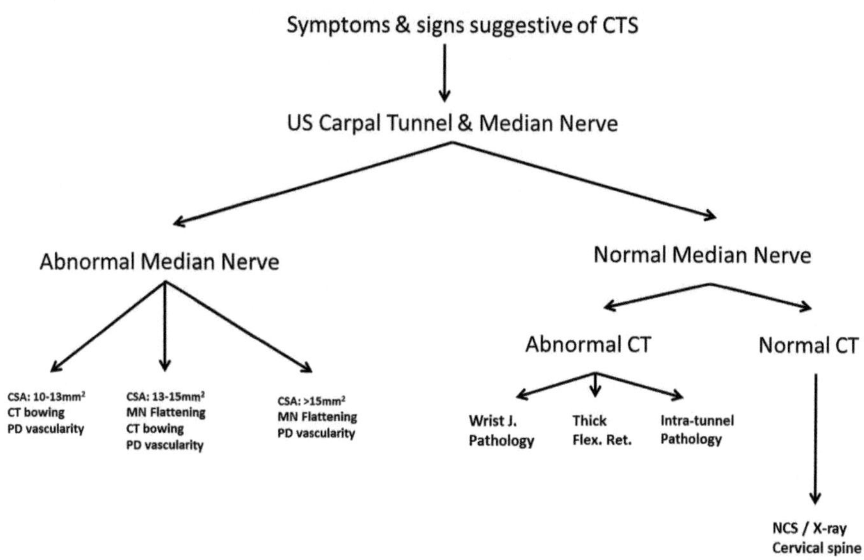

Fig. 10.12 Algorithm for carpal tunnel syndrome diagnosis using ultrasonography. *CSA* Cross sectional area, *CT* Carpal tunnel, *PD* Power Doppler, *MN* Median nerve, *J* joint, *Flex. Ret.* flexor retinaculum, *NSC* nerve conduction study

US As a Biomarker for CTS Progression and Management

There are various CTS treatment options which can be stratified into conservative approach and surgical decompression. As research studies revealed comparative results of both treatment modalities [123], the question every treating physician would face, would be "what is the most appropriate treatment modality advised for the patient?" Another challenge is to assess changes in the median nerve and CT following treatment. A valid biomarker should be objectively measured, indicatory of normal biology as well as the pathologic process, good indicator of response to therapy and prognosis. It should also be a valid indicator of modification of the pathology process and should help to identify (in early cases) the patients who are going to respond quickly to therapy with a vision to tailor management approach to the patient status [124]. A recent research studied the use of a combination of US outcome measure in setting up a treatment approach tailored to the patient's condition [65]. According to the study results, the sonographic changes of the median nerve in CTS can be divided into three phases according to disease duration: (1) stage of hypervascularity, (2) nerve edema, and (3) nerve flattening. In CTS patients presenting early in the disease course, median nerve-enhanced vascularity tends to be perineural. Later in the disease course, enhanced vascularity extends to be intraneural (Fig. 10.13). Using a combination of US outcome measures, it was possible to stratify the patients presenting with CTS manifestations based on the severity of the nerve compression. The risk of a poor outcome was significantly higher in the patients with high median nerve flattening ratio at the CT inlet (relative risk 3.3, 95 % confidence interval (CI) 1.73–6.43, $P=0.0004$). This risk was most marked in the cohort with nerve flattening associated with longer duration of illness (relative risk 4.3, 95 % CI 1.82–10.29, $P=0.006$) and low PD signal (relative risk 4.1, 95 % CI 1.71–9.47, $P=0.005$). There was also significant inverse relation ($r=-0.372$, $P<0.03$) between PD and disease duration. Assessing response to treatment, whether conservative or surgical (Fig. 10.14), PD was the first US outcome measure to show significant change in response to treatment as early as 1 week. This improvement continued till 1 month of management, then the process became slower/plateaued. On the other hand, while the CSA and FR improved significantly (back to normal range in patients with mild or mild-moderate nerve compression), they remained unchanged, or showed slight improvement, in the severe median nerve compression cases. In contrast with NCS which did not show any significant correlation with the treatment response at 6 months, there was significant correlation ($P<0.01$) between baseline US variables, namely CSA, FR, and PD, and the percentage of improvement at first week, first month, and 6 months of management. These data represent a major transformation in the assessment and management of CTS patients. In addition to the diagnostic value of sonography in CTS, US can help in tailoring a treatment approach to individual patients and was reported to be a fit radiologic biomarker for CTS.

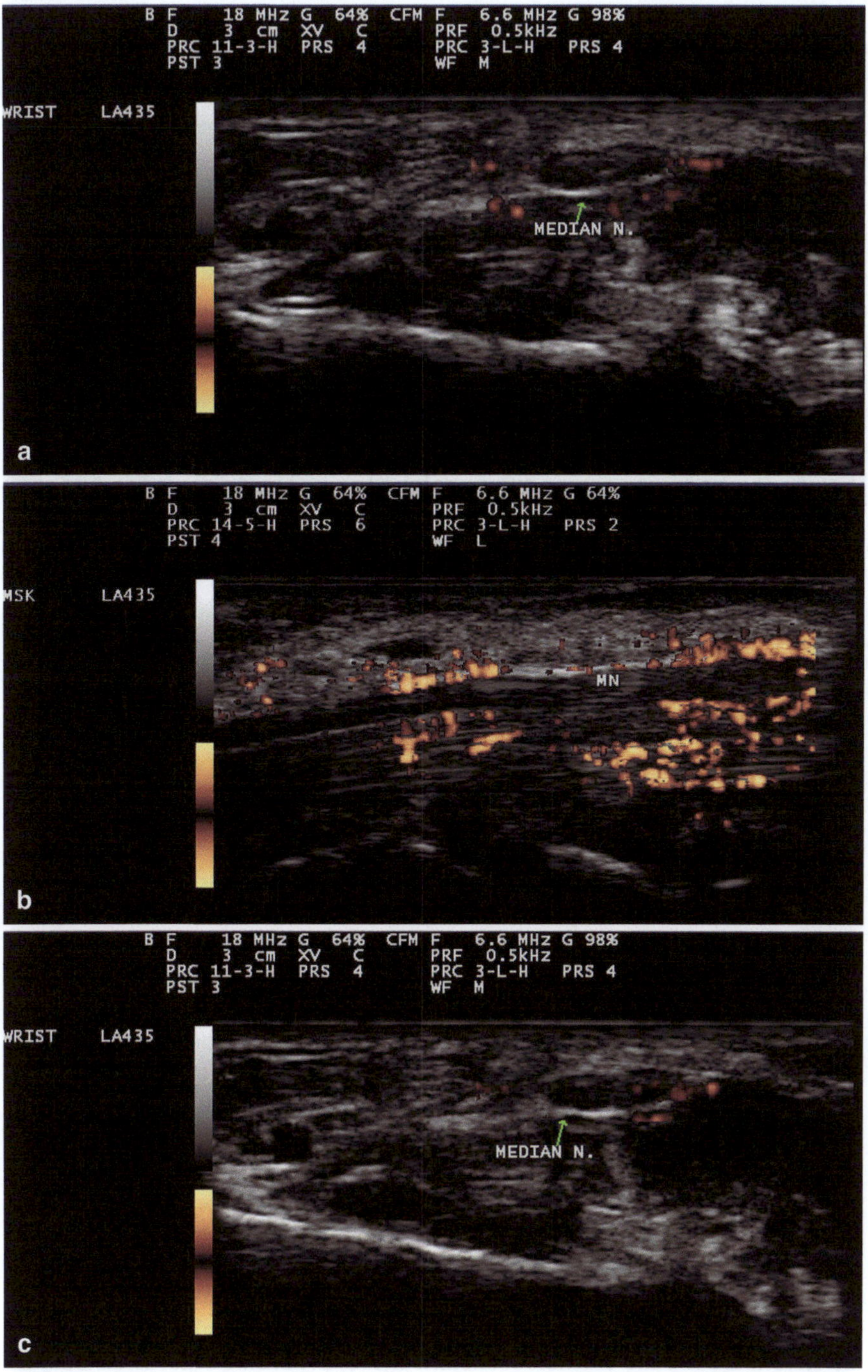

Fig. 10.13 Gray scale US using Esaote Mylab 25 system — **a.** Palmar transverse view at the proximal inlet of the carpal tunnel showing perineurial enhanced vascularity of the median nerve secondary to tenosynovitis; **b.** Palmar longitudinal view at the proximal inlet of the carpal tunnel showing perineurial enhanced vascularity of the median nerve; **c.** Palmar transverse view at the proximal inlet of the carpal tunnel showing endoneurial enhanced vascularity of the median nerve. MN Median nerve

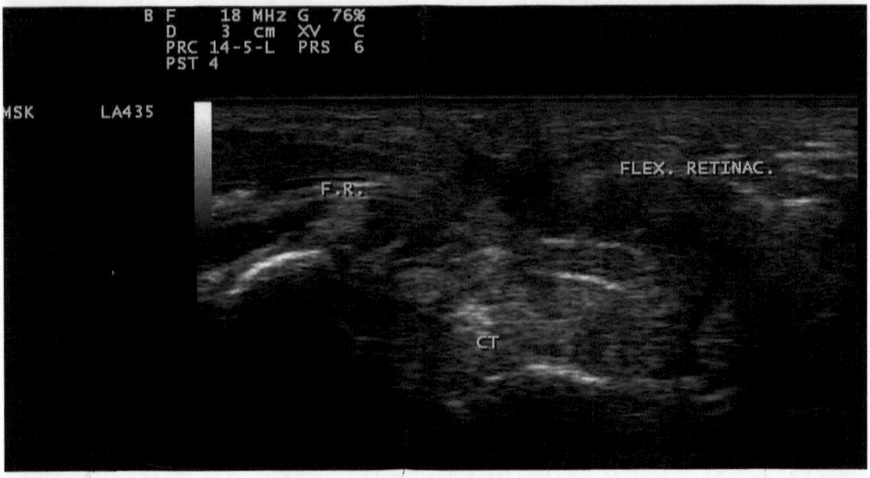

Fig. 10.14 Gray scale US using Esaote Mylab 25 system: Palmar transverse view of the carpal tunnel showing postoperative decompression of the carpal tunnel with excision of the middle section of the transverse carpal ligament. *Flex. Retinac.* flexor retinaculum, *CT* carpal tunnel, *F.R.* Flexor retinaculum

US-Guided Carpal Tunnel Injection

When nonsurgical treatment is agreed, clinicians may elect to inject steroids into the CT to reduce symptoms of pain and paresthesias [125]. On the basis of visible and palpable anatomic landmarks, multiple approaches to CT injections have been described [126, 127]. Although each approach targets minimum patient discomfort and to avoid median nerve injury, while achieving the desired clinical outcomes, insufficient evidence exists to document the superiority of one technique over the others [128, 129]. Regardless of the chosen technique, a therapeutic response may not be achieved if the injection is not successfully placed in the CT [130]. Furthermore, even an accurately placed needle within the CT following one of the previously described techniques may injure the median nerve because of well-described anatomic variations (e.g., a bifid median nerve or a median nerve in an abnormal location) [131].

In view of the encouraging outcomes of earlier studies showing the value of ultrasound-guided intra-articular injections, clinicians have begun to explore the potential therapeutic role of sonography in patients with CTS. However, in contrast to the increased emphasis on the use of US in CTS patients, few published studies pertaining to its use in guiding therapeutic CT injections in appropriately selected patients [132, 133]. Although the optimal therapeutic location for a CT injection has not been determined, the aim of any local injection is to infuse the steroid around the edematous median nerve, producing what is called "target sign" as well as to infiltrate the subsynovial connective tissue [134]. In concordance with the standard US-guided injection techniques, a proximal-to-distal needle approach, similar to the non-guided techniques previously described, has been suggested [132]. Using this approach, some authors prefer to visualize the needle in a short-axis view relative to the transducer, [22] whereas others prefer the long-axis (in-plane) view [133], while Doppler imaging can be used to confirm the position of blood vessels if necessary. In either scenario, US allows the treating doctor to visualize the needle shaft/tip throughout the procedure, and provides the flexibility to precisely target any desired location within the CT. The standard injection consists of 1 mL of 40 mg/mL methylprednisolone and 1 mL of 1 % lidocaine. US-guided ulnar approach to the tunnel has been studied with good outcomes [129].

Conclusion

US as a tool for CTS diagnosis can be universally employed when a standardized protocol is adopted. Several studies documented that the obtained measurements are reproducible and reliable. The use of a combination of US outcome measures and provocative testing may assist the clinician in sorting out symptomatic patients who either currently have negative findings or are already at a severe stage. In addition, it would help in setting up a treatment approach tailored to the patient's condition.

References

1. Klauser AS, Faschingbauer R, Bauer T, et al. Entrapment neuropathies II: carpal tunnel syndrome. Semin Musculoskelet Radiol. 2010;14:487–500.
2. Roll SC, Case-Smith J, Evans KD. Diagnostic accuracy of ultrasonography vs. electromyography in carpal tunnel syndrome: a systematic review of literature. Ultrasound Med Biol. 2011;37:1539–53.
3. Steinberg DR, Szabo RM. Anatomy of the median nerve at the wrist—open carpal tunnel release—classic. Hand Clinics. 1996;12:259–69.
4. Cobb TK, Dalley BK, Posterato RH. Anatomy of the flexor retinaculum. J Hand Surg. 1993;18:91–9.
5. Yugueros P, Berger R. Anatomy of the carpal tunnel, Chapter 2. In: Carpal Tunnel Syndrome. Luchetti R, Amadio P, editors. Springer. 2002. pp. 10–12.
6. Joshi SD, Joshi SS, Athavale SA. Lumbrical muscles and carpal tunnel. J Anat Soc. 2005;54(1):12–5.
7. Siegel DB, Kuzma G, Eakins D. Anatomic investigation of the role of lumbricals muscle in carpal tunnel syndrome. J Hand Surg. 1995;20(5):860–3.
8. Cobb TK, An KN, Cooney WP, Berger RA. Lumbrical muscles incursion into the carpal tunnel during finger flexion. J Hand Surg. 1994;19(4):434–8.
9. Cobb TK, An KN, Cooney WP. Effect of lumbrical muscle incursion within the carpal tunnel on carpal tunnel pressure: a cadaveric study. J Hand Surg. 1995;20(2):186–92.
10. Bezerra AJ, Carvalho VC, Nucci A. An anatomical study of the palmar cutaneous branch of the median nerve. Surg Radiol Anat. 1986;8:183–8.
11. Poisel S. Ursprung und verlauf des r.muscularis des nervus digitalis palmaris communis I (N. medianus). Chir Praxis. 1974;18:471–4.

12. Pecket P, Gloobe H, Nathan H. Variations in the arteries of the median nerve with special considerations on the ischemic factor in the carpal tunnel syndrome. Clin Orthop. 1973;97:144–7.
13. Blunt M. The vascular anatomy of the median nerve in the forearm and hand. J Anat. 1959;93(1):15–22.
14. Denny-Brown D, Brenner C. Paralysis of nerve induced by direct pressure and by tourniquet. Arch Neurol Psychiat. 1944;51:1–26.
15. Lanz U. Anatomical variations of the median nerve in the carpal tunnel. J Hand Surg Am. 1977;2(1):44–53.
16. Al-Qattan MM, Al-Zahrani K, Al-Omawi M. The bifid median nerve re-visited. J Hand Surg Eur. 2009;34(2):212–4.
17. Jones DP. Bilateral palmaris profundus in association with bifid median nerve as a cause of failed carpal tunnel release. J Hand Surg Am. 2006;31(5):741–3.
18. Ogino T, Ohno K. A case of bipartite median nerve at the wrist. J Hand Surg Br. 1991;16(1):96–7.
19. Kessler I. Unusual distribution of the median nerve at the wrist. A case report. Clin Orthop Relat Res. 1969;(67):124–6.
20. Amadio PC. Bifid median nerve with double compartment within the transverse carpal canal. J Hand Surg Am. 1987;12(3):366–8.
21. Szabo RM, Pettey J. Bilateral median nerve bifurcation with an accessory compartment within the carpal tunnel. J Hand Surg Br. 1994;19(1):22–3.
22. Mackinnon SE. Pathophysiology of nerve compression. Hand Clin. 2002;18(2):231–41.
23. Ibrahim I, Khan W, Goddard N, Smitham P. Carpal tunnel syndrome: a review of the recent literature. Open Orthop J. 2012;6(Suppl 1: M8):69–76.
24. Werner RA, Andary M. Carpal tunnel syndrome: pathophysiology and clinical neurophysiology. Clin Neurophysiol. 2002;113(9):1373–81.
25. Alfonso C, Jann S, Massa R, Torreggiani A. Diagnosis, treatment and follow-up of the carpal tunnel syndrome: a review. Neurolog Sci. 2010;31(3):243–52.
26. Burns TM. Mechanisms of acute and chronic compression neuropathy. In: Dyck PJ, Thomas PK, editors. Peripheral neuropathy, 4th ed. Amsterdam: Elsevier; 2005. pp. 1391–402.
27. Sud V, Tucci MA, Freeland AE, Smith WT, Grinspun K. Absorptive properties of synovium harvested from the carpal tunnel. Microsurgery. 2002;22:316–9.
28. Tucci M, Freeland A, Mohamed A, Benghuzzi H. The role of proteoglycans in idiopathic carpal tunnel syndrome. Biomed Sci Instrum. 2005;41:141–6.
29. Wehbe MA, Schlegel JM. Nerve gliding exercies for thoracic outlet syndrome. Hand Clin. 2004;20(1):51–5.
30. Millesi H, Zoch G, Rath T. The gliding apparatus ofperipheral nerve and its clinical significance. Ann Chir Main Memb Super. 1990;9:87–97.
31. MacDermid JC, Doherty T. Clinical and electrodiagnostic testing of carpal tunnel syndrome: a narrative review. J Orthop Sports Phys Ther. 2004;34(10):565–88.
32. Lundborg G. Intraneural microcirculation. Orthop Clin North Am. 1988;19(1):1–12.
33. Yoon JH, Halper J. Tendon proteoglycans: biochemistry and function. J Musculoskeletal Neuronal Interact. 2005;5:22–4.
34. Hirata H, Nagakura T, Tsujii M. The relationshiot of VEGF and PGE2 expression to extracellular matrix remodelling of the tenosynovium in the carpal tunnel syndrome. J Pathol. 2004;205(5):606–12.
35. Ettema AM, Amadio PC, Zhao C. A histological and immunohistochemical study of the subsynovial connective tissue in idiopathic carpal tunnel syndrome. J Bone and Joint Surgery Am. 2004;86-A(7):1458–66.
36. Lundborg G. Nerve entrapment. In Lundborg G, editor. Nerve injury and repair. Philadelphia: Churchill Livingstone; 1988. pp. 102–48.
37. Colombini D, Occhipinti E, Cairoli S. Le affezioni muscoloscheletriche degli arti superiori e inferiori come patologie professionali; quali e a quali condizioni. Documento di consenso di un gruppo di lavoro nazionale. Med Lav. 2003;94(3):312–29.

38. Duncan I, Sullivan P, Lomas F. Sonography in the diagnosis of carpal tunnel syndrome. Am J Roentgenol. 1999;173:681–4.
39. El Miedany YM, Aty SA, Ashour S. Ultrasonography versus nerve conduction study in patients with carpal tunnel syndrome: substantive or complementary tests? Rheumatology. 2004;43:887–95.
40. Koyuncuoglu HR, Kutluhan S, Yesildag A, Oyar O, Guler K, Ozden A. The value of ultrasonographic measurement in carpal tunnel syndrome in patients with negative electrodiagnostic tests. Eur J Radiol. 2005;56:365–9.
41. Wiesler ER, Chloros GD, Cartwright MS, Smith BP, Rushing J, Walker FO. The use of diagnostic ultrasound in carpal tunnel syndrome. J Hand Surg Am. 2006;31:726–32.
42. Wong SM, Griffith JF, Hui AC, Tang A, Wong KS. Discriminatory sonographic criteria for the diagnosis of carpal tunnel syndrome. Arthritis Rheum. 2002;46:1914–21.
43. Gelberman RH, Eaton R, Urbaniak JR. Peripheral nerve compression. Instr Course Lect. 1994;43:31–53.
44. Sernik RA, Abicalaf CA, Pimentel BF, Braga-Baiak A, Braga L, Cerri GG. Ultrasound features of carpal tunnel syndrome: a prospective case-control study. Skeletal Radiol. 2008;37(1):49–53.
45. Yucel A, Yilmaz O, Babaoglu S, Acar M, Degirmenci B. Sonographic findings of the median nerve and prevalence of carpal tunnel syndrome in patients with Parkinson's disease. Eur J Radiol. 2008;67(3):546–50.
46. Duncan I, Sullivan P, Lomas F. Sonography in the diagnosis of carpal tunnel syndrome. Am J Roentgenol. 1999;173(3):681–4.
47. Buchberger W, Schon G, Strasser K, Jungwirth W. High-resolution ultrasonography of the carpal tunnel. J Ultrasound Med. 1991;10(10):531–7.
48. Chen P, Maklad N, Redwine M, Zelitt D. Dynamic high-resolution sonography of the carpal tunnel. Am J Roentgenol. 1997;168(2):533–7.
49. Martinoli C, Bianchi S, Gandolfo N, Valle M, Simonetti S, Derchi LE. US of nerve entrapments in osteofibrous tunnels of the upper and lower limbs. Radiographics. 2000;20:199–213.
50. Buchberger W, Judmaier W, Birbamer G, Lener M, Schmidauer C. Carpal tunnel syndrome: diagnosis with high-resolution sonography. Am J Roentgenol. 1992;159(4):793–8.
51. Klauser AS, Halpern EJ, De Zordo T, Feuchtner GM, Arora R, Gruber J, et al. Carpal tunnel syndrome assessment with US: value of additional cross-sectional area measurements of the median nerve in patients versus healthy volunteers. Radiology. 2009;250:171–7.
52. Moran L, Perez M, Esteban A, Bellon J, Arranz B, del Cerro M. Sonographic measurement of cross-sectional area of the median nerve in the diagnosis of carpal tunnel syndrome: correlation with nerve conduction studies. J Clin Ultrasound. 2009;37:125–31.
53. Pinilla I, Martín-Hervás C, Sordo G, Santiago S. The usefulness of ultrasonography in the diagnosis of carpal tunnel syndrome. J Hand Surg Eur. 2008;33:435–9.
54. Hobson-Webb LD, Massey JM, Juel VC, Sanders DB. The Ultrasonographic wrist-to-forearm median nerve area ratio in carpal tunnel syndrome. Clin Neurophysiol. 2008;119:1353–7.
55. Nakamichi K, Tachibana S. Ultrasonographic measurement of median nerve cross-sectional area in idiopathic carpal tunnel syndrome: diagnostic accuracy. Muscle Nerve. 2002;26:798–803.
56. Karadag YS, Karadag O, Cicekli E, et al. Severity of carpal tunnel syndrome assessed with high frequency ultrasonography. Rheumatol Int. 2010;30(6):761–5.
57. Mondelli M, Filippou G, Gallo A, Frediani B. Diagnostic utility of ultrasonography versus nerve conduction studies in mild carpal tunnel syndrome. Arthritis Rheum. 2008;59:357–66.
58. Sernik RA, Abicalaf CA, Pimentel BF, Braga-Baiak A, Braga L, Cerri GG. Ultrasound features of carpal tunnel syndrome: a prospective case-control study. Skeletal Radiol. 2008;37(1):49–53.
59. Yucel A, Yilmaz O, Babaoglu S, Acar M, Degirmenci B. Sonographic findings of the median nerve and prevalence of carpal tunnel syndrome in patients with Parkinson's disease. Eur J Radiol. 2008;67(3):546–50.

60. Buchberger W, Schon G, Strasser K, Jungwirth W. High-resolution ultrasonography of the carpal tunnel. J Ultrasound Med. 1991;10(10):531–7.
61. Chen P, Maklad N, Redwine M, Zelitt D. Dynamic high-resolution sonography of the carpal tunnel. Am J Roentgenol. 1997;168(2):533–7.
62. Martins RS, Siqueria MG, Simplicio H, Agapito D, Medeiros M. Magnetic resonance imaging of idiopathic carpal tunnel syndrome: correlation with clinical findings and electrophysiologic investigation. Clin Neurol Neurosurg. 2008;110:38–45.
63. Kunze NM, Goetz JE, Thedens DR, Baer TE, Lawler EA, et al. Localized measures of tendon impingement on the Median nerve within the carpal tunnel. 34th Annual meeting of the American society of biomechanics, 2010; August 18–21, Providence, Rhode Island. Abstract ID: 262. Poster Presentation. Tissue Mechanics Session.
64. Monagle K, Dai G, Chu A. Quantitative MR imaging of carpal tunnel syndrome. Am J Roentgenol. 1999;172(6):1581–6.
65. El Miedany Y, El Gaafary M, Youssef S, Ahmed I, Nasr A. Ultrasound assessment of the median nerve: a biomarker that can help in setting a treat to target approach tailored for carpal tunnel syndrome patients. Springerplus. 2015 Jan 13;4:13. doi: 10.1186/s40064-014-0779-4.
66. Somay G, Somay H, Cevik D, Sungur D, Berkman Z. The pressure angle of the Median nerve as a new magnetic resonance imaging parameters for the evaluation of carpal tunnel. Clin Neurol Neurosurg. 2009;111:28–33.
67. Keles I, Kendi ATK, Aydın G, Zo¨g G, Orkun S. Diagnostic precision of ultrasonography in patients with carpal tunnel syndrome. Am J Phys Med Rehabil. 2005;84:443–50.
68. Altinok T, Baysal O, Karakas HM, Sigirci A, Alkan A, Kayhan A. Ultrasonographic assessment of mild and moderate idiopathic carpal tunnel syndrome. Clin Radiol. 2004;59:916–25.
69. KeleŞ I, Karagülle Kendi AT, Aydin G, Zöğ SG, Orkun S. Diagnostic precision of ultrasonography in patients with carpal tunnel syndrome. Am J Phys Med Rehabil. 2005;84:443–50.
70. Yesildag A, Kutluhan S, Sengul N, Koyuncuoglu HR, Oyar O, Guler K, et al. The role of ultrasonographic measurements of the median nerve in the diagnosis of carpal tunnel syndrome. Clin Radiol. 2004;59:910–915.
71. Kim M, Jeon H, Park S, et al. Value of ultrasonography in the diagnosis of carpal tunnel syndrome: correlation with electrophysiological abnormalities and clinical severity. J Korean Neurosurg Soc. 2014;55:78–82.
72. Dejaco C, Stradner M, Zauner D, et al. Ultrasound for diagnosis of carpal tunnel syndrome: comparison of different methods to determine median nerve volume and value of power Doppler sonography. Ann Rheum Dis. 2012;0:1–6.
73. Miller TT, Reinus WR. Nerve entrapment syndromes of the elbow, forearm, and wrist. Am J Roentgenol. 2010;195(3):585–94.
74. Klauser AS, Halpern EJ, De Zordo T, et al. Carpal tunnel syndrome assessment with US: value of additional cross-sectional area measurements of the median nerve in patients versus healthy volunteers. Radiology. 2009;250:171–7.
75. Hobson-Webb LD, Massey JM, Juel VC, et al. The ultrasonographic wrist-to-forearm median nerve area ratio in carpal tunnel syndrome. Clin Neurophysiol. 2008;119:1353–7.
76. Hobson-Webb LD, Padua L. Median nerve ultrasonography in carpal tunnel syndrome: findings from two laboratories. Muscle Nerve. 2009;40:94–7.
77. Choi WK, Kang YK, Kim YH, Park EM. Diagnosis of carpal tunnel syndrome by diagnostic ultrasound. J Korean Acad Rehabil Med. 2001;25:134–9.
78. Buchberger W, Judmaier W, Birbamer G, Lener M, Schmidauer C. Carpal tunnel syndrome: diagnosis with high-resolution sonography. Am J Roentgenol. 1992; 159:793–8.
79. Kim H, Joo S, Han Z, et al. The nerve/tunnel index: A new diagnostic standard for carpal tunnel syndrome using sonography. J Ultrasound Med. 2012;31:23–9.
80. Bianchi S, Martinoli C. Wrist. In: Bianchi S, Martinoli C editors, Ultrasound of the musculoskeletal system. Heidelberg: Springer; 2007. pp. 445–67.
81. Millesi H, Zoch G, Rath T. The gliding apparatus of peripheral nerve and its clinical significance. Ann Chir Main Memb Super. 1990;9:87–97.

82. MacDermid JC, Doherty T. Clinical and electrodiagnostic testing of carpal tunnel syndrome: a narrative review. J Orthop Sports Phys Ther. 2004;34(10):565–88.
83. LaBan MM, MacKenzie JR, Zemenick GA. Anatomic observations in carpal tunnel syndrome as they relate to the tethered median nerve stress test. Arch Phys Med Rehabil. 1989;70(1):44–6.
84. van Doesburg M, van der Molen A, Henderson J, et al. Sonographic measurements of subsynovial connective tissue thickness in patients with carpal tunnel syndrome. J Ultrasound Med. 2012;31(1):31–6.
85. Ettema AM, Belohlavek M, Zhao C, Oh SH, Amadio PC, An KN. High-resolution ultrasound analysis of subsynovial connective tissue inhuman cadaver carpal tunnel. J Orthop Res. 2006;24:2011–20.
86. Mallouhi A, Pu¨ lzl P, Trieb T, Piza H, Bodner G. Predictors of carpal tunnel syndrome: accuracy of gray-scale and color Doppler sonography. Am J Roentgenol 2006;186(5):1240–5.
87. Abdel Ghaffar M, El-Shinnawy M, Fawzy H, Eldessouki S. Gray scale and color Doppler sonography in the diagnosis of carpal tunnel syndrome. Egy J Radiology Nuc Med. 2012;43:581–7.
88. Joy V, Therimadasamy AK, Chan YC, Wilder-Smith EP. Combined Doppler and B-mode sonography in carpal tunnel syndrome. J Neurol Sci 2011; 308(1–2):16–20.
89. Evans K, Roll S, Volz K, et al. Relationship between intraneural vascular flow measured with sonography and carpal tunnel syndrome diagnosis based on electrodiagnostic testing. J Ultrasound Med. 2012;31:729–36.
90. Sarria L, Cabada T, Cozcolluela R, Martinez-Berganza T, Garcia S. Carpal tunnel syndrome: usefulness of sonography. Eur Radiol. 2000;10:1920–5.
91. Wong SM, Griffi th JF, Hui AC, Lok SK, Fu M, Wong KS. Carpal tunnel syndrome: diagnostic usefulness of sonography. Radiology. 2004;232(1):93–9.
92. Yesildag A, Kutluhan S, Sengul N, et al. The role of ultrasonographic measurements of the median nerve in the diagnosis of carpal tunnel syndrome. Clin Radiol. 2004;59(10):910–5.
93. Wiesler ER, Chloros GD, Cartwright MS, Smith BP, Rushing J, Walker FO. The use of diagnostic ultrasound in carpal tunnel syndrome. J Hand Surg Am. 2006;31(5):726–32.
94. Sernik RA, Abicalaf CA, Pimentael BF, Braga-Baiak A, Braga L, Cerri GC. Ultrasound features of carpal tunnel syndrome: a prospective case-control study. Skeletal Radiol. 2008;37:49–53.
95. Mondelli M, Filippou G, Gallo A, Frediani B. Diagnostic utility of ultrasonography versus nerve conduction studies in mild carpal tunnel syndrome. Arthritis Rheum. 2008;59(3):357–66.
96. Nakamichi K, Tachibana S. Ultrasonographic measurement of median nerve cross-sectional area in idiopathic carpal tunnel syndrome: diagnostic accuracy. Muscle Nerve. 2002;26:798–803.
97. Kele H, Verheggen R, Bittermann HJ, Reimers CD. The potential value of ultrasonography in the evaluation of carpal tunnel syndrome. Neurology. 2003;61:389–91.
98. Altinok T, Baysal O, Karakas HM, et al. Ultrasonographic assessment of mild and moderate idiopathic carpal tunnel syndrome. Clin Radiol. 2004;59(10):916–25.
99. Visser LH, Smidt MH, Lee ML. High resolution sonography versus EMG in the diagnosis of carpal tunnel syndrome. J Neurol Neurosurg Psychiatry. 2008;79(1):63–7.
100. Swen WA, Jacobs JW, Bussemaker FE, de Waard JW, Bijlsma JW. Carpal tunnel sonography by the rheumatologist versus nerve conduction study by the neurologist. J Rheumatol. 2001;28:62–9.
101. Kwon BC, jung KI, Baek GH. Comparison of sonography and electrodiagnostic testing in the diagnosis of carpal tunnel syndrome. J Hand Surg Am. 2008;33(1):65–71.
102. Pastare D, Therimadasamy AK, Lee E, Wilder-Smith E. Sonography versus nerve conduction studies in patients referred with a clinical diagnosis of carpal tunnel syndrome. J Clin Ultrasound. 2009;37(7):380–93.
103. Seror P. Sonography and electrodiagnosis in carpal tunnel syndrome, an analysis of the literature. Eur J Radiol. 2008;67:146–52.

104. El Miedany Y, Ashour S, Youssef S, Mehanna A, Meky FA. Clinical diagnosis of carpal tunnel syndrome: Old tests-new concepts. Joint Bone Spine. 2008;75(4):451–7.
105. El Miedany Y, Ashour S, Youssef S, Mehanna A, Meky F. Diagnosis of carpal tunnel syndrome: How far can we rely on clinical assessment. Arthritis Rheum. 2006;54(9):S651.
106. US Department of Health and Human Services, Agency for Healthcare Research and Quality. National guideline clearinghouse website. http://www.guideline.gov/. Accessed 6 Nov 2011.
107. American Association of Electrodiagnostic Medicine. Practice parameter for electrodiagnostic studies in carpal tunnel syndrome: summary statement. American Association of Electrodiagnostic Medicine website. http://www.aanem.org/getmedia/7ddc9ef9-ee91-4b48-9c1a-53454313001e/CTS.pdf.aspx. Accessed 6 Nov 2011.
108. Jordan R, Carter T, Cummins C. A systematic review of the utility of electrodiagnostic testing in carpal tunnel syndrome. Br J Gen Pract. 2002;52:670–3.
109. Strickland JW, Gozani SN. Accuracy of in-office nerve conduction studies for median neuropathy: a meta-analysis. J Hand Surg Am. 2011;36:52–60.
110. Lew HL, Date ES, Pan SS, Wu P, Ware PF, Kingery WS. Sensitivity, specificity, and variability of nerve conduction velocity measurements in carpal tunnel syndrome. Arch Phys Med Rehabil. 2005;86:12–6.
111. Salerno DF, Franzblau A, Werner RA, et al. Reliability of physical examination of the upper extremity among keyboard operators. Am J Ind Med. 2000;37:423–30.
112. Kuhlman KA, Hennessey WJ. Sensitivity and specificity of carpal tunnel syndrome signs. Am J Phys Med Rehabil. 1997;76:451–7.
113. Mondelli M, Passero S, Giannini F. Provocative tests in different stages of carpal tunnel syndrome. Clin Neurol Neurosurg. 2001;103:178–83.
114. Atroshi I, Gummesson C, Johnsson R, Ornstein E. Diagnostic properties of nerve conduction tests in population-based carpal tunnel syndrome. BMC Musculoskelet Disord. 2003;4:9–14.
115. Koyuncuoglu HR, Kutluhan S, Yesildag A, Oyar O, Guler K, Ozden A. The value of ultrasonographic measurement in carpal tunnel syndrome in patients with negative electrodiagnostic tests. Eur J Radiol. 2005;56(3):365–9.
116. Pastare D, Therimadasamy AK, Lee E, Wilder-Smith E. Sonography versus nerve conduction studies in patients referred with a clinical diagnosis of carpal tunnel syndrome. J Clin Ultrasound. 2009;37(7):380–93.
117. Descatha A, Huard L, Duval S. Letter to the editor: the sensitivity and specificity of ultrasound for the diagnosis of carpal tunnel syndrome: A meta-analysis. Clin Orthop Relat Res. 2011;469:901–2.
118. Roll SC, Case-Smith J, Evans KD. Diagnostic accuracy of ultrasonography vs electromyography in carpal tunnel syndrome: a systematic review of literature. Ultrasound Med Biol. 2011;37:1539–53.
119. Wang LY, Leong CP, Huang YC, Hung JW, Cheung SM, Pong YP. Best diagnostic criterion in high-resolution ultrasonography for carpal tunnel syndrome. Chang Gung Med J. 2008;31:469–76.
120. Mallouhi A, Pultzl A, Trieb T, Piza H, Bodner G. Predictors of carpal tunnel syndrome: accuracy of gray-scale and color Doppler sonography. Am J Roentgenol. 2006;186:1240–5.
121. Rahmani M, Ghasemi Esfe AR, Vaziri-Bozorg SM, Mazloumi M, Khalilzadeh O, Kahnouji H. The ultrasonographic correlates of carpal tunnel syndrome in patients with normal electrodiagnostic tests. Radiol Med. 2011;116:489–96.
122. Evans K, Roll S, Volz K, Freimer M. Relationship between intraneural vascular flow measured with sonography and carpal tunnel syndrome diagnosis based on electrodiagnostic testing. J Ultrasound Med. 2012; 31(5):729–36.
123. Ly-Pen D, Andréu J, de Blas G, et al. Surgical decompression versus local steroid injection in carpal tunnel syndrome. Arthritis Rheum. 2005;52(2):612–9.
124. Ferraccioli G, Alivernini S, Gremese E. Biomarkers of joint damage in rheumatoid arthritis: where are we in 2013? J Rheumatol. 2013;40:1244–6.

125. Katz JN, Simmons BP. Clinical practice: carpal tunnel syndrome. N Engl J Med. 2002;346:1807–12.
126. Racasan O, Dubert T. The safest location for steroid injection in the treatment of carpal tunnel syndrome. J Hand Surg. 2005;30:412–4.
127. Graham RG, Hudson DA, Solomons M, Singer M. A prospective study to assess the outcome of steroid injections and wrist splinting for the treatment of carpal tunnel syndrome. Plast Reconstr Surg. 2004;113:550–6.
128. Dubert T, Racasan O. A reliable technique for avoiding the median nerve during carpal tunnel injections. Joint Bone Spine. 2006;73:77–9.
129. Smith J, Wisniewski S, Finnoff J, Payne J. Sonographically guided carpal tunnel injections: the ulnar approach. J Ultrasound Med. 2008;27:1485–90.
130. Wood MR. Hydrocortisone injections for carpal tunnel syndrome. Hand. 1980;12:62–4.
131. Propeck T, Quinn TJ, Jacobson JA, et al. Sonography and MR imaging of bifid median nerve with anatomic and histologic correlation. Am J Roentgenol. 2000;175:1721–5.
132. Grassi W, Farina A, Filipucci E, Cervini C. Intralesional therapy in carpal tunnel syndrome: a sonographic-guided approach. Clin Exp Rheumatol. 2002;20:73–6.
133. McNally E. Musculoskeletal interventional ultrasound. In: McNally E, editor. Practical musculoskeletal ultrasound. New York: Elsevier; 2005. p. 293.
134. Oh S, Belohlavek M, Zhao C, et al. Detection of differential gliding characteristics of the flexor digitorum superficialis tendon and subsynovial connective tissue using color Doppler sonographic imaging. J Ultrasound Med. 2007;26:149–55.

125. Katz JN, Simmons BP. Clinical practice. Carpal tunnel syndrome. N Engl J Med. 2002;346:1807–12.

126. [illegible] C, [illegible] P. The safest location for steroid injection in the treatment of carpal tunnel syndrome. J Hand Surg. 2003;30:412–4.

127. [illegible]

128. [illegible]

129. [illegible]

Chapter 11
Soft Tissue Rheumatism

Adham Aboul Fotouh Khalil MD, MSc and Cristina Hernández-Díaz MD, MS

The term "soft tissue rheumatism" refers to a wide range of musculoskeletal disorders or syndromes characterized by non-joint-related pain and limited motion, and is usually classified as *diffuse* or *local*. Local soft tissue rheumatism includes tendon, tendon sheath, muscle, ligament, bursa, fascia, and subcutaneous tissue pathology. Its etiology is usually unknown [1]. Ultrasound is an excellent imaging technique to assess these pathologies, especially because it confers the additional advantage of being dynamic, a very important advantage in assessing periarticular structures in real time. It also provides high structural detail in axial resolution in gray scale, and when needed with color/power Doppler.

Knowledge of basic anatomy and pathology of joints and soft tissues are essential for the evaluation of soft tissue lesions [2].

Classification of these entities involves articular regions, structures, or mechanisms of damage [3]. For the purpose of this chapter, ultrasound evaluation of soft tissue rheumatism is divided into specific structures, sonographic changes, and assessment key points of the main pathology.

Normal sonoanatomy knowledge is a must. Tendon pathology is an area where ultrasound excels.

Tendon

Tendons are highly specialized structures that attach muscle to bone, allowing joint movement and providing for stabilization of motion. Its histological composition of collagen fibers provides high tensile strength. Consequently, the mechanism of

A. A. F. Khalil (✉)
New Kasr El Aini Teaching Hospital, Cairo University, Cairo, Egypt
e-mail: dr_adham@hotmail.com

C. Hernández-Díaz
Laboratorio de Ultrasonido Musculoesquelético y Articular, Instituto Nacional de Rehabilitación, Mexico City, Mexico
e-mail: cristy_hernandez@hotmail.com

Y. El Miedany (ed.), *Musculoskeletal Ultrasonography in Rheumatic Diseases*,
DOI 10.1007/978-3-319-15723-8_11

lesion is complex. Noninflammatory damage includes impingement, hypoxia, and rupture [4]. Tendons are divided into two types: type I is surrounded by paratenon and type II is surrounded by a tendon sheath with a cell lining identical to synovium. Based on the angle and mobility, enthesis (tendon attachment to bone) can be fibrous or fibrocartilaginous [5]. All these characteristics define pathology.

Tendon Instability

Tendon instability normally occurs in tendons with synovial sheaths due to injury of the surrounding structures, such as ligaments, retinacula, or annular pulleys—mechanisms are vast and regionally dependent.

Instability could lead to subluxation (permanent or intermittent) or luxation (dislocation). Ultrasound may reveal the degree of damage and associated findings, such as effusion and tears. We see that a dynamic examination is a must to enhance lesions when possible. Short-axis scanning is preferred, but both planes should be assessed especially for beginners [6, 7].

The main locations of instability are the shoulder, hand, and ankle (Table 11.1).

In the shoulder, the long head of the biceps tendon is the most representative lesion in association with a supraspinatus tear and coracohumeral ligament tear or

Table 11.1 Tendon instability

Location	Structure	Main findings	Gamuts
Shoulder	Long head bicipital tendon	Empty groove Intraarticular portion of the long head of the biceps tendon is the most frequent location Coracohumeral ligament could be torn alone or in association with supraspinatus tear	Two biceps tendons that really represent a longitudinal tear The absence due to complete tendon tear Tendinopathy
Hand	Flexor tendons	Empty anatomical site Snapping sensation Flexor contracture	Partial distal tears Painful flexion with dynamic scanning due to associated trigger finger
	Extensor carpi ulnaris tendon	Dislocation is normally volar side There is no empty groove sign Dynamic scan should include supination, ulnar deviation, and flexion movements	Dorsal subluxation directs to think about distal radius fracture
Ankle	Peroneal tendons	Always dislocate anteriorly Partial or complete retinaculum tear Intra-sheath peroneal tendon luxation	Longus and brevis peroneal tendons can reverse their anatomical position during dynamic scan Peroneus quartus is a frequent accessory muscle in the ankle (22 %) Look for avulsion near the retinaculum or malleolus

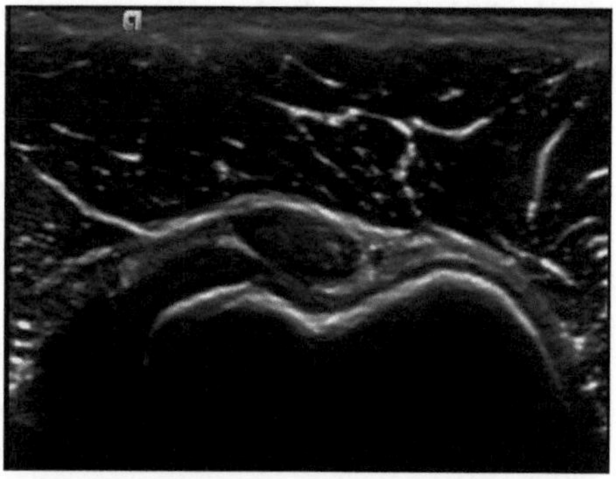

Fig. 11.1 Shallow bicipital groove and tendency of biceps tendon subluxation. Note the disappearance of the hypoechoic triangle that lies between the tendon and the medial border (*left* side of the figure) of the upper portion of the groove

lesion. The most frequent location is the intra-articular portion due to its curvilinear course and reflection over the humeral head. The sonographer should look for hypoechoic triangle that lies between the tendon and the medial border of the upper portion of the groove that in the case of disappearance, indirectly means subluxation [8], the demonstration of which requires dynamic external rotation (Fig. 11.1).

In the hand, damage of the A2 and A4 annular pulleys in the flexor tendons are the most common sites. The ring and middle fingers are commonly involved. Excessive traction on the flexor tendons when the tendons are flexed causes anterior bowstringing which no longer lines against the bone plane; if this rupture is not diagnosed, flexion contractures can be present alongside other tendon lesions [9].

Other sites of lesions are the dorsal carpal bones between the carpometacarpal interline and the extensor tendons at the level of the radius or fingers [10]. Extensor compartments contribute widely to subluxation and wrist instability; sixth compartment is the most commonly involved due to its particular anatomy. The extensor carpi ulnaris tendon is normally held between a fibro-osseous sheath and instability may result in either subluxation or intermittent dislocation. Dynamic maneuver should be performed doing progressive supination of the forearm with ulnar deviation and flexion while the probe is in the short axis at the ulnar styloid [11, 12].

Ankle pathology involves peroneal tendons in patients with chronic ankle sprains and sport injuries. Normally, laxity of the superior peroneal retinaculum, valgus, and accessory tendons are involved in the etiology; although it can be voluntary or could secondarily involve disruption of the superior peroneal retinaculum in various grades. Clinically, it is revealed with forced dorsiflexion and eversion of the foot, showing a palpable snapping or discomfort with pain. Ultrasound examination in the short axis should be dynamic to look for abnormal tendon movement over the malleolus or intra-sheath luxation [13].

Tendinopathy: Tears

The damage produced in the collagen fibers of the tendon is usually referred to as *tendinopathy*. This term refers to specific changes in the appearance of the tendon, histological changes in the collagen and tendon disruption that makes collagen fibers thinner with the loss of the hierarchical structure with consequent wound and subsequent healing. With ultrasound, the primary findings are tendon thickening, heterogeneous hypoechogenicity, and the loss of parallel superficial and deep tendon borders. These first changes, referred to as *tendinosis,* can persist for months or even years, but if damage continues, it will lead to hypoxia and myxoid degeneration that cause ischemia and focal zones of hypovascularization, resulting in tendon rupture [14, 15].

Normally tears are not produced unless external and severe trauma or excessive loading forces involve the musculotendinous system. When a tendon tear appears, it typically does so at the insertion near the bone or at the myotendinous junction. Tears can be partial, complete, or intrasubstance, depending on the particular region or the mechanism that caused it [16–18] (Table 11.2.).

Table 11.2 Tendon tears

Mechanism	Tendon	Main findings	Gamuts
Overuse damage	Supraspinatus tendon Long head of the biceps tendon Extensor and flexor tendons of the elbow Patellar tendon Achilles tendon	Tendinosis is focal or diffuse, poorly demarcated Double cortex appearance and cortical irregularities Spur formation in epicondyle, patella, tibia, and calcaneus Non-insertional paratenonitis with partial tears in Achilles tendon injury could be associated Bursal fluid often present	Compression with the probe to obliterate space when there is fluid with the tear Intrasubstance tears are hypoechoic areas within the tendon not related to articular or bursal surfaces Biceps tendon, complete tears, tracing the muscle belly in the arm differentiates from dislocation Tendon calcification could be present Ultrasound scanning in the knee with slight flexion might be necessary In Achilles tendon distal tears, plantaris tendon insertional portion could be intact
Excessive frictional forces Steroid treatment	All regions	Tendinosis Paratenonitis Tears	Avoid anisotrophy Initial stages only focal regions of low echogenicity Steroids can be seen as intrasubstance hyperechoic spots depending on the time of injection
Inflammatory and metabolic systemic diseases	Quadriceps tendon Patellar tendon Extensor and flexor digitorum tendons Posterior tibial tendon	Bursitis could be present Enlargement or thickening of the tendons Loss of definition of tendon margins Anechoic ring at the extensor retinaculum or the malleolus	Doppler signal could be present in active diseases

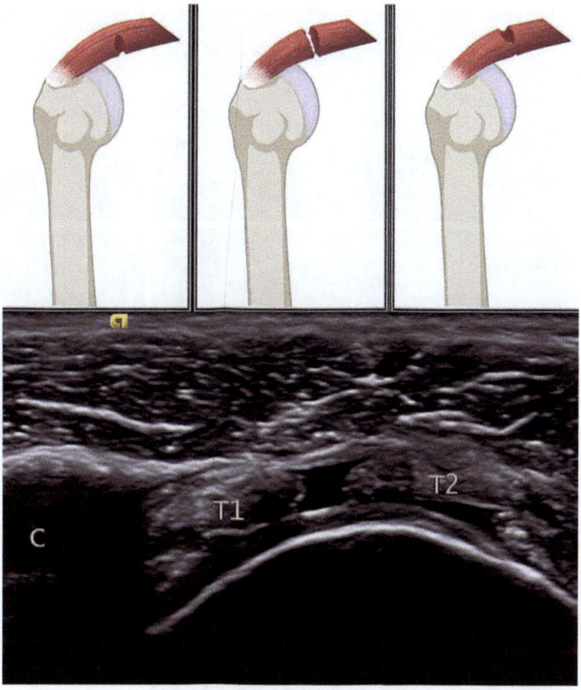

Fig. 11.2 Supraspinatus tendon tears. The drawing shows different examples of partial and complete tears. The ultrasonography image shows complete tear with tendon edges (T1 and T2) retraction. *C*=coracoid process. Note the anechoic fluid replacing the tendon. Also of relevance is the double cortex sign (cartilage interface sign)

Partial tears appear as a region devoid of tendon fibers while complete tears cause tendon disruption with visible tendon stumps. Dynamic imaging shows approximation (Fig. 11.2).

Paratendinitis

Paratendinitis refers to an inflammatory process that involves the vascular paratenon which surrounds the patellar and Achilles tendons mainly. It represents a continuous inflammatory process that causes high-grade hypoxic tendinosis. Ultrasound features include thickening of the paratenon, irregular tendon margin, and fluid, with or without color/power Doppler. Bursitis is frequently associated and reflects irritation or local overload and work stress. Soleus and gastrocnemius muscle injuries could be present in association with Achilles paratendinitis (Fig. 11.3).

Tenosynovitis

In tendons with a synovial sheath, inflammation is secondary to mechanical factors (repetitive microtrauma, overuse, or osseous friction), foreign bodies, infection, tumors, or inflammatory entities (e.g., rheumatoid arthritis). Acutely, inflammation

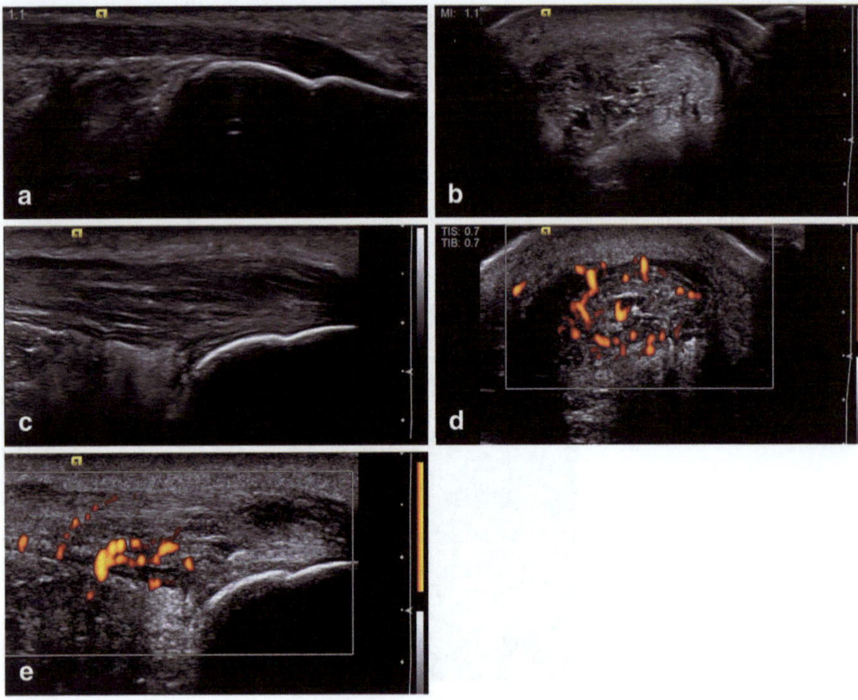

Fig. 11.3 Achilles tendinitis and paratendinitis. (**a**) Normal Achilles tendon in long axis. (**b**) Short axis. (**c**) Long-axis gray scale with increased thickness, hypoechoic changes, fibril separation, edema, and loss of normal tendon boundary. (**d**) Short axis. (**e**) Long-axis power Doppler for the same findings with positive signal in and around the tendon

is characterized by fluid encircling the tendon forming a ring or halo around it; chronic disease causes tendon sheath thickening that may cause tendon entrapment and/or thicker retinacula or pulley (e.g., de Quervain's disease, trigger finger of flexor and extensor tendons, peroneal entrapment) [9, 19] (Figs. 11.4, 11.5).

In the hand, effusion is easy to identify when it is proximal to the metacarpal head. In the fingers, effusion assumes a lobulated appearance and creates a discontinuous array of the pulleys. In infection processes, there is no characteristic image but tenosynovitis seems to be more hyperechoic and the subcutaneous tissue appears hyperechoic with thickening.

Special Conditions: Trigger Finger

Trigger finger is the common term for stenosing tenosynovitis, usually affecting the flexor tendons in the thumb and index finger and involves A1 pulley pathology, although it can affect other fingers and other pulleys as well. It is a painful condition in which the tendon locks when flexed and is difficult to extend due to inflammation of its tendon sheath, provoking a click sound when the tendon passes through it. The normal pulley, visualized by ultrasound, appears as a hypoechoic band superficial

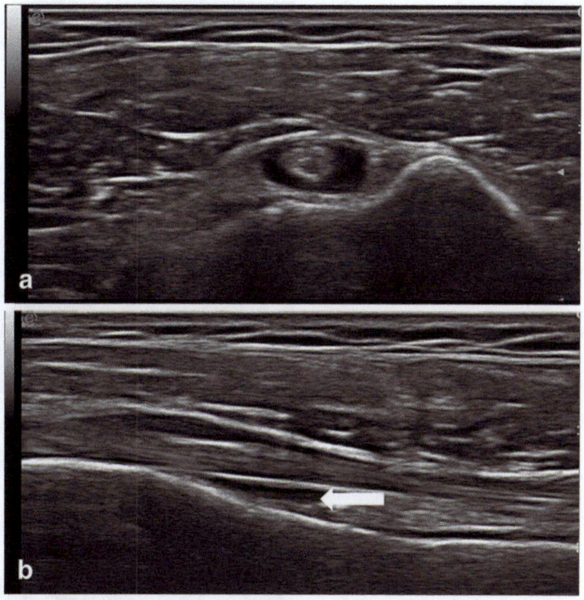

Fig. 11.4 Tenosynovitis of the biceps tendon. (**a**) Short- and (**b**) in long-axis views demonstrate effusion (*arrow*)

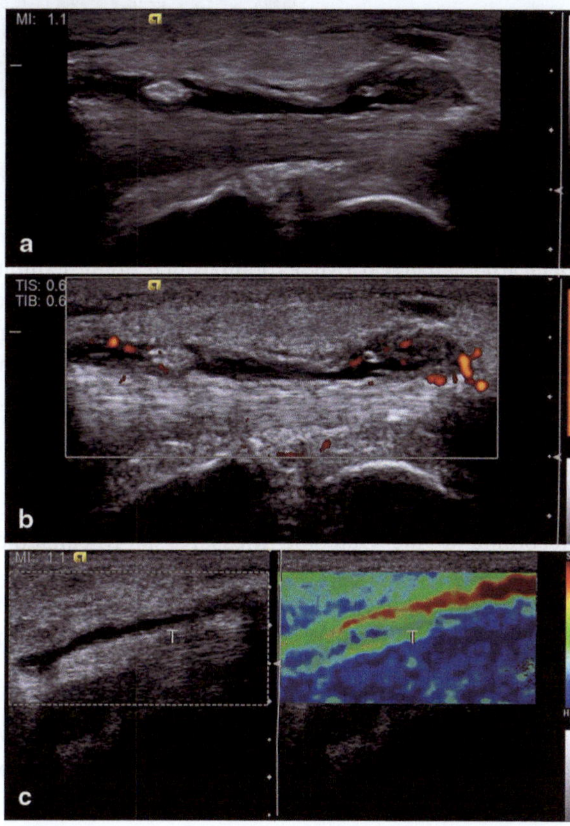

Fig. 11.5 Tenosynovitis in different ultrasonographic modalities. (**a**) In gray scale showing anechoic effusion together with echogenic synovial hypertrophy. (**b**) With power Doppler confirming synovial hypertrophy as well as activity. (**c**) With elastography in which synovial effusion appears softer than the tendon

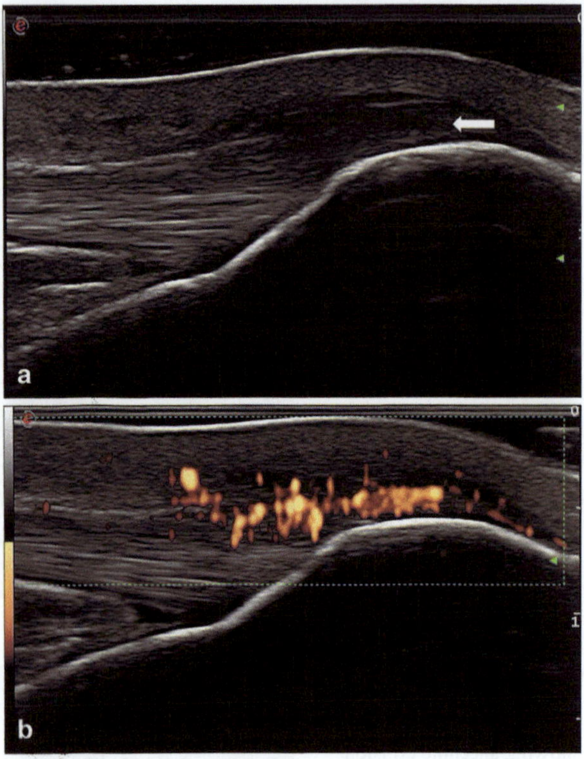

Fig. 11.6 Enthesopathy of the distal patellar tendon in long axis. (**a**) Gray scale shows increased thickness and hypoechoic changes (*arrow*). (**b**) Power Doppler signal in the same hypoechoic area

to the flexor tendon sheath. In the short axis, lateral expansion looks hypoechoic causing an anisotropic artifact that could be reduced by tilting the probe. In trigger finger, the pulley looks like a diffuse hypoechoic thickened band at the level of the metacarpophalangeal (MCP) joint which might exhibit Doppler signal. Dynamic scanning allows for the detection of changes in the shape of the synovial sheath and tendon pathology such as peritendinous effusion, tendinosis, or tenosynovitis. Additionally, ultrasound allows for the guidance of local infiltration therapy [19, 20].

Enthesopathy (Noninflammatory)

According to the Outcome Measures in Rheumatoid Arthritis Clinical Trials' (OMERACT) definition, enthesopathy is an "abnormally hypoechoic (loss of normal fibrillar architecture) and/or thickened tendon or ligament at its bony attachment (may occasionally contain hyperechoic foci consistent with calcification), seen in 2 perpendicular planes that may exhibit Doppler signal and/or bony changes including enthesophytes, erosions, or irregularity" [21].

In non inflammatory lesions, the most common injury mechanism is sport related or direct trauma. Affected regions are the elbow, hand, knee, and ankle (Fig. 11.6).

Surgical Changes in Tendon

Surgical treatment in any tendon has the main objective of restoring function—indications vary but the appearance depends on the various processes tendons undergo to heal. These conditions are related to histological healing, complete tendon fiber restitution, and clinical and functional evolution [22]. Histological changes are related to cell infiltration and neovascularization that could be visualized with color Doppler signal present at the site of healing or at the most painful site when clinical exam is performed. With the use of novel techniques such as contrast-enhanced ultrasound, it is possible to visualize the increased vascular flow in the areas of interest, where Doppler might fail [23].

Depending on the history of the tendon injury, its location, time between injury and surgery, surgical technique performed, postoperative complications and course are the findings we have to look for in the tendon. It is important to note that sutures, surgical wires, or intratendinous surgical material could be found during the ultrasound exam. We also have to remember that, depending on the time after surgery, the tendon can present with different lesions as shown in Table 11.3 (Fig. 11.7).

It is important to note that in tendons with tendon sheaths, the main lesion is tenosynovitis; in tendons without it, lesions such as tendinopathy, tendonitis, and paratendinitis could be present. Tendon tears after surgery deserve a special mention—these lesions can appear in the repaired tendon or within the substance around the lesion of importance and needs to be confirmed in longitudinal and transverse views because anechoic spots do not necessarily mean rupture and could be part of the healing process. In the operated tendon, complete tears can occur in the first month and it is mandatory to look at the other areas, probably away from the injury site and hematoma. Structures such as retinaculum, muscle, and ligaments at the peripheral surface surrounding the affected tendon can be damaged [24, 25].

In general, when a tendon is damaged, it loses its mechanical properties—ultrasound gives information of the lesion in many aspects, both in gray scale and Doppler, but new techniques such as elastography improves the diagnostic capability of conventional ultrasound. This technique involves manual compression of the tendon to generate tissue deformations. The strain distribution of the involved region could be compared before and after this tissue deformation (axial elastography). While this technique could be highly operator dependent and erratic regarding reproducibility, the most recent shear elastography, which depends on the intrinsic tissue elasticity shown, seems promising (Fig. 11.8).

The most studied tendon with elastography has been the Achilles followed by the common extensor tendon in the elbow. In the Achilles, the normal tendon is harder and more homogeneous when compared with surgically repaired tendons which appear softer and more heterogeneous. Although elastography is still in the process of being validated for use in rheumatology and musculoskeletal specialties, the technique has proven useful in demonstrating changes in tendon architecture and is a good method to improve the assessment of other structures as well [26].

Table 11.3 Characteristics of operated tendons

Characteristics	Ultrasound features	Description
Morphology	Thickness and width	Loss of the normal pattern Thicker the first 3 months, irreversible Tendon stumps cause focal thickness or callus Re-rupture: Thinner when re-ruptured Loss of tendon pattern and/or continuity of the tendon Anechoic image between the opposite parts of the tear and/or hematoma
	Contour	Circumferential, hypoechoic, peritendinous area after surgical repair up to 3 months Hypoechoic–anechoic halo surrounding the repaired tendon
	Continuity	Anechoic images in the repaired zone A re-rupture could appear in the repair zone
Structure	Internal	Loss of fibrillar pattern Heterogeneous structure Hypoechoic areas surrounding the stitches and metal devices if present
Vascularization	Intratendinous Peritendinous	No vascularization immediate to surgeryIntrasubstance the first month, small or few Doppler spotsHypervascularization next 3 months No Doppler after 6 months In peritendinous areas could be no Doppler signal at any time
Dynamic scan	Active Passive	Immediate after surgery, there is no movement at all The first 6 months, reduction of movement is observed and depends on the type of surgery performed and the repaired surface
Other	Associated lesion	Synovitis Enthesitis Cortical irregularities Intratendinous ossification or calcification Surgical artifacts: sutures, surgical wires, or intratendinous surgical material Adhesions
	Elastography	Soft Heterogeneous
	Contrast-enhanced ultrasound	Displayed the maximum intensity of vascularization in suture anchor and/or the tendon peripheral structures (bursae, muscle, peripheral bursal fat stripe) after surgery

Muscle

Skeletal muscle is the largest tissue in the human body, composed of parallel fibers surrounded by connective tissue that is organized into group of bundles called fascicles which are in turn surrounded by tissue called the perimysium and finally enclosed in the epimysium. Due to its function and anatomic position, these fibers

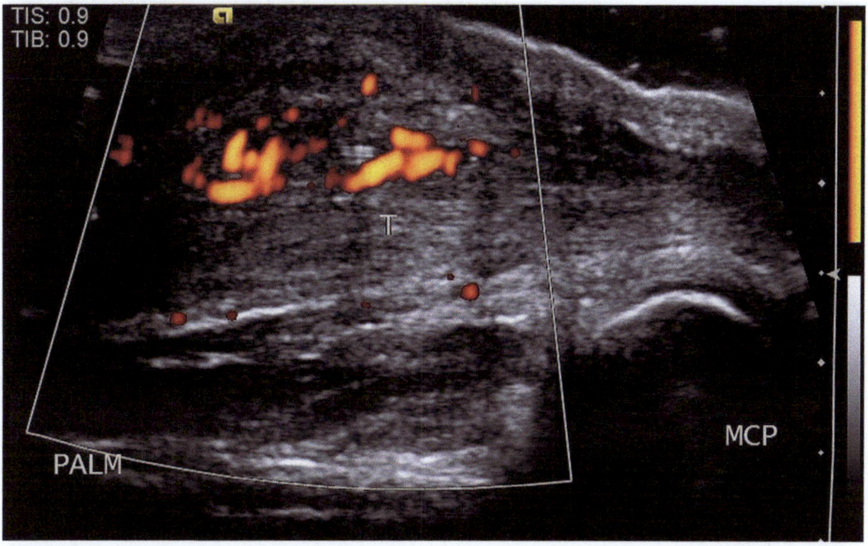

Fig. 11.7 Postoperative tendon (*T*). The flexor tendon is thickened and surrounded by thick echogenic tissue that exhibits Doppler signal, representing adhesions

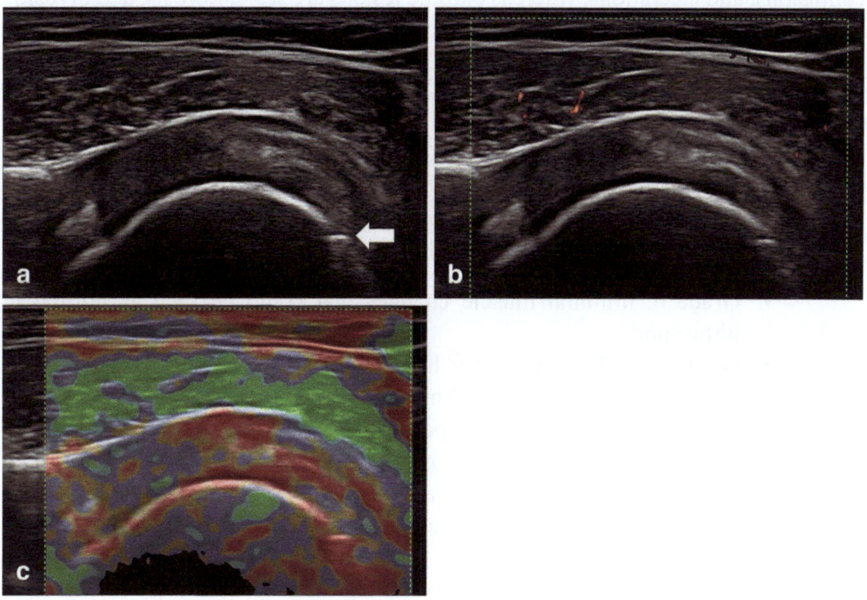

Fig. 11.8 **(a)** Supraspinatus tendon 2 years postsurgical repairing, the image is in longitudinal view and gray scale exhibit as hypoechoic blurred structure. **(b)** Doppler signal is located in the muscle. **(c)** Elastography shows a marked red area in the zone of the repair (*soft*); an anchor is located distally (*arrow*).

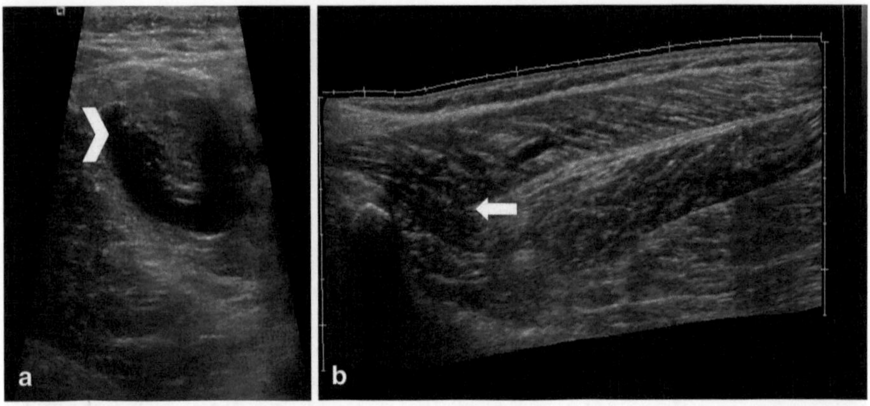

Fig. 11.9 Muscle showing the central aspect of the muscle tear (*arrow head*) in short axis, and B grade II tear (*arrow*) in long axis

are susceptible to lesions. Ultrasound appearance of the muscle is affected by many factors including angulation of the ultrasound beam. Sport is the most common cause of injury [27].

Classification of muscle lesions is *acute* and *chronic*. Acute lesions are tears, contusions, and lacerations, whereas chronic lesions are scars, fibrosis, and myositis ossificans:

1. Strain, tears, and lacerations commonly result from over elongation; signs are localized and cause tender points that increase with active movements [28].

    a) The main objective of ultrasound in acute lesions is to determine the severity and extent of the lesion. There are several systems for classifying lesions though none of them are accurate or globally accepted. Ultrasound features can however be graded into:

        - Grade 0: minimal muscle changes that could be seen as normal in ultrasound.
        - Grade I: focal or generalized hyperechoic areas located within a bundle, represents $\leq 5\%$ of the total muscular area affected.
        - Grade II: loss of muscle or perimysial striation or discontinuity of the fibers with vascularization; loss of echogenicity and a hyperechoic halo surrounding the myotendinous junction or hematoma. Lesion involves more than 5 % of the muscle area. Dynamic ultrasound could reveal the bell clap sign (Fig. 11.9).
        - Grade III: complete disruption of the muscle and loss of longitudinal fiber integrity; associated with hematoma and/or muscle retraction (Fig. 11.10).

    b) *Hematoma* healing starts immediately after the lesion: firstly, the hematoma undergoes a process of liquefaction, with subsequent resorption during which showing a very heterogeneous appearance with areas of

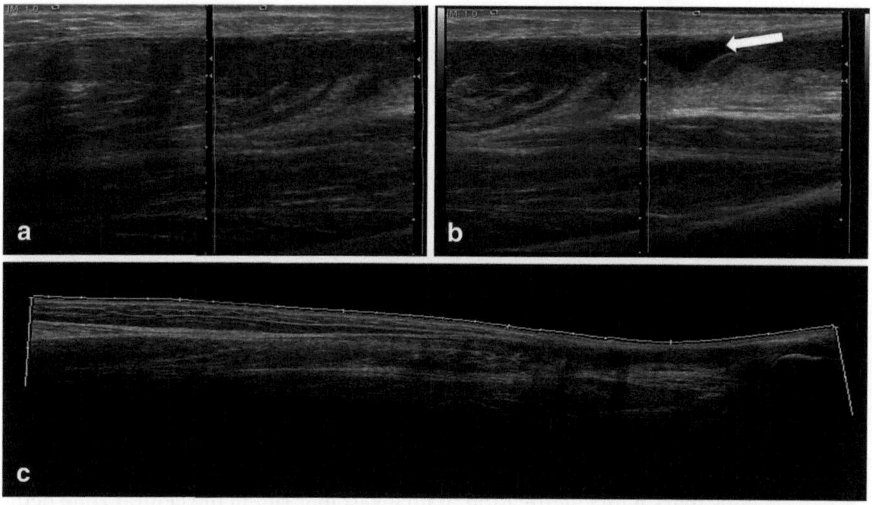

Fig. 11.10 **(a)** Medial gastrocnemius muscle long axis, in gray scale with muscle grade II tear. **(b)** double screen view in long axis of the medial soleus muscle with a hematoma (*arrow*), and (c) panoramic view of the Achilles tendon and soleus muscle in which tendon rupture is shown and muscle tear is depicted

reflective material or fibrous septae. Over time, these areas disappear and muscle fibers reorganize and appear normal—complete recovery can take weeks or months. With repeated trauma, granulomatous tissue forms and a scar appears. If the hematoma fails to resolve, a cyst forms and a discrete area devoid of reflectivity with acoustic shadowing is seen [29].

c) *Contusion* is the lesion that results from direct blunt impact and is most frequently associated with sport. Muscle contusion represents a capillary and muscle fiber disruption with microhemorrhage, dissecting fibers and leading to inflammation—ultrasound shows mild or focal muscle swelling against the background of the undamaged muscle tissue, secondary to hematoma in the first 24–48 h that normally resolves leaving no permanent damage. Over the next 2–3 days, collections become hypo- or anechoic increasing in echogenicity as time passes. After few weeks, the hematoma organizes and focal scar tissue develops if the contusion is large [30].

d) *Acute compartmental syndrome* arises following the trauma due to hematoma formation and muscle swelling—ultrasound can be helpful in identifying focal collections which, if drained, can decompress the compartment.

2. Chronic muscle injury is the result of abnormal healing or extended lesions as a consequence of untreated acute lesions. The main chronic lesions are:

   a) *Scars* that appear as linear or irregular echogenic structures surrounded by hypoechoic zones that are usually found at the fascial interface or myotendinous junction.

b) *Muscle herniation* results from fascial defects that allow extra compartmental herniation, commonly in the lower limb, which become prominent with muscle contraction. Ultrasonographically, there is normal muscle through the focal epimysial defect; longitudinal images show perimysium bowed into the defect. If the defect is large, bell clap sign is observed with active motion.
c) *Atrophy* is related to chronic inflammation; fatty infiltration and decreased muscle fibers result with an increased echogenicity of the muscle involved.
d) *Myositis ossificans* is a benign non-neoplastic condition, often seen in young people, following muscle trauma and is defined as heterotopic bone or cartilage formation in or adjacent to muscle. There are four types of the condition but the most common is post-traumatic. Phase changes can be detected by ultrasound; during early phase (precalcified), ultrasound can show a vascular solid mass with Doppler signal present at the periphery, but caution and vigilance are required because this image can also be seen in sarcoma. Late phase (calcified) appears as a hypoechoic or heterogeneous mass with calcification, with a vascular rim in the central zone demonstrating enhanced Doppler signal. Progressively, Doppler disappears and an acoustic shadowing appears due to peripheral ossification [31].

3. Inflammatory entities can injure muscle with infection and autoimmune disease being the most common causative factors:

   a) In *infectious myositis*, mix echogenicity is seen because of the different types which include purulent, tubercular, and viral. The muscle is edematous, thickened, and tender with swelling of surrounding tissues. Anechoic or hypoechoic fluid can be present but it is not specific; septations may be present between the collections and Doppler signal secondary to vascularization may be increased in the periphery. In chronic infections, loss of muscular pattern due to atrophy is visualized [32].
   b) *Inflammatory myopathy* show consistent but nonspecific patterns. Inflammatory changes include hyperechogenic muscle fibers, hypoechogenic fibradipose septae with increased muscle diameter. While atrophic criteria include increased echogenicity of the muscle with reduced diameter of the fibers. Polymyositis hyperechogenicity was more prominent than in dermatomyositis and other conditions. Elastography results correlated well with decreased elasticity, probably due to fibrosis and atrophy. While in few cases, increased elasticity with *soft* look might be due to fatty infiltration [33] (Fig. 11.11).

Ligament

Ultrasound is a good technique to visualize ligaments. Ligaments are fibrous structures that attach bone to bone and that could be intrinsic or extrinsic, intra or extracapsular. On ultrasound examination, ligaments appear as band-like structures, usually trilaminar and hyperechoic, being slightly more compact when compared to

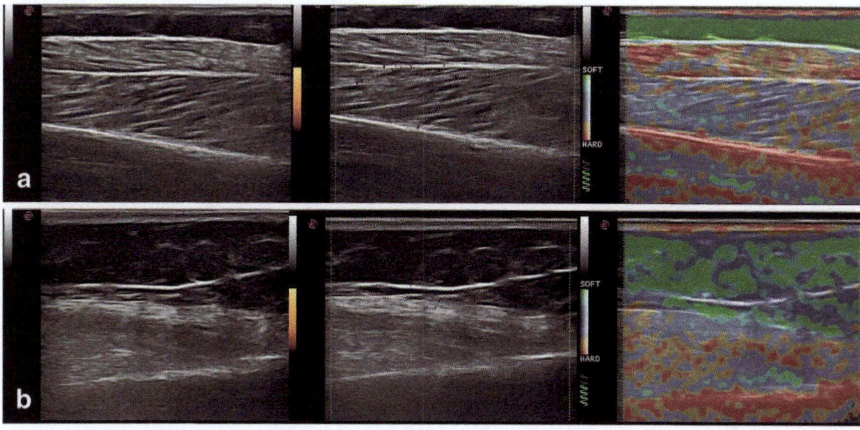

Fig. 11.11 Myositis in gray scale, Doppler, and elastography. **(a)** Hyperechoic gastrocnemius muscle fibers with hypoechoic fibroadipose septae, reduced diameter of the fibers, and reduced muscle elasticity. **(b)** Deltoid muscle with loss of muscle pattern and increased elasticity due to fatty infiltration

tendons. All regions have ligaments but not all can be seen by ultrasound and most of their lesions are related to sports injuries. Ligaments that are usually involved in noninflammatory or sport pathology are the collateral lateral ligament in the knee, collateral ligament in the elbow, collateral ulnar ligament in the first MCP joint, and the ligaments in the ankle. We summarize the most important lesions not related to sports for each region in Table 11.4 [34–44].

When ligaments rupture, lesions present in three stages all of which can be seen by ultrasound. At the beginning, the ligament looks hypoechoic and thickened but continuous. If second stage is present, discontinuity is seen with hypoechoic and thickened appearance. In the final stage, full discontinuity is present and signs of hemorrhage could be observed. In acute lesions, Doppler signal can be present. Dynamic examination is recommended [7] (Fig. 11.12).

Bursae

Bursae are normally seen as hypoechoic bands between two layers of hyperechoic fat. Bursitis refers to a distention of the band, usually beyond 2 mm, seen as a hyperechoic thickening of the walls. In noninflammatory conditions, it is related to trauma, hemorrhage, infection, or metabolic diseases. The most affected bursae are the communicating ones. Bursae are classified as deep or superficial according to its location, and ultrasound can visualize most of them in pathologic conditions. The main pathologies depicted are summarized in Table 11.5 [45–47].

Ultrasound shows distension, thickening and/or synovial hypertrophy with/without septae, and can exhibit Doppler signal but it is not frequent unless an inflammatory process is associated.

Table 11.4 Ligament lesions

Region	Ligament	Normal US view	Pathology, US aspect
Shoulder	Coracoacromial ligament	Thin ligament is divided into two major bands (Y shape): is part of the reflection pulley and can visualize in the interval scanning	Lesion makes the ligament to look like the surrounding fat (slightly hyperechoic) Thickening suggests adhesive capsulitis Its lesion is related to impingement or bursitis
	Gleno-humeral ligament: superior, medial, inferior	Scanning position: maximal abduction, neutral rotation Superior glenohumeral lesion is part of the reflection pulley The two bands of the inferior glenohumeral ligament are bit thicker than the axillary pouch	Superior glenohumeral ligament lesion causes glenohumeral joints instability
	Coraco-humeral ligament	Linear hypoechoic band surrounded by hyperechoic fat	Shortened and thickened in adhesive capsulitis
Elbow	Medial collateral ligament	Triangular shape: anterior, posterior transverse bundles Hyperechoic compared to the common flexor tendon	Luxation Tear Instability
	Lateral collateral ligament	"Y"-shaped structure composed by the lateral ulnar collateral ligament (normally hyperechoic), radial collateral ligament and the annular ligament (normally hypoechopic band indistinguishable from the humeral cartilage Triangular shape in ligament insertion and radial head	Luxation Tears Snapping elbow
Hand	Pisotriquetral joint ligaments	Pisohamate ligament looks slighthly concave, hyperechoic, and fibrillar; insertion is hypoechoic Pisometacarpal ligament is a fibrous band, slightly hypoechoic Pisotriquetral ligament is seen as a hypoechoic fibrillar thickening of the capsule, below the extensor retinaculum that becomes hyperechoic with passive deviation	TearInstability (in osteoarthritis)
	Intrinsic ligaments	Scapholunate and lunotriquetral ligaments are C-shaped ligaments with palmar, intermediate, and dorsal portions that look fibrillar and echoic. The scapholunate appears as triangular Capsular ligaments are six ligaments in the palmar side and three dorsally, not always visible by US	Lesion of the scapholunate ligament appears as widening and loss of the regular echoic aspect with partial thickness tear Instability Loss of the radio–ulnocarpal congruence

Table 11.4 (continued)

Region	Ligament	Normal US view	Pathology, US aspect
	Extrinsic ligaments	Radiocarpal/ulnocarpal collateral ligaments are part of the capsular ligaments and not completely visible on US Triangle fibrocartilaginous complex includes the triangular ligament, that looks hyperechoic, formed by the articular disk, that look less echoic but more homogeneous, and the radio–ulnar ligaments surrounding the disk and attaching to the ulna	
Knee	Medial collateral ligament	The layers are well identified along the medial part (deep fascia, ligament, deep meniscofemoral and meniscotibial bands) that looks homogeneous parallel hyperechoic bands with loose hypoechoic tissue imposed between them	Thickened, heterogeneous, hypoechoic can be associated with edema, hemorrhage, and fluid collection Isolated injuries result from valgus stress without rotation, with knee in flexion Tear that could present into three grades
	Fibular collateral ligament	Cord-like structure, fibrillar pattern	Distension due to related bursitis
Ankle and foot	Anterior tibiofibular ligament	Several band-like structures that looks fibrillar hyperechoic and exhibit anisotropy	In acute injuries, ligament is replaced by amorphous hypoechoic mass with evident tears; chronic lesions show absence or thickend ligament Severe lesion could exhibit avulsion fracture of the lateral malleolus Overuse disorders Trauma Sport lesions
	Anterior talofibular ligament	Band-like structure extending from the distal fibula to the talar bone, fibrillar hyperechoic and exhibit anisotropy	Involved in 95 % of sprains
	Posterior talofibular ligament	Fibrillar hyperechoic and exhibit anisotropy	Normally, it is not usually injured unless there is a frank dislocation or fracture of the ankle
	Calcaneofibular ligament	Fibrillar hyperechoic	Tears Overuse disorders Trauma Sport lesions

Table 11.4 (continued)

Region	Ligament	Normal US view	Pathology, US aspect
	Posterior deltoid ligament	It can be seen only partially, both the superficial and deep bands are hypoechoic structures	
	Anterior deltoid ligament	A long part of the tibiocalcaneal ligament is seen, but not always	
	Spring ligament	Superomedial, inferoplantar, and medioplantar calcaneonavicular ligaments Superomedial looks like a fibrillar hyperechoic structure	Increasing in the thickening and loss of normal fibrillar pattern in the distal portion visualized in chronic lesion confine to the superomedial portion along with chronic dysfunction of peroneal tendons; could exhibit Doppler signal

US ultrasound

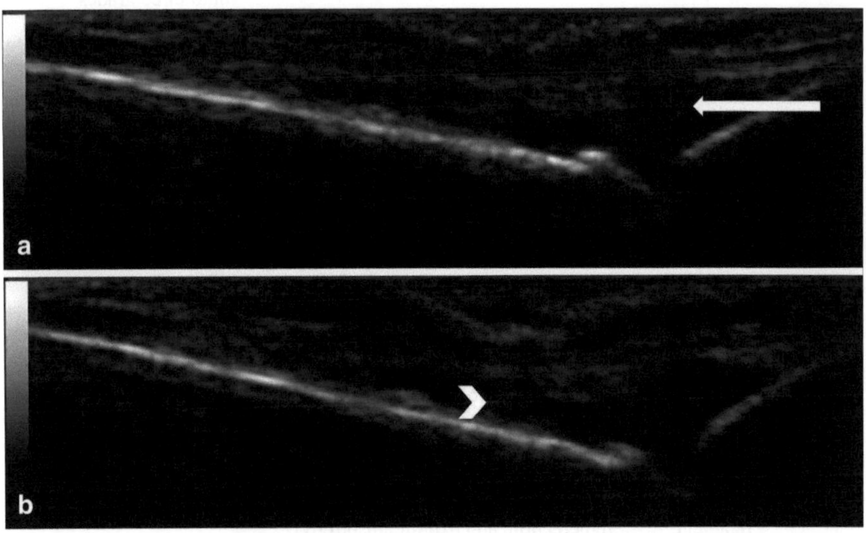

11.12 (**a**) Collateral ulnar ligament in the first MCP joint in neutral position (*arrow*). (**b**) Strained ligament in extension in which a hypoechoic appearance and loss of normal pattern are seen (*arrow head*)

In acute bursitis, bursae distension causes pain and limited range of motion, normally related to friction and repetitive trauma. Fluid is anechoic or slightly hypoechoic and compressible. Sometimes a posterior acoustic shadow can be seen and shows little or no change with compression (Fig. 11.13).

Table 11.5 Bursae location and main pathology mechanism

Bursae location		Pathology mechanism
Shoulder	Subacromial–subdeltoid (SA-SD) bursae Coracoid bursae Subscapular bursae	Acute, occasionally it is related to rotator cuff injury. When it turns chronic, ultrasound can detect the loss of cortical regularity associated with bursal changes Chronic or inflammatory Chronic or inflammatory
Elbow	Olecranon bursae	Acute due to repetitive trauma Chronic, related to inflammatory or infectious diseases Can be recurrent Sometimes the bursae can exhibit septae
Hip	Iliopsoas bursae Trochanteric Bursae	Acute or chronic, related to inflammatory disease, avascular necrosis, osteoarthritis, trauma, pigmented villonodular synovitis, crystal deposition diseases, and tumor. Differential diagnosis includes inguinal hernia, tumors, ganglia; it is important to consider that this bursae can reach abdomen Superficial bursae, related to obesity, trauma, and inflammatory diseases. It is the largest bursae in the greater trochanter
Knee	Prepatellar bursae Suprapatellar bursae Superficial infrapatellar bursae Deep infrapatellar bursae Anserine bursae Gastrocnemius–semimembranosus bursae	Acute or chronic, it is distended over the patellar superficial margin and can reach the anterior region in the knee. It is tricompartmental bursae. Main cause is repetitive trauma from kneeling It is communicating bursae, when distended is related to joint effusion and its main cause is inflammatory diseases (e.g., rheumatoid arthritis) and osteoarthritis Chronic, related to overuse and repetitive trauma (occupational kneeling and repetitive trauma) Acute, usually is related to overuse and sports injuries due to repetitive trauma Chronic, related to repetitive trauma, stress, or anatomic knee structural changes. It can be related to medial collateral ligament injury Acute or chronic, communicating bursae also known as Baker's cyst, it the most affected bursae in the knee joint
Ankle	Superficial calcaneal bursae (subcutaneous Achilles bursae) Precachilles bursae (deep retrocalcaneal bursae)	Acute or chronic, usually related to poorly fitting shoes, overtraining, or Haglund's disease Acute, chronic, and frequently related to inflammatory diseases
Foot	Metatarsal bursae	Acute, chronic, and related to inflammatory diseases and shoe fitting

Chronic bursitis presents with an irregular thickening of the bursae walls, fluid becomes echoic and is less or not compressible. Sometimes, calcifications and septae can be visualized; hyperechoic spots with or without acoustic shadow are seen.

Metabolic diseases, infections, bone fragments, and steroids can be seen as spots and a detailed history is needed, especially in chronic cases. Septic bursitis can be

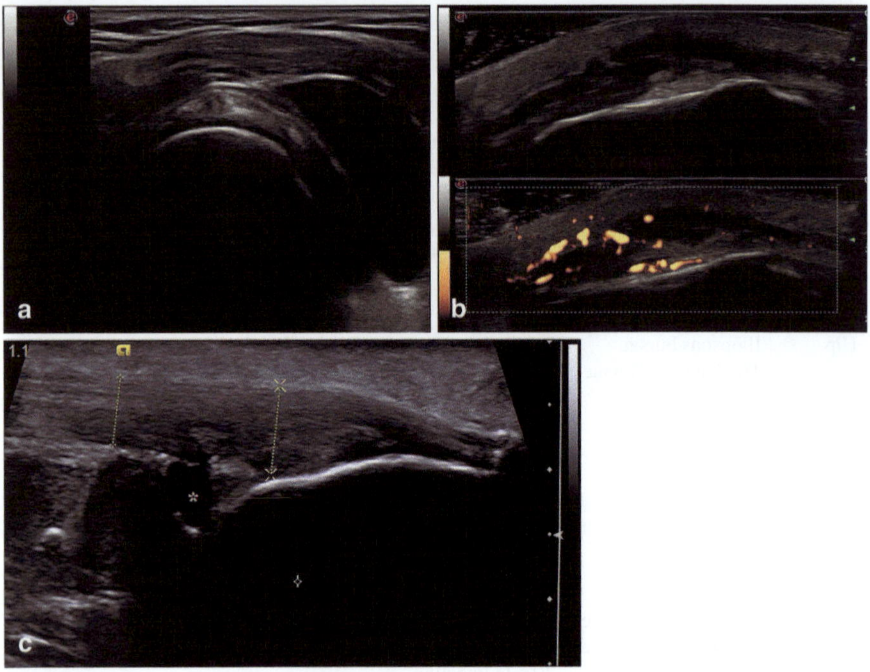

Fig. 11.13 (**a**) Subscapular bursae, (**b**) suprapatellar bursa in gray scale and with power Doppler signal, and (**c**) deep retrocalcaneal bursae

confused or misdiagnosed if an inflammatory disease is present—some differences can distinguish one from another. Septic bursitis gas can be visualized as posterior acoustic shadowing, walls are hyperechoic and bursae content is cloudy and hypoechoic and could exhibit Doppler signal (Fig. 11.14).

Fascia

Fascia is the most important tissue in the body in musculoskeletal anatomy and is composed of three layers (superficial, deep, and visceral). The deep layer is the most important because it is part of most tissues. Retinacula is the thickening of this deep fascia that prevents tendon migration in osseofibrous canals such as flexor retinaculum of the wrist, medial ankle flexor retinaculum, peroneal retinaculum, and the annular pulleys in the fingers. The main pathology related to aponeurosis and pulleys was previously described.

Palmar fascia is a connective tissue located superficially in the hand, also known as palmar aponeurosis—its inflammation is known as palmar fibromatosis or

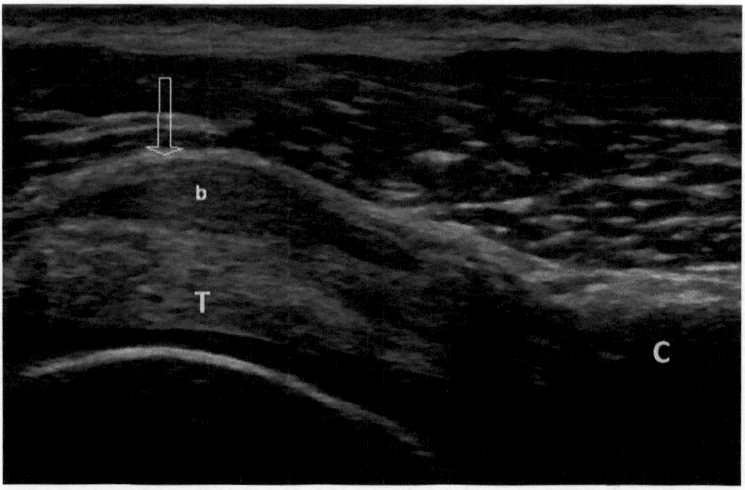

Fig. 11.14 Chronic subacromial–subdeltoid bursitis (*b*). The tendon (*T*) is distinctly demarcated. *C*=coracoid process. The bursal wall is thickened (*arrow*). The contents are echogenic. Findings denote chronicity

Dupuytren's disease and is characterized by hypoechoic nodular thickening and variable retraction of the palmar aponeurosis that might exhibit Doppler signal, and if severe can limit hand movement and cause flexion contraction of the fingers. It is sometimes associated with neoplasia or metabolic diseases like diabetes (Figs. 11.15, 11.16).

Plantar fascia is a thickened fibrous sheet of connective tissue originating in the medial tubercle of calcaneus and attached to the plantar surface of the metatarsophalangeal joints. Its most common injury is plantar fasciitis which is easily visualized by ultrasound—its appearance is thickened and hypoechoic with surrounding edema (usually thickness is more than 4 mm) [48] (Fig. 11.17).

Plantar fibromatosis or Ledderhose disease appears as a fusiform, hypoechoic, or mixed echogenic mass located in the middle or distal fascia, easily located by clinical examination.

Subcutaneous Tissue

Subcutaneous tissue appears as discrete hypoechoic layers with hypoechoic background fat and hyperechoic linear echoes of the connective tissue septae. Pathology is vast and ultrasound technology, including Doppler and elastography, has improved the study of this tissue. Pathology is mostly related to dermatology, but

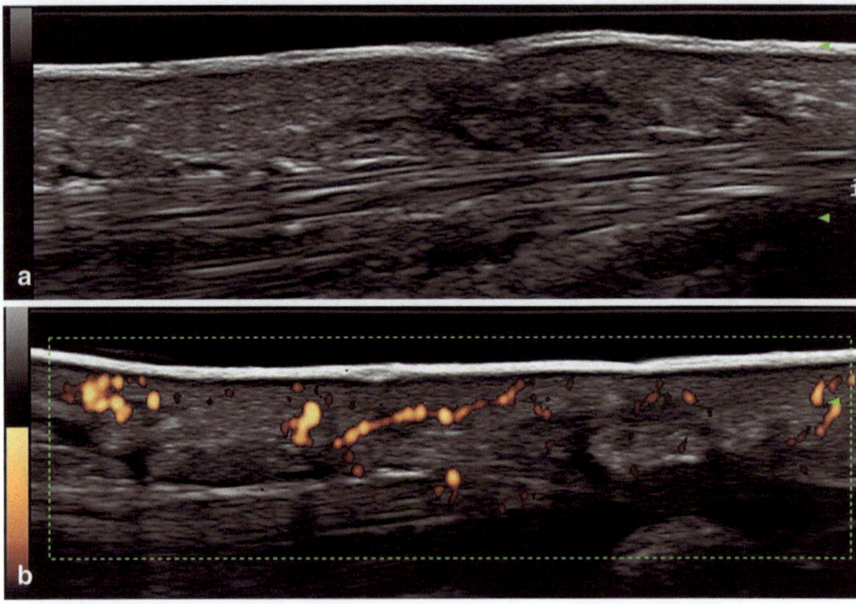

Fig. 11.15 (**a**) Palmar fascia with hypoechoic nodular thickening. (**b**) Positive power Doppler signal

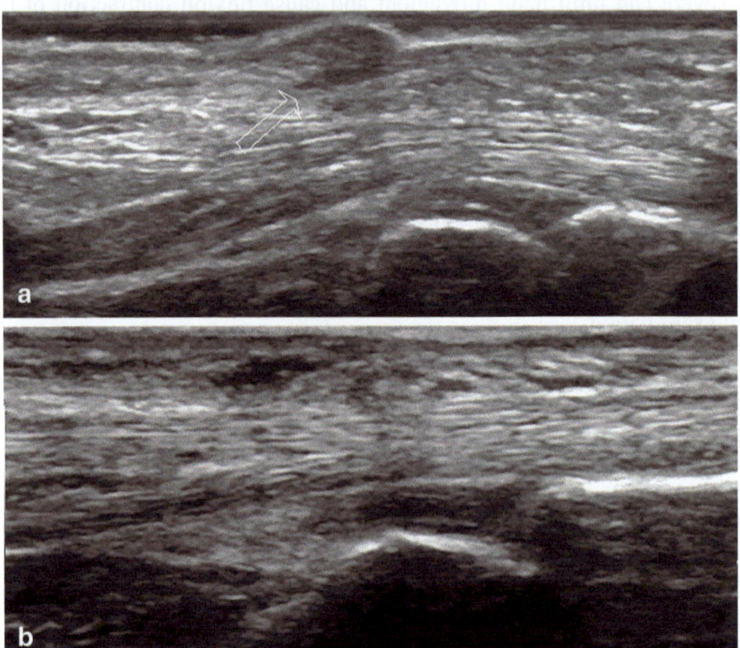

Fig. 11.16 Palmar fibromatosis. (**a**) Thickened palmar fascia superficial to the flexor tendon with nodular formation (*arrow*) causing skin corrugation above. (**b**) Post local injection, thickening is less, the skin is more flat, and changes are less marked

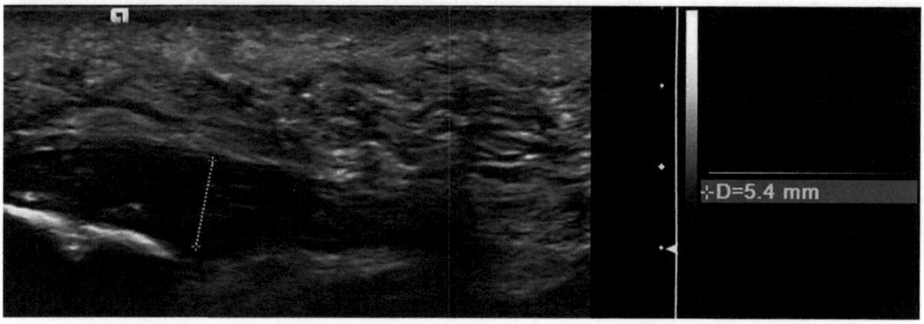

Fig. 11.17 Plantar fascia shows thickening, hypoechoic changes with heterogeneous pattern

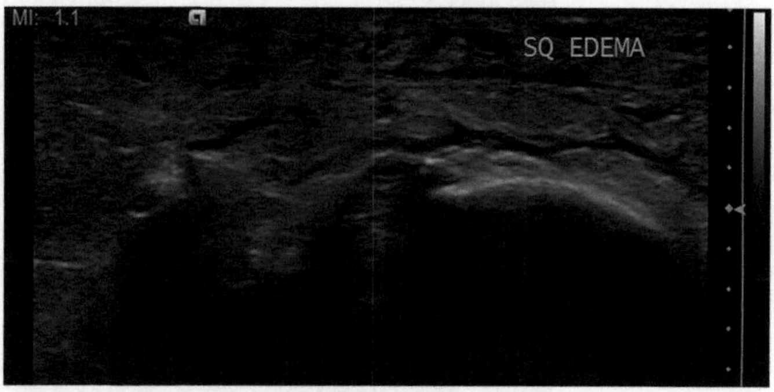

Fig. 11.18 Grayscale image depicting connective septae enlargement and anechoic strands until fat lobules separated from one to another due to anechoic fluid representing subcutaneous edema

rheumatologists and other specialists are not exempt from knowing of it. The most frequent are described below.

Subcutaneous edema can be caused by trauma or inflammatory diseases and is visualized as hyperechoic appearance of fat lobules. The early stages involve the deep layers that are seen as hypoanechoic due to fluid accumulation. Progression relates to connective tissue septae enlargement and anechoic strands until fat lobules are separated from one another due to the anechoic fluid. When the cutaneous layers thicken and anechoic fluid localizes between the fatty lobules of the hypodermis, it is called lymphedema [49] (Fig. 11.18).

Cellulitis involves skin and subcutaneous tissue and is related to infectious diseases—differentiation from abscesses is very important. Ultrasound appearance of cellulitis is not specific, but demonstrates skin thickening, subcutaneous edema, and an irregular hyperechoic appearance of fat with blurred tissue planes and associated edema of deep layers. Doppler shows hypervascularity, and echogenic foci with dirty shadowing that may be related to air or gas [50] (Fig. 11.19).

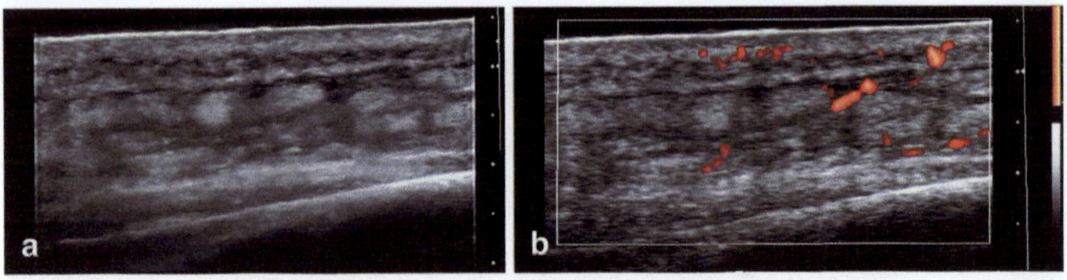

Fig. 11.19 (**a**) Zoom image of the skin with thickening, subcutaneous edema and irregular hyperechoic appearance of fat with blurred tissue planes with associated edema of deep layers. (**b**) Power Doppler signal shows hypervascularity

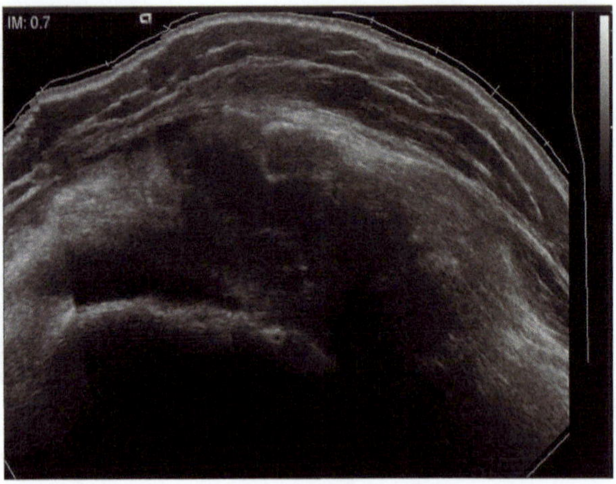

Fig. 11.20 Hypoechoic area in the hip in short axis with characteristic posterior acoustic enhancement containing echogenic spots and an homogeneously hypoechoic mass that illustrates an abscess

Panniculitis involves adipose tissue of the subcutaneous layer and is associated with systemic diseases or local injuries such as trauma. On ultrasound visualization, septal panniculitis shows thickening and hypoechoic septae with increased echogenicity of fatty tissue, while lobular panniculitis presents with increased echogenicity and blurred fatty tissue, and sometimes anechoic or hypoechoic round-shaped pseudocyst structures can be seen from the resulting liquefaction.

Abscess is demonstrated as an irregular fluid-filled hypoechoic or isoechoic area with characteristic posterior acoustic enhancement containing echogenic spots (pus) that, very often, fluctuate with slight pressure with the probe. Doppler shows blood flow in the abscess walls and surrounding tissues; deep abscess may exhibit hypoechoic subperiostial fluid. When the abscess is well formed, it appears as a homogeneously anechoic or hypoechoic mass containing gas; the ongoing liquefaction visualized like an echogenic fluid without mass and focal changes. It is important to search for communicating tracts to the skin or deep layers (muscular/articular) [51] (Fig. 11.20).

Table 11.6 Study appearance of soft tissue tumors

Characteristic	Description
Location	Superficial: skin, subcutaneous tissue, fascia Deep: muscle, joint, cortical Central/peripheral body location
Appearance	Size, shape, capsule Limits: well defined, blurred Internal echogenicity: hypo/hyper/iso/anechoic Internal architecture: linear echoes, septae, speckles, cysts, acoustic shadowing, calcifications, gas/air
Consistency	Soft, firm, hard Compressibility
Doppler signal	Present or absent Peripheral or central Pattern: organized or chaotic
Surrounding tissues	Related to skin, tendon, muscle, nerve, bone
Other lesions	Edema Swelling of surrounding soft tissues

Tumors

Soft tissue tumors present in a wide spectrum from benign to malignant mostly located in the subcutaneous tissue and varying in shapes and sizes. Ultrasound is useful to determine the presence of the tumor, its anatomical location, extent, consistency, relation to other structures, vascular involvement, and for the follow-up of diagnostic and treatment procedures [52]. The ultrasound appearance of most of the soft tissue tumors is known and allows for an accurate diagnosis.

In general, benign tumors are present for a long time, without a change of size, (demonstrating a slow growth), superficial in location, and do not necessarily exhibit a Doppler signal, whereas malignant tumors present in a short time, with rapid growth, deep location are vascularized. Other characteristics shown in Table 11.6 [53] must be considered in the ultrasound examination for tumors.

Lipomas are the most common soft tissue solid masses. They present as well-defined oval/round-shaped structures that follow the axis of the skin layers and exhibit hypovascularity. When located in risky areas such as near the temporal artery, carotid arteries, jugular veins, and brachial artery, a vascular view must also be performed. When located superficially its aspect is well defined with a thin wall, hypoechoic, and homogeneous (Fig. 11.21).

Fibrolipomas are hyperechoic and intramuscular lipomas that are usually located in lower limbs. They present as well-circumscribed or non-infiltrative types. They are seen as a well-defined ovoid, noncompressible mass contained inside a muscle usually without Doppler signal. Infiltrative type substitutes muscle with fat, and muscle fiber separation is seen with heterogeneous aspect and not very well-defined walls but this does not represent a sign of malignancy unless increased Doppler signal is

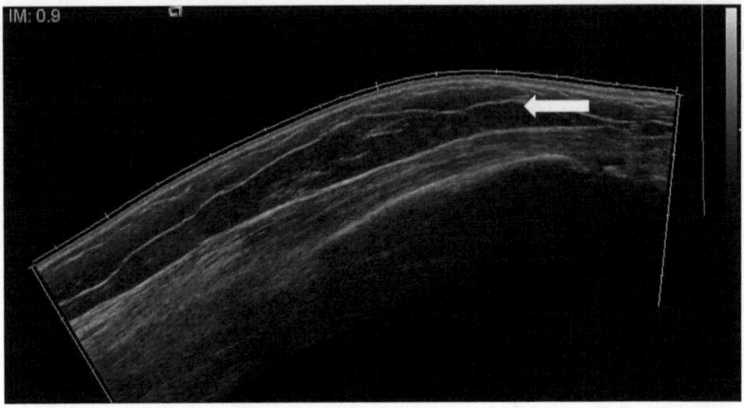

Fig. 11.21 Well-defined subcutaneous tissue with a thin wall, hypoechoic, and homogeneous soft tissue mass representing a lipoma (*arrow*)

Table 11.7 Ganglion and synovial cyst

Characteristics	Ganglion	Synovial cyst
Shape	Rounded Unilocular or multilocular	Rounded or not defined Simple, multiloculated, or septated
Walls	Well defined, hyperechoic, and related to connective tissue lining Can contain sharply defined septae	Defined, hypo- or hyperechoic related to synovial cells lining and communicate with joint space
Location	Hand and feet Juxta-articular, intra-articular, periosteal	Hand, hip, and knee
Appearance	Mucinous collection Variable echogenicity (anechoic, hypoechoic, or mixed)	Fluid collection Heterogeneous, mostly hypo- or anechoic
Compressibility	Hardly compressible	Compressible
Doppler signal	Absent	May be present

present, which might reveal liposarcomas or other type of lesions. In this case, magnetic resonance imaging or computed tomography should be done [5, 49, 54].

Another type of soft tissue masses is *ganglion* or *synovial cysts*. Its main characteristics are depicted in Table 11.7.

There is no consensus about the differences between both lesions and in general literature refers indistinctly to them. The main distinction is related to whether they do or do not communicate with the joint space [55–57] (Fig. 11.22, 11.23).

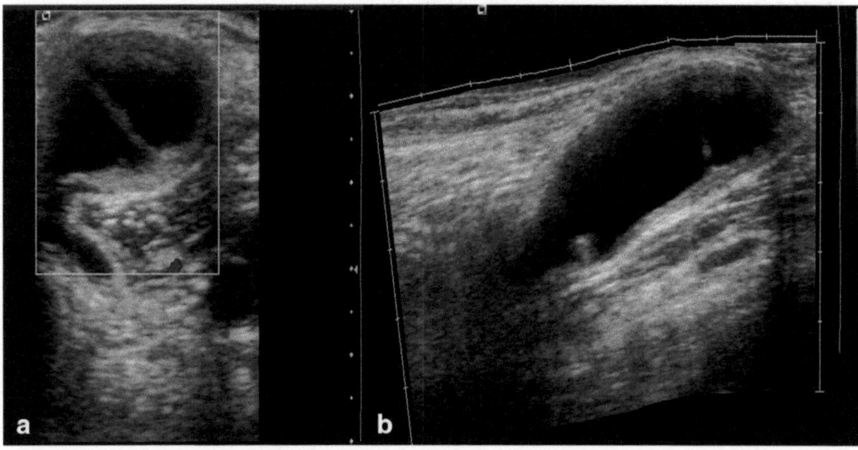

Fig. 11.22 **(a)** Semimembranosus bursae in short axis with a central septae and hypoechoic areas. **(b)** Same bursae in long axis

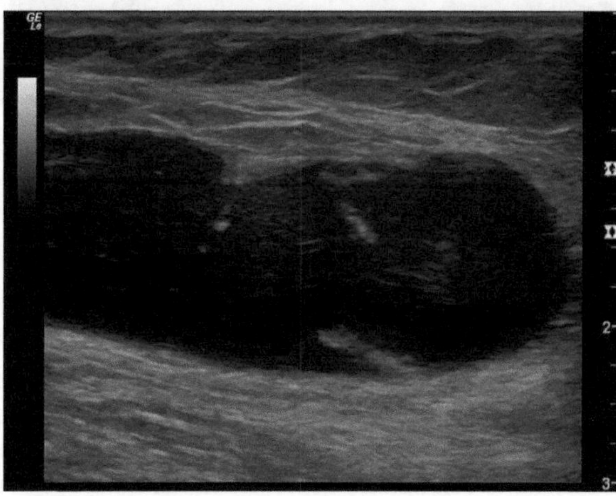

Fig. 11.23 Large Baker's cyst with hypoechoic content

Foreign Bodies

Ultrasound is a good imaging technique to identify foreign bodies in subcutaneous tissue. There are several types of foreign bodies and each can be characterized by this method. Its characteristics are depicted in Table 11.8.

In the case of a suspected foreign body, it is important to take into account the size, material, location, related symptoms (infection, mass, granuloma), duration of symptoms, migration, and associated therapeutic decisions [58–61] (Fig. 11.24).

Table 11.8 Ultrasound characteristics of foreign bodies

Type of foreign body	US appearance
Wood or another organic material (bee stinger, rose splinter)	Radiolucent Linear, hyperechoic with underlying acoustic clean shadowing Shadow may be either complete or partial depending on the angle of insonation Hypoechoic halo can be seen, representing edema, abscess, or granulation tissue
Glass	Hyperechoic with underlying dirty acoustic shadowing Comet tail artifact can be present
Metal	Hyperechoic with underlying acoustic shadowing and comet tail artifact Reverberation

US ultrasound

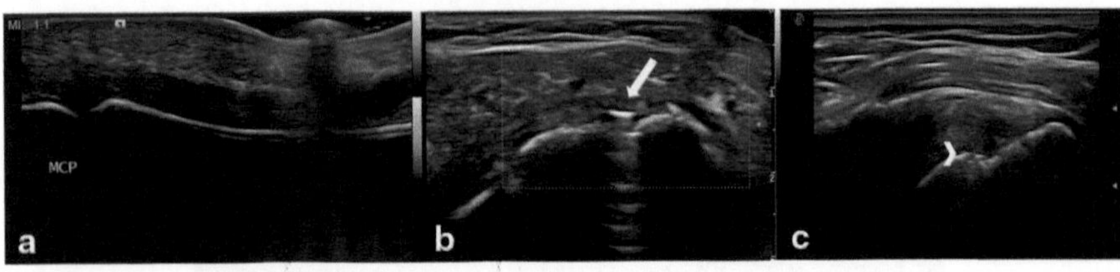

Fig. 11.24 Different types of artifacts. (**a**) Reverberating artifact image cause by a ring. (**b**) Post-surgical Achilles tendinopathy, sutures (*arrow*) still visible. (**c**) Subscapular tendon in which the head of a screw (*arrow*) is depicted. (**d**) Anchor (*arrow head*) in a supraspinatus tendon

Conclusion

Musculoskeletal ultrasonography is a quickly achieving major advancement and enhancement. This, consequently, is clearly helping in an area like soft tissue rheumatism, where not only anatomy shows wide variety but also pathology is vast and sometimes difficult to distinguish. Musculoskeletal ultrasonographist must familiarize himself/herself with normal sonoanatomy, get an in-depth reading and/or hands on experience, and learn how to precisely and objectively describe and differentiate different findings. The use of new techniques such as elastography should be mastered. The result will highly pinpoint the issue of concern. Proper history, clinical aspects, and other imaging modalities are always of necessity.

References

1. Álvarez-Nemegyei J, Canoso JJ. Name and classification of soft-tissue rheumatism. Reumatol Clín. 2007;3(4):151–2.
2. Toprak H, Kilic E, Serter A, Kocakoc E, Ozgocmen S. Ultrasound and Doppler US in evaluation of superficial soft-tissue lesions. J Clin Imag Sci. 2014;4:12.

3. Katz JN, Buchbinder R. Soft tissue syndromes. Bailliere's Clin Rheumatol. 1995;9(3):585–98.
4. Cawston TE, Riley GP, Hazleman BL. Tendon lesions and soft tissue rheumatism—great outback or great opportunity? Ann Rheum Dis. 1996;55(1):1–3.
5. Zamorani MP, Valle M. Muscle and tendon. In: Bianchi S, Martinoli C, editors. Ultrasound of the musculoskeletal system. 1st ed. Berlin: Springer; 2007. p. 45–96.
6. Gaitini D. Shoulder ultrasonography: performance and common findings. J Clin Imag Sci. 2012;2:38.
7. Möller I, Bong D, de Miguel Mendieta E. Soft tissue rheumatism. In: Wakefield RJ, D´Agostino MA, editors. Essential applications of musculoskeletal ultrasound in rheumatology. 1st ed. Philadelphia: Saunders, Elsevier; 2010. p. 219–35.
8. Brasseur JL. The biceps tendons: from the top and from the bottom. J Ultrasound. 2012;15(1):29–38.
9. Bianchi S, Martinoli C, de Gautard R, Gaignot C. Ultrasound of the digital flexor system: Normal and pathological findings(). J Ultrasound. 2007;10(2):85–92.
10. Fantino O, Borne J, Bordet B. Conflicts, snapping and instability of the tendons. Pictorial essay. J Ultrasound. 2012;15(1):42–9.
11. Pratt RK, Hoy GA, Bass Franzcr C. Extensor carpi ulnaris subluxation or dislocation? Ultrasound measurement of tendon excursion and normal values. Hand Surg: Int J Devoted H and Upper Limb Surg Related Res: J Asia-Pacific Fed Soc Surg Hand. 2004;9(2):137–43.
12. Lee KS, Ablove RH, Singh S, De Smet AA, Haaland B, Fine JP. Ultrasound imaging of normal displacement of the extensor carpi ulnaris tendon within the ulnar groove in 12 forearm-wrist positions. AJR Am J Roentgenol. 2009;193(3):651–5.
13. Raikin SM, Elias I, Nazarian LN. Intrasheath subluxation of the peroneal tendons. J Bone Joint Surg Am Vol. 2008;90(5):992–9.
14. Kaux JF, Forthomme B, Goff CL, Crielaard JM, Croisier JL. Current opinions on tendinopathy. J Sports Sci Med. 2011;10(2):238–53.
15. Smith J, Maida E. Musculoskeletal ultrasound of tendons. Oper Tech Orthop. 2013;23(2):50–5.
16. Lew HL, Chen CP, Wang TG, Chew KT. Introduction to musculoskeletal diagnostic ultrasound: examination of the upper limb. Am J Phys Med Rehabil/Assoc Acad Physiatr. 2007;86(4):310–21.
17. Chew K, Stevens KJ, Wang TG, Fredericson M, Lew HL. Introduction to diagnostic musculoskeletal ultrasound: part 2: examination of the lower limb. Am J Phys Med Rehabil/Assoc Acad Physiatr. 2008;87(3):238–48.
18. Mathieson JR, Connell DG, Cooperberg PL, Lloyd-Smith DR. Sonography of the Achilles tendon and adjacent bursae. AJR Am J Roentgenol. 1988;151(1):127–31.
19. Vuillemin V, Guerini H, Bard H, Morvan G. Stenosing tenosynovitis. J Ultrasound. 2012;15(1):20–8.
20. Guerini H, Pessis E, Theumann N, Le Quintrec JS, Campagna R, Chevrot A, et al. Sonographic appearance of trigger fingers. J Ultrasound Med: Off J Am Inst Ultrasound Med. 2008;27(10):1407–13.
21. Wakefield RJ, Balint PV, Szkudlarek M, Filippucci E, Backhaus M, D'Agostino MA, et al. Musculoskeletal ultrasound including definitions for ultrasonographic pathology. J Rheumatol. 2005;32(12):2485–7.
22. Cohen M. US imaging in operated tendons. J Ultrasound. 2012;15(1):69–75.
23. Adler RS, Johnson KM, Fealy S, Maderazo A, Gallo RA, Gamradt SC, et al. Contrast-enhanced sonographic characterization of the vascularity of the repaired rotator cuff: utility of maximum intensity projection imaging. J Ultrasound Med: Off J Am Inst Ultrasound Med. 2011;30(8):1103–9.
24. Waitches GM, Rockett M, Brage M, Sudakoff G. Ultrasonographic-surgical correlation of ankle tendon tears. J Ultrasound Med: Off J Am Inst Ultrasound Med. 1998;17(4):249–56.
25. Blankstein A, Israeli A, Dudkiewicz I, Chechik A, Ganel A. Percutaneous Achilles tendon repair combined with real-time sonography. Isr Med Assoc J. 2007;9(2):83–5.
26. Ooi CC, Malliaras P, Schneider ME, Connell DA. "Soft, hard, or just right?" Applications and limitations of axial-strain sonoelastography and shear-wave elastography in the assessment of tendon injuries. Skelet Radiol. 2014;43(1):1–12.

27. Vera-Pérez E, Ventura-Ríos L, Gallegos-Nava S, Bernal A, Hernández-Díaz C, Pineda C. Son las Lesiones Musculares Adecuadamente Diagnosticadas por Ultrasonografía? Rev Chil Reumatol. 2013;29(4):226–31.
28. Woodhouse JB, McNally EG. Ultrasound of skeletal muscle injury: an update. Semin Ultrasound, CT, MR. 2011;32(2):91–100.
29. Campbell RSD, Wood J. Ultrasound of muscle. Imaging. 2002;14(3):229–40.
30. Koh ES, McNally EG. Ultrasound of skeletal muscle injury. Sem Musculoskelet Radiol. 2007;11(2):162–73.
31. Abate M, Salini V, Rimondi E, Errani C, Alberghini M, Mercuri M, et al. Post traumatic myositis ossificans: sonographic findings. J Clin Ultrasound. 2011;39(3):135–40.
32. Campbell SE, Adler R, Sofka CM. Ultrasound of muscle abnormalities. Ultrasound Q. 2005;21(2):87–94; quiz 150, 3–4.
33. Botar-Jid C, Damian L, Dudea SM, Vasilescu D, Rednic S, Badea R. The contribution of ultrasonography and sonoelastography in assessment of myositis. Med Ultrasonogr. 2010;12(2):120–6.
34. Michelin P, Delarue Y, Duparc F, Dacher JN. Thickening of the inferior glenohumeral capsule: an ultrasound sign for shoulder capsular contracture. Eur Radiol. 2013;23(10):2802–6.
35. Homsi C, Bordalo-Rodrigues M, da Silva JJ, Stump XM. Ultrasound in adhesive capsulitis of the shoulder: is assessment of the coracohumeral ligament a valuable diagnostic tool? Skelet Radiol. 2006;35(9):673–8.
36. Teixeira PA, Omoumi P, Trudell DJ, Ward SR, Lecocq S, Blum A, et al. Ultrasound assessment of the lateral collateral ligamentous complex of the elbow: imaging aspects in cadavers and normal volunteers. Eur Radiol. 2011;21(7):1492–8.
37. Merolla G, Del Sordo S, Paladini P, Porcellini G. Elbow ulnar collateral ligament reconstruction: clinical, radiographic, and ultrasound outcomes at a mean 3-year follow-up. Musculoskelet Surg. 2014;98(Suppl 1):S87–S93
38. Moraux A, Vandenbussche L, Demondion X, Gheno R, Pansini V, Cotten A. Anatomical study of the pisotriquetral joint ligaments using ultrasonography. Skelet Radiol. 2012;41(3):321–8.
39. Renoux J, Zeitoun-Eiss D, Brasseur JL. Ultrasonographic study of wrist ligaments: review and new perspectives. Semin Musculoskelet Radiol. 2009;13(1):55–65.
40. De Maeseneer M, Marcelis S, Boulet C, Kichouh M, Shahabpour M, de Mey J, et al. Ultrasound of the knee with emphasis on the detailed anatomy of anterior, medial, and lateral structures. Skelet Radiol. 2014;43(8):1025–39.
41. Kleinbaum Y, Blankstein A. Mild to moderate medial collateral ligament (Mcl) injuries of the knee: sonographic findings and sonographic valgus stress test. J Musculoskelet Res. 2008;11(01):9–14.
42. De Maeseneer M, Marcelis S, Jager T, Shahabpour M, Van Roy P, Weaver J, et al. Sonography of the normal ankle: a target approach using skeletal reference points. AJR Am J Roentgenol. 2009;192(2):487–95.
43. Mansour R, Teh J, Sharp RJ, Ostlere S. Ultrasound assessment of the spring ligament complex. Eur Radiol. 2008;18(11):2670–5.
44. Mansour R, Jibri Z, Kamath S, Mukherjee K, Ostlere S. Persistent ankle pain following a sprain: a review of imaging. Emerg Radiol. 2011;18(3):211–25.
45. Holsbeeck MTv, Introcaso JH. Ecografía de las bolsas articulares. In: Holsbeeck MTv, Introcaso JH, editors. Ecografia musculoesqueletica. 2nd ed. Madrid: Marban Libros; 2002. p. 131–69.
46. Blankstein A, Ganel A, Givon U, Mirovski Y, Chechick A. Ultrasonographic findings in patients with olecranon bursitis. Ultraschall Med. 2006;27(6):568–71.
47. Hirji Z, Hunjun JS, Choudur HN. Imaging of the bursae. J Clin Imag Sci. 2011;1(1):22.
48. Girish G, Finlay K, Landry D, O'Neill J, Popowich T, Jacobson J, et al. Musculoskeletal disorders of the lower limb—ultrasound and magnetic resonance imaging correlation. Can Assoc Radiol J. 2007;58(3):152–66.

49. Wortsman X, Carreño L, Morales C. Inflammatory diseases of the skin. In: Wortsman X, Jemec GBE, editors. Dermatologic ultrasound with clinical and ultrasound correlations. New York: Springer; 2013. p. 73–117.
50. Turecki MB, Taljanovic MS, Stubbs AY, Graham AR, Holden DA, Hunter TB, et al. Imaging of musculoskeletal soft tissue infections. Skelet Radiol. 2010;39(10):957–71.
51. Chau CL, Griffith JF. Musculoskeletal infections: ultrasound appearances. Clin Radiol. 2005;60(2):149–59.
52. Cardinal E, Bureau N, Chem RK. Masas del Tejido Blando: Un Enfoque Algorítmico. In: Cardinal E, Chem RK, editors. Ecografía Musculoesquelética: Pautas y gamuts. Buenos Aires: Ediciones Journal; 1999.
53. Hung EH, Griffith JF. Pitfalls in ultrasonography of soft tissue tumors. Semin Musculoskelet Radiol. 2014;18(1):79–85.
54. Verdugo M. Ultrasonido en el estudio de tumores de partes blandas. Rev Chil Radiol. 2009;15:5–18.
55. Beaman FD, Peterson JJ. MR imaging of cysts, ganglia, and bursae about the knee. Radiol Clin N Am. 2007;45(6):969–82, vi.
56. Wang G, Jacobson JA, Feng FY, Girish G, Caoili EM, Brandon C. Sonography of wrist ganglion cysts: variable and noncystic appearances. J Ultrasound Med: Off J Am Inst Ultrasound Med. 2007;26(10):1323–8; quiz 30–1.
57. Cardinal E, Buckwalter KA, Braunstein EM, Mih AD. Occult dorsal carpal ganglion: comparison of US and MR imaging. Radiology. 1994;193(1):259–62.
58. Vargas B, Wildhaber B, La Scala G. Late migration of a foreign body in the foot 5 years after initial trauma. Pediatr Emerg Care. 2011;27(6):535–6.
59. Leung A, Patton A, Navoy J, Cummings RJ. Intraoperative sonography-guided removal of radiolucent foreign bodies. J Pediatr Orthop. 1998;18(2):259–61.
60. Gibbs TS. The use of sonography in the identification, localization, and removal of soft tissue foreign bodies. J Diagn Med Sonogr. 2006;22(1):5–21.
61. Mohammadi A, Ghasemi-Rad M, Khodabakhsh M. Non-opaque soft tissue foreign body: sonographic findings. BMC Med Imag. 2011;11:9.

Chapter 12
Inflammatory Allied Conditions

Margarida Alexandre Oliveira MD and Anna Ciechomska MD, PhD

Abbreviations

ACPA	Anti-citrullinated protein antibodies
ACR	American College of Rheumatology
CD	color Doppler
CDUS	color Doppler ultrasonography
CRP	C-reactive protein
CT	computed tomography
CTA	computed tomography—angiography
ESR	erythrocyte sedimentation rate
EULAR	European league against rheumatism
FM	fibromyalgia
GCA	giant-cell arteritis
HIV	human immunodeficiency virus
LV-GCA	large-vessel giant-cell arteritis
MRA	magnetic resonance angiography
MRI	magnetic resonance imaging
PET	positron emission tomography
PIH	parenchymal inhomogeneity
PMR	polymyalgia rheumatica
PSV	peak systolic velocity
RA	rheumatoid arthritis
RF	rheumatoid factor

M. A. Oliveira (✉)
Department of Rheumatology, Centro Hospitalar Cova da Beira, Covilhã, Portugal
e-mail: mar.a@sapo.pt

A. Ciechomska
Department of Rheumatology, Wishaw General Hospital, Wishaw, UK
e-mail: anna.ciechomska@lanarkshire.scot.nhs.uk

Y. El Miedany (ed.), *Musculoskeletal Ultrasonography in Rheumatic Diseases,*
DOI 10.1007/978-3-319-15723-8_12

RI	resistive index
SAPHO	synovitis–acne–pustulosis–hyperostosis–osteomyelitis/osteitis
SCJ	sternoclavicular joint
SLE	systemic lupus erythematosus
SM	sono-myography
SME	sonomyoelastography
SOV	single-organ vasculitis
SS	Sjögren's syndrome
SVV	small-vessel vasculitis
TMJ	temporomandibular joint
TP	tender points
US	ultrasound

Introduction

Inflammatory nonarthritic musculoskeletal conditions encompass a wide spectrum of disorders affecting various anatomical structures. In spite of the fact that the term musculoskeletal "inflammatory allied conditions" may sound less popular than inflammatory arthritic diseases, they rank as one of the leading causes of morbidity and/or disability. However, there are still gaps in our knowledge of the different diseases included under the title of inflammatory allied musculoskeletal conditions. This chapter discusses the value of US in the diagnosis as well as management of this inhomogeneous group of diseases, including vasculitis, fibromyalgia (FM), polymyalgia rheumatica (PMR), inflammatory myopathies, parasternal joint disease, Sjögren's syndrome (SS) as well as the temporomandibular joint (TMJ).

Vasculitis

Large-Vessel Vasculitis: Clinical Features and Sonographic Pattern

Systemic vasculitis are a group of disorders characterized by inflammation of vessel walls, that may affect large, medium, or small vessels, often leading to serious organ damage. They may appear as a primary process or secondary to another underlying disease. In vasculitis, the inflammatory process is initiated in the blood vessel itself as inflammatory leucocytes accumulate in the vessel walls, leading to swelling and damage of the mural structures, subsequently narrowing of the vessels' lumen which can cause ischemia and necrosis [1]. An updated nomenclature categorizes vasculitis as: large-vessel vasculitis, medium-vessels vasculitis, small-vessels vasculitis (SVV), vasculitis of variable size, single-organ vasculitis

(SOV), vasculitis secondary to a systemic disease, and vasculitis with suspected etiology [2].

Large-vessel vasculitis refers to the vasculitis affecting the aorta and its major branches: giant-cell arteritis (GCA) and Takayasu's arteritis [2]. It includes classic temporal arteritis (cranial GCA), large-vessel giant-cell arteritis (LV-GCA), describing patients with extracranial artery involvement, Takayasu's arteritis, and isolated aortitis [2–4].

The incidence of GCA increases with age and peaks during the eighth decade, been the mean age of diagnosis of about 75 years. GCA is 2–3 times more frequent in women than in men and is more common in populations from Northern European countries than in Mediterranean countries. A cyclic variation in disease incidence has been observed in several studies, supporting the hypothesis that the disease is triggered by an infectious agent [5].

In these pathologies, color Doppler (CD) ultrasonography (CDUS) was reported as a sensitive, noninvasive technique, of value to diagnose vasculitis of the large-vessels vasculitis [6]. Temporal arteritis is a good example of how US can be of help, not only for the disease diagnosis but also for monitoring response to therapy as well as identifying flare up of the inflammatory process. In temporal arteritis, there is a characteristic circumferential wall thickening, "halo sign" (a hypoechoic, dark area around the lumen), in the affected arteries which seems to disappear within 2 days to 6 months after starting treatment with corticosteroids [7–9]. However, the halo sign reappears in GCA patients suffering a flare [7].

Giant Cell Arteritis/Temporal Arteritis

Temporal arteritis is an important cause of secondary headache in older adults. According to the 1990 American College of Rheumatology criteria, GCA is diagnosed if the patient fulfills three of a core of five features. These include age of 50 or older, at onset; a new-onset headache or changed pattern of headache; a clinical temporal artery abnormality; elevated erythrocyte sedimentation rate (ESR) of at least 50 ml/h; and abnormal temporal artery biopsy. GCA is generally regarded as a cranial arteritis, although 20–30 % of the patients will also have involvement of the aortic trunk and branches [10].

Clinically, temporal arteritis patients tend to complain of localized headache (72–74 % of the cases), usually accompanied by swollen and tender temporal arteries, with reduced pulse (64 % of the cases). In addition, jaw claudications are usually reported in 37 % of the cases. The headache is typically acute or subacute in onset and is also associated with scalp tenderness, in either temporal or occipital artery branches [11]. It is classically persistent, but intermittent headaches have also been reported [12]. About 32 % of the patients may have ophthalmological complications, such as double vision or anterior ischemic optic neuropathy. Blindness is the most feared complication of GCA, occurring in approximately 15–20 % of patients [11]. Systemically, patients usually lose weight and feel ill, with possible night sweats. ESR is higher than 50 mm/h in 85 % of the cases. C-reactive protein (CRP) is also commonly elevated. While about 50 % of the patients have concomitant PMR [3], the pathophysiologic relationship between these conditions continues to be a matter of debate.

Ultrasound in Temporal Arteritis

US has the highest resolution of all the imaging techniques used for the diagnosis of vasculitis. High-frequency probes of more than 10 MHz provide axial and lateral resolution of 0.1 mm. The temporal arteries are localized about 4 mm below the skin surface. They are easily accessible with US, and this image technique allows the assessment of the whole length of the two branches of the temporal arteries. US depicts the artery wall and provides information about blood-flow characteristics within the artery.

In acute temporal arteritis, CDUS of temporal arteries shows hypoechoic edematous wall swelling, "halo sign" (Fig. 12.1a, Fig. 12.1b). This tissue is not compressible [13], and it disappears with glucocorticoid treatment after 2–3 weeks, in most patients. The dimensions of the halo sign seem to be very important with maximum widths ≥ 1 mm being correlated with high specificity (93 %) [14]. A meta-analysis concluded that a unilateral halo sign had sensitivity for diagnosis of 68 % and a specificity of 91 %. The specificity is even greater when bilateral halo sign is identified and the sensitivity is improved by assessing large vessels (common carotid and axillary arteries) in addition to the temporal arteries [6, 15]. In a specialized US clinic, US showed a sensitivity of 88 % and a specificity of 96 % with regard to clinical diagnosis and a sensitivity of 95 % as well as specificity of 99.5 % with regard to positive temporal artery biopsy [4]. When stenosis occurs, it is represented by a reduction in lumen width associated with an increase in blood-flow velocity; in this case, CDUS would show a mixture of colors with persisting color signals in the diastolic phase. Occlusion is represented by a US image of the artery with absence of color Doppler (CD).

US Technique to Examine the Two Branches of the Temporal Arteries

To perform temporal US, the patient lies supine with the head towards the opposite side of the artery that the sonographer examines. The sonographer can sit in front or behind the patient. Place the probe longitudinally in front of the ear. After detecting the common superficial temporal artery, the probe is moved continuously forward to the parietal branch. Then the probe is placed transversely to examine the parietal branch and the common superficial temporal artery in short axis. From the bifurcation, the frontal branch is followed longitudinally and transversely [4, 16].

The sonographer needs to avoid putting too much pressure, to avoid compressing the temporal arteries, and more gel should be applied in areas with hair. The

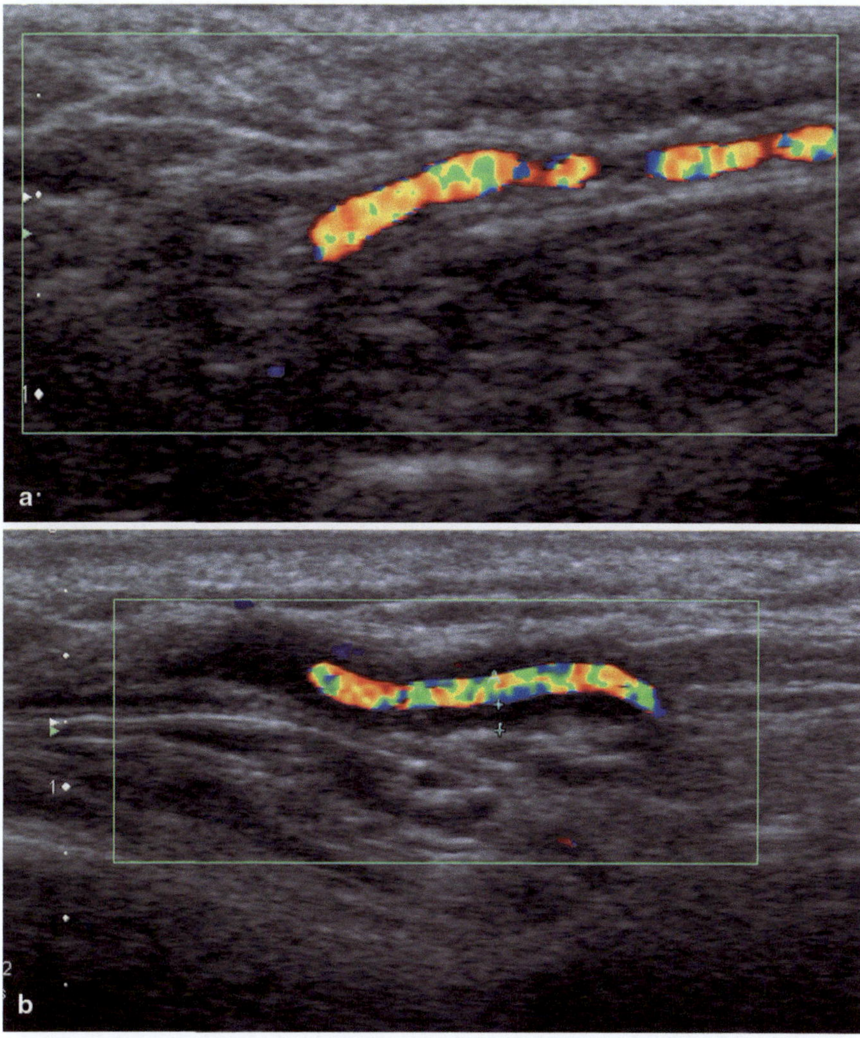

Fig. 12.1 a Longitudinal scan of a normal temporal artery; 12.1b Longitudinal scan of a temporal artery from a patient with temporal arteritis. Notice the halo sign

frequency of the linear probe should be as high as possible, preferably 15 MHz or higher. Adjust the color gain, because if it is too low, only the center of the lumen will show color signals leaving an anechoic rim between the lumen and the wall, mimicking the pathological halo sign; if the color gain is too high, the inflamed area may be covered and disease may be missed. The pulse repletion frequency should be set at about 2.5 kHz. The sonographer uses CD mode and angles the box so that the blood flow is not completely parallel to the probe. Color should cover the artery lumen completely, but should not continue over the artery wall. Both sides should be examined. Temporal arteries should be investigated as completely as possible [17].

The facial arteries are mainly involved in patients with jaw claudication. In patients presenting with retro-auricular pain, consider to examine the occipital arteries, which are exclusively involved in some patients [18].

As mentioned before, in this pathological condition, stenosis of short segments tend to occur. The CD shows a blurring mixture of colors (aliasing) together with persistent blood flow in the diastole. Pulsed-wave Doppler then displays at least a twofold increase in maximum systolic blood-flow velocity in the stenosis, compared with the blood-flow velocity in the area before or behind the stenosis [17].

Many studies have been performed comparing temporal artery US with histology and clinical diagnosis of temporal arteritis. Three meta-analyses have been published [15, 19, 20]. The sensitivity of temporal artery duplex US was 87 % with regard to clinical diagnosis, and specificity was 96 % in one of the meta-analyses [20]. As the halo of the temporal arteries disappears quickly and recurs only in severe flares, routine US examinations are not necessary. Despite this, biopsy of the temporal artery is still considered the gold standard in the diagnosis of GCA [21].

Large-Vessel Giant-Cell Arteritis

Patients with LV-GCA may present as the aforementioned classical cranial temporal arteritis, pure PMR, intermittent arm claudication, paresthesias and Raynaud's phenomenon or pyrexia of unknown origin [3, 4]. Extracranial arterial involvement is much more common than previously assumed. US facilitated the examination of extracranial arteries, such as the femoral and popliteal arteries, the aorta as well as the proximal arm arteries. Clinically, in addition to auscultation of the axillary region, peripheral pulses should be palpated and blood pressure measured bilaterally; if the pulses are not palpable, the arteries of both arms and legs should be examined.

When comparing patients with classical cranial temporal arteritis with patients with proximal arm artery involvement in LV-GCA, the first patients' cohort tend to be older (72 vs. 66 years), and there are lesser females (66 vs. 83 %). Time between onset of symptoms and diagnosis was longer in LV-GCA (7 vs. 2 months) [3]. Patients with LV-GCA less commonly develop anterior ischemic optic neuropathy [10, 22], more frequently complain of PMR, and often lack typical cranial symptoms [23]. Critical limb ischemia does not usually occur [22]. The axillary arteries are more commonly involved than the subclavian and brachial arteries [4]. Bilateral involvement is usually present [4, 24]. In contrast with GCA, in LV-GCA, the wall swelling remains much longer and the dark appearance of the wall becomes more echogenic because of less edema and increasing fibrosis [4] (Table 12.1).

US in LV-GCA

About 50 % of newly diagnosed GCA patients have axillary artery involvement [4, 25]. Only 60 % of the patients with LV-GCA have temporal artery involvement.

Table 12.1 Comparison of patients with extracranial and cranial GCA. (Results from 110 patients with established diagnosis of GCA. Adapted from Czihal et al. 2012) [23]

Variable	Extracranial GCA (n=59)	Cranial GCA (n=51)
Age (years)	65±7.6	73.7±7.0
Time to diagnosis	28.7±25.1	6.5±6.5
Cranial symptoms	32 (54.2)	51 (100)
Headache	19 (32.2)	40 (78.4)
Jaw claudication	12 (20.3)	35 (68.6)
Persistent visual impairment	3 (5.1)	27 (52.9)
Polymyalgia rheumatic	29 (49.2)	14 (27.5)
Constitutional symptoms	40 (67.8)	21(42)
Temporal artery abnormality	9 (15.3)	35 (68.6)

GCA giant-cell arteritis

To examine the axillary arteries, place the probe longitudinally in the axilla along the humeral head and neck, like the scan for detecting glenohumeral joint effusions [16, 26]. The axillary artery is located at the level of the humerus or 1–2 cm medial to it, and proximal to the circumflexa humeri artery. The color box needs to be steered to avoid a blood flow perpendicular to the sound waves. Like evaluating the temporal artery wall, the grayscale image probe and vessel should be as parallel as possible. A normal vessel shows an intima–media complex of less than 1 mm. A bright line represents the interface between artery lumen and vessel wall, followed by a dark line representing wall tissue, and another bright line representing the interface between media and adventitia (Fig. 12.2).

In the large-vessel vasculitis, the artery wall is thickened, usually more than 1 mm. A homogeneous wall swelling, preferably circumferential, of 1.5 mm or more is pathognomonic for the diagnosis of axillary vasculitis [3]. The edematous

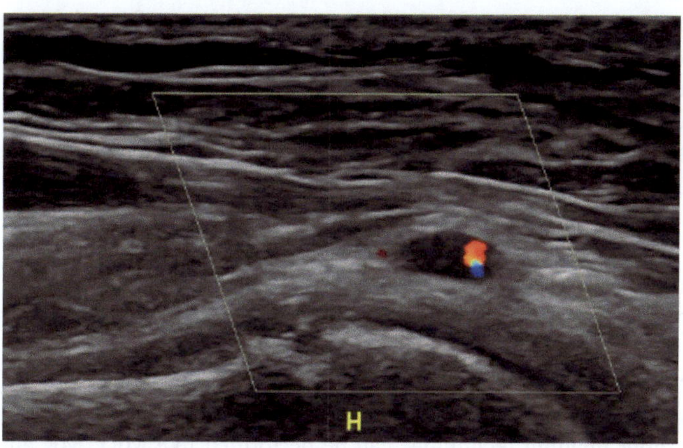

Fig. 12.2 Color Doppler image of the axillary artery of a woman aged 66 years, with temporal arteritis and no palpable radial pulse in the right arm

wall swelling, hypoechoic in untreated patients, frequently persists at follow-up and becomes brighter with treatment, because of fibrosis.

Other arteries like subclavian, common carotid and vertebral arteries can also be examined by US. However, these arteries are rarely affected in GCA, without the involvement of either the temporal or the axillary arteries.

Takayasu's Arteritis

In contrast to LV-GCA patients in which the age of onset is usually more than 50 years, Takayasu's arteritis usually occurs before the age of 40 years. Arteries involved include, the subclavian arteries which are the most commonly involved (93 %), followed by the aorta (65 %), and the common carotid arteries (58 %) [17]. Aortitis results in markedly increased risk for developing thoracic aortic aneurysms and dissections [27]. Characteristically, temporal arteries are never affected in Takayasu's arteritis. Though the arterial involvement pattern may be similar for both LV-GCA and Takayasu's arteritis, in this last one, arteritis affection is usually less symmetrical, affecting more frequently left subclavian and common carotid arteries and less frequently the axillary arteries [24].

Ultrasound in Takayasu's Arteritis

In Takayasu's arteritis, US can be very useful in early diagnosis, to monitor disease progression and the effect of therapy. In general, the US findings in Takayasu's arteritis are similar to those in GCA. CD assessment reveals hypoechoic, concentric wall thickening, brighter than in temporal arteritis, sometimes termed "macaroni sign". The reason for this difference in echogenic pattern lies, probably, in the more chronic course of Takayasu's arteritis and less wall edema [14]. In suspected cases, it is advisable to examine the carotid, subclavian, vertebral arteries and abdominal aorta. The renal arteries should also be examined in case of arterial hypertension. Vasculitis can be easily differentiated from arteriosclerotic lesions, which are heterogeneous and irregular with calcifications.

However, US and angiography are considered complementary in the evaluation of Takayasu's arteritis patients and, unfortunately, in most cases, the diagnosis of Takayasu's arteritis is not established before stenosis occurs.

Isolated Aortitis

This vasculitis is listed as an SOV in the new Chapel Hill nomenclature [2]. Clinically, many patients present with pyrexia of unknown origin. If US findings are not clear, other imaging techniques can be performed like MRI, MR angiography

(MRA), positron emission tomography (PET), computerized tomography angiography (CTA), or thoracic and abdominal CT scan.

Polymyalgia Rheumatica

PMR is a common inflammatory condition involving the shoulders, neck and pelvic girdle, of elderly people. The etiology of PMR is currently unknown although genetic and environmental factors are thought to be contributing factors [28]. Symptoms may also involve the proximal upper and lower extremities. Pain usually limits active and passive range of motion, and the stiffness is marked in the morning, causing patients difficulty in getting out of bed and performing activities of daily living, such as washing and dressing. Pain often interferes with sleep at night and results in significant daytime fatigue. The shoulders are more often involved than the hips, and symptoms may be unilateral early but tend to become bilateral and symmetrical. Distal musculoskeletal manifestations, which are present in up to 50 % of PMR patients, may include carpal tunnel syndrome, swelling of hands and feet with pitting edema and nonerosive peripheral arthritis. About 40 % of these patients can also experience weight loss, fatigue, and low-grade fever.

PMR shares the same pattern of age and sex distribution as GCA. The diagnosis is usually based in clinical and laboratory findings. Synovitis, vasculitis, and bursitis have all been demonstrated in PMR [29], and some authors say that PMR and GCA are different facets of the same disease. About 40–60 % of patients with GCA have PMR symptoms, and about 9–20 % of patients with PMR also have GCA. However, patients with PMR associated with GCA often have more severe disease, with significant abnormalities in most laboratory variables, including higher ESR and platelet counts, and lower hemoglobin values, than patients with isolated PMR. Patients with PMR should be carefully interviewed and examined to detect features suggestive of GCA [5].

Radiographs of the affected joints in PMR are often normal and are only useful in excluding other conditions like degenerative or inflammatory joint diseases or crystal arthropathies. The presence of bony erosions on radiographs is more consistent with rheumatoid arthritis (RA) [30].

US in Polymyalgia Rheumatica

Shoulder and hip US increase the specificity for diagnosing PMR; therefore, musculoskeletal US has been incorporated into the new European League against Rheumatism/American College of Rheumatology (EULAR/ACR) classification criteria [31]. Based on US and MRI findings, PMR predominantly affects periarticular structures. The findings of bilateral subacromial/subdeltoid bursitis (Fig. 12.3), tendinosis of the long head of the biceps (Fig. 12.4), and bilateral trochanteric bursitis are particularly consistent. Glenohumeral and hip joint effusions and synovitis can also be seen, but are less specific [32]. Entheseal involvement and effusions of the

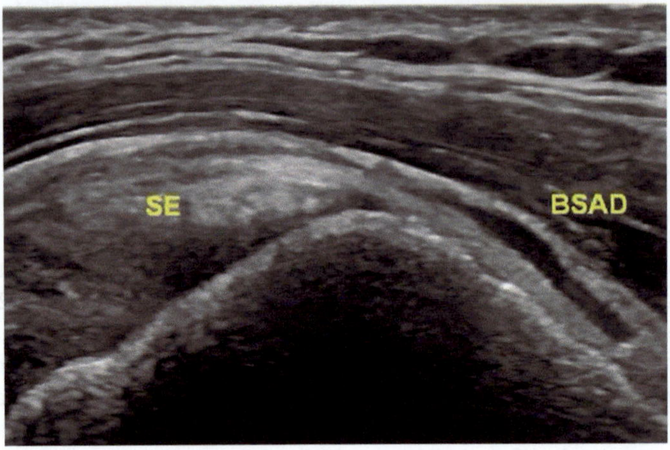

Fig. 12.3 Longitudinal scan of the right shoulder showing a subacromial/subdeltoid bursitis (BSAD) in a patient with PMR. *SE*—supraspinatus tendon

knees, elbows, and radiocarpal joints also have been described [14]. An intra- and inter-reader exercise, performed as a basis for the ACR/EULAR PMR classification criteria study, showed 87 % overall agreement.

A score of 4 or more is categorized as PMR in the algorithm without US and a score of 5 or more is categorized as PMR in the algorithm with US (Table 12.2). In the EULAR/ACR classification study, US had high specificity in discriminating PMR from other shoulder conditions (89 %), but not in discriminating PMR from RA (70 %) [31].

The clinical manifestations of GCA and PMR respond usually within 12–48 h of starting corticosteroid treatment. The initial dose for GCA is oral prednisolone or equivalent of 40–60 mg/day, and for patients with PMR a dosage of 15–20 mg/day is usually sufficient. A prolonged monitoring is necessary, and corticosteroids are gradually tapered, guided by regular clinical evaluation and ESR and/or CRP measurement. In PMR, the initial dose is usually maintained for 2–4 weeks. Sub-

Fig. 12.4 Transverse scan of the long head of the biceps tendon showing a tenosynovitis in a patient with PMR

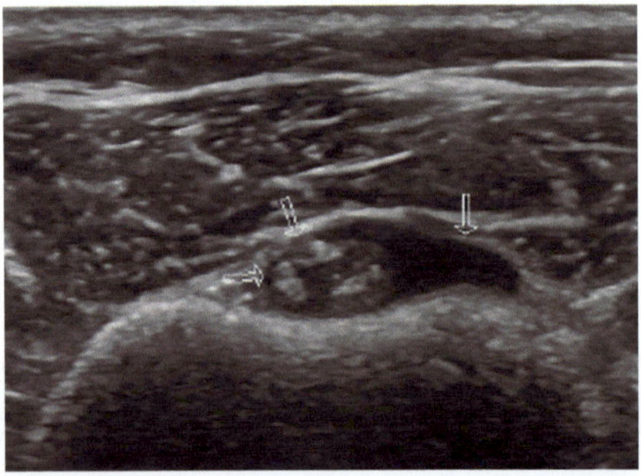

Table 12.2 PMR classification criteria scoring algorithm-required criteria: age ≥50 years; bilateral shoulder aching; abnormal CRP and/or ESR. (Adapted from 2012 provisional classification criteria for polymyalgia rheumatic: a EULAR/ACR collaborative initiative) [31]

	Points without US (0–6)	Points with US (0–8)
Morning stiffness >45 min	2	2
Hip pain or limited range of motion	1	1
Absence of RF or ACPA	2	2
Absence of other joint involvement	1	1
If at least one shoulder with subdeltoid bursitis and/or biceps tenosynovitis and/or glenohumeral synovitis (posterior or axillary) and at least one hip with synovitis or trochanteric synovitis		1
Both shoulders with subdeltoid bursitis, biceps tenosynovitis or glenohumeral synovitis		1

PMR polymyalgia rheumatica, *CRP* C-reactive protein, *ESR* erythrocyte sedimentation rate, *EULAR* European League Against Rheumatism, *ACR* American College of Rheumatology, *RF* rheumatoid factor, ACPA anti-citrullinated peptide antibody, *US* ultrasound

sequently, the dose is decreased by 2.5 mg every 2–4 weeks until the patient is at 10 mg daily; following this, the dose is tapered by 1 mg a month [33, 34]. Methotrexate may be useful as corticosteroid-sparing agent in GCA and PMR in patients with corticosteroid-related toxicity or patients with frequent disease relapses [32]. Retrospective studies favor aspirin (acetylsalicylic acid) as an effective adjuvant treatment for reducing the ischemic complications of GCA [5]. Recent anecdotal reports suggest interleukin (IL)-6 may be a helpful therapeutic target in PMR and large-vessel vasculitis, but additional studies are needed [32].

Inflammatory Myopathies

MRI and US are regarded as the tools of choice for imaging muscle trauma and disease, bearing in mind that each one has its own specific advantages. MRI is considered the gold standard for muscle diseases imaging. On the other hand, the morphology of normal muscles can be clearly depicted on US, a technique with a superior accessibility and ability to obtain real-time dynamic images.

Sono-anatomy studies revealed that individual muscle fibers are surrounded by a thin fascial layer called *endomysium* (which cannot be seen with US). The fibers are packed together in bundles (fascicles) which are surrounded by *perimysium,* and the fascicles are packed together to form muscles, which in turn are surrounded by a thicker *epimysium*. Muscle fascicles appear relatively hypoechoic compared to the hyperechoic encasing linear septations and sheath. In longitudinal scanning, these hypoechoic fascicles are separated by multiple hyperechoic long lines which represent perimysium or fibroadipose septa. This is called pennate appearance. On short-axis images, muscle septations appear as white dots on a hypoechoic back-

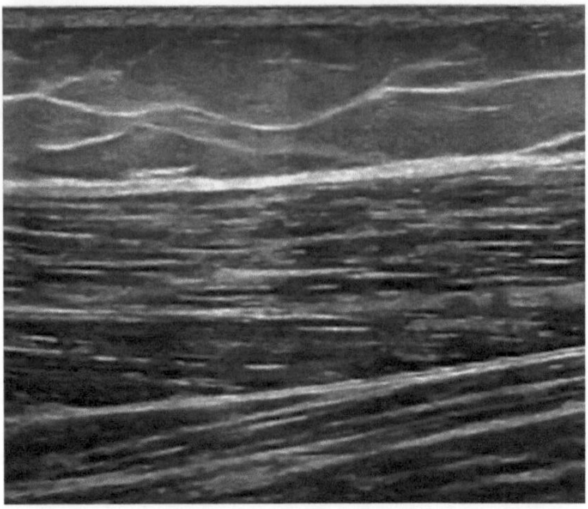

Fig. 12.5 Longitudinal scan of the muscles of the thigh showing the normal pennate appearance

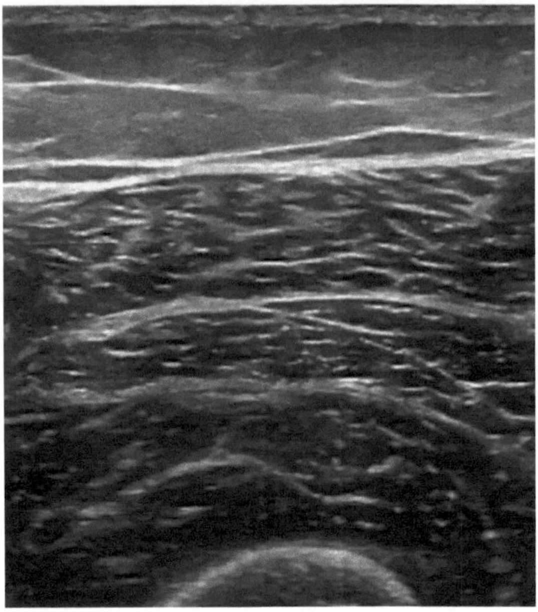

Fig. 12.6 Transverse scan of the normal muscles of the thigh

ground (Figs. 12.5, 12.6). Within the muscle itself, the muscle fibres attach to a fibrous aponeurosis that eventually becomes a tendon [35].

Muscle tissue inflammation can arise secondary to infection, trauma, vasculitis or autoimmune diseases, including RA, lupus, syndromes, and scleroderma. Polymyositis, dermatomyositis, and inclusion-body myositis are the major entities of a group of skeletal muscle diseases called idiopathic inflammatory myopathies.

MRI permits the identification of muscle edema in fat-suppressed T2-weighted MRI, and allows a detailed examination of a large anatomic area, which is very

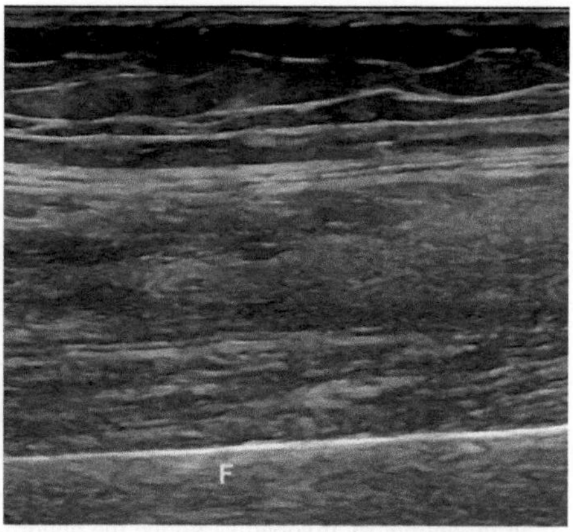

Fig. 12.7 Longitudinal scan of the muscles of the thigh in a patient with acute myositis, *F* femoral bone

useful as symptoms are usually difficult to localize to one muscle group. In T1-weighted MRI images, fatty atrophy of the musculature can be seen reflecting the chronic phase of the disease [36]. By using MRI, it is possible to identify the best area to do a muscle biopsy.

US in Inflammatory Myopathies

In rheumatology, US has been mainly used for inflammatory muscle conditions. Before performing a US muscle examination, a clinical history and physical examination are essential. Linear transducers or curved-array transducers (which are preferable especially for obese or very muscular patients, particularly to examine gluteal region or the proximal thigh) can be used. With this technique, it is possible and preferable to do a dynamic and real-time comparison of the contralateral region. To demonstrate a pathology, it is useful to use, if possible, a panoramic or extended view. In early acute stage of myositis, the muscle belly becomes edematous, resulting in diffuse enlargement and relative decrease in overall echogenicity; therefore, US features of acute myositis include normal or increased muscle size, lower echogenicity and enhanced perfusion of the affected muscles [37] (Fig. 12.7). In chronic disease stage, muscle size and perfusion are reduced (Fig. 12.8). The increased Doppler activity within the muscle may be more sensitive to change than gray-scale and have a predictive potential [35]. A study based on contrast-enhanced US scanning showed significantly higher blood-flow velocity, blood flow, and blood volume in patients with acute myositis than in normal volunteers [38]. Calcifications can easily be detected because of high echo intensity and acoustic shadowing on US imaging. Being widely available and cheap, muscular US became a useful tool for studying and monitoring patients with myositis. Both MRI and US are also able to identify fluid collections, adipose infiltration, atrophy, and fibrosis.

Ultrasonic elastography (sonoelastography) measures tissue deformation as a response to an external force, assuming that the deformation is lower in rigid tissues,

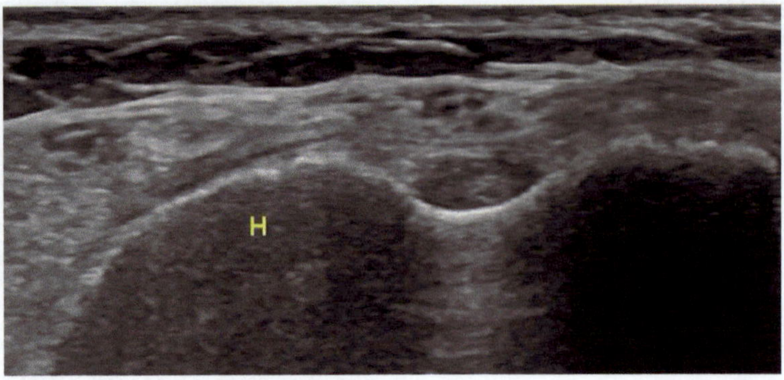

Fig. 12.8 Transverse scan of the upper part of the shoulder from a patient with chronic evolution of an inflammatory myopathy. Notice the abnormal deltoid pattern and echogenicity, *H* humeral bone

compared to elastic soft tissues. This method is based on comparing the radiofrequency of ultrasonic waves obtained before and after an easy made compression with a conventional transducer, using a free hand technique. As muscular elasticity is affected, hence modified in myopathies, sonoelastography became a useful method for its assessment. A study involving 24 patients with muscular pathology, namely polymyositis, dermatomyositis, mixed connective tissue disease, systemic lupus erythematosus (SLE) and RA myopathy, analyzed the utility of US in the assessment of inflammatory myopathies, and of sonoelastography in the assessment of the elasticity of skeletal muscle in myositis. Despite all the limitations, the authors concluded that analysis of the color information from elastography could be a reliable method for the management of patients with idiopathic inflammatory myopathies [39].

Fibromyalgia

FM is a common form of noninflammatory rheumatism, with symptoms often mimicking those of arthritis or muscle disorders, and a widespread pain of idiopathic origin. Because of its very wide range of symptoms, including fatigue, headache, irritable bowel syndrome, paresthesias, Raynaud's syndrome, muscle weakness, bladder dysfunction, sleep disorders, decreased cognitive functioning, depression, and anxiety, it can be confused with other rheumatic and non-rheumatic diseases.

Its prevalence increases with age and affects women more than men. According to the 1990 ACR diagnostic criteria, the diagnosis of FM requires the presence of chronic widespread pain and tenderness in at least 11 out of 18 tender points (TP) when applying a pressure of 5 kg. However, in 2010, a new set of criteria has been proposed by the ACR that does not require a TP examination but includes a subjective measure of the number of painful body regions and a somatic symptom severity scale.

Ultrasound in Fibromyalgia

Arthralgic symptoms in the hip region are commonly mentioned by FM patients. A study about sonographic assessment of the hip in FM patients failed to demonstrate significant pathological US abnormalities in the majority of the patients [40].

A study used sono-myography (SM) and sonomyoelastography (SME) to analyse morphology, stiffness and blood flow in the TP of a group of FM patients, in comparison to a healthy control group. There was no significant difference in stiffness, nor number of localized hypoechoic areas at TP. Similarly, there was no significant difference between FM women and healthy female controls using SME. Furthermore, local blood-flow assessment did not appear to reveal any differences between both the patients and control groups [41].

In another study, which included ten FM patients as well as ten control subjects, using contrast-enhanced US, results revealed reduced muscular vascularization in FM patients when compared with controls [42].

Regional Pain Syndrome

Regional pain is one of the most common complaints encountered in the rheumatology clinic. Various conditions manifest as regional pain, including ligamentous and tendinous structure abnormalities, nerve entrapments, and tumors, muscles and bone pathologies. Clinical examination may be sufficient for the diagnosis; however, imaging modalities are useful in unclear cases. Reproducing patient's symptoms by movements or palpation while scanning (sonopalpation), make ultrasonography one of the most useful imaging modalities for differential diagnosis of regional pain [43].

Some of the conditions leading to regional pain were described in detail in other parts of this book. The aim of this section is to help understand the concept of scanning in regional pain syndromes and present less common or less familiar conditions which may manifest as regional pain.

In general, clinical examination should always precede US scanning. Differential diagnosis of potential causes of pain should then be systematically revised during US examination, starting with most common pathologies and easy-to-assess structures, followed by less common causes of pain and more profound assessment. In the majority of cases, pathological changes result in decreased echogenicity, thickening of involved structures, and/or irregular outlines (enthesitis, tendinopathy, stenosing tenosynovitis, nerve entrapment) (Figs. 12.9, 12.10).

Detection of focal lesions, compressible (i.e. bursae) or not (benign and malignant neoplasms, or ganglion cysts), usually does not present difficulties. In smaller lesions, a comparison to the contralateral side may be necessary. Sonopalpation helps to correlate the ultrasound findings to the patient's symptoms. This is confirmed on direct compression of the localized lesion which would elicit the anticipated type of pain. Dynamic examination may reveal abnormalities not apparent on static images, like

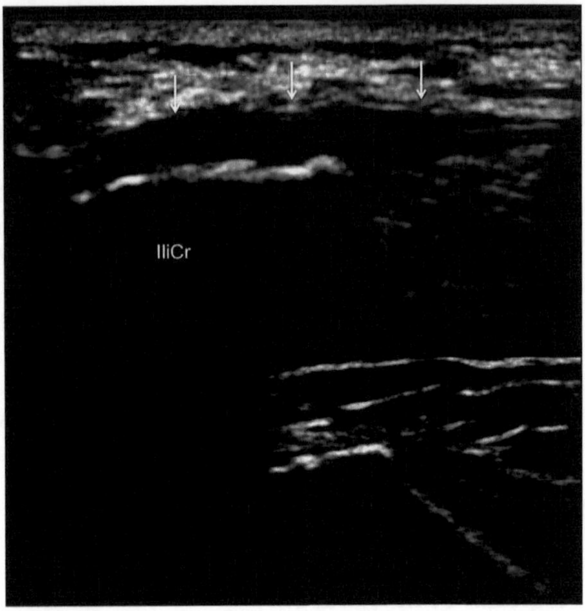

Fig. 12.9 Longitudinal scan of the iliac crest (*IliCr*) of the patient with spondyloarthopathy and left iliac crest pain. Thickening and decreased echogenicity of the fascia insertion (*arrows*), and irregular outline of the iliac crest reflect enthesitis. The lesion is clearly distinguishable by its low echogenicity in comparison with the normal isoechoic entheses in adjacent areas (*inset*)

adhesions following previous inflammation, ligamentous laxity, tendon subluxation, entrapment, impingement, and snapping syndromes (Figs. 12.11, 12.12).

Unexpected findings may be occasionally seen, confirming the value of US as the rheumatology bedside procedure (Figs. 12.13, 12.14).

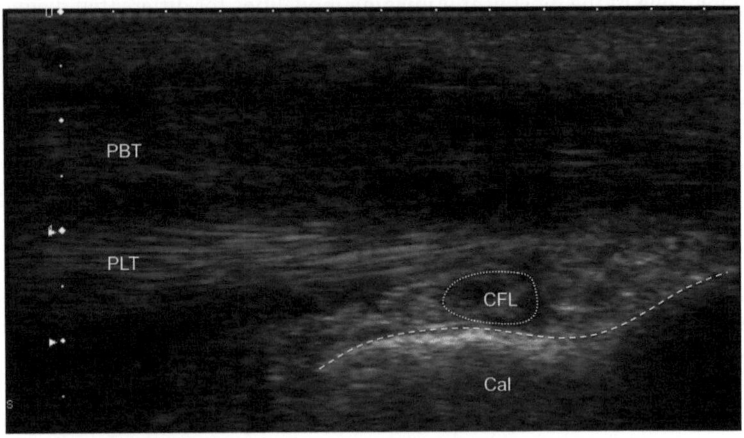

Fig. 12.10 Longitudinal scan of the peroneus tendons below the lateral malleolus. The patient was complaining of pain behind the lateral ankle exacerbated by activities. Ultrasound examination shows normal structure of peroneus longus tendon (*PLT*) and hypoechoic, blurred structure of the peroneus brevis tendon (*PBT*). Diagnosis: peroneus brevis tendonitis. *CFL* calcaneofibular ligament, *Cal* calcaneus

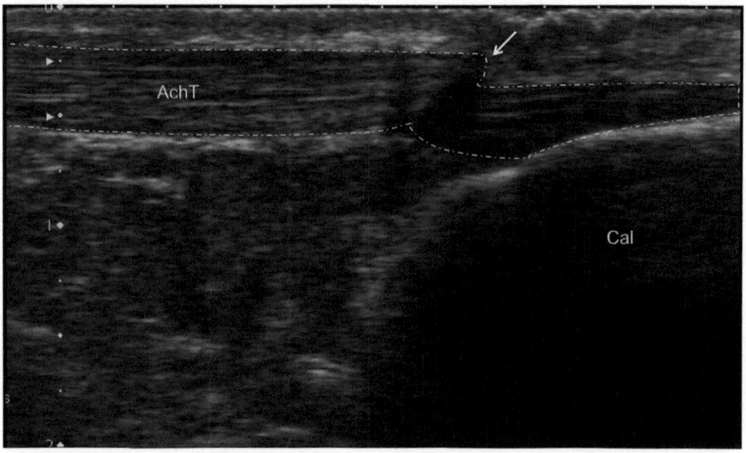

Fig. 12.11 Longitudinal scan of Achilles tendon (*AchT*) of patient complaining of chronic Achilles tendon pain. Thickness and echostructure of Achilles do not showed abnormality. On plantar flexion, apparent folding of Achilles (*arrow*) revealing adhesions within retrocalcaneal bursa as the result of previous bursitis, *Cal* calcaneus

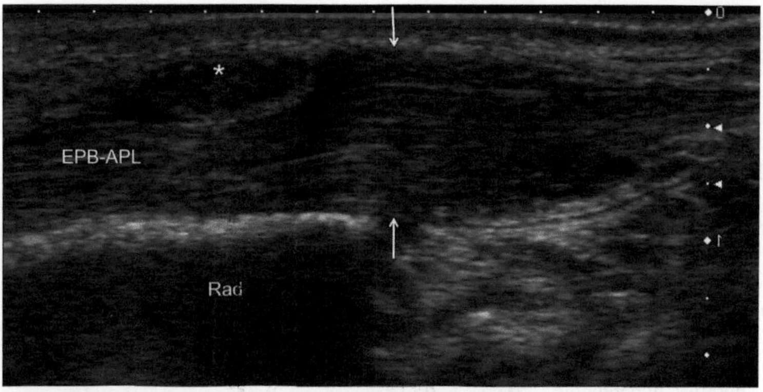

Fig. 12.12 Longitudinal scan on the radial side of the wrist of patient with de Quervain's disease. Thickening of the retinaculum (*) causing entrapment of the tendons. During passive abduction of the thumb, ultrasound revealed crowding of the tendons before the retinaculum (*arrows*). The patient was not able to actively abduct the thumb because of tight stenosis of the first extensor's compartment. *EPB–APL* extensor pollicis brevis and abductor pollicis longus tendons, *Rad* radius

Retinacula

Normal tendon retinacula appear as fibrous bands surrounding the tendons and attach to the underlying bone [44]. Normal thickness of retinacula of the ankle ranges between 0.9 and 1.5 mm [45]. Chronic overuse in the presence of additional intrinsic factors (i.e., diabetes) may lead to degeneration and thickening of the retinaculum, restraining the tendons' sliding (retinaculum impingement) [46]. Morphologically, thickening of the retinaculum was found to be the result of fibrosis, hyaline

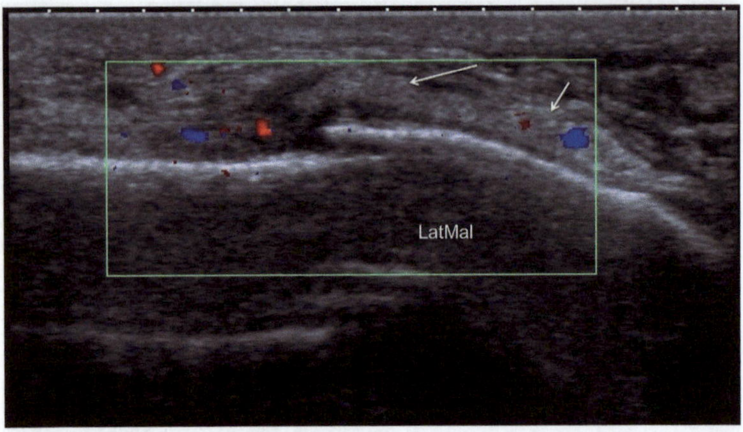

Fig. 12.13 Longitudinal scan of the lateral ankle of the patient with rheumatoid arthritis, who complained of ankle pain on weight bearing, ankle swelling, and restricted range of movements. The pain did not respond to blind steroid injection to tibiotalar joint. Ultrasound scan showed fractures of lateral (*LatMal*) and medial malleolus: cortical discontinuity, thickening of periosteum (*arrows*), swelling of subcuteneous tissue, and hypervascularity

degeneration and vascular proliferation [47]. Thickening of the tendon sheath and fibrosis of the tendon are commonly associated findings [48]. Stenosing tenosyno-

Fig. 12.14 Ultrasound scan of the patella of patient complaining of the knee pain on movements. Ultrasound showed the destruction of the patella by the pathological lesion (*Les*). Patient presented also with small subcutaneous nodules, hyperechoic on ultrasound scan. Biopsy confirmed the diagnosis of angiosarcoma

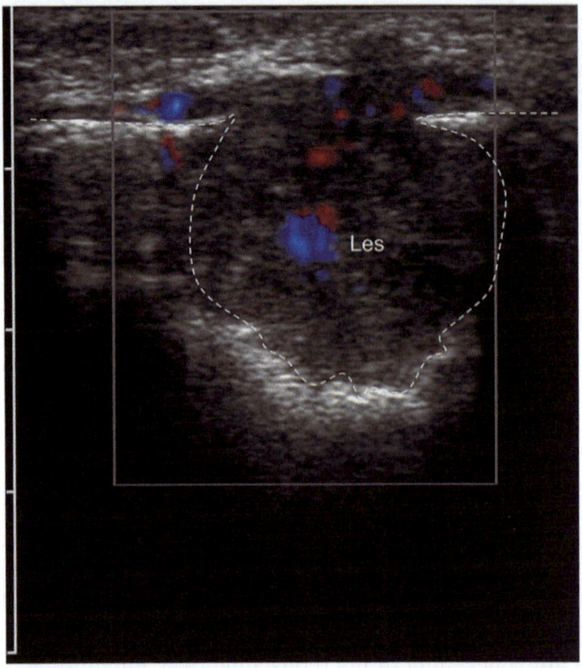

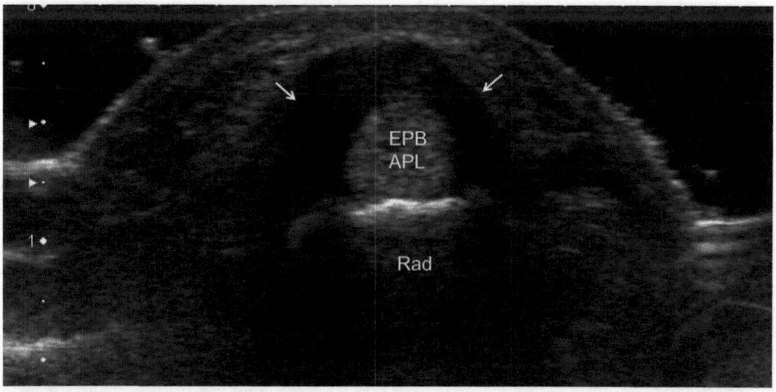

Fig. 12.15 Transverse scan of the radial side of the wrist. Patient presented with pain and swelling on the radial side of the wrist exacerbated by adduction of the thumb. Thickening of the retinaculum (*arrows*) over first extensors compartment (extensor pollicis brevis—*EPB*, abductor pollicis longus—*APL*) in keeping with the diagnosis of de Quervain's disease. *Rad* radial bone

vitis of finger flexors and de Quervain's disease are common examples of retinaculum impingement (Figs. 12.12, 12.15) [49].

Muscle Anatomic Variants

Muscle anatomic variants may appear as swelling and cause pain or discomfort during activities. Accessory muscles can be easily mistaken with bursitis, tenosynovitis, tendonitis or neoplasm, and may be a cause of nerve compression [50].

Selected muscular anomalies and their clinical implications are listed [50–53]:

- reversed, digastric or duplicated palmaris longus (prevalence 7 %)—may be confused with flexor tenosynovitis and may cause median nerve compression
- Extensor digitorum brevis manus (2–3 %)—may be confused with extensor tenosynovitis (Fig. 12.16)
- Anconeus epitrochlearis (11 %)—may cause ulnar nerve entrapment in cubital tunnel
- Accessory abductor digiti minimi (24 %)—may cause ulnar nerve compression in Guyon's canal
- Accessory soleus muscle (0.7–5.5 %)—may cause localized compartment syndrome and be confused with soft tissue tumor
- Flexor digitorum accessorius longus (6–8 %)—may cause tarsal tunnel syndrome
- Peroneus quartus (13–26 %)—may cause ankle pain and instability

If not anticipated, additional muscles may cause confusion during US examination (Fig. 12.16.). Awareness of such anatomical variation type helps to avoid unnecessary therapeutic actions.

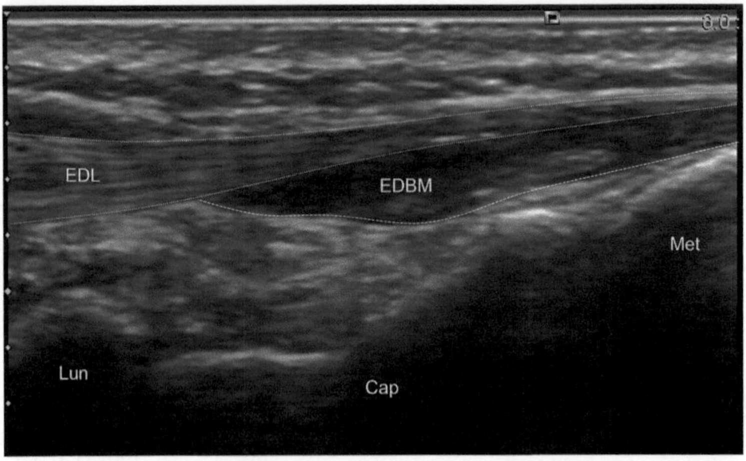

Fig. 12.16 Longitudinal scan of the wrist. Patient presented with swelling on the dorsum of the hand and pain on activities. Ultrasound scan shows hypoechoic, poorly compressible area adjacent to the extensor tendons. Extensor digitorum brevis manus (*EDBM*) may be easily confused with tenosynovitis of extensor digitorum longus (*EDL*). *Lun* lunate, *Cap* capitate, *Met* metacarpal bones

Nerve Entrapments

Entrapment neuropathies typically occur within osteofibrous tunnels. Sonographically, swelling and decrease echogenicity of the nerve, blurring of internal structure, thickening and increased echogenicity of perineurium may be seen proximally and distally to the site of compression [54]. Simple principle may be accommodated while diagnosing entrapment neuropathies: swollen, easy to spot and homogenously hypoechoic nerve often indicate nerve pathology (Fig. 12.17).

Nerve Tumors

On US, nerve tumors usually present a fusiform shape with slightly tapered ends, oriented longitudinally in the nerve axis. Neurofibromas develop centrally, whereas schwannomas are located eccentrically in relation to the nerve (Fig. 12.18). Most lesions appear as hypoechoic masses and have well-defined margins. Schwannomas commonly are heterogeneous tumors, well vascularized, and may contain internal cystic changes [54].

Neuromas are commonly formed following nerve injury as the result of the reparative process and fibrosis (Fig. 12.19). Local swelling of the nerve with decreased echogenicity, disorganized fibrillar pattern and poorly defined borders may be the result of partial damage of the nerve fibres, while bulb shape of proximal ending of the nerve at the site of the injury is the effect of complete nerve disruption [54].

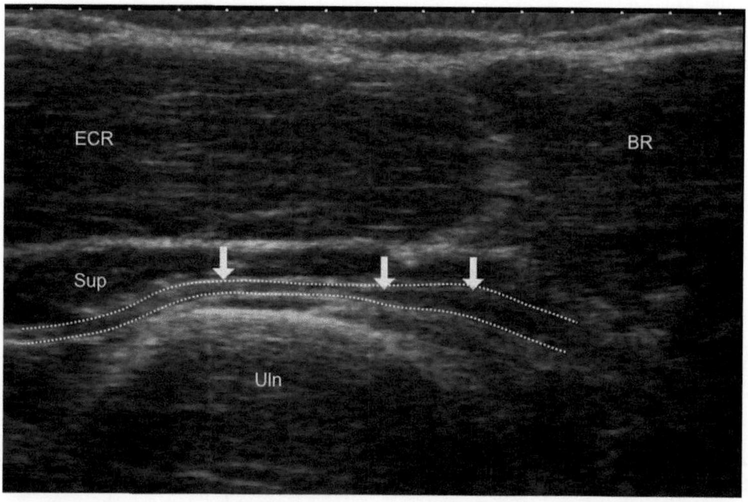

Fig. 12.17 Oblique scan of upper lateral forearm. Patient complained of pain at and below the lateral epicondyle exacerbated by activity. Ultrasound scan shows swelling and decreased echogenicity of deep branch of the radial nerve (*thick arrows*) proximally to its compression within supinator (*Sup*) muscle. *ECR* extensor carpi radialis, *BR* brachoradialis muscle, *Uln* ulna

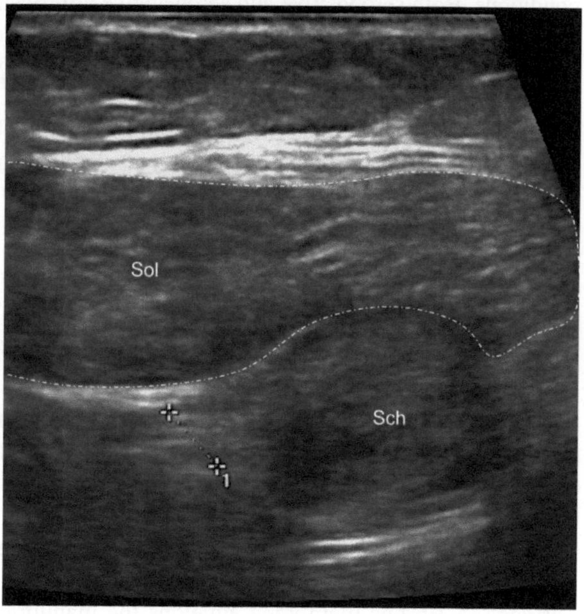

Fig. 12.18 Transverse scan of the calf in its distal third. Patient with anxiety and fibromyalgia presented with pain and paresthesia of the right lower leg. Ultrasound scan at the point of greatest tenderness showed ovoid mass adjacent to the tibial nerve (measure), located below the soleus muscle (*Sol*). MRI scan confirmed the diagnosis of schwannoma (*Sch*). *MRI* magnetic resonance imaging

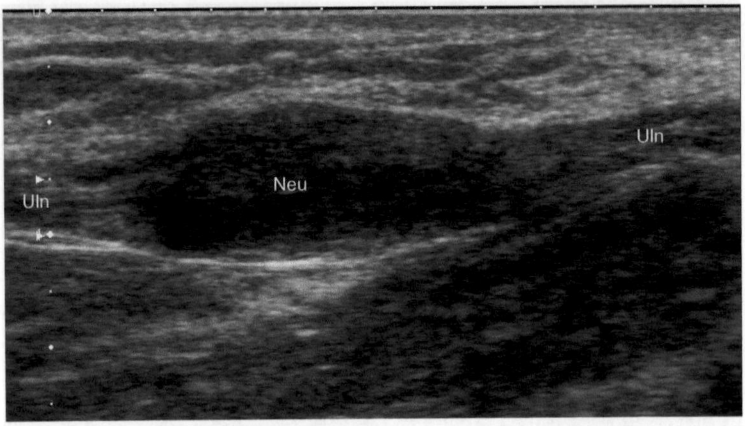

Fig. 12.19 Neuroma (*Neu*) formed following complete ulnar nerve (*Uln*) disruption treated with reconstruction of the nerve: local nerve thickening and decreased echogenicity

Parasternal Joints

The pathologies involving parasternal joints are not uncommon in the rheumatic conditions. Both seropositive and seronegative arthropathies as well as degenerative conditions can affect sternoclavicular, manubriosternal and sternocostal joints. The diagnosis in the anterior chest pain syndrome is essentially clinical with imaging evaluation used optionally to increase diagnostic confidence. Plain radiography, CT, and magnetic resonance imaging (MRI) are commonly used. US was proposed as a useful method of assessment of the parasternal joints in the rheumatology practice [55, 56].

Anatomy

The sternoclavicular joint (SCJ) is a saddle-type diarthrodial joint between medial end of the clavicle, sternal manubrium, and first costal cartilage (Fig. 12.20). The articular surface of the clavicle is disproportionally large in relation to the shallow socket with which it articulates. The articular disc made of fibrocartilage separates the articular surfaces and divides the joint space into two synovial cavities [57].

The manubriosternal joint is a symphysis connecting the manubrium and sternal body. The bony surfaces are covered by hyaline cartilage and connected by a fibrocartilage. A fibrous membrane envelops the manubriosternal joint and the entire bone [58].

The sternocostal joints connect costal cartilages of the first to seventh rib with the sternum. Except for the first rib, the sternocostal articulations are arthrodial joints containing a small synovial cavity. The joint cavity eventually obliterates, and the costal cartilages become continuous with the sternum in older individuals [59]. The cartilage of first rib is directly united with the sternum, forming a synarthrodial joint.

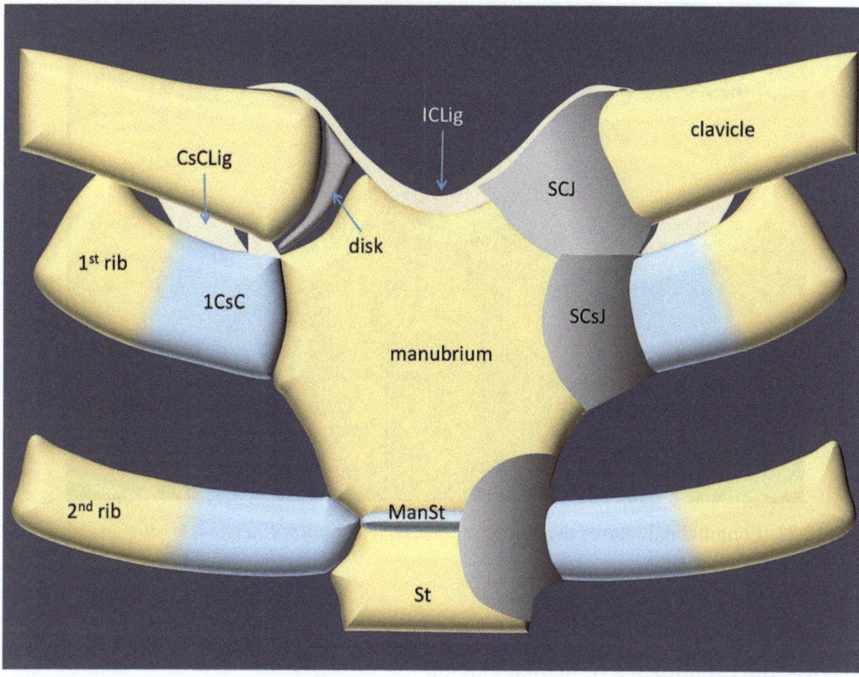

Fig. 12.20 Anatomy of the parasternal joints. *BS* body of sternum, *ManSt* manubriosternal symphysis, *SCJ* sternoclavicular joint, *ICLig* interclavicular ligament, *SCsJ* sternocostal joint, *CsCLig* costoclavicular ligament, *1CsC* first costal cartilage

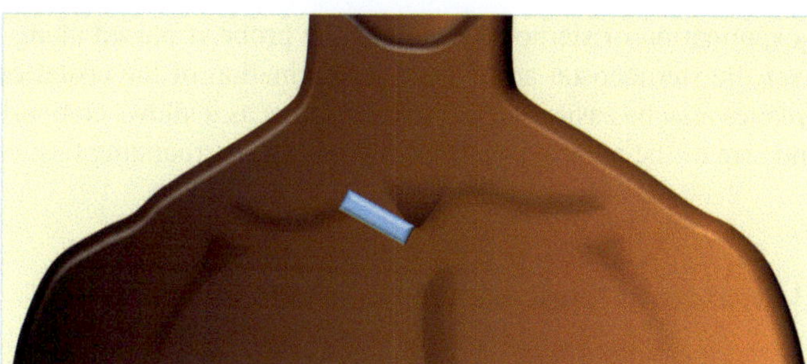

Fig. 12.21 Position of the probe for scanning of the sternoclavicular joint

US Examination

US examination of parasternal joints is conducted with patient lying supine or in half-lying position. High-frequency 7–13 MHz linear-array transducers are used [57].

For the examination of SCJ, the probe is placed parallel to the long axis of the medial clavicle, and medial pole of the probe is moved medially and inferiorly towards the manubrium (Fig. 12.21) [57]. The joint is identified as the hypoechoic

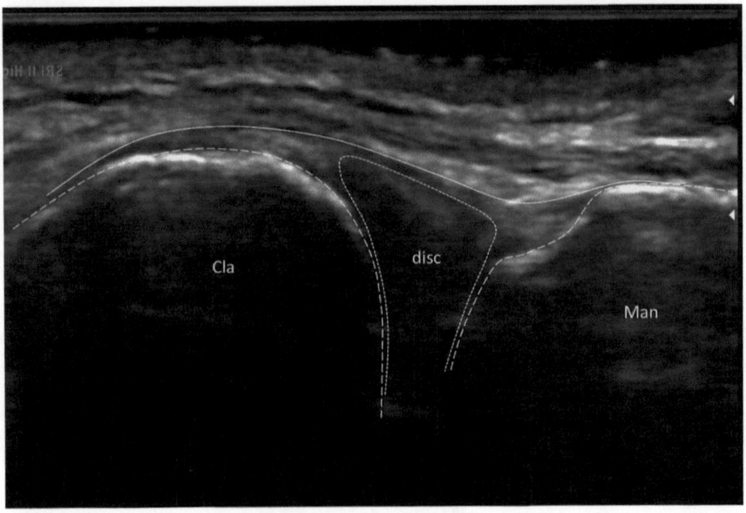

Fig. 12.22 Longitudinal scan of the normal sternoclavicular joint. *Cla* clavicle, *disc* articular disc, *Man* manubrium

gap in between the two bony structures. A generous amount of gel may be required to cover the prominence of medial clavicle.

Examination of manubriosternal joint requires scans in longitudinal and transverse planes. The probe is placed on the sternal angle and moved from central towards lateral aspects of the sternum in sagittal planes and from cranial to caudal in axial planes.

For the examination of sternocostal joints, the probe is placed along the costal cartilage over the sternocostal articulation. Examination of the costal cartilage in transverse planes may be easier and more informative as it allows comparison of the structure and size of the adjacent costal cartilages and surrounding tissues.

Normal Ultrasound Appearance

On US examination of the SCJ in oblique axial planes, the medial clavicle and sternal manubrium are seen as hyperechoic lines with posterior acoustic shadowing (Fig. 12.22). The medial part of the clavicle is positioned more anterior and superior with respect to the sternum and has a rounded configuration [55]. In between clavicle and sternum, the hypoechoic articular disc may be seen. The joint capsule overlies the medial clavicle and manubrium and stretches over the joint space. In transverse planes, the oval shape of hypoechoic disc is seen in the central part of the joint.

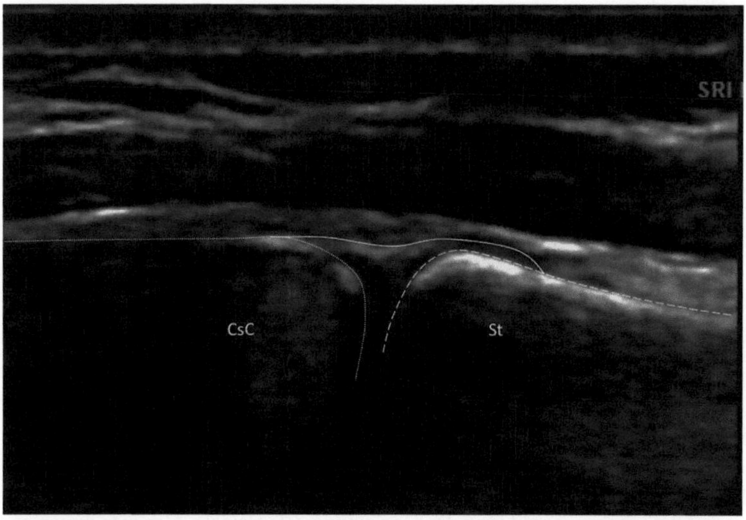

Fig. 12.23 Longitudinal scan of normal sternocostal joint. *CsC* costal cartilage, *St* body of sternum

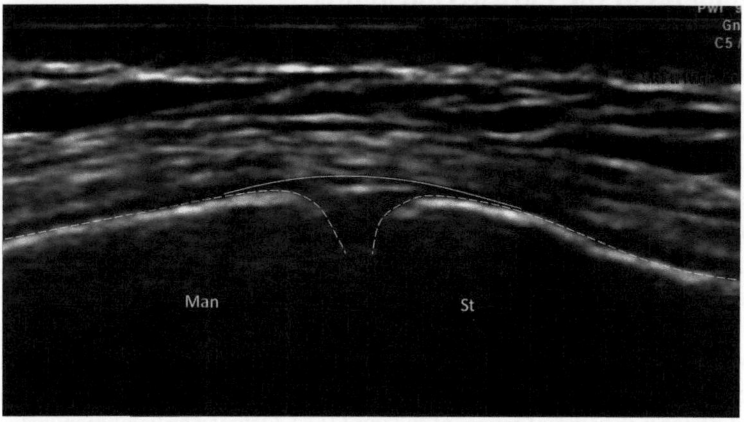

Fig. 12.24 Longitudinal scan of normal manubriosternal joint. *Man* manubrium, *St* body of sternum

The sternocostal joint appears as a junction between the bony outline of the sternum and the homogenous, hypoechoic costal cartilage. The thin hyperechoic line of the joint capsule overlies the junction (Fig. 12.23). In sagittal planes, the costal cartilage appears as an oval-shaped hypoechoic structure [55, 60].

Manubriosternal symphysis appears as a gap between the smooth outlines of the manubrium and sternal body (Fig. 12.24). In individuals above the age of 30 years, the joint may appear fused [61].

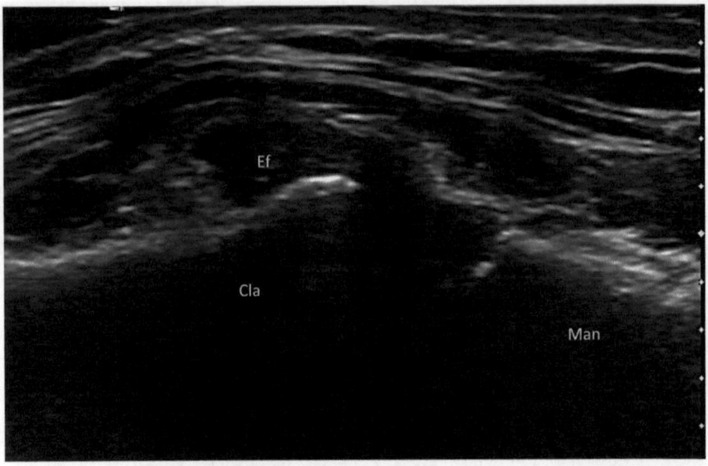

Fig. 12.25 Longitudinal scan of sternoclavicular joint of patient with osteoarthritis. Effusion (*Ef*), narrowing of the joint space, and irregular bone outline. *Cla* clavicle, *Man* manubrium

Pathologies of the Parasternal Joints

During the US examination of parasternal joint, the outlines of the bones, thickness of the joint capsule, echogenicity of the joint content, thickness of the costal cartilage, and the cartilaginous echogenicity and presence of CD signal are evaluated.

Osteoarthritis of SCJ manifests by narrowing of the joint space, osteophytes formation, and articular surface irregularities (Fig. 12.25) [57]. The articular disc may become hyperechoic and inhomogeneous. The joint capsule may be thickened and hypoechoic and, occasionally, the increased CD signal may be found. The capsule extends over the clavicle but not over the manubrium, and its distension usually does not exceed 5 mm [62].

Degenerative changes in the sternocostal joints are characterized by osteophytes extending from the sternal aspect of the joint [63]. Osteoarthritis of manubriosternal joint demonstrates by irregularity (erosions and osteophytes) of the outlines of the bones-forming symphysis [64].

Although classically involved in spondyloarthropathies, synovitis of SCJ can be found on US in around 40% of patients with RA [56]. US is more sensitive than clinical examination in detecting SCJ involvement in RA. In one series of patients, US detected twofold times more SCJ with synovitis than clinical examination [56]. On US, synovitis of SCJ manifests by distension of the joint capsule by hypoechoic intra-articular tissue with or without CD signal (Fig. 12.26). In patients with RA, increased CD signal correlates with the activity of the disease [56].

On US, inflammation in the manubriosternal joint manifests by thickening of the joint capsule and irregular outlines of the bone. Liquefaction of the joint fibrocartilage may mimic presence of the synovial fluid (Fig. 12.27).

In inflammatory arthropathies, synovitis and erosions may also be detected in sternocostal joints [65].

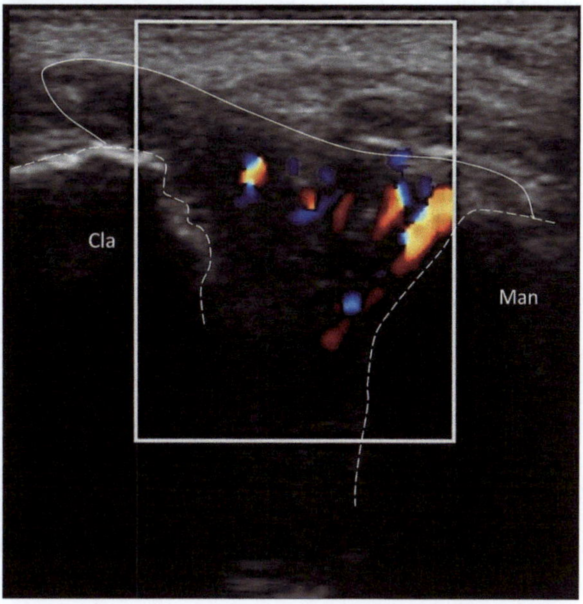

Fig. 12.26 Patient with spondyloarthropathy and clinically silent inflammation in sternoclavicular joint. Irregular outline of medial end clavicle, hyperemia within joint space. *Cla* clavicle, *Man* manubrium

US findings in septic arthritis of sternoclavicular joint include distension of the joint capsule of more than 10 mm, extending over both clavicle and sternum, and hyperemia on CD [62].

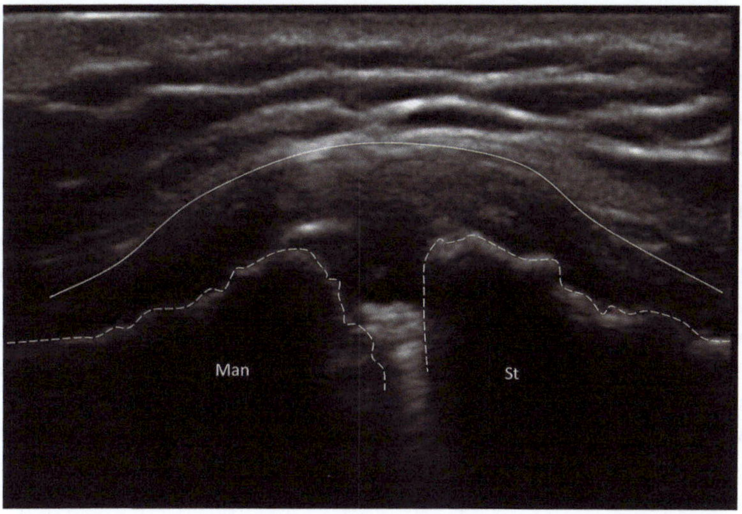

Fig. 12.27 Patient with rheumatoid arthritis presenting with swelling and tenderness of manubriosternal joint. Longitudinal scan of manubriosternal joint shows irregular outlines of the bones, thickening of joint capsule, and fibrous membrane. Liquefaction of fibrocartilage, strong echo of cartilage debris within the joint space. *Man* manubrium, *St* body of sternum

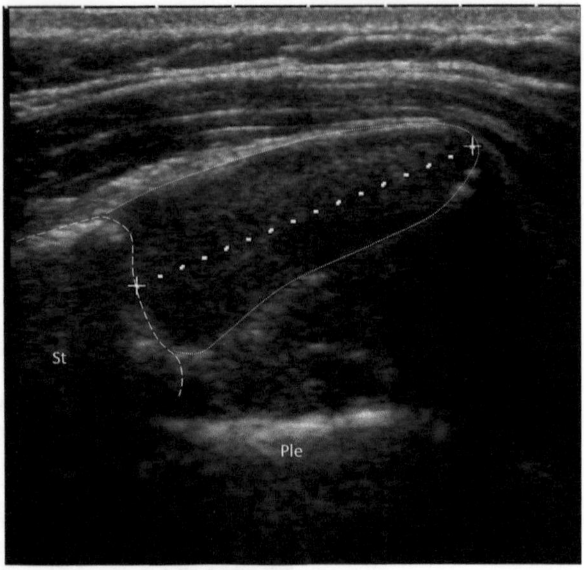

Fig. 12.28 Ultrasound scan of anterior chest wall of patient complaining of swelling and tenderness on the left side of sternum. In axial plane, accessory costal cartilage is visible (measure), located in third intercostal space, and attached to the sternum just below the third sternocostal joint. *St* body of sternum, *Ple* pleura

Tietze syndrome is a benign and self-limiting disorder characterized by swelling and tenderness of one or more costal cartilages. Most commonly, unilateral second or third costal cartilage is affected.

Thickening of the costal cartilage in comparison to the opposite normal side, inhomogeneous, dotty-like increase of the echogenicity of the cartilage and intense broad posterior acoustic shadow are typical US features of Tietze syndrome [60, 66]. Corticosteroid injection leads to decrease in size of the affected costal cartilage and resolution of the acoustic shadow within a week following the injection [60].

The diagnosis of Tietze syndrome is based on clinical examination. US can help to exclude other conditions mimicking Tietze syndrome as neoplasms, lipomas, muscular changes and chest wall deformities (Fig. 12.28).

Synovitis–acne–pustulosis–hyperostosis–osteomyelitis/osteitis (SAPHO) syndrome is a rare condition, which typically manifests by aseptic osteitis and hyperostosis of the anterior chest wall in association with neutrophilic dermatoses. The acronym SAPHO describes most common signs associated with this condition (synovitis–acne–pustulosis–hyperostosis–osteomyelitis/osteitis) [67]. Any component of the sternocostoclavicular region can be involved. The early stage of the disease affects costoclavicular ligament and may manifest as enthesopathy. In the following stages, arthropathy of SCJ develops with osteolytic and osteosclerotic changes of the medial end of clavicle, sternum, first rib, and costal cartilage [68]. In the active phase of osteitis, periosteal reaction and new bone formation may manifest on ultrasonography as cortical irregularities, soft tissue swelling, and increased CD signal surrounding the involved bone [69]. Calcifications/ossifications of pericartilage at the bone–cartilage border, and at the first rib, were also described in SAPHO syndrome [70]. Erosive arthritis of SCJ and manubriosternal joint may be present [70].

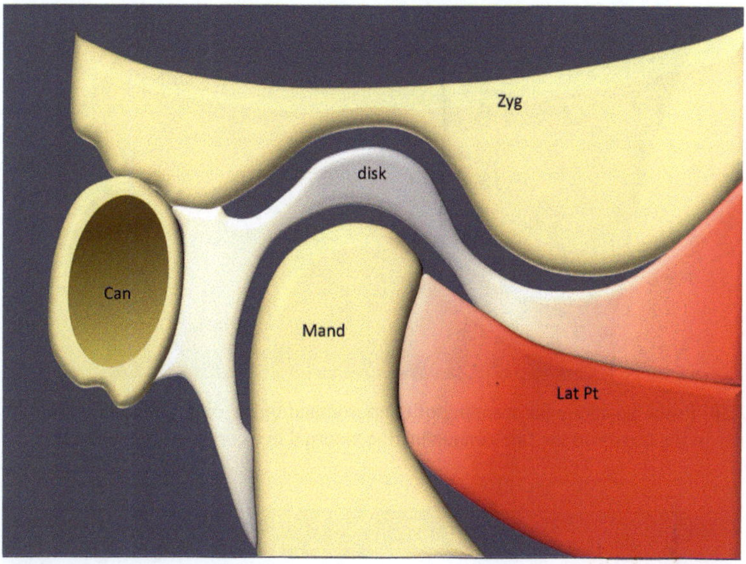

Fig. 12.29 Anatomy of temporomandibular joint. *Zyg* zygomatic arc, *Mand* mandible condyle, *Can* auditory canal, *Lat Pt* lateral pterygoid muscle

Temporomandibular Joint

Anatomy

TMJ is formed by the mandibular condyle and articular eminence of the temporal bone (Fig. 12.29). Articular disc located between the condyle and articular eminence splits the joint into two separate synovial cavities [71]. During the initial phase of mouth opening, the condyle rotates in the mandibular fossa. Past 15° of the rotation, the head of the mandible glides forward onto the articular tubercle [71].

Scanning Technique

For the US examination, linear, high-frequency probes are used (7.5–15 MHz). The most optimal images are obtained with 9 MHz [72].

The patient is positioned lying supine with the head turned a little to the opposite side. The probe is placed along longitudinal axis of ramus of mandible anteriorly to the tragus (Fig. 12.30) [73]. The scans are obtained both in the open-mouth and in the closed-mouth position. Transverse scans are obtained with the probe placed below and parallel to the zygomatic arch.

Despite superficial location of TMJ, acoustic shadow of the zygomatic arch limits the assessment of the joint, in particular, the evaluation of the articular disc [73–75]. Dynamic examination while opening the mouth, tilting or rotating of the probe can help with optimization of the US image [55, 72, 76–78].

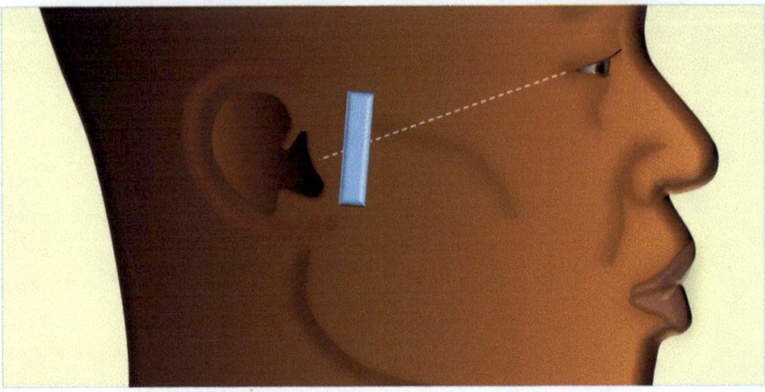

Fig. 12.30 Probe positions for scanning of temporomandibular joint. The joint is located 15 mm anteriorly to the tragus on the line connecting the external eye corner with the tragus

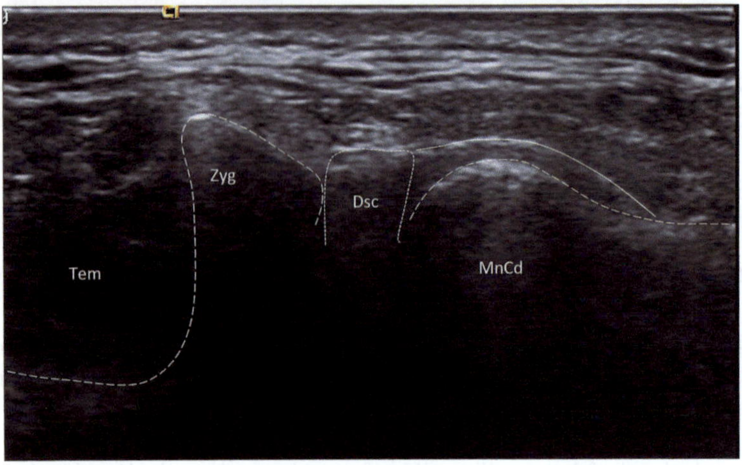

Fig. 12.31 Longitudinal scan of the temporomandibular joint. *Tem* temporalis muscle, *Zyg* zygomatic arc, *Dsc* articular disc, *MnCd* mandibular condyle

Normal Ultrasound Appearance

With the probe parallel to the ramus of mandible and mouth closed, top of zygomatic arch and mandible condyle can be seen as hyperechoic lines (Fig. 12.31). The gap above the top of the condyle indicates the joint space. When visible, disc appears as a hypoechoic or isoechoic band with a subtle hypoechoic halo, positioned above the condyle [72, 78, 79]. The joint capsule forms a thin hyperechoic line overlying the mandibular condyle.

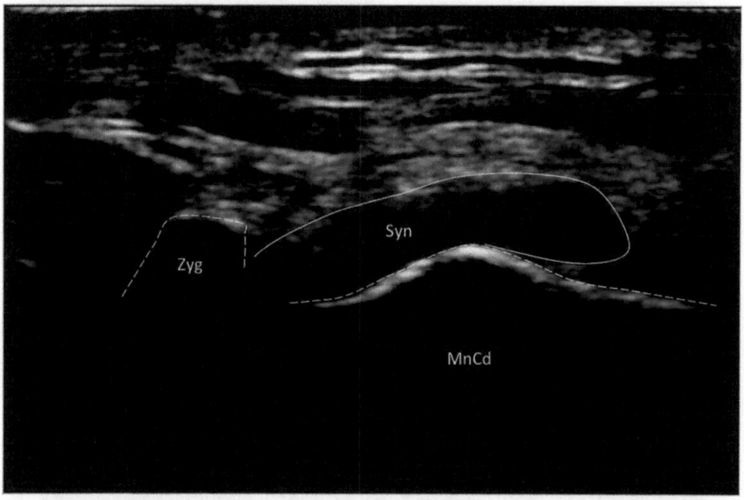

Fig. 12.32 Rheumatoid arthritis, new onset of temporomandibular joint pain. Longitudinal scan of TMJ shows distension of the joint capsule (*Syn*). *Zyg* zygomatic arc, *MnCd* mandibular condyle

Pathologies of the TMJ

Effusion and synovitis cause the distension of the joint capsule (Fig. 12.32). Two ways of measurement are proposed for the assessment of the joint capsule thickening and effusion: between the joint capsule and latero-superior surface of the condyle (normal up to 3 mm) and between the joint capsule and the most lateral point of the condyle (normal up to 2 mm) [72, 79, 80]. Effusion in TMJ is not disease specific and may be found in patients with inflammatory and noninflammatory arthropathies involving TMJ [81].

Erosions can be seen on the anterior and lateral aspect of the condyle. When compared with autopsy specimens, sensitivity of US in detecting condylar erosions was found to be 95 % and specificity 90 %, indicating high reliability of US in detecting erosions [82].

The osteoarthritic changes are demonstrated by mandible condyle flattening, irregular outline of the condyle and osteophyte formation (Fig. 12.33) [83, 84]. Pieces of osteophytes can appear as loose calcified bodies within the joint space [84].

The articular disc displacement is defined as lack of the disc echo above the postero-superior aspect of mandible condyle at the closed-mouth position [81, 85]. Because of inconstant visualization of the disc in all sonograms, lateral and anterior capsule–condyle distances have been proposed as indirect US signs to determine disc positions [72, 86, 87].

Ultrasound-Guided TMJ Injection

US guidance steroid injection to TMJ has been investigated in patients with juvenile idiopathic arthritis with good results [88]. For this purpose, the probe was placed

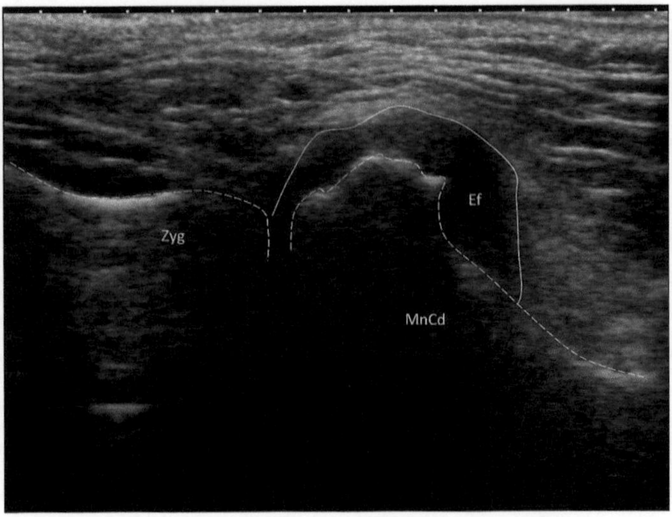

Fig. 12.33 Patient with spondyloarthropathy and complaints of TMJ pain. Longitudinal scan of TMJ shows erosions and osteophytes of mandibular condyle and distension of the joint capsule (*Ef*). TMJ temporomandibular joint, *Zyg* zygomatic arc, *MnCd* mandibular condyle

in coronal plane, and a needle is inserted under sterile conditions from under the inferior pole of the transducer, targeting the joint space [88].

Salivary Glands and Sjögren's Syndrome

US examination of the salivary glands is not a new concept. US has been used for diagnosing of various salivary glands' conditions including benign and malignant tumors, ductal calculi, acute and chronic sialadenitis since the early 1980s [89–91]. Bradus et al. first described US examination of salivary glands in SS in 1988 [92].

SS is a chronic, inflammatory autoimmune disease characterized by lymphocytic infiltration of the exocrine glands leading to their destruction. The typical clinical presentation of SS is dryness of the eyes and mouth but as this feature is not specific, additional criteria have to be fulfilled to confirm the diagnosis [93, 94].

US examination of the salivary gland allows noninvasive and easy-to-perform assessment of the glands' parenchyma and has been suggested as a promising replacement of sialography and minor salivary gland biopsy in the diagnostic criteria of the SS [95–99].

Anatomy

There are three pairs of the major salivary glands: parotid, submandibular and sublingual (Fig. 12.34).

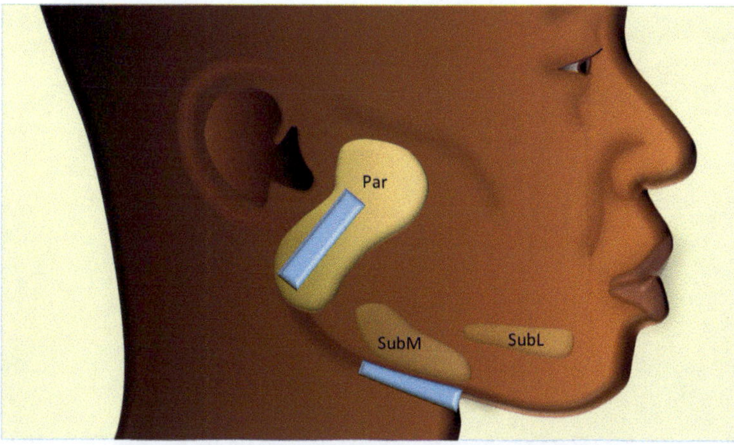

Fig. 12.34 Location of major salivary glands and position of the probe for scanning of the parotid and submandibular glands. *Par* parotid, *SubM* submandibular, *SubL* sublingual

The parotid gland is located along and behind the mandibular process, overlying the posterior part of the masseter muscle [90, 100, 101].

The parotid gland measures approximately 46 mm in its vertical and 37 mm in its horizontal dimension [102, 103]. The size of normal salivary glands is proportional to body weight, and thus may be large in obese patients [102, 103].

The submandibular gland is located in a posterior part of the triangle formed by the body of mandible and two bellies of the digastric muscle. The dimensions of the submandibular gland are in average 33 × 35 × 14 mm [102, 103].

Sublingual gland lies in direct relation to the body of the mandible, between the muscles of the oral cavity floor.

Scanning Technique

For the assessment of salivary glands, the patient is positioned supine with the head turned a little to the opposite side (parotid) and neck extended (submandibular).

The linear array probes of frequency 7.5–12 MHz are used.

The parotid and submandibular glands are examined in longitudinal and transverse planes. Oblique planes are useful in examining the deep lobe of the parotid gland around the angle of the mandible [96] (Fig. 12.34).

Normal Ultrasound Appearance

Figure 12.35 presents normal US appearance of the submandibular gland.

The echostructure of major salivary glands is homogeneous, and the echogenicity is comparable to that of the thyroid gland [100, 101]. It can vary from markedly

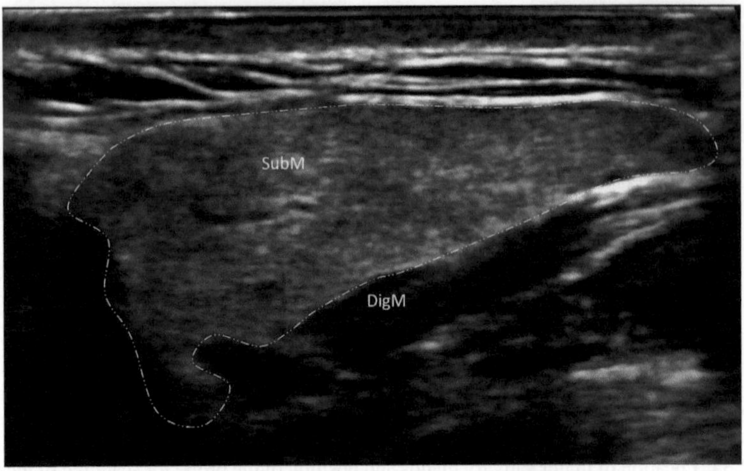

Fig. 12.35 Longitudinal scan of normal submandibular gland (*SubM*). The gland is hyperechoic in comparison with the adjacent digastric muscle and has homogeneous echostructure. *DigM* digastric muscle

hyperechoic to only slightly hyperechoic in comparison to adjacent muscles. This variation depends on the amount of the intraglandular adipose tissue [100].

Within the parotid gland, the lymph nodes can be found, localized most commonly in the area of upper and lower poles of the gland. Normal lymph nodes are hypoechoic and oval in shape with hyperechoic hilum. In the short axis, normal lymph node is not greater than 5–6 mm [104].

Sjögren's Syndrome

The inflammatory and destructive process in SS affects all major salivary glands. In acute phase, swelling and decreased echogenicity can be found [105, 106] (Fig. 12.36). Irregular echostructure is the most common finding (Fig. 12.37). In long-lasting SS, the glands are usually small, hypoechoic, inhomogeneous and difficult to delineate (Fig. 12.38) [98, 101, 107].

The majority of studies indicate parenchymal inhomogeneity (PIH) of both parotid glands as the most specific and sensitive finding [105, 106, 108, 109]. Inhomogeneity refers to the presence of hypoechoic spots and echogenic lines. A direct correlation between pathological changes and salivary gland images has not yet been established. Hypoechoic or anechoic areas are believed to represent lymphocytic infiltration, damaged salivary parenchyma and dilated ducts, while the echogenic lines represent fibrosis and the accumulation of adipose tissues [110, 111].

The degree of PIH varies depending on the degree of the gland damage. This led to the development of the descriptive and quantitative assessment of the salivary glands in SS [97, 106, 109, 112, 113].

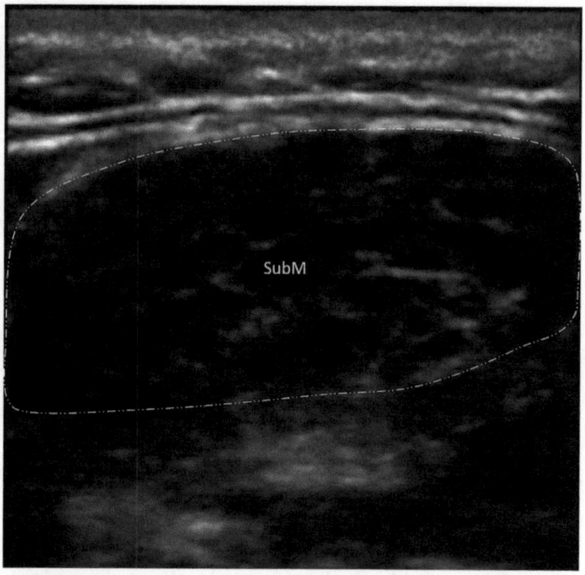

Fig. 12.36 Longitudinal scan of submandibular gland (*SubM*) in early Sjögren's syndrome shows decreased echogenicity and increased volume of the gland. Hyperechoic lines separate hypoechoic areas giving the gland patchy, inhomogeneous appearance

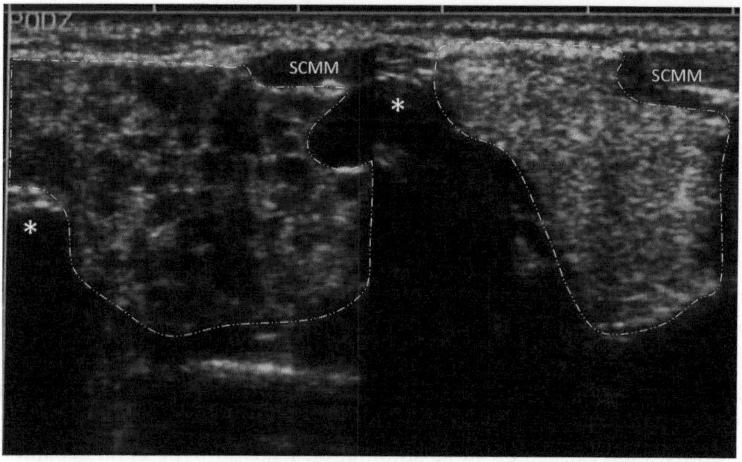

Fig. 12.37 Transverse scan of submandibular gland in Sjögren's syndrome (*left*) and normal salivary gland for comparison (*right*). Parenchymal inhomogeneity: multiple hypoechoic spots. Ramus of mandible (*), *SCMM* sternocleidomastoid muscle

Grade 0: Normal salivary gland.
Grade 1: Mild PIH. A diffuse microareolar structure, the borders of the hypoechoic areolae are blurred and the areolae are generally less than 2 mm in diameter.
Grade 2: Evident PIH. The diffuse hypoechoic areas are larger (2–6 mm), with sharp borders.
Grade 3: Gross PIH. Large, more than 6-mm circumscribed hypoechoic areas.

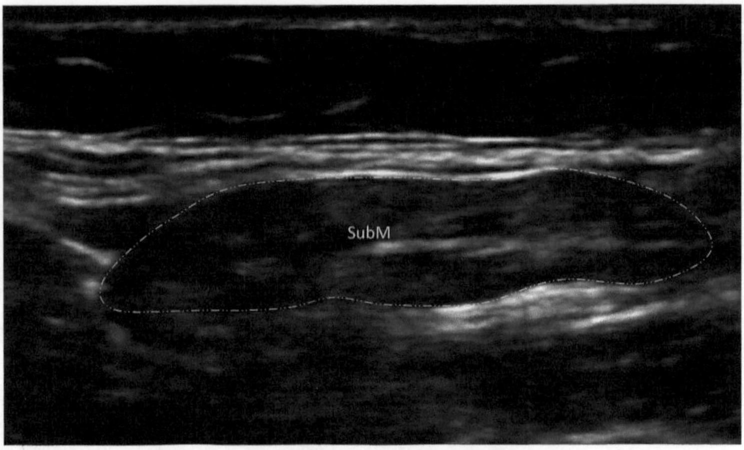

Fig. 12.38 Small, hypoechoic and difficult-to-delineate submandibular gland (*SubM*) in Sjögren's syndrome

Grade 1 in parotids can be regarded only as a warning sign of the possibility of SS. Grade 2 and 3 are considered to be of true diagnostic value and displayed a good correlation with the results of sialography, scintigraphy, MRI and histology of the minor salivary glands [96, 97, 106, 108]. Depending on the methodology and cohort of patients, sensitivity of ultrasonography has been reported between 59 and 90 % and specificity between 73 and 100 % [96, 97, 106, 109, 110, 113–117].

Color Doppler Assessment

Inflammatory process in salivary glands leads to hyperemia. CD sonograms show a higher number of flow signals in patients with higher grades PIH [118].

Peak systolic velocity (PSV) and resistive index (RI) can quantitatively assess blood flow to salivary glands in facial artery (principal branch tributary of submandibular gland) and external carotid artery (principal branch tributary of parotid gland) [119].

In healthy subjects, lemon juice stimulation leads to hyperemia accompanying the secretion of saliva. This process results in significant increase in PSV.

Patients with SS have significantly higher PSV at baseline (representing salivary glands hyperemia) but there is no significant change in PSV after lemon juice stimulation [119]. No change was detected in RI in both SS and control groups before and after lemon juice stimulation; therefore, RI seems to be of limited value in assessment of salivary glands in SS.

Differential Diagnosis

Various other conditions may lead to multiple cyst-like lesions of the salivary gland. Acute bacterial infection can cause inhomogeneity and reduced echogenicity of the parenchyma, but changes are usually unilateral [100, 101]. In chronic sialadenitis, resulting from recurrent infections, the gland appears hypoechoic with irregular margins due to atrophy and replacement of normal gland parenchyma with fibrotic tissue [103].

Viral infections usually involve both parotids causing their enlargement and decreased echogenicity. Although clinical presentation is sufficient for diagnosis, it may be difficult or impossible to differentiate viral parotitis from SS basing on US examination alone [101]. Bilateral cystic changes in parotid glands have been described as an early sign of HIV (human immunodeficiency virus) infection. Cysts are generally larger, and cervical adenopathy is more common than in SS [97]. Sarcoidosis of parotid glands can mimic the clinical presentation of SS but the changes of parenchyma generally appear on ultrasonography as solid lesions [110].

Follow-up

Patients with primary SS have increased risk of developing lymphoma and need regular follow-ups with careful examination of salivary glands on palpation and, when required, also with imaging modality [96]. Change in the US appearance of salivary gland may be a sign of malignant transformation. In lymphoma, large, ill-defined, hypoechoic area can be found within the parotid or submandibular glands (Fig. 12.39) [108]. MRI is regarded as the most reliable technique for detection of salivary gland tumors and should be considered in all suspected cases.

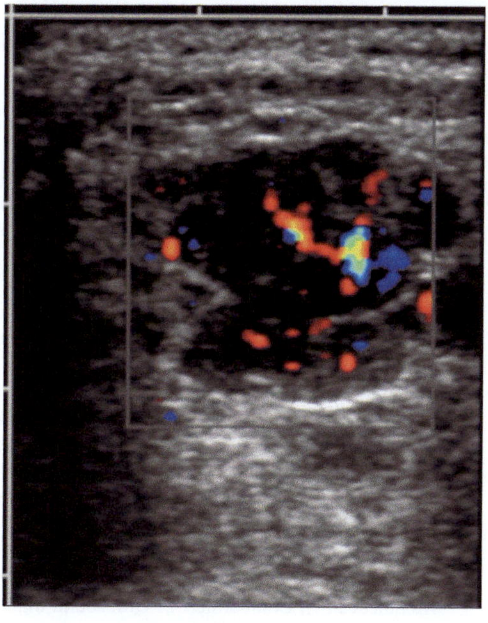

Fig. 12.39 Parotid gland in Sjögren's syndrome and accidental finding of the hypoechoic, lobulated lesion with pathological vessels, indicating malignant transformation

Concluding Remarks

US offers noninvasive assessment in various inflammatory diseases including GCA, PMR, myositis and SS, and can be used as a first-line investigation supplementing clinical diagnosis, to determine the extent of the disease and in follow-up. Although many studies confirm the value of US in diagnosing the above diseases, US cannot replace histopathology examination, and the results of the US should be carefully interpreted in relation to clinical presentation and other diagnostic tests.

In regional pain, FM US is particularly useful in the differential diagnosis. Sonopalpation, dynamic examination and comparison with contralateral side are irreplaceable techniques.

Clinical assessment of parasternal and TMJs is limited and US examination of these areas appears to be useful to rule out clinically silent synovitis, or to differentiate causes of chest wall swelling.

References

1. Watts R, Scott D. Recent developments in the classification and assessment of vasculitis. Best Pract Clin Rheumatol. 2009;23:429–43.
2. Jennette J, et al. Revised international Chapel Hill consensus conference nomenclature of vasculitides. Arthritis Rheum. 2013;65:1–11.
3. Schmidt WA. Role of ultrasound in the understanding and management of vasculitis. Ther Adv Musculoskel Dis. 2014;6(2):39–47.
4. Schmidt W, Seifert A, Gromnica-Ihle E, Krause A, Natusch A. Ultrasound of proximal upper extremity arteries to increase the diagnostic yield in large-vessel giant cell arteritis. Rheumatology (Oxford). 2008c;47:96–101.
5. Schmidt J, Warrington KJ. Polymyalgia rheumatica and Giant cell arteritis in older patients. Drugs Aging. 2011;28(8):651–66.
6. Diamantopoulus AP, et al. Diagnostic value of color Doppler ultrasonography of temporal arteries and large vessels in giant cell arteritis: a consecutive case series. Arthritis Care Res. 2014;66:113–9.
7. De Miguel E, Castillo C, Rodriguez A, De Agustin J. Learning and reliability of colour Doppler ultrasound in giant cell arteritis. Clin Exp Rheumatol. 2009;27:S53–8.
8. Perez LJ, Solans LR, Bosch GJ, Molina CC, Huguet RP, Vilardell TM. Colour-duplex ultrasonography of the temporal and ophthalmic arteries in the diagnosis and follow-up of giant cell arteritis. Clin Exp Rheumatol. 2009;27:S77–82.
9. Santoro L, D`Onofrio F, Bernardi S, Gremese E, Ferraccioli G, Santoliquido A. Temporal ultrasonography findings in temporal arteritis: early disappearence of halo sign after only 2 days of steroid treatment. Rheumatology (Oxford). 2013;52:622.
10. Schmidt WA, Moll A, Seifert A, Schicke B, Gromnica-Ihle E, Krause A. Prognosis of large-vessel giant cell arteritis. Rheumatology (Oxford). 2008b;47:1406–8.
11. Smith JH, Swanson JW. Giant cell arteritis. Headache. 2014.
12. Solomon S, Cappa KG. The headache of temporal arteritis. J Am Geriatr Soc. 1987;35:163–5.
13. Aschwanden M, Daikeler T, Kesten F, Baldi T, Benz D, Tyndall A, et al. Temporal artery compression sign—a novel ultrasound finding for the diagnosis of giant cell arteritis. Ultraschall Med. 2013;34:47–50.
14. Delle SA, Riente L, Filippucci A, Iagnocco A, Meenagh G, Epis O, Grassi W, Valesini G, Montecucco C, Bombardieri S. Ultrasound imaging for the rheumatologist. XV Ultrasound imaging in vasculitis. Clin Exp Rheumatol. 2008;26:391–4.

15. Arida A, Kyprianou M, Kanakis M, Sfikakis P. The diagnostic value of ultrasonography-derived edema of the temporal artery wall in giant cell arteritis: a second meta-analysis. BMC Musculoskelet Disord. 2010;11:44.
16. Bruyn G, Schmidt WA. Introductory guide to musculoskeletal ultrasound for the rheumatologist. Houten: Bohn Stafleu van Loghum/Springer; 2011.
17. Schmidt WA. Technology insight: the role of color and power Doppler ultrasonography in rheumatology. Nat Clin Pract Rheumatol. 2007 Jan;3(1):35–42.
18. Pfadenhauer K, Weber H. Duplex sonography of the temporal and occipital artery in the diagnosis of temporal arteritis. A prospective study. J Rheumtol. 2003;30:2177–81.
19. Ball E, Walsh S, Tang T, Gohil R, Clarke J. Role of ultrasonography in the diagnosis of temporal arteritis. Br J Surg. 2010;97:1765–71.
20. Karassa F, Matsagas M, Schmidt W, Ioannidis J. Meta-analysis: test performance of ultrasonography for giant-cell arteritis. Ann Int Med. 2005;142:359–69.
21. Diamantopoulos AP, Myklebust G. Long-term inflammation in the temporal artery of a giant cell arteritis patient as detected by ultrasound. Ther Adv Musculoskel Dis. 2014;6(3):102–3.
22. Schmidt WA, Krause A, Schicke B, kuchenbecker J, Gromnica-Ihle E. Do temporal artery duplex ultrasound findings correlate with ophthalmic complications in giant cell arteritis? Rheumatology (Oxford). 2009;48:383–5.
23. Czihal M, Rademacher A, Tato F, Kuhlencordt PJ, Schulze-Koops H, Hoffmann U, et al. Sonographic and clinical pattern of extracranial and cranial giant cell arteritis. Scand J Rheumatol. 2012;41:231–6.
24. Grayson P, et al. Distribution of arterial lesions in Takayasu`s arteritis and giant cell arteritis. Ann Rheum Dis. 2012;71:1329–34.
25. Schmidt W. Imaging in vasculitis. Best Pract Res Clin Rheumatol. 2013;27:107–18.
26. Blockmans D, de Ceuninck L, Vanderschueren S, Knockaert D, Mortelmans L, Bobbaers H. Repetitive 18F-fluorodeoxyglucose positron emission tomography in giant cell arteritis: a prospective study of 35 patients. Arthritis Rheum. 2006;55:131–7.
27. Kermani T, Warrington K. Polimyalgia rheumatic. Lancet 2013;381:63–72.
28. Camellino D, Cimmino MA. Imaging of polymyalgia rheumatic: indications on its pathogenesis, diagnosis and prognosis. Rheumatology. 2012;51:77–86.
29. Pease C, Haugberg G, Montague B, Hensor E, Bahkta B, Thomson W, et al. Polymyalgia rheumatica can be distinguished from late onset rheumatoid arthritis at baseline: results of a 5-yr prospective study. Rheumatology. 2009;48:123–7.
30. Dasgupta B, Cimmino M, Maradit-Kremers H, Schmidt W, Schirmmer M, Salvarani C, et al. Provisional classification criteria for polymyalgia rheumatica: a European League against Rheumatism/American College of Rheumatology collaborative initiative. Ann Rheum Dis. 2012;71:484–92.
31. Kermani T, Warrington K. Advances and challenges in the diagnosis and treatment of polymyalgia rheumatica. Ther Adv Musculoskel Dis. 2014;6(1):8–19.
32. Hernandez-Rodriguez J, Cid M, Lopez-Soto A, Espigol-Frigole G, Bosch X. Treatment of polymyalgia rheumatic: a systematic review. Arch Intern Med. 2009;169:1839–50.
33. Dasgupta B, Borg F, Hassan N, Barraclough K, Bourke B, Fulcher J, et al. BSR and BHPR guidelines for the management of polymyalgia rheumatic. Rheumatology. 2010;49:943–54.
34. Robinson P, Wakefield RJ. Essential applications of musculoskeletal ultrasound in rheumatology. Muscle. 2010:137–50 (Saunders Elsevier).
35. Iaccarino L, Ghirardello A, Bettio S, Zen M, Gatto M, Punzi L, Doria A. The clinical features, diagnosis and classification of dermatomyositis. J Autoimmun. 2014:1–6.
36. Connell MJ, Wu TS. Bedside musculoskeletal ultrasonography. Crit Care Clin. 2014;30:243–73.
37. Weber MA, Krix M, Jappe U, et al. Pathological skeletal muscle perfusion in patients with myositis: detection with quantitative contrast-enhanced US. Initial results. Radiology. 2005;238: 640–9.
38. Botar-Jid C, Damian L, Dudea SM, Vasilescu D, Rednic S, Badea R. The contribution of ultrasonography and sonoelastography in assessment of myositis. Med Ultrasonogr. 2010;12(2):120–6.

39. Meenagh G, Sakellariou G, Iagnocco A, Delle Sedie A, Riente L, Filippuci E, Di Geso L, Grassi W, Bombardieri S, Valesini G, Montecucco C. Ultrasound imaging for the rheumatologist XXXIX. Sonographic assessment of hip in fibromyalgia patients. Clin Exp Rheumatol. 2012;30(3):319–21.
40. Muro-Culebras A, Cuesta-Vargas AI. Sono-myography and sono-myoelastography of the tender points of women with fibromyalgia. Ultrasound Med Biol. 2013;39(11):1951–7.
41. Elvin A, Siosteen AK, Nilsson A, Kosek E. Decreased muscle blood flow during standardizing muscle exercise: a contrast media enhanced colour Doppler study. Eur J Pain. 2006;10:137–44.
42. Bruyn GA, Moller I, Klauser A, Martinoli C. Soft tissue pathology: regional pain syndromes, nerves and ligaments. Rheumatology (Oxford). 2012 Dec;51(Suppl 7):vii22–5.
43. Lee JC, Healy JC. Normal sonographic anatomy of the wrist and hand. Radiographics. 2005 Nov–Dec;25(6):1577–90.
44. Numkarunarunrote N, Malik A, Aguiar RO, Trudell DJ, Resnick D. Retinacula of the foot and ankle: MRI with anatomic correlation in cadavers. AJR Am J Roentgenol. 2007;188(4):W348–54.
45. VanHeest AE, Luger NM, House JH, Vener M. Extensor retinaculum impingement in the athlete: a new diagnosis. Am J Sports Med. 2007 Dec;35(12):2126–30.
46. Takahashi Y, Hashizume H, Inoue H, Ogura T. Clinicopathological analysis of de Quervain's disease. Acta Med Okayama. 1994 Feb;48(1):7–15.
47. Trevino S, Gould N, Korson R. Surgical treatment of stenosing tenosynovitis at the ankle. Foot Ankle. 1981;2(1):37–45.
48. Lee KH, Kang CN, Lee BG, Jung WS, Kim do Y, Lee CH. Ultrasonographic evaluation of the first extensor compartment of the wrist in de Quervain's disease. J Orthop Sci. 2014;19(1):49–54.
49. Sookur PA, Naraghi AM, Bleakney RR, Jalan R, Chan O, White LM. Accessory muscles: anatomy, symptoms, and radiologic evaluation. Radiographics. 2008;28(2):481–99.
50. Bianchi S, Abdelwahab IF, Oliveri M, Mazzola CG, Rettagliata P. Sonographic diagnosis of accessory soleus muscle mimicking a soft tissue tumor. J Ultrasound Med. 1995;14(9):707–9.
51. Pai MM, Prabhu LV, Nayak SR, Madhyastha S, Vadgaonkar R, Krishnamurthy A, Kumar A. The palmaris longus muscle: its anatomic variations and functional morphology. Rom J Morphol Embryol. 2008;49(2):215–7.
52. Ranade AV, Rai R, Prabhu LV, Rajanigandha V, Prakash, Janardhanan JP, Ramanathan L, Prameela MD. Incidence of extensor digitorum brevis manus muscle. Hand (N Y). 2008;3(4):320–3.
53. Martinoli C, Bianchi S, Dahmane M, Pugliese F, Bianchi-Zamorani MP, Valle M. Ultrasound of tendons and nerves. Eur Radiol. 2002;12(1):44–55.
54. Delle SA, Riente L, Iagnocco A, Filippucci E, Meenagh G, Valesini G, Grassi W, Bombardieri S. Ultrasound imaging for the rheumatologist. VI. Ultrasonography of the elbow, sacroiliac, parasternal, and temporomandibular joints. Clin Exp Rheumatol. 2006;24(6):617–21.
55. Rodríguez-Henríquez P, Solano C, Peña A, León-Hernández S, Hernández-Díaz C, Gutiérrez M, Pineda C. Sternoclavicular joint involvement in rheumatoid arthritis: clinical and ultrasound findings of a neglected joint. Arthritis Care Res (Hoboken). 2013;65(7):1177–82.
56. Ferri M, Finlay K, Popowich T, Jurriaans E, Friedman L. Sonographic examination of the acromioclavicular and sternoclavicular joints. J Clin Ultrasound. 2005;33(7):345–55.
57. Standring S. Gray's Anatomy. 40th edn. Churchill: Elsevier; 2008.
58. Arent CF. Ultrasound of the shoulder. Master Medical Books; 2013. http://www.shoulderus.com/ultrasound-of-the-shoulder/sternocostal-joint.
59. Kamel M, Kotob H. Ultrasonographic assessment of local steroid injection in Tietze's syndrome. Br J Rheumatol. 1997;36(5):547–50.
60. Jurik AG, Justesen T, Graudal H. Radiographic findings in patients with clinical Tietze syndrome. Skelet Radiol. 1987;16(7):517–23.
61. Johnson MC, Jacobson JA, Fessell DP, Kim SM, Brandon C, Caoili E. The sternoclavicular joint: can imaging differentiate infection from degenerative change? Skeletal Radiol. 2010;39(6):551–8.

62. Schils JP, Resnick D, Haghighi P, Trudell D, Sartoris DJ. Sternocostal joints. Anatomic, radiographic and pathologic features in adult cadavers. Invest Radiol. 1989;24(8):596–603.
63. Ehara S. Manubriosternal joint: imaging features of normal anatomy and arthritis. Jpn J Radiol. 2010;28(5):329–34.
64. Guglielmi G, Cascavilla A, Scalzo G, Salaffi F, Grassi W. Imaging of sternocostoclavicular joint in spondyloarthropaties and other rheumatic conditions. Clin Exp Rheumatol. 2009;27(3):402–8.
65. Martino F, D'Amore M, Angelelli G, Macarini L, Cantatore FP. Echographic study of Tietze's syndrome. Clin Rheumatol. 1991;10(1):2–4.
66. Benhamou CL, Chamot AM, Kahn MF. Synovitis-acne-pustulosis hyperostosis-osteomyelitis syndrome (SAPHO). A new syndrome among the spondyloarthropathies? Clin Exp Rheumatol. 1988;6(2):109–12.
67. Nguyen MT, Borchers A, Selmi C, Naguwa SM, Cheema G, Gershwin ME. The SAPHO syndrome. Semin Arthritis Rheum. 2012;42(3):254–65.
68. Bianchi G, Marinelli A, Frizziero A, Mercuri M. Hyperostosis and osteitis in Sapho syndrome: conservative or surgical treatment. Chir Organi Mov. 2004;89(1):45–9.
69. Dihlmann W, Dihlmann SW. Acquired hyperostosis syndrome: spectrum of manifestations at the sternocostoclavicular region. Radiologic evaluation of 34 cases. Clin Rheumatol. 1991;10(3):250–63.
70. Gilroy AM, MacPherson BR, Ross LM. Atlas of Anatomy. 2nd edn. Thieme: Thieme Medical Publishers, Inc. 2012. pp. 672.
71. Elias FM, Birman EG, Matsuda CK, Oliveira IR, Jorge WA. Ultrasonographic findings in normal temporomandibular joints. Braz Oral Res. 2006;20(1):25–32.
72. Melis M, Secci S, Ceneviz C. Use of ultrasonography for the diagnosis of temporomandibular joint disorders: a review. Am J Dent. 2007;20(2):73–8.
73. Elias FM, Birman EG, Jorge WA, Homsi C. Ultrasonography of the temporomandibular joint: where is the disc? J Oral Maxillofac Surg. 2002;60(11):1381.
74. Li C, Su N, Yang X, Yang X, Shi Z, Li L. Ultrasonography for detection of disc displacement of temporomandibular joint: a systematic review and meta-analysis. J Oral Maxillofac Surg. 2012;70(6):1300–9.
75. Emshoff R, Bertram S, Rudisch A, Gassner R. The diagnostic value of ultrasonography to determine the temporomandibular joint disk position. Oral Surg Oral Med Oral Pathol Oral Radiol Endod. 1997;84(6):688–96.
76. Motoyoshi M, Kamijo K, Numata K, Namura S. Ultrasonic imaging of the temporomandibular joint: a clinical trial for diagnosis of internal derangement. J Oral Sci. 1998;40(2):89–94.
77. Dupuy-Bonafé I, Picot MC, Maldonado IL, Lachiche V, Granier I, Bonafé A. Internal derangement of the temporomandibular joint: is there still a place for ultrasound? Oral Surg Oral Med Oral Pathol Oral Radiol. 2012;113(6):832–40.
78. Tognini F, Manfredini D, Melchiorre D, Zampa V, Bosco M. Ultrasonographic vs magnetic resonance imaging findings of temporomandibular joint effusion. Minerva Stomatol. 2003;52(7–8):365–70, 370–2.
79. Manfredini D, Tognini F, Melchiorre D, Zampa V, Bosco M. Ultrasound assessment of increased capsular width as a predictor of temporomandibular joint effusion. Dentomaxillofac Radiol. 2003;32(6):359–64.
80. Manfredini D, Tognini F, Melchiorre D, Bazzichi L, Bosco M. Ultrasonography of the temporomandibular joint: comparison of findings in patients with rheumatic diseases and temporomandibular disorders. A preliminary report. Oral Surg Oral Med Oral Pathol Oral Radiol Endod. 2005;100(4):481–5.
81. Rudisch A, Emshoff R, Maurer H, Kovacs P, Bodner G. Pathologic-sonographic correlation in temporomandibular joint pathology. Eur Radiol. 2006;16(8):1750–6.
82. Brandlmaier I, Bertram S, Rudisch A, Bodner G, Emshoff R. Temporomandibular joint osteoarthrosis diagnosed with high resolution ultrasonography versus magnetic resonance imaging: how reliable is high resolution ultrasonography? J Oral Rehabil. 2003;30(8):812–7.

83. Hussain AM, Packota G, Major PW, Flores-Mir C. Role of different imaging modalities in assessment of temporomandibular joint erosions and osteophytes: a systematic review. Dentomaxillofac Radiol. 2008;37(2):63–71.
84. Bas B, Yılmaz N, Gökce E, Akan H. Diagnostic value of ultrasonography in temporomandibular disorders. J Oral Maxillofac Surg. 2011;69(5):1304–10.
85. Landes C, Walendzik H, Klein C. Sonography of the temporomandibular joint from 60 examinations and comparison with MRI and axiography. J Craniomaxillofac Surg. 2000;28(6):352–61.
86. Hayashi T, Ito J, Koyama J, Yamada K. The accuracy of sonography for evaluation of internal derangement of the temporomandibular joint in asymptomatic elementary school children: comparison with MR and CT. AJNR Am J Neuroradiol. 2001;22(4):728–34.
87. Habibi S, Ellis J, Strike H, Ramanan AV. Safety and efficacy of US-guided CS injection into temporomandibular joints in children with active JIA. Rheumatology (Oxford). 2012;51(5):874–7.
88. Schwerk WB, Schroeder HG, Eichhorn T. High-resolution real-time sonography in salivary gland diseases. I: Inflammatory diseases. HNO. 1985;33(11):505–10.
89. Gritzmann N. Sonography of the salivary glands. AJR Am J Roentgenol. 1989;153(1):161–6.
90. Bartlett LJ, Pon M. High-resolution real-time ultrasonography of the submandibular salivary gland. J Ultrasound Med. 1984;3(10):433–7.
91. Bradus RJ, Hybarger P, Gooding GA. Parotid gland: US findings in Sjögren syndrome. Work in progress. Radiology. 1988;169(3):749–51.
92. Shiboski SC, Shiboski CH, Criswell L, Baer A, Challacombe S, Lanfranchi H, Schiødt M, Umehara H, Vivino F, Zhao Y, Dong Y, Greenspan D, Heidenreich AM, Helin P, Kirkham B, Kitagawa K, Larkin G, Li M, Lietman T, Lindegaard J, McNamara N, Sack K, Shirlaw P, Sugai S, Vollenweider C, Whitcher J, Wu A, Zhang S, Zhang W, Greenspan J, Daniels T. Sjögren's International Collaborative Clinical Alliance (SICCA) Research Groups. American College of Rheumatology classification criteria for Sjögren's syndrome: a data-driven, expert consensus approach in the Sjögren's International Collaborative Clinical Alliance cohort. Arthritis Care Res (Hoboken). 2012;64(4):475–87.
93. Vitali C, Bombardieri S, Jonsson R, Moutsopoulos HM, Alexander EL, Carsons SE, Daniels TE, Fox PC, Fox RI, Kassan SS, Pillemer SR, Talal N, Weisman MH. European Study Group on Classification Criteria for Sjögren's Syndrome. Classification criteria for Sjögren's syndrome: a revised version of the European criteria proposed by the American-European Consensus Group. Ann Rheum Dis. 2002;61(6):554–8.
94. Cornec D, Jousse-Joulin S, Marhadour T, Pers JO, Boisramé-Gastrin S, Renaudineau Y, Saraux A, Devauchelle-Pensec V. Salivary gland ultrasonography improves the diagnostic performance of the 2012 American College of Rheumatology classification criteria for Sjogren's syndrome. Rheumatology (Oxford). 2014;53(9):1604–7.
95. El Miedany YM, Ahmed I, Mourad HG, Mehanna AN, Aty SA, Gamal HM, El Baddini M, Smith P, El Gafaary M. Quantitative ultrasonography and magnetic resonance imaging of the parotid gland: can they replace the histopathologic studies in patients with Sjogren's syndrome? Joint Bone Spine. 2004;71(1):29–38.
96. Makula E, Pokorny G, Rajtár M, Kiss I, Kovács A, Kovács L. Parotid gland ultrasonography as a diagnostic tool in primary Sjögren's syndrome. Br J Rheumatol. 1996;35(10):972–7.
97. Takagi Y, Kimura Y, Nakamura H, Sasaki M, Eguchi K, Nakamura T. Salivary gland ultrasonography: can it be an alternative to sialography as an imaging modality for Sjogren's syndrome? Ann Rheum Dis. 2010;69(7):1321–4.
98. Milic V, Petrovic R, Boricic I, Radunovic G, Marinkovic-Eric J, Jeremic P, Damjanov N. Ultrasonography of major salivary glands could be an alternative tool to sialoscintigraphy in the American-European classification criteria for primary Sjögren's syndrome. Rheumatology (Oxford). 2012;51(6):1081–5.
99. Bialek E, Jakubowski W, Zajkowski P, Szopinski KT, Osmolski A. US of the major salivary glands: anatomy and spatial relationships, pathologic conditions and pitfalls. RadioGraphics. 2006;26:745–63.

100. Gritzmann N, Rettenbacher T, Hollerweger A, Macheiner P, Hübner E. Sonography of the salivary glands. Eur Radiol. 2003 May;13(5):964–75. Epub 2002 Sep 5.
101. Dost P, Kaiser S. Ultrasonographic biometry in salivary glands. Ultrasound Med Biol. 1997;23:1299–303.
102. Zengel P, Schrötzlmair F, Reichel C, Paprottka P, Clevert DA. Sonography: the leading diagnostic tool for diseases of the salivary glands. Semin Ultrasound CT MR. 2013 June;34(3):196–203.
103. Ying M, Ahuja A, Metreweli C. Diagnostic accuracy of sonographic criteria for evaluation of cervical lymphadenopathy. J Ultrasound Med. 1998 July;17(7):437–45.
104. De Clerck LS, Corthouts R, Francx L, Brussaard C, de Schepper A, Vercruysse HA, Stevens WJ. Ultrasonography and computer tomography of the salivary glands in the evaluation of Sjögren's syndrome. Comparison with parotid sialography. J Rheumatol. 1988 Dec;15(12):1777–81.
105. Salaffi F, Argalia G, Carotti M, Giannini FB, Palombi C. Salivary gland ultrasonography in the evaluation of primary Sjögren's syndrome. Comparison with minor salivarygland biopsy. J Rheumatol. 2000 May;27(5):1229–36.
106. Salaffi F, Carotti M, Iagnocco A, Luccioli F, Ramonda R, Sabatini E, De Nicola M, Maggi M, Priori R, Valesini G, Gerli R, Punzi L, Giuseppetti GM, Salvolini U, Grassi W. Ultrasonography of salivary glands in primary Sjögren's syndrome: a comparison with contrast sialography and scintigraphy. Rheumatology (Oxford). 2008;47(8):1244–9.
107. Makula E, Pokorny G, Kiss M, Vörös E, Kovács L, Kovács A, Csernay L, Palkó A. The place of magnetic resonance and ultrasonographic examinations of the parotid gland in the diagnosis and follow-up of primary Sjögren's syndrome. Rheumatology (Oxford). 2000;39(1):97–104.
108. De Vita S, Lorenzon G, Rossi G, Sabella M, Fossaluzza V. Salivary gland echography in primary and secondary Sjögren's syndrome. Clin Exp Rheumatol. 1992;10(4):351–6.
109. Takashima S, Morimoto S, Tomiyama N, Takeuchi N, Ikezoe J, Kozuka T. Sjogren syndrome: comparison of sialography and ultrasonography. J Clin Ultrasound. 1992;20(2):99–109.
110. Tzioufas AG, Moutsopoulos HM. Ultrasonography of salivary glands: an evolving approach for the diagnosis of Sjögren's syndrome. Nat Clin Pract Rheumatol. 2008;4(9):454–5.
111. Ariji Y, Ohki M, Eguchi K, Izumi M, Ariji E, Mizokami A, Nagataki S, Nakamura T. Texture analysis of sonographic features of the parotid gland in Sjögren's syndrome. AJR Am J Roentgenol. 1996;166(4):935–41.
112. Hocevar A, Ambrozic A, Rozman B, Kveder T, Tomsic M. Ultrasonographic changes of major salivary glands in primary Sjogren's syndrome. Diagnostic value of a novel scoring system. Rheumatology (Oxford). 2005;44(6):768–72.
113. Kawamura H, Taniguchi N, Itoh K, Kano S. Salivary gland echography in patients with Sjögren's syndrome. Arthritis Rheum. 1990;33(4):505–10.
114. Yonetsu K, Takagi Y, Sumi M, Nakamura T, Eguchi K. Sonography as a replacement for sialography for the diagnosis of salivary glands affected by Sjögren's syndrome. Ann Rheum Dis. 2002;61(3):276–7.
115. Niemelä RK, Takalo R, Pääkkö E, Suramo I, Päivänsalo M, Salo T, Hakala M. Ultrasonography of salivary glands in primary Sjogren's syndrome. A comparison with magnetic resonance imaging and magnetic resonance sialography of parotid glands. Rheumatology (Oxford). 2004;43(7):875–9. Epub 2004 April 27.
116. Chikui T, Shimizu M, Kawazu T, Okamura K, Shiraishi T, Yoshiura K. A quantitative analysis of sonographic images of the salivary gland: a comparison between sonographic and sialographic findings. Ultrasound Med Biol. 2009;35(8):1257–64.
117. Martinoli C, Derchi LE, Solbiati L, Rizzatto G, Silvestri E, Giannoni M. Color Doppler sonography of salivary glands. AJR Am J Roentgenol. 1994;163(4):933–41.
118. Carotti M, Salaffi F, Manganelli P, Argalia G. Ultrasonography and colour doppler sonography of salivary glands in primary Sjögren's syndrome. Clin Rheumatol. 2001;20(3):213–9.
119. Zengel P, Schrötzlmair F, Reichel C, Paprottka P, Clevert DA. Sonography: the leading diagnostic tool for diseases of the salivary glands. Semin Ultrasound CT MR. 2013;34(3):196–203.

Chapter 13
Sports Medicine

Zbigniew Czyrny MD

Introduction

Musculoskeletal injuries constitute the largest class of athletic injuries sustained in sports. By definition, any injury that occurs to a skeletal muscle, tendon, ligament, joint, nerve, or a blood vessel that services the musculoskeletal system or any related tissues is a musculoskeletal injury. The majority of these injuries tend to resolve without significant long-term sequelae. The most common cause of musculoskeletal injury is a combination of physical overloads created by overtraining/collision while playing or by the repetitive use of a joint or a particular muscle group. The distribution of the frequency of the different types of musculoskeletal injuries is relatively equal between male and female athletes and approximately 25 % of all athletes will expect to sustain a musculoskeletal injury in a 12-month period.

In sports medicine, musculoskeletal ultrasound (US) has become a crucial imaging technique for the evaluation of soft tissue and joint-related damage. With multi-planar abilities, the feasibility of using movement as well as compression to dynamically assess structures and compare affected and unaffected sides can easily be achieved. Therefore, it became an indispensable imaging tool for the diagnosis and monitoring of athletes with joint/soft tissue pain or masses.

To achieve a more accurate diagnosis, it is integral first to get familiar with the tendon/muscle normal histology and anatomy. Bochenek and Reicher [1] described the tendon as: "a fibrous structure built of compact connective tissue of white-silver color and very strong." When stretched, the tendon can elongate without injury to its fibers by 4 % of its length. Muscle's tendon can be divided into its naked or bare part and the part that is distributing fibers at the muscle belly level (Fig. 13.1) [2]. Compact connective tissue tendinous fibers do not "switch" onto a muscle fiber but are being "unpacked" to the form of loose connective tissue (endomysium and perimysium—EP) at both ends of every myofiber (Figs. 13.2, 13.3, and 13.4). All

Z. Czyrny (✉)
Private Practice, Diagnostic Imaging, Warsaw, Poland
e-mail: zbigniew.czyrny@wp.pl

Y. El Miedany (ed.), *Musculoskeletal Ultrasonography in Rheumatic Diseases*,
DOI 10.1007/978-3-319-15723-8_13

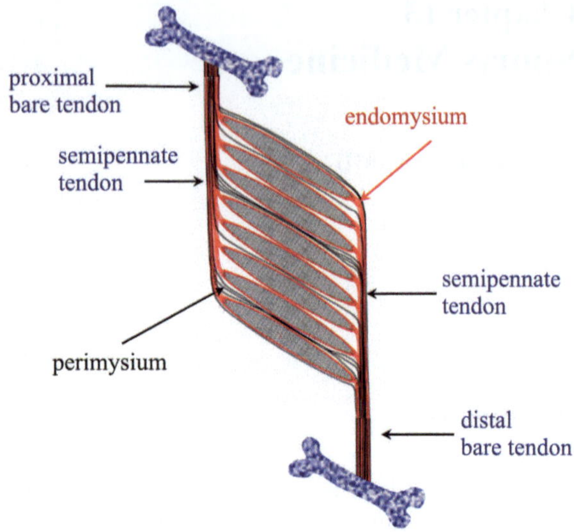

Fig. 13.1 A detailed scheme of a most simple muscle structure with both tendons—semipennate/superficial. Based on a figure from [2]

together—tendons and endomysium + perimysium form the tendinous system of a muscle which is a continuous structure. Some tendons do not have a bare part, that is, they distribute fibers which become endomysium and perimysium along all their length.

At the terminal myotendinous junction, tendinous collagen fibrils are inserted into deep recesses formed by myocyte processes, allowing the tension generated by centrally located myofibrils to be transmitted to the collagen fibrils of the terminal myotendinous junction (Fig. 13.2). Sarcomeres attach to endomysium (the tendinous system) at every Z membrane level along the whole length of the myofiber—forming parietal myotendinous junctions (PMTJ) (Fig. 13.2). Human muscles have 500 PMTJ every 1 mm. It is the main force transmission device of the myofiber and of the muscle. At the beginning/end of a myofiber, its myofibrils are separated from each other by endomysium and form terminal myotendinous junction (TMTJ; Fig. 13.2). This is a very important proprioceptive zone as Golgi apparatus as well as Vatter-Paccini and Ruffini corpuscules are largely located here [3]. This complex architecture reduces the tensile stress exerted on the tendons during muscle contraction. It is wrongly understood that the terminal myotendinous junction is the weakest link of a muscle as a whole. In fact, it is supported by both superficial endomysium (Fig. 13.3) and, additionally, by tendinous fibers arising from the TMTJ (Fig. 13.4). What is thought to be a "myotendinous junction" tear is a tendinous tear with frequent involvement of endomysium and perimysium including the level of the TMTJ and PMTJ too. We must remember that 1 mm of the tendon myofiber has already attached to endomysium at 500 levels; close but quite away in terms of myofiber built. At the muscle belly level, tendons may distribute their fibers on one side—these are called semipennate (Fig. 13.1), or on all sides—pennate (Fig. 13.5). This means that semipennate tendon is located on the surface of the muscle and pennate tendon is located within a muscle belly. Many muscles have tendons which run

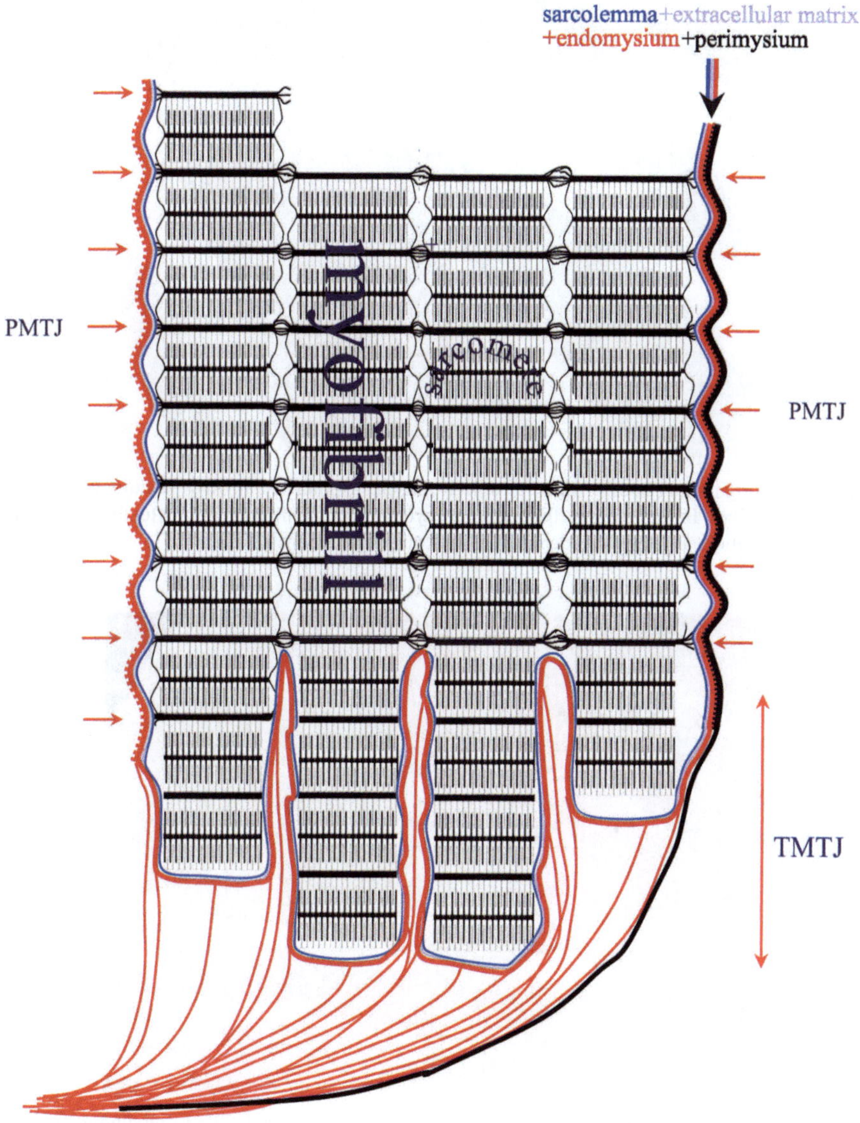

Fig. 13.2 A scheme of the tendinous system of the myofiber. Based on a figure from [2]

superficially (semipennate) and at some level become pennate—"dive" inside the muscle belly (Fig. 13.5). This may be wrongly interpreted as a tendon end when, in fact, the tendon disappears from the muscle surface [2].

Tendons respond to repetitive overload beyond the physiological threshold with either inflammation of their sheath or tearing with possible degeneration of scars of their body or both. Tendon or ligament tears are followed by a repair process

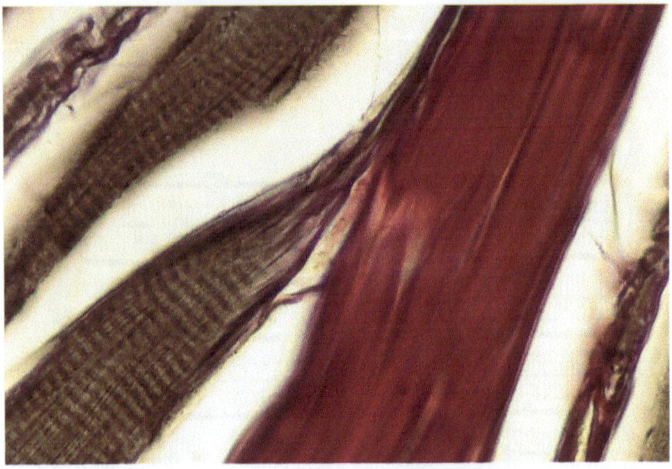

Fig. 13.3 Longitudinal section of a rat soleus muscle, Gomori stain. *Red*—tendinous system, *brown*—myofiber. In this section, the terminal myotendinous junction with its tendinous inserts between myofibrils are very well depicted. Based on a figure from [2]

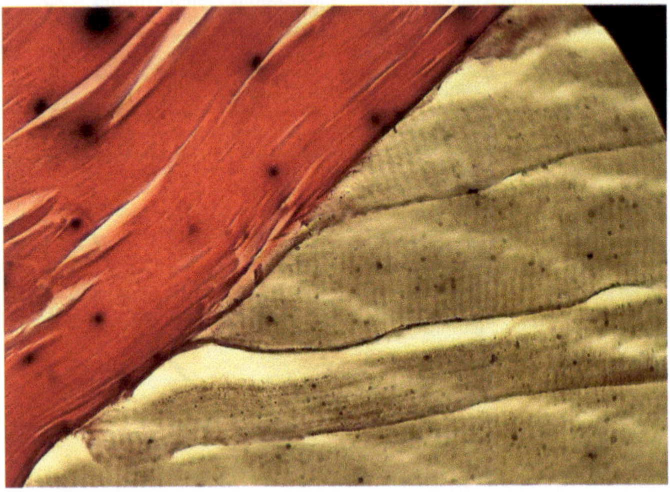

Fig. 13.4 Longitudinal section of a rat soleus muscle, Gomori stain. *Red*—tendinous system, *brown*—myofiber. In this section, endo-/perimysial "coat" of the myofiber is very well depicted. Based on a figure from [2]

which may differ in quality and duration. However, tendon damage may even occur from stresses within the physiological limits as frequent cumulative microtrauma may not allow enough time for repair. Frank inflammatory lesions and granulation tissues are infrequent and are mostly associated with tendon ruptures or other associated chronic inflammatory conditions. The presence of inflamed synovial or fatty tissue (paratenon, bursas, sheaths, and their fatty tissues) may disturb the scar healing process.

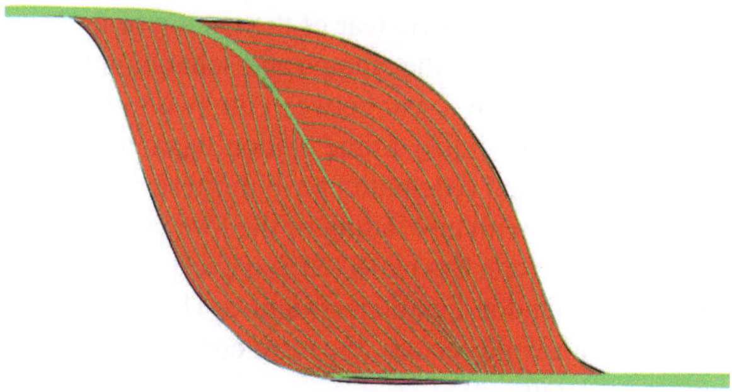

Fig. 13.5 A scheme of a muscle with proximal semipennate and then pennate tendon, distal semipennate tendon. *Red*—muscle belly, *green*—tendons and EP. By permission from [4]. *EP* endomysium and perimysium

Tendon healing occurs in three overlapping phases. Initially, in the inflammatory phase, the erythrocytes and inflammatory cells (particularly neutrophils) enter the site in the first 24 h. This is followed by the proliferative phase which begins after 5 days. Synthesis of type III collagen peaks during this stage and lasts for few weeks. After about 6 weeks, the remodeling phase starts with decreased cellularity as well as reduced collagen synthesis.

Classically, in many cases, tendons with intratendinous lesions detected on US or magnetic resonance imaging (MRI) are not painful. However, pain may be present in patients with tendinopathy, which is usually attributed to inflammation. Pain may originate from a composite of both mechanical and biochemical factors. Anti-inflammatory treatment is usually helpful in such cases and, clinically, it is justified by pain and edema.

Sonographic Findings

Tendons and Ligaments

Tendons and ligaments are those structures that tend to tear first in acute or chronic injuries. The normal fibrillar pattern of tendons and ligaments appears as several long bright echoes surrounded by hypoechoic/anechoic matter. Tendons and ligamentous tears can be stratified into three grades:

Grade I—deformation of a tendon without injury to the collagen fibers. Such deformation may elongate the tendon by 4–6 %. The only visible sonographic feature of the injury is mild edema which makes the fibrillar pattern look a bit darker compared to contralateral side which should be used for measurement reference.

Grade II—some fibers completely torn, some with grade I, whereas some other fibers remain normal. In its early stages (first week or two), a tear appears as hypo-/anechoic focal lesion (reflecting fiber tear and blood metabolites) within the tendon/ligament. At 2–3 weeks, the scar is forming and initially does not have much structure. At 4–6 weeks, the scar tissue is well structured and this scar is called microfibrillar as it is built of finer collagen fibers than the original.

Grade III—a complete tear of the tendon or ligament. Depending on the tendon/ligament and the site, the stumps and separating hematoma/fluid can be seen. Later stumps tend to form a scar between the two torn ends unless they are retracted.

In general, any injury of the tendon/ligament collagen fibers whether in its early stages, i.e., grade I (edematic) or in more severe conditions (grade II or III), is usually associated with unwinding of the torn sites and twisted structures, thus gaining volume. Therefore, it is expected that every tear is most likely associated with thickening of the tendon/ligament structure. That is the least specific feature of a tear.

In the following section, the most common specific tendon injuries are discussed in further detail.

The Shoulder

Rotator cuff injury and inflammation of the subdeltoid bursa are one of the most common causes of shoulder pain. Rotator cuff consists of the tendons of subscapularis, supraspinatus, infraspinatus as well as teres minor muscles in addition to the capsulo-ligamentous complex of the shoulder joint. The strongest part of the capsulo-ligamentous complex is a superior complex and antero-inferior complex. Superior complex [5, 6] is located under both supraspinatus and infraspinatus tendons and consists of superior gleno-humeral ligament, coraco-humeral ligament, and superior–posterior gleno-humeral ligament. The thickness of the complex varies and is of similar thickness to the tendinous part of the cuff [7, 8] (Figs. 13.6 and 13.7). It must be stressed here that the supraspinatus and infraspinatus tendons do not form two layers. The two layers that we see in US (and MRI) are tendinous and capsulo-ligamentous ones [4].

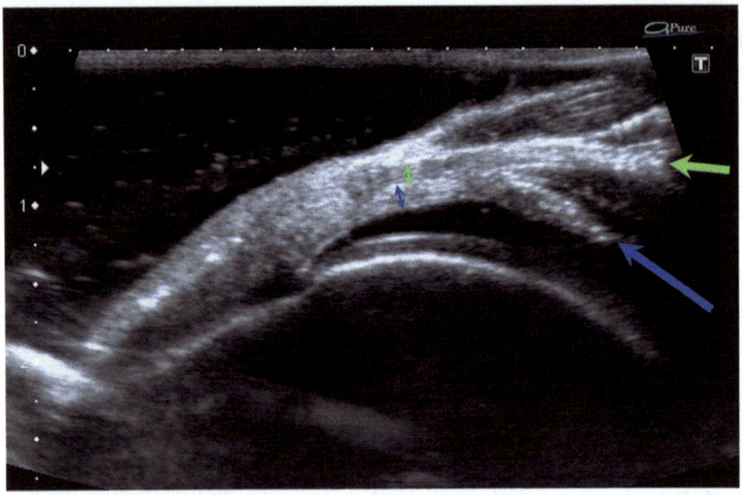

Fig. 13.6 Longitudinal scan of the supraspinatus zone of the normal rotator cuff fresh specimen in a water bath. Tendinous (*green arrows*) layer, previously bluntly, separated from capsulo-ligamentous layer (*blue arrows*) on the length of approx. 10 mm

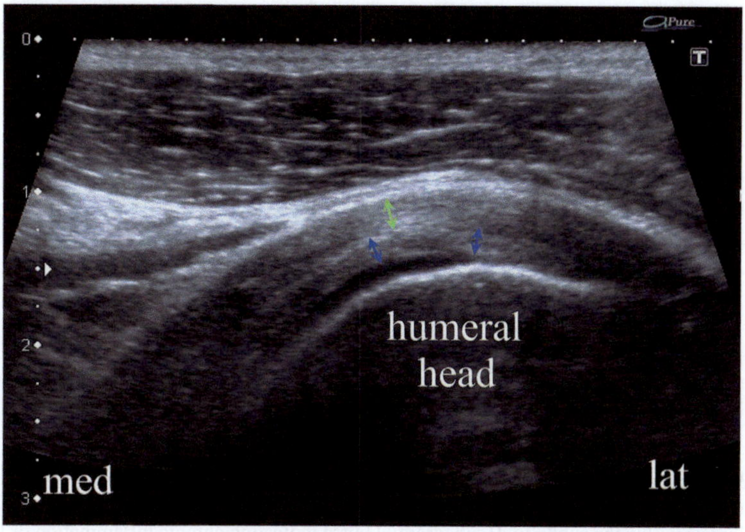

Fig. 13.7 Longitudinal scan of the infraspinatus zone of the normal rotator cuff. Tendinous layer—*green arrows*, capsulo-ligamentous layer—*blue arrows*

The antero-inferior complex consists of the medial and inferior gleno-humeral ligaments connected distally by a transversely oriented fasciculus obliquus [6]. This zone also features two layers comparable in thickness. The width of capsulo-ligamentous and tendinous insertion width varies from front to back and fits between 3.5 and 9.1 mm for a capsulo-ligamentous layer, and 3.5–9.7 mm for a tendinous one [8]. When there is diffuse fibrosis, layers cannot be distinguished; however, it can be predicted which layer is damaged, particularly in view of the finding that superior and antero-inferior complex forms at least 1/3–1/2 of the rotator cuff thickness and, in places, reaches 2/3.

Rotator Cuff

The most common type of the rotator cuff injury is tearing, whether full thickness or partial (Figs. 13.8, 13.9, 13.10, and 13.11). Common tear location is enthesis. Regardless of the initial factor, which may be inflammatory or mechanical, it may lead to imbalance between the endurance of the enthesal cortex and the load on either tendon or ligament. Enthesis site is commonly known as "enthesis zone". Enthesis zone is the enthesis ± 10 mm.

There are three main ways how the enthesis zone tears [7]:

1. Without injury to the bone
2. With erosive and cystic destruction of the bone
3. With mineralization or even ossification of the enthesis cartilage and tendinous/ligamentous scar.

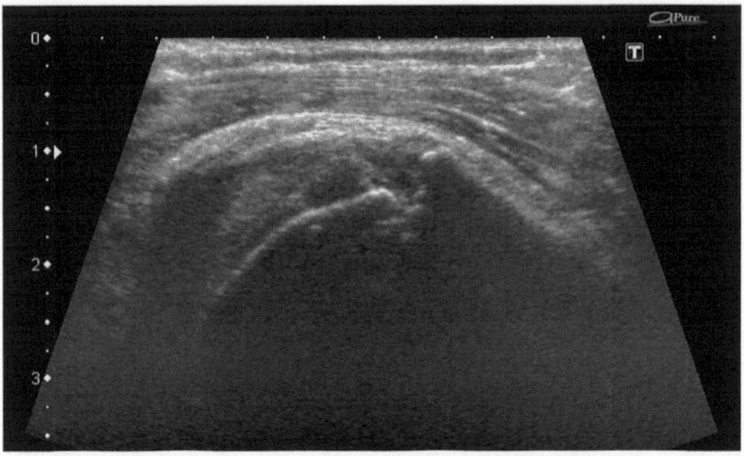

Fig. 13.8 Supraspinatus zone of the rotator cuff, longitudinal scan. Enthesopathy, a scar-filled erosion of the enthesis of the tendinous layer—a chronic tear with bone destruction. Note quite some fibrosis within the subdeltoid–subacromial bursa indicating chronic inflammation or fibrosis directly due to healing process within the tendon

There may also be a mixed type.

The primary sonographic finding of full thickness tears is volume loss and focal tendon nonvisualization. Most full thickness tears occur in the distal enthesis of the supraspinatus, but some may arise more proximally and leave a distal tendon stump attached to the greater tubersoity. In long-standing tears, fluid is usually absent and the gap is filled by fibrous tissue and proliferating synovium. Graded compression must be liberally used to depict full thickness tears of either layer of the rotator cuff because complex fluid, fibrous tissue, and proliferating synovium may all mimic a heterogenous but continuous tendon on static images. Dynamic compression induces shift of these materials from the tear and allows for accurate diagnosis.

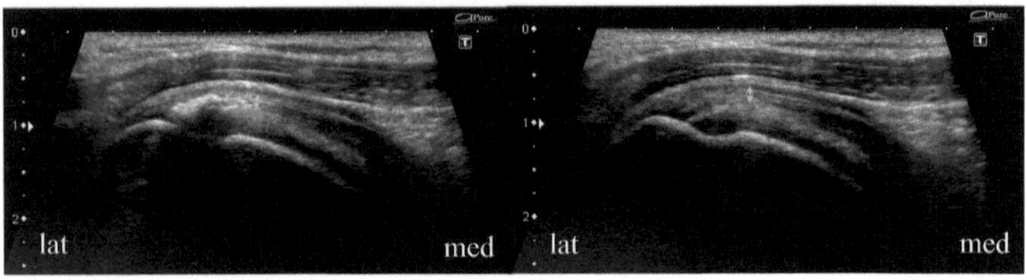

Fig. 13.9 Supraspinatus zone of the rotator cuff, longitudinal scans of two adjacent layers. Enthesopathy with scar mineralization mainly within tendinous layer (*left*) and on the surface of the superior complex (*right*). Such mineral cavities within a scar may become quite large, mainly those following tears right at the border of insertion of the tendinous and capsulo-ligamentous complex where shearing forces are generated

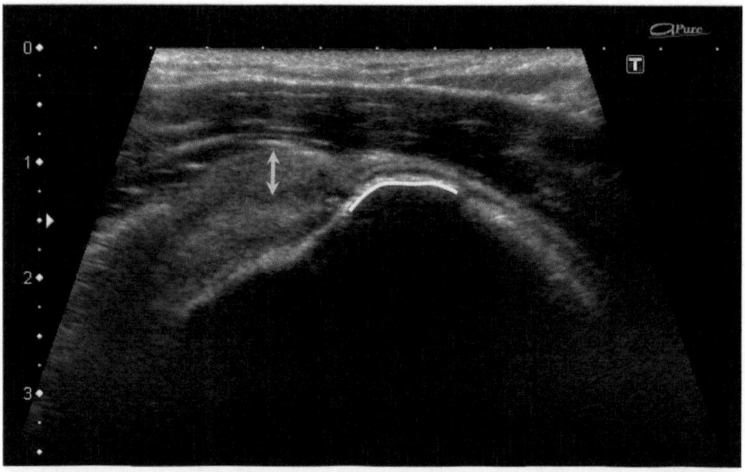

Fig. 13.10 Supraspinatus zone of the rotator cuff, longitudinal scan. A total tear of the supraspinatus tendon (*green arrow*) with preserved continuity of the capsulo-ligamentous layer (*blue arrow*) allowing little retraction of the tendon. *White line* is the enthesis from which the tendon was torn off

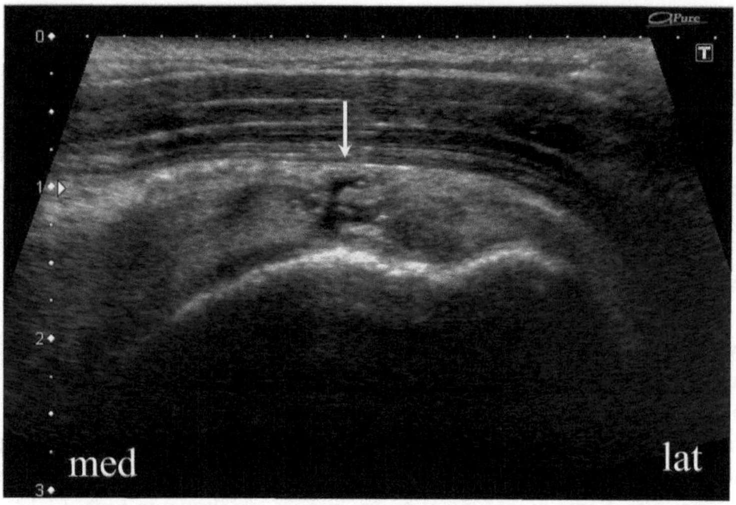

Fig. 13.11 Supraspinatus zone of the rotator cuff, longitudinal scan. A total tear of the supraspinatus tendon and superior complex. *Arrow* shows fluid separating the stumps. Little retraction of the torn structures indicates a small tear

Elbow

Contrary to expectations, the elbow zone is relatively a common site for tendon damage. Several forms of tendon/ligamentous affection may occur in the elbow

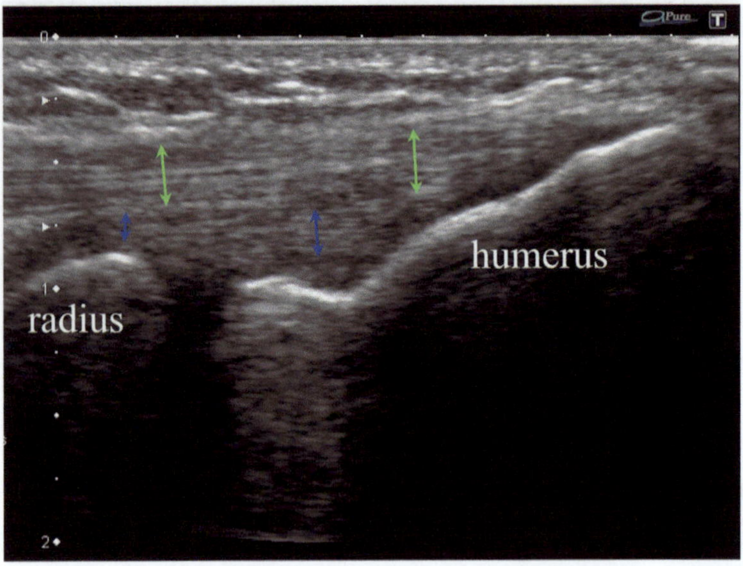

Fig. 13.12 Longitudinal scan of the common extensor tendon (*green arrows*) and radial collateral and anular ligaments (*blue arrows*)

region including partial tears of the common extensor tendon or/and radial collateral ligament [9]. Less frequent is the injury to the common flexor tendon and ulnar collateral ligament. Triceps brachii tears occur nearly always at the enthesis zone.

Common extensor tendon with radial collateral and anular ligaments is a classic example of fibrillar echo pattern (Fig. 13.12).

Common Extensor Tendon

The symptoms of common extensor tendon and radial collateral ligament tears are diagnosed clinically as "tennis elbow." Unlike in the shoulder, enthesopathy at the common extensor origin is usually limited to the thickness of fibrocartilage of the enthesis—it becomes fibrotic and cannot be measured. It is unlikely to see an erosion or a cyst at the affected site. Scar mineralization is relatively common at the enthesis injury site. Some most typical injury images are shown in Figs. 13.13, 13.14, and 13.15.

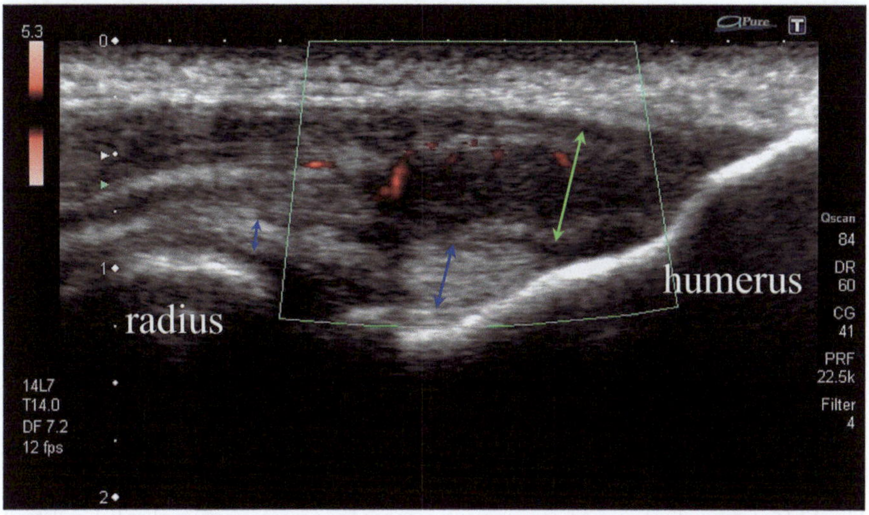

Fig. 13.13 Longitudinal scan of the common extensor tendon (*green arrow*) and radial collateral and anular ligaments (*blue arrow*). A partial tear of the common extensor tendon—hypoechoic scar with blurred microfibrillar tissue pattern and hypervascularity indicating healing process, not inflammation. The scar may be around 3–4 weeks old, it can also be much older as it may be a chronic tear. Patient's history is crucial

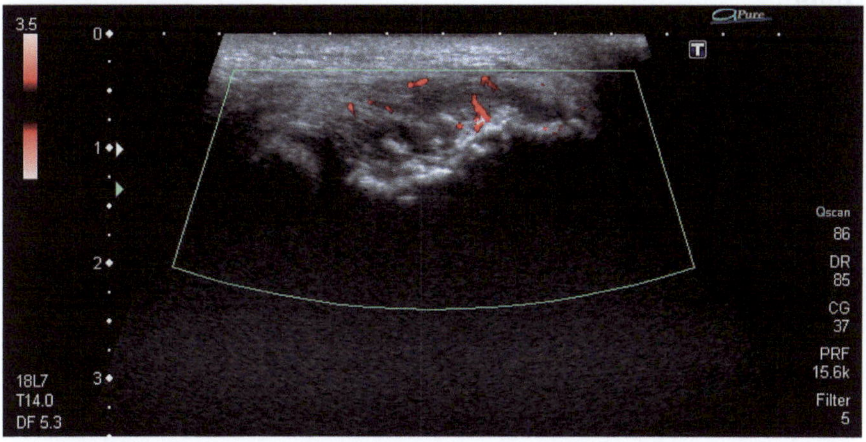

Fig. 13.14 Patient after three steroid injections to the tendon/ligament. Common extensor tendon and radial collateral and anular ligaments with multifocal tears, diffused fibrosis, areas with distorted and amorphous material in both layers which are difficult to distinguish. Severe destruction of the enthesis zone—multiple erosions and mineralized scars. Hyperperfusion indicates healing and how difficult it is to destroy a living tissue

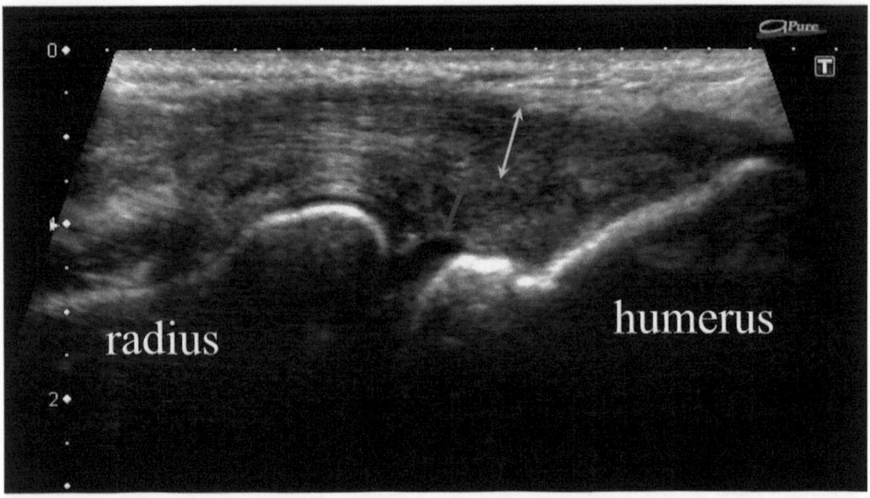

Fig. 13.15 Longitudinal scan of common extensor tendon (*green arrow*) and radial collateral and anular ligaments (*blue arrow*). A complete tear of the common extensor tendon and radial collateral ligament near humeral enthesis. When pressed, the scar is soft and sliding of layers near the enthesis is seen. Little retraction of stumps due to strong bond between the fascia and common extensor tendon

Hand

Sport injuries of the hand and wrist are common and include a variety of diagnoses. This is attributed to the several tendons present in the hand region with their sheaths and retinacula. Repetitive stress injury causes inflammation in the tendons' synovial sheath lining and its irritable chronic nature brings some inflammatory factor to the tendons region. Trigger fingers and de Quervain's disease are good examples of inflammatory-related conditions. Although, in most of the cases, it starts secondary to mechanical etiology, it is relatively prevalent in association with metabolic disorders such as diabetes mellitus. Inflammation of the tendinous sheath leads to retinacular fibrosis followed by contracture and stiffening (Fig. 13.16); that leads to tendon compression, hypovascularity, and change of morphology. The pain is generated by ill tendon pressing against ill retinaculum during movements.

Apart from inflammatory-related diseases, hand diagnostics include some contusions that are not related to inflammation or even irritation (Figs. 13.17, 13.18, and 13.19).

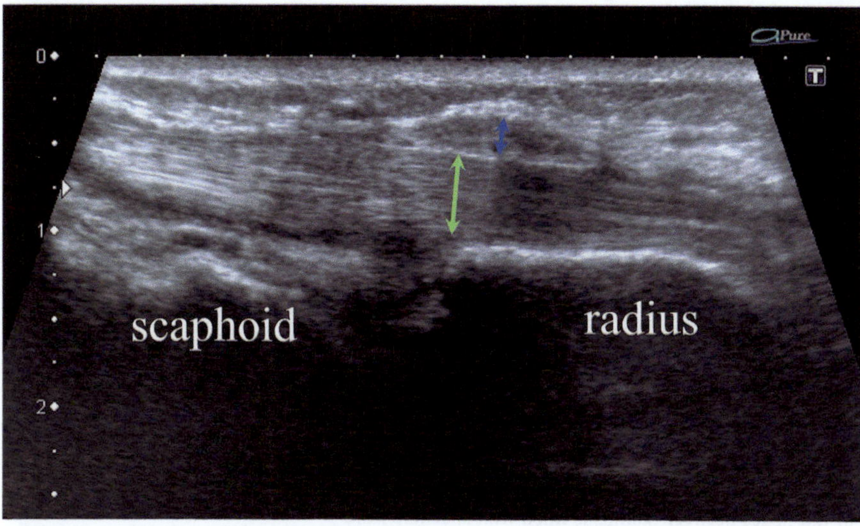

Fig. 13.16 de Quervain's disease. First extensors (*green*) compartment at the level of its pulley (*blue*). Severe fibrosis of the pulley in the course of mechanically induced inflammation of the tendinous sheath combined with constriction and stiffening of the pulley causes in severe cases tendon blocking against the pulley

Hip and Thigh

The area of the hip is covered with muscles and their tendons which makes the hip region able to absorb acute stresses, hence less likely to sustain acute injuries, but

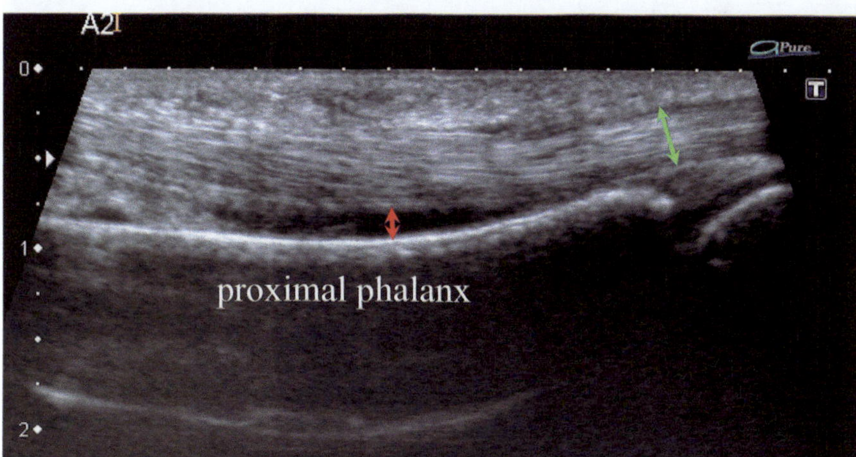

Fig. 13.17 Longitudinal scan of the flexors (*green arrow*) at the level of the A2 pulley. Flexors of D2–D5 look the same. Image of A2 pulley tear/insufficiency—*red arrow* shows the distance between the phalanx and the tendon which appears when there is no pulley. The distance is larger during dynamic evaluation (flexion with resistance). This space normally does not exist, the tendon is compressed against the bone. When fresh tear, some more edema would be seen

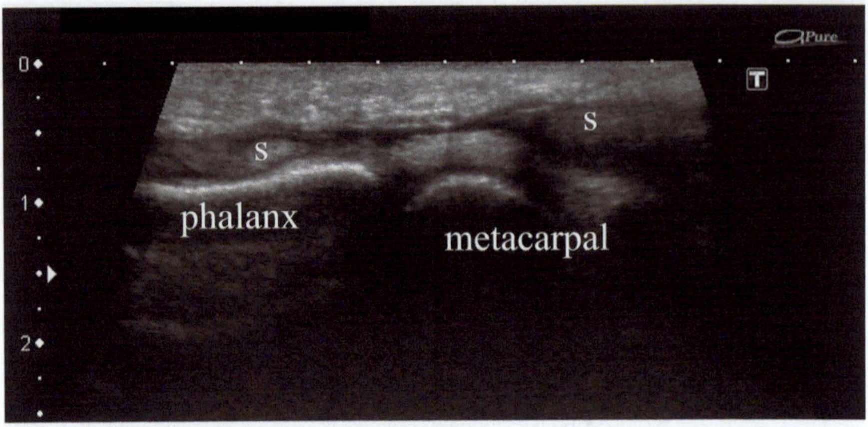

Fig. 13.18 Longitudinal scan over the flexors (flexors of D2–D5 look alike) at the level of distal metacarpal and proximal phalanx. S—retracted stumps of completely torn flexors tendons. The slit-like space between stumps is filled with anechoic fluid

more frequently chronic ones. Typical injury is within or close to enthesis with mineralized tendon scars at the enthesis zone. This is relatively common after chronic injuries of the rectus femoris and gluteal muscles. Rectus femoris proximal tendon tear away from enthesis is the most common in the anterior thigh, in particular,

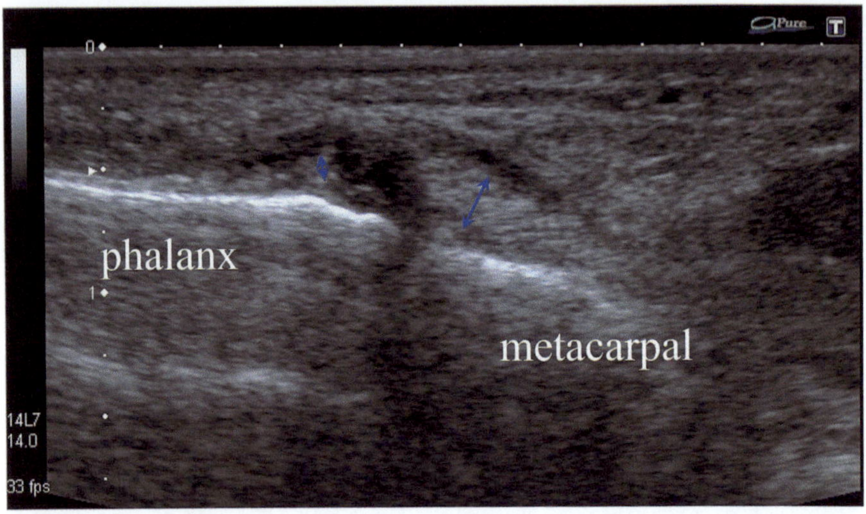

Fig. 13.19 Longitudinal scan of the MCP joint ulnar collateral ligament of the thumb (*blue arrows*). A complete tear with stumps separated by a zone of fluid and edematic torn tissue. MCP metacarpophalangeal

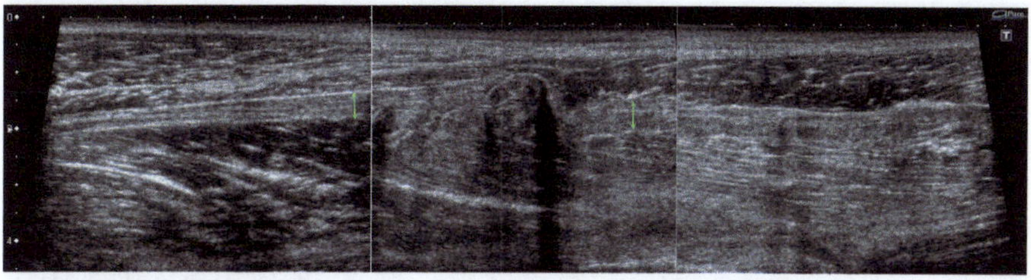

Fig. 13.20 Longitudinal image made of three. Rectus femoris proximal pennate tendon complete tear. The tendon (*green arrows*) is completely torn in the area of the zigzag. Both stumps sustained multilevel interstitial partial tears, the distal stump on the whole length, proximal on approx. 2 cm distance

among footballers. The tendon is proximally bare, then semipennate and finally pennate. It can tear on all levels, most spectacular is the pennate stretch (Fig. 13.20).

In addition to tendons and ligaments, a new category can also be affected at the hip region—the cartilage, namely fibrocartilage and hyaline cartilage. Hyaline cartilage is well exposed to the US but only on the anterior surface of the femoral head. The labrum of the hip joint's acetabulum tears most frequently in its anterior and superior parts where it is best visualized by the US (Fig. 13.21).

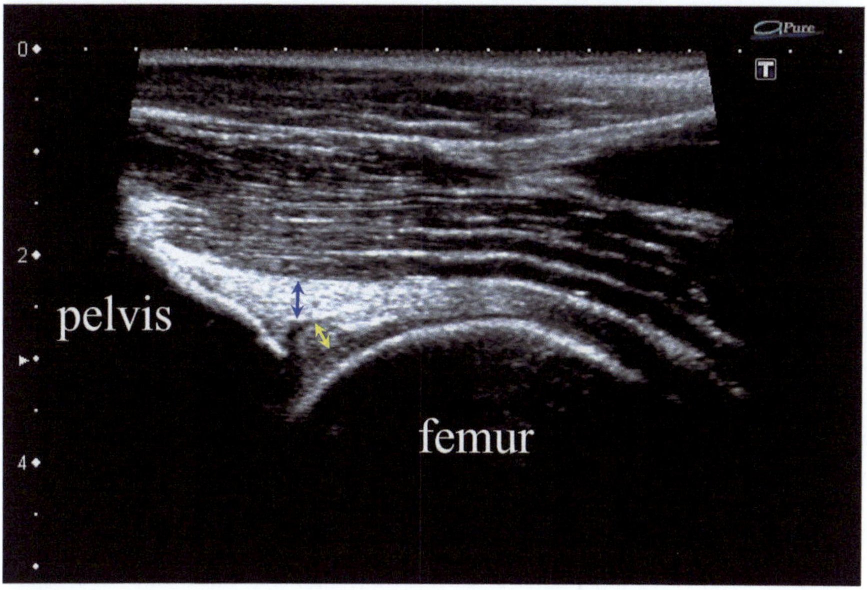

Fig. 13.21 Longitudinal scan over the anterior hip joint. *Blue*—anterior capsulo-ligamentous complex, *yellow*—labrum torn off from the acetabular rim. The tear level is indicated by *blue arrow*

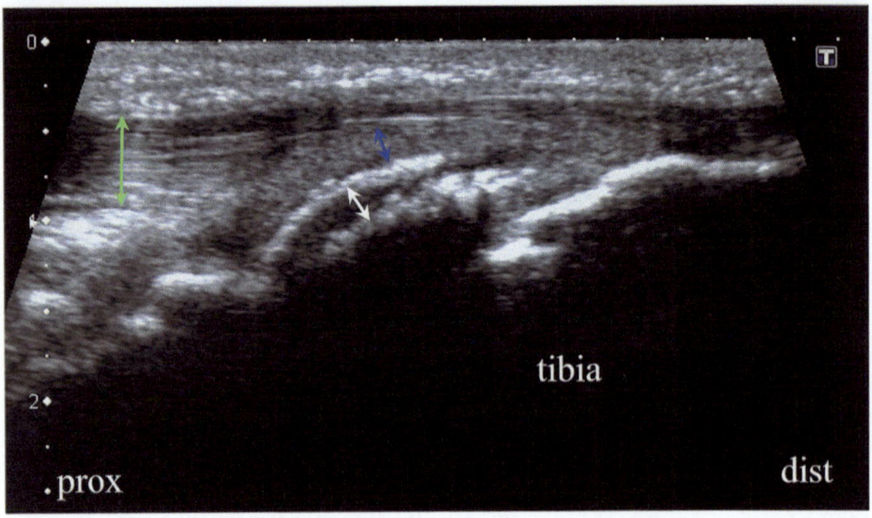

Fig. 13.22 Longitudinal scan over adolescent patellar ligament (*green*) and tibial tuberosity. OSD type I—closed delamination fracture of the ossification center of the anterior tibial epiphysis. *Blue arrow*—cartilage around the ossification center, *white arrow*—the delamination fracture distance

Knee

As the knee combines all possible structures, it is considered a universal model for sport soft tissue injury.

US turns out to be the method of choice for the diagnosis and follow-up of adolescent patient injuries such as Sinding-Larsen-Johansson disease (SLJD) and Osgood-Schlatter disease (OSD). US brings the most important elements to front—the cartilage, patellar ligament, and bone surface.

There are three types of OSDs [10, 11]:

Type I—pure delamination tear/fracture of the ossification center of the tibial tuberosity (Fig. 13.22).

Type II—delamination tear/fracture of the ossification center with fracture of the overlying cartilage situated outside (usually proximal) the patellar ligament's footprint, but extending into the deep infrapatellar bursa (Fig. 13.23).

Type III—delamination/tear of the ossification center with the cartilage fracture line within the patellar ligament's footprint, causing delamination injury to the patellar ligament; in most cases, the fracture line also extended into the deep infrapatellar bursa (Fig. 13.24).

In adults, the most frequent site of tear is the proximal stretch and insertion of the patellar ligament. Clinical symptoms are frequently called "jumper's knee" (Fig. 13.25).

Among ligaments of the knee, medial or tibial collateral ligament is the most commonly torn (Fig. 13.26). This ligament heals very well, sometimes without any treatment.

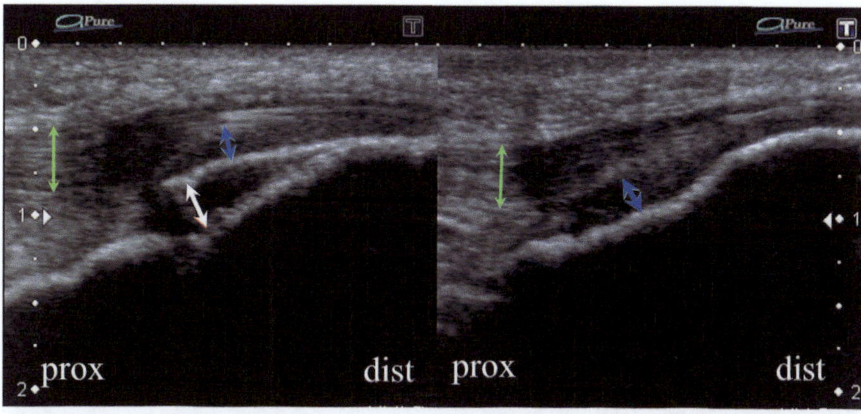

Fig. 13.23 Longitudinal scan over adolescent patellar ligament (*green*) and tibial tuberosity. Right—normal, *left*—OSD type II—delamination fracture of the ossification center with cartilage fracture proximally to the ligament's insertion to the anterior tibial epiphysis. *Blue arrow*—cartilage around the ossification center, *white arrow*—the delamination and fracture distance

Similar to the hip joint, fibrocartilage and hyaline cartilage affection has been reported in knee joints' painful conditions.

Menisci appear triangular in shape and produce bright echoes when healthy. Tears and different quality scars usually cause decreased echogenicity; however, it is difficult to say how many well-healed scars we regard as normal meniscus (Fig. 13.27). Many small asymptomatic tears can be seen occasionally in the meniscus. Hyaline cartilages appear as hypoechoic or anechoic, with a smooth surface where we look for defects (Fig. 13.28).

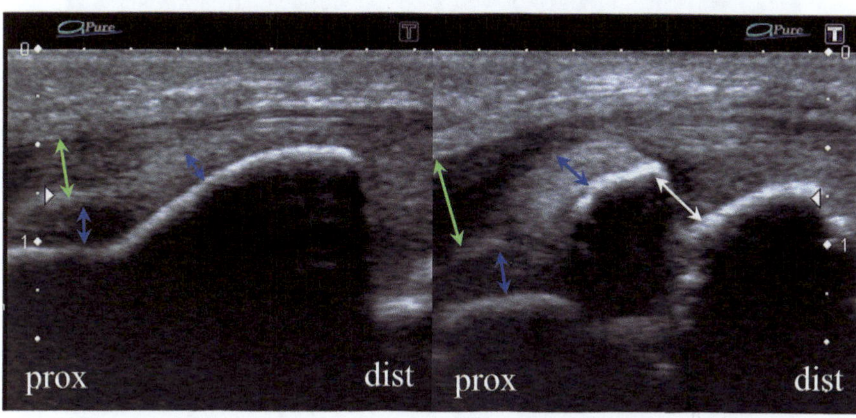

Fig. 13.24 Longitudinal scan over adolescent patellar ligament (*green*) and tibial tuberosity. *Left*—normal, *right*—OSD type III—delamination fracture of the ossification center with cartilage fracture within the patellar ligament's insertion to the anterior tibial epiphysis. *Blue arrow*—cartilage around the ossification center, *white arrow*—the delamination and fracture distance

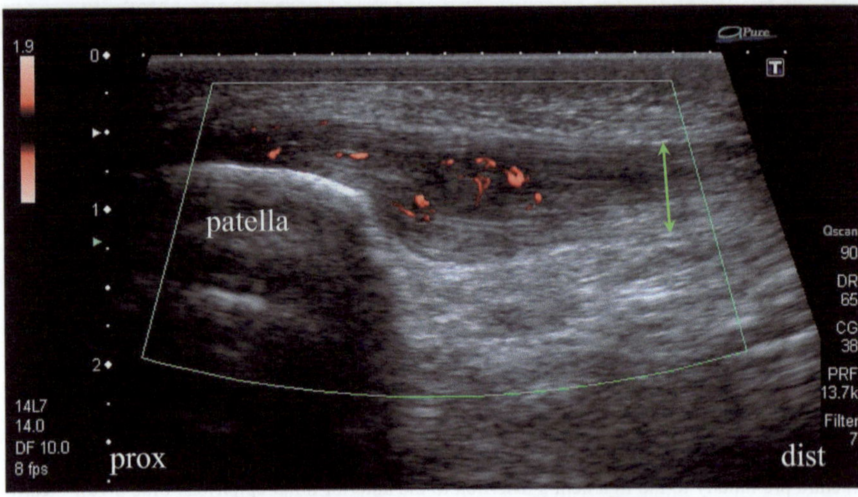

Fig. 13.25 Longitudinal scan of the mid-proximal patellar ligament. A partial tear with enhanced vascularity within and around the hypoechoic scar indicating healing process

Foot

The foot and the Achilles tendon are a good example of all kinds of tendinous and ligamentous structures. Here, like the hands, tendinous pathologies may be complicated by an inflammatory process in the bursa or tendon sheath. However, microtears are relatively common in the feet; therefore, an equilibrium between healing

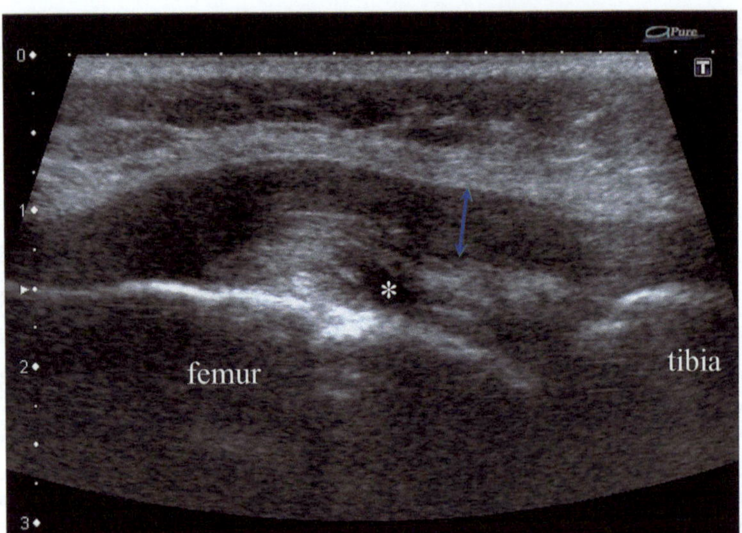

Fig. 13.26 Longitudinal scan of the mid-proximal medial collateral ligament (*blue*). A interstitial multilevel partial tear of the anterior part of the ligament. *Asterisk* shows fluid separating stumps of the completely torn menisco-femoral ligament

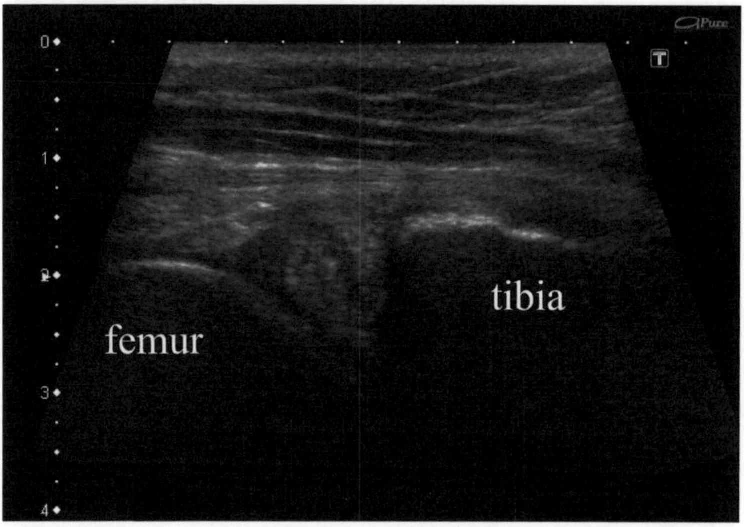

Fig. 13.27 Medial meniscus, posterior horn. A cleavage, horizontal tear reaching tibial surface of the meniscus

and anti-inflammatory treatment should be obtained. Furthermore, it is particularly dangerous to treat tendinous pathologies of the foot as inflammatory without US diagnostics. Steroid anti-inflammatory treatment suppresses the healing processes. Tibialis anterior tendon (Fig. 13.29) is a good example of mechanical tendon pathology. It is very well exposed and very thick. Sometimes, in chronic tears, there might not even be any inflammation in the tendon sheath.

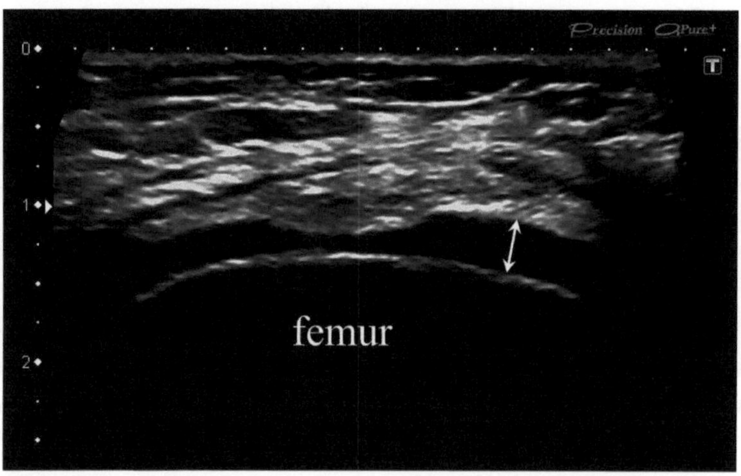

Fig. 13.28 Femoral cartilage (*white arrow*) defect reaching half thickness of the cartilage

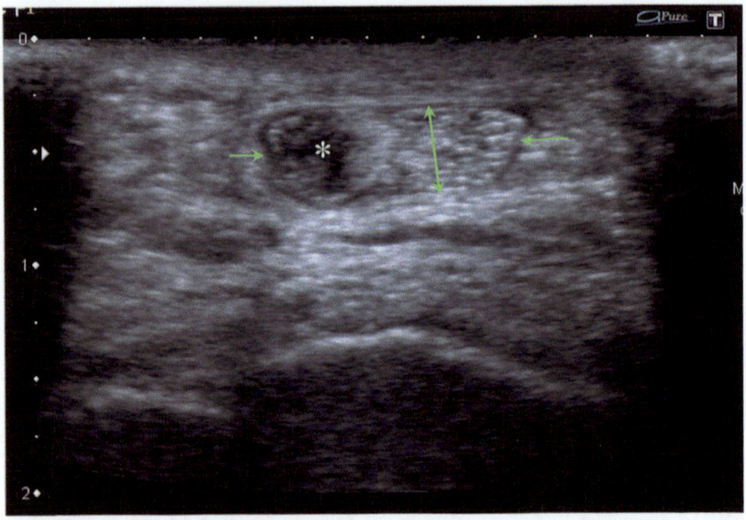

Fig. 13.29 Transverse scan of the tibialis anterior tendon (*green arrows*) at the level of approx. Chopart joint. A fresh or chronic degenerated partial tear (anechoic zone marked by an *asterisk*) of the tendon. There is little synovial irritation within the tendinous sheath

By far, the most often torn ligament of the foot is anterior talo-fibular ligament (Fig. 13.30). It is important to early diagnose the tears of this ligament because it "likes" to heal.

Among ligamentous structures of the plantar side of the foot, we cannot miss the plantar aponeurosis (Fig. 13.31). Frequently, patients are told that the pain comes from a spur. It is not true. The spur is a form that builds up and ossifies for months and years following asymptomatic micro-tears of the proximal tendon of the flexor digitorum brevis. Most often, the pain on the plantar side of the calcaneus is a partial tear of the aponeurosis. The spur is only this what an X-ray can show, not what hurts the patient.

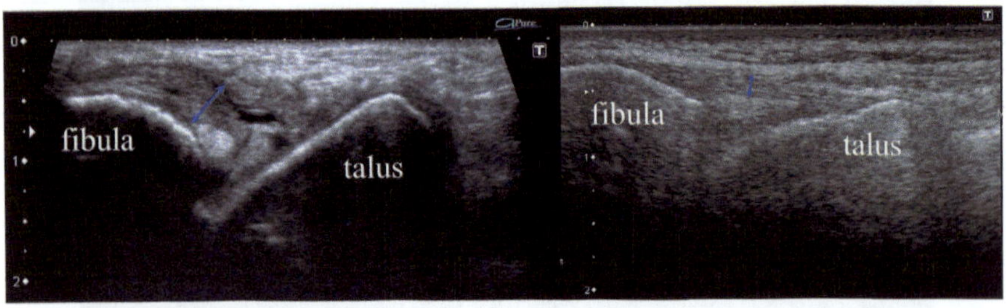

Fig. 13.30 Longitudinal scan of the anterior talo-fibular ligament (*blue arrows*), *right*—normal, *left*—a complete tear from the talar insertion with retracted interstitially partially torn stump lying deep in the talo-fibular joint

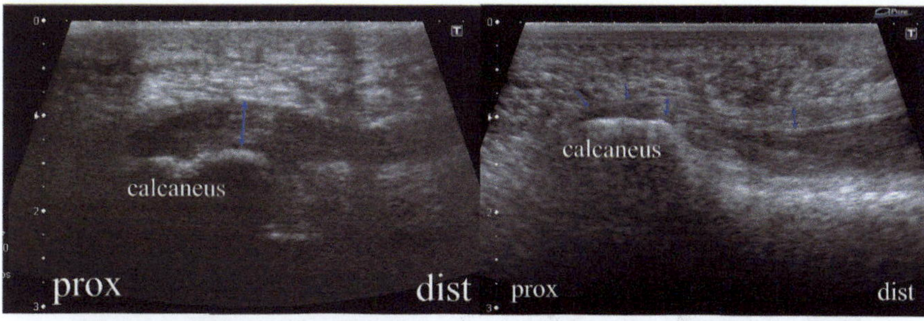

Fig. 13.31 Longitudinal scan of the proximal plantar aponeurosis (*blue*), *right*—normal, *left*—partial tear of the enthesis zone and proximal stretch. Hypoechoic, thick scar diffused within the whole volume of the aponeurosis indicating a chronic tear built from many micro-tears. An erosion of the enthesis shows destructive enthesopathy. It is an extreme rarity to see aponeurosis' enthesal scars mineralize or ossify

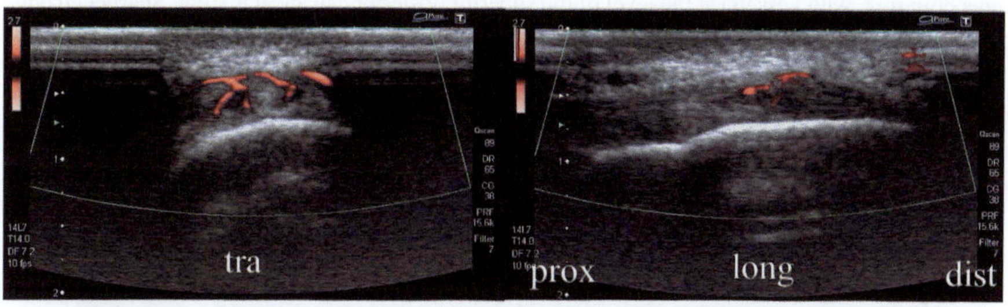

Fig. 13.32 Transverse left and longitudinal right scans of the Achilles tendon. A scar after a partial tear of the tendon within and near the enthesis (hypoechoic zone with vessels). Hypervascularity is a good sign of healing process, not inflammation. Some of such scars mineralize or ossify and form structures called spurs. So a spur is actually a product of multiple, chronic tears mineralizing and/or ossifying in the end stage of healing

Achilles tendon tears similarly and differently depending on the level (Figs. 13.32, 13.33, and 13.34). The differences are based on adjacent synovial (paratenon and Achilles tendon bursa) and bony structures [12]. Tears at the enthesis site may be calcified or ossified forming a structure called a spur which, in fact, is a moved enthesis. The level of the calcaneal upper-posterior margin and Achilles tendon bursa is a common site for tears which may be complicated with inflammation of the Achilles tendon bursa. That makes the treatment of a tear at this site more complex and difficult.

Above the Achilles tendon bursa is the mid-portion of the tendon which, in fact, is the proximal part of the bare tendon (Fig. 13.34). Tears at this area are usually uncomplicated, with no inflammation, unless infection is involved.

A form of the Achilles tendon tear is a tear of the distal tendon of the medial gastrocnemius muscle (Fig. 13.35). This tendon forms the distal tendon of the triceps surae—the Achilles tendon.

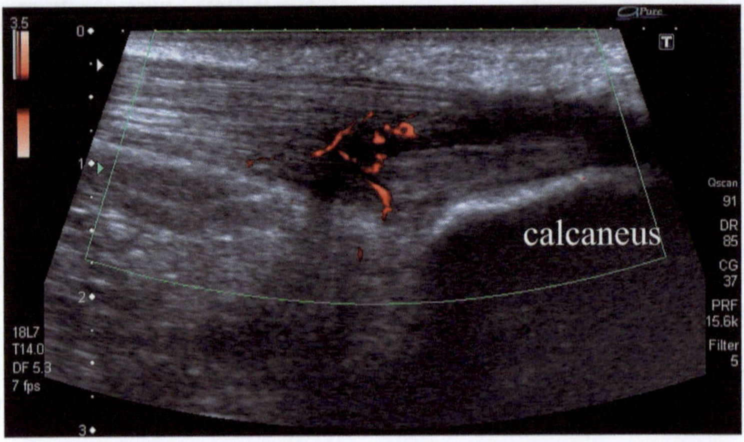

Fig. 13.33 A longitudinal scan of the Achilles tendon. Image of a partial interstitial tear (hypo-/anechoic zone with vessels) at the level of the superior posterior calcaneal margin. Sharpened edge of the calcaneus indicates that there is high probability of mechanical conflict as a reason for this tear location. This pathology may be complicated by inflammatory process in the Achilles tendon bursa due to proximity and common vascularity

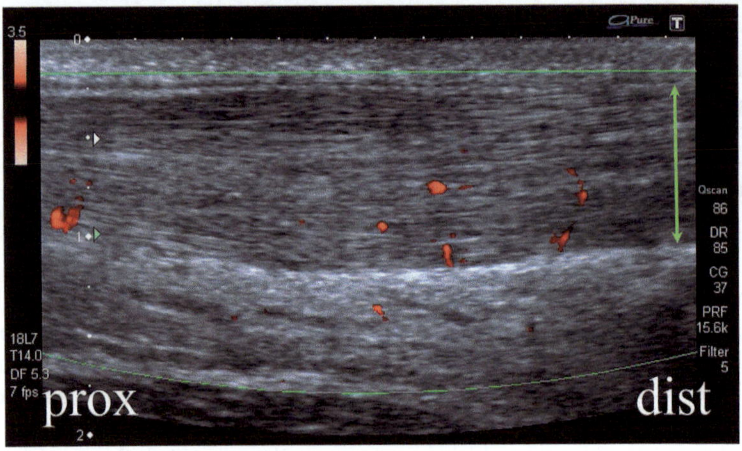

Fig. 13.34 Longitudinal scan over the midportion of the Achilles tendon (*green arrow*). Diffused thickening and blurring of the fibrillar tendinous pattern with low-level of hyperemia—a typical example of the healing process induced by an interstitial partial tendon tear located within all three bundles of the Achilles tendon. Hypoechoic zone in the superficial layer represents the least advanced in healing/newest partial tear site

In conclusion, apart from joint cavities with cartilaginous surfaces, labra and menisci, it is nearly all tendinous systems of muscles and ligaments that are mainly diagnosed in sports-related musculoskeletal contusions. The key to understanding the sonographic pattern of variable regions is identifying their different (which at

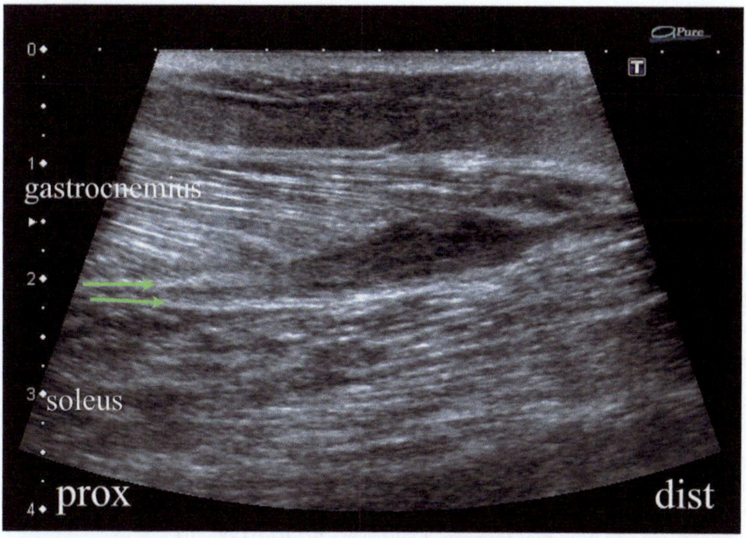

Fig. 13.35 Longitudinal scan over the distal medial gastrocnemius and mid-soleus muscles, muscles which tendons form the Achilles tendon. *Green arrows* are drawn in the location of the distal tendons of the muscles. Medial gastrocnemius distal tendon's partial tear—thickening/edema of the torn tendon with nearly anechoic zones of the torn tendinous tissue. Just below it the *bright line* of the distal soleus tendon which runs on the posterior surface of the muscle

times may appear similar) anatomy. By knowing basic principles of tendinous/ligamentous tears and their heal/repair mechanisms, one can easily apply this knowledge to new anatomical regions individually.

It is also important to identify the pathological background, whether inflammation or a tear, or both. The key point is the difference in the management strategy. Inflammation means rest and anti-inflammatory therapy including local and/or systemic approach. Similarly, a tear and repair may give identical symptoms and require rest too; however, nonsteroid anti-inflammatory drug therapy should be limited in both quantity and time as it suppresses the healing performance of a torn structure. Steroids, especially locally administered, should be avoided at all costs as they are a very strong suppressor of the repair processes. If steroids are used, the ill area should be immobilized to protect the healing tissue.

The challenging scenario is when a lesion of a mixed nature occurs, e.g., a tear associated with inflammation of the surrounding tissues. The clue might be in thorough assessment of the affected area bearing in mind its anatomical landmarks. In some cases, the inflammation may be induced by the mechanical trauma or even patient's habit. If not, then special care, including immobilization, should be considered in particular while administering strong anti-inflammatory drug therapy.

References

1. Bochenek A, Reicher M. Anatomia człowieka; PZWL, Warszawa, wydanie X, Tom I; 1990.
2. Czyrny Z. Muscles–histology, micro/macroanatomy and US anatomy, a brand new perspective. Ultrasonografia 2012;12:48.
3. DiGiacomo G, Pouliart N, Costanini A, De Vita Andrea. Atlas of functional shoulder anatomy. Milan: Springer-Verlag; 2008.
4. Czyrny Z. O systemach ścięgnistych mięśni (On tendinous systems of muscles). Spectrum Media: Warszawa; 2014. ISBN 978-83-938977-0–4.
5. Pouliart N, Somers K, Eid S, Gagey O. Variations in the superior capsulo-ligamentous complex and description of a new ligament. J Shoulder Elbow Surg. 2007;16:821–36.
6. Pouliart N, Somers K, Gagey O. Arthroscopic glenohumeral folds and microscopic glenohumeral ligaments: the fasciculus obliquus is the missing link. J Shoulder Elbow Surg. 2008;17:418–30.
7. Czyrny Z. Diagnostic anatomy and diagnostics of enthesal pathologies of the rotator cuff. J Ultrason. 2012;12:178–87.
8. Nimura A, Kato A, Yamaguchi K, Mochizuki T, Okawa A, Sugaya H, Akita K. The superior capsule of the shoulder joint complements the insertion of the rotator cuff. J Shoulder Elbow Surg. 2012;21:867–72
9. Dębek A, Nowicki P, Czyrny Z. Ultrasonographic diagnostics of pain in the lateral cubital compartment and proximal forearm. J Ultrason. 2012;12:188–201.
10. Czyrny Z, Greenspan A. Osgood-Schlatter disease: a new perspective and classification based on ultrasonography. Ultrasonografia. 2009;9:38.
11. Czyrny Z. Osgood-Schlatter disease in ultrasound diagnostics—a pictorial essay. Med Ultrason. 2010;12(4):323–35.
12. Czyrny Z. Ultrasonography of the Achilles tendon—basics of anatomy and pathology. J Orthop Trauma Surg Relat Res. 5/2011;17–28.

Chapter 14
Ultrasound-Guided Interventional Maneuvers

Mihaela C. Micu MD and Daniela Fodor MD, PhD

Introduction

Interventional musculoskeletal ultrasound (MSUS)-guided maneuvers refer to a variety of invasive procedures, performed percutaneously, covering both diagnostic and therapeutic indications. This include simple procedures such as joints, tendon sheaths, or other periarticular structures injection as well as more complex maneuvers such as biopsies, foreign bodies removal, or intratumoral therapeutic injections.

Intra-articular injections, with corticosteroid (CS) preparations, have been used for several decades in the management of different inflammatory or degenerative joint conditions when first-line conservative management approach fails to provide adequate pain relief. Despite the significant developments achieved in the last years, in the treatment of inflammatory/degenerative rheumatic diseases, intra- and periarticular injections remained an important tool in daily clinical practice.

The current European League Against Rheumatism (EULAR), the American College of Rheumatology (ACR), Assessment of SpondyloArthritis International Society (ASAS), and the Osteoarthritis Research Society International (OARSI) therapeutic guidelines include intra-articular CS injections in the recommended treatment protocols [1–8]. Indeed, more data, emerging from the recent literature, support the extension of intra- and periarticular injections with other different drugs, like viscoelastic and biologic agents, radioactive isotopes, etc. [9].

Aspiration and examination of the joint fluid, a routine maneuvers in clinical practice, still remains of paramount importance as it may rapidly differentiate septic or

M. C. Micu (✉)
Department of Rehabilitation II, Rheumatology Division, Rehabilitation Clinical Hospital Cluj-Napoca, Cluj-Napoca, Romania
e-mail: mcmicu@yahoo.com

D. Fodor
Department of Internal Medicine, "Iuliu Hatieganu" University of Medicine and Pharmacy, Cluj-Napoca, Romania
e-mail: dfodor@umfcluj.ro

Y. El Miedany (ed.), *Musculoskeletal Ultrasonography in Rheumatic Diseases*,
DOI 10.1007/978-3-319-15723-8_14

crystal deposition arthritis from other inflammatory joint conditions such as rheumatoid arthritis (RA), spondyloarthropathies (SpA), osteoarthritis (OA), etc. [10–11].

When performing these kinds of interventional maneuvers, it is always desirable to adopt an imaging needle-guided technique. Considering efficacy, economic impact on health-care and safety profile, Ultrasonography comes first as the most feasible and practical imaging tool [9, 12–16].

The interest in minimally invasive guided maneuvers is rapidly mounting. In the past few years there has been an exponential growth of number of studies published in this field reflecting the growing interest to implement these techniques as part of the standard clinical practice and rheumatology curriculum in many countries. Unfortunately, this procedure remains insufficiently exploited among rheumatologists and, according to a recent survey, only 10% of the rheumatologists routinely perform MSUS-guided joint/ periarticular structures aspirations or injections in most European countries [17].

This chapter describes the basic techniques of the MSUS-guided joints and periarticular structures aspiration, injections, and biopsy. It will also provide an update to the therapeutic standards, indications and contraindications, efficacy and safety profile of the methods.

General Approach

Basics

The decision for performing a certain interventional maneuver is considered following an algorithm which incudes: the patient's history (trigger conditions, disease onset, other comorbidities, current or chronic treatment), general/local clinical assessment in addition to systematic MSUS assessment of the target region (according to published guidelines and protocols) [18–20].

A complex set of information regarding anatomical structures, starting from skin to cortical bone, can be obtained by using MSUS grayscale and power Doppler assessment. Whenever indicated, multiplanar dynamic assessment may be also carried out. Some of the pathological conditions may also require the complementary radiographic evaluation with focus on the bony structure and possible anatomical variants.

Initially, in clinical practice, preinterventional MSUS assessment is important to confirm the primary clinical decision. In some cases, the MSUS findings may alter the preliminary decision of whether to perform the planned interventional maneuver or change the targeted structure(s) [21–23].

The probe type and the operating frequencies are adapted according to different anatomical areas. Linear probes (5–18 MHz) are usually preferred, though convex probes are considered the best option for deep areas such as hip or shoulder, with lower frequency (2.5–5 MHz) and higher penetration. After the detection of the target lesion, the shortest and safest way (avoiding neurovascular proximity structures) from skin to target lesion is visualized using a multiplanar MSUS region evaluation.

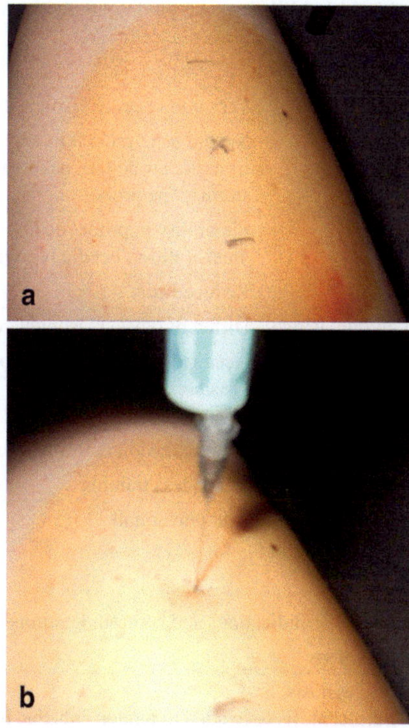

Fig. 14.1 After ultrasound lesion depiction, the skin is marked at the probe ends and center. The needle is penetrating perpendicular to the skin in the center of the target lesion skin projection

Proper patient positioning is essential in order to obtain a relaxed posture during the procedure and to assure an easy approach for the performing physician. In general, needles should be kept outside the patient's visual field and every effort should be made to minimize the possibility of the patient sustaining a vagal reaction. Specific details for each anatomical area will be discussed in more details later in the chapter.

MSUS Puncture Methods

Technically, there are two MSUS-guided interventional methods: (1) using a needle-guiding device attached to the probe and (2) the "free-hand" technique. The second one can be carried out in either of two ways: the indirect (semiguided, skin marking) and direct methods [16, 24].

In the indirect technique, the semiguided method, MSUS is used only for identification and evaluation of the lesion and to mark the skin and measure the lesion's depth (calipers). The needle is then inserted blindly, perpendicular to the skin reaching the previously measured depth. The method is easily performed but the needle progression cannot be tracked (Fig. 14.1).

The direct MSUS-guided technique is a superior approach as it offers a real time and continuous visual monitoring of the needle penetration and advancement, through every tissue layer, till the targeted area. Needle-guiding devices attached to the probe may help in achieving a maximum precision for hitting the target

Table 14.1 MSUS needle-guiding methods. Advantages and disadvantages of direct "free-hand" and "device-guided" needle techniques

Method	Advantages	Disadvantages
Free hand	Very good needle visualization	Difficulties in keeping the needle and lesion in the same plane
	Needle penetrates skin at distance from the probe (not necessary to sterilize the probe)	
	Flexibility in probe/ needle movements (allows needle reorientation)	Needs more training and longer learning curve
	Reduced cost	
Guiding device	Different types, reasonable cost	Difficulties in needle manipulation
	Needle direction preselection is possible	Needle penetrates close to the probe; needs sterile probe cover and sterile gel
	High precision in hitting the target	Higher costs (probe, guiding device, sterile equipment, more personnel)
	Needs less training	

MSUS musculoskeletal ultrasound

Table 14.2 Indications and contraindications in MSUS-guided interventional maneuvers

Indications
For diagnosis: Arthrocentesis (extended to tendon sheaths, bursae) and biopsy
For therapy: Decompression (effusion under pressure, hematoma, abscess) Medication (corticosteroids, viscosupplementation, radioactive substance, etc.), lavage
Contraindications
Relative, temporary: Infection, wounds, psoriasis, skin vasculitis (large areas)!! advantage in using nonconventional puncture areas under MSUS guidance Anticoagulant drugs: INR > 3.5 in the last 72 h; attention to combination with NSAIDs Hemophilia: when factor VIII, IX is < 20 % Diabetes mellitus, high blood pressure, glaucoma
Allergy to medication or drug composition

MSUS musculoskeletal ultrasound, *NSAID* nonsteroidal anti-inflammatory drugs, *INR* international normalized ratio

structure, but the predefined direction of the needle must be kept throughout the entire procedure. In contrast to this, "free-hand" technique, direct MSUS-guided technique is highly flexible, giving the facility of various needle insertion positions. Table 14.1 shows the main advantages and disadvantages of the two techniques.

Similar to any other conventional therapeutic procedure, MSUS-guided approach has its pros and cons (Table 14.2). Possible untoward side effects linked to local drug deposition or postinterventional complications are summarized in Table 14.3 [24, 25].

Table 14.3 Potential side effects, postinterventional course recommendations

Side effects/complications
Increase of local pain: secondary to traumatic act (needle insertion), microcrystalline synovitis induced by CS, irritation produced by viscosupplementation drugs
Facial, chest rash
Local skin depigmentation
Local skin atrophy
Severe infections: iatrogenic (antiseptic rules), sepsis in the layers the needle penetration takes place
Tendon rupture, necrotizing fasciitis (intra-tendon/ fascia injection)
High blood pressure, high glycemia
Postinterventional course
No need for bed rest
Current treatment with anticoagulant drugs—reduction of walking effort some days

CS corticosteroid

Table 14.4 General conditions for performing MSUS-guided invasive interventions

1.	Re-examination of the region is mandatory for decision reconfirmation, optimizing patient and doctor position
2.	The patient must be informed regarding the necessity of the chosen invasive method; the informed consent must be signed
3.	The performing physician must collect information regarding coagulopathies, hemorrhagic accidents in the past, INR value, number of thrombocytes, allergies, intolerance to some drugs
4.	Easy access to emergency equipment; intravenous branula for selected cases
5.	Disinfection of anatomic area/ probe—simple/ complex
6.	Protection measures for the doctor—sterile/nonsterile gloves, mask/more complex equipment
7.	Local anesthetic—lidocaine 1 %, bupivacaine, etc. are used in biopsies, percutaneous drainage, sedation
8.	Avoid vagal reactions—manipulate needles outside the patients visual field
9.	Needle type selection—length, caliber
10.	Postinterventional—in outpatient clinic—patient must report the occurrence of fever/ persistence >48–72 h; hospital surveillance for those with complex interventional maneuvers, drainage catheter

MSUS musculoskeletal ultrasound

General rules must be constantly applied when any of these intervention techniques is considered. Agreed protocols should be adopted. Table 14.4 entails some of the basic steps.

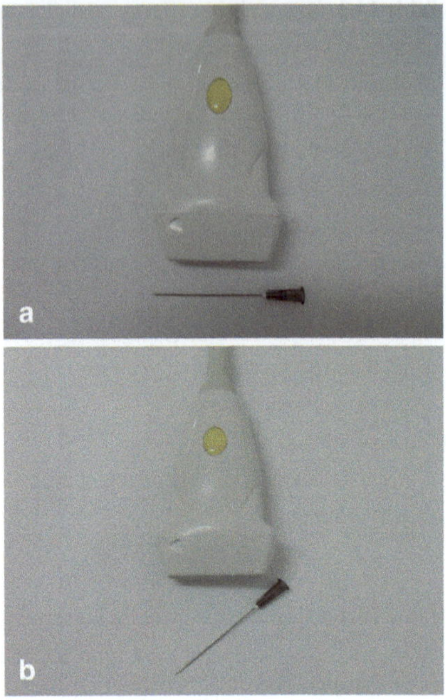

Fig. 14.2 When needle is placed at angulation of 90° and perpendicular to the ultrasound (US) beam, optimal visualization is achieved (hyperechoic, continuous line). When needle is placed at angulation between 45° and 90°, visualization is possible (hyperechoic, discontinuous line)

Free-Hand Direct MSUS-Guided Interventional Maneuvers

Apart from its flexibility in choosing the site of needle penetration, distance in relation to the probe varies from minimum 0.5 cm for small areas up to 2–3 cm for bigger areas). Needle angulation to the ultrasound (US) beam (90–45°) can be adapted according to the anatomic region and targeted structure. If necessary, needle redirection can be considered at any moment and depth, subject to local factors.

The needle visualization inside a body tissue or fluid is possible due to its physical characteristics (metal, hyper-reflectogen, glittering hyperechogen structure); however, this is also subject to its positioning inside the US beam. Higher caliber needles are better visualized in comparison to thin ones. The use of needle-guiding systems attached to the probe allows a perfect needle penetration, perpendicular to the US beam and parallel to the transducers margins, with optimal needle visualization. Though "free-hand" technique needle visualization may be less accurate, it is good enough for identifying the needle tip and the penetration of soft tissue layers (Figs. 14.2 and 14.3).

Skin and Probe Disinfection

Skin and probe preparation protocols may vary from simple protocols set according to local rheumatology settings regulations and/or personal experience up to

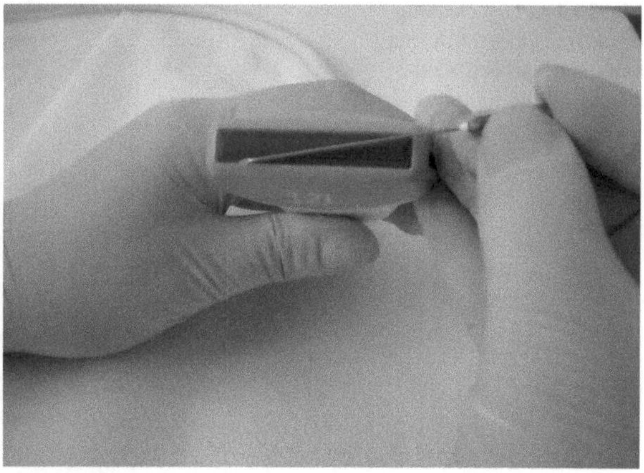

Fig. 14.3 When needle is placed at angulation of 90° and oblique to the ultrasound (US) beam, visualization is possible (hyperechoic, discontinuous line)

complex rules (use of sterile probe envelope, sterile gel, sterile cleaning solutions, alcoholic solutions, surgical drapes, etc.) [9, 26, 27].

In our experience, there is no need for using sterile probe envelope or sterile gel. Rigorous, dry cleaning of the probe prior to each interventional maneuvers accompanied by rigorous skin disinfection (similar to conventional blind injections) with bethadine should be enough if the security distance between the probe and puncture site is respected. This simplifies the entire technique and reduces substantially the costs. A recent study showed that adopting these simple rules ensured a very good safety profile and satisfactory outcomes [14].

Alcoholic cleaning solutions, if used for probe cleaning, must be applied strictly to the plastic cover and not on the footprint, in order to avoid its damage [28]. The use of bethadine solution for skin disinfection has some further advantages. In addition to the fact that bethadine is less irritating to the skin in comparison to alcoholic disinfection swaps, its brown color will highlight the disinfected area, hence it will be easier to trace the probe position.

Puncture Preparation

For puncture preparation, a small quantity of nonsterile gel will be applied on the footprint in order to avoid the overflow of excessive gel to the puncture site and compromise the sterile maneuvers status (Fig. 14.4).

Multiplanar scanning technique allows the optimal identification of the targeted lesion. Afterwards, the best scanning position for a lesion exposure will be memorized. For beginners, it is helpful to mark the transducer's footprint position by using a permanent marker. After final probe positioning, the needle will be inserted. There are two modalities for insertion: "in plane"—the needle will enter the tissues parallel to the long transducer axes and the whole needle length will be visualized or "out of plane"—the needle will enter perpendicular or oblique to the short axes and it will be visualized as a hyperechoic dot or it will be partially detected (only the

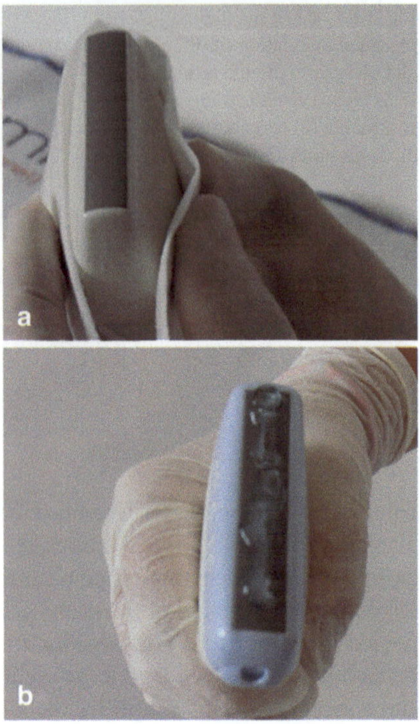

Fig. 14.4 **a** Dry cleaning of the transducer. **b** A small amount of gel is placed strictly on the footprint

portion inside the US beam will be identified; Figs. 14.5 and 14.6). Once the needle insertion is finalized, minimal sliding or tilting maneuvers to adjust the probe position is allowed for optimal image acquisition.

Though in MSUS, there are usually good ultrasonographic windows for almost all anatomic regions, there are still some limitations, in particular for accurate deep structures imaging, caused by artifacts such as osteophytes, bony fragments, calcifications, or due to the presence of air.

Basic Rules for Joint, Tendon Sheath, Cysts, and Bursa Puncture

In standard rheumatology practice, MSUS-guided procedures usually include joints, tendon sheaths, cysts, or bursae aspiration/injections. In general, for optimal needle and target visualization, "in-plane" insertion is usually the recommended approach, if possible.

The joint is a triangular cavity delineated by two bones and capsule. The needle penetration in a joint cavity can be made in a longitudinal or transversal approach (Fig. 14.7). On the other hand, the tendon sheath injection approach varies subject to the anatomical region as well as the operator preferences. The puncture can be carried out by using a transversal/longitudinal scanning view, with "in-plane" needle insertion (Fig. 14.8). For cysts and bursae, it is left open for the performing physician to

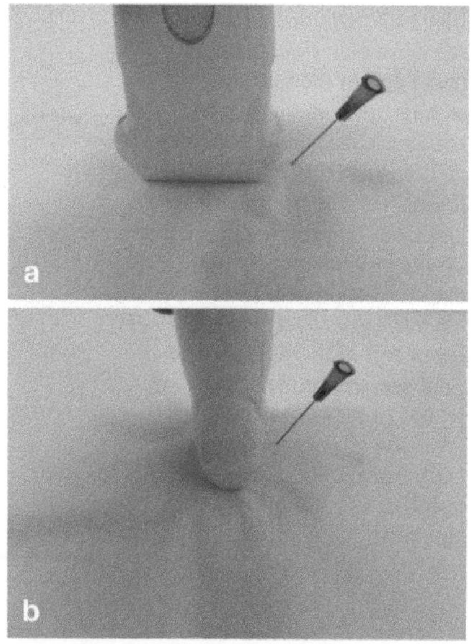

Fig. 14.5 **a** "In-plane" needle insertion. **b** "Out-of-plane" needle insertion

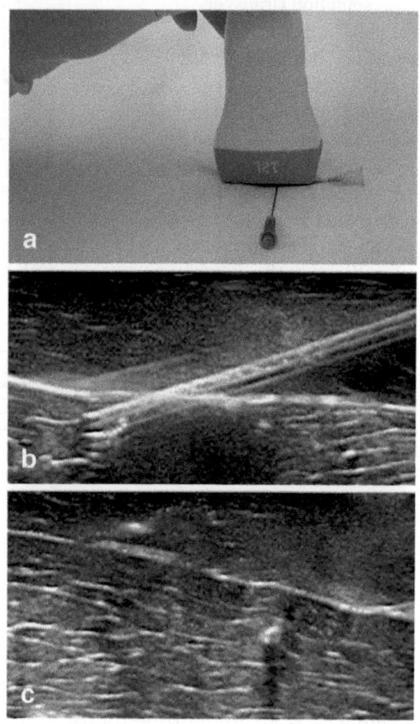

Fig. 14.6 **a** "In-plane" and "out-of-plane" needle insertion. **b** "In-plane" technique—detection of the needle as a hyperechoic *line* generating reverberation artifact. **c** "Out-of-plane" technique—detection of the needle as hyperechoic *dot* generating acoustic shadow

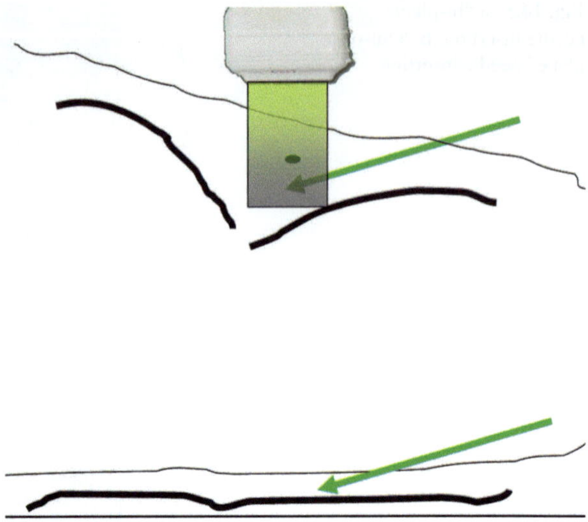

Fig. 14.7 Schematic joint representation (longitudinal, upper part of the figure)—triangular space delineated by two bones (*black thick lines*) and capsule (*thin black line*). *Arrow*—needle penetration "in-plane" approach, *dot*—needle visualization—"out-of-plane" approach. Joint representation (transversal, lower part of the figure)—the joint space is delineated by one of the bones cortex (*thick black line*) and capsule (*thin black line*). *Arrow*—needle penetration "in-plane" approach

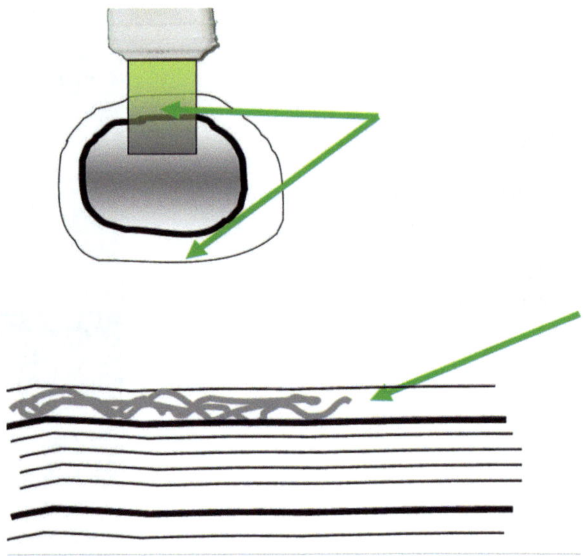

Fig. 14.8 Schematic tendon representation (transversal view, upper part of the figure)—"in-plane" needle (*arrow*) penetration inside the tendon sheath. Tendon representation (longitudinal view, lower part of the figure)—"In-plane" needle (*arrow*) penetration inside the tendon sheath and drug deposition (*gray lines*)

choose the most convenient scanning view, preferably adopting the "in-plane" needle insertion. Optimal needle angulation in relation to the US beam should be 90° for superficial structures and can vary between 90° and 45° for more profound structures. Beyond this angulation limit, there would be difficulty in detecting the needle. When using a 4–5-cm-long probe footprint for visualizing small structures, we recommend the target lesion to be displayed on the right side of the monitor for right-handed operators and on the left side of the screen for left-handed operators. In this way, we shorten the distance from the skin penetration area to the target lesion, offering the possibility for using shorter, thinner, and less traumatic needles (Fig. 14.9).

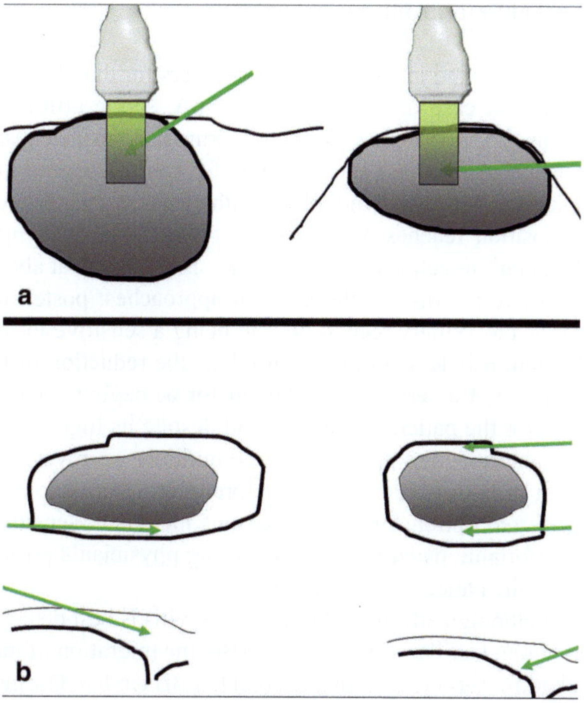

Fig. 14.9 **a** Schematic cyst representation. "In-plane" needle approach—right side— a superficial structure and needle can penetrate with angulation of 90°, left side— a deeper structure and needle is penetrating with angulation < 90°. **b** Schematic representation of the target structure on the monitor for right-handed (structure is displaced at the right side of the screen) and left-handed (structure is displaced at the left side of the screen) performing physicians

Needles caliber must be adapted to the desired procedure. It is recommended to use 16–21 gauge needles for aspiration and viscoelastic drug injection and thinner needles (22–24 gauge) for other medications like CS, anesthetic drug, etc. For profound areas like hip or glenohumeral joint, we recommend the spinal puncture needles.

Applications for Anatomic Regions

This section focuses on different types of interventional approaches, in different anatomic areas, using the most suitable and simple technique, designed for one performing doctor, and considering the most important pathologies encountered in clinical practice.

Shoulder Region

Several target areas are identified in the shoulder region: the glenohumeral and acromioclavicular joints, the subacromial–subdeltoid bursa, the long head of the biceps brachialis (BB) tendon, intratendinous calcifications.

Glenohumeral Joint

MSUS-guided interventional maneuvers for the glenohumeral joint are a treatment option in case of synovitis (RA, SpA, septic arthritis, microcrystalline arthritis, trauma), OA (viscous lubricating drugs), adhesive capsulitis (repeated capsule distension procedures) and a method for contrast agent injection (magnetic resonance imaging (MRI) arthrography). With visual assistance, contrast agent and other drug deposition reaches an accuracy of 94–100 % in comparison to classic anatomic landmark injections where accuracy is estimated at about 50 % [14, 29–32].

There are three different joint approaches: posterior, anterior, and axillary. As far as the axillary recess, despite being a sensitive location for joint pathology detection, it is less commonly used, as the reduction of the joint range of motion is considered a serious impediment for achieving a comfortable examination position for the patient. Similarly, and despite its high accuracy, a major limitation for the anterior joint approach is the multiple neurovascular structures located in the joint proximity [24, 33]. Therefore, the posterior joint approach is considered the preferred option for several reasons including: sensitivity of pathology detection, comfortable patient's and performing physician's position, and preparations away from the patient's visual field.

Pathologically, glenohumeral synovitis is best detectable with MSUS at the posterior and axillary recess level. Also, the migration of intra-articular effusion can be identified around the long head of the BB tendon. During the shoulder external rotation maneuvers, exposure of a small quantity of effusion/synovial hypertrophy (not visible in neutral arm position) can be seen inside the posterior joint recess [33].

Injection Technique and Positioning It is recommended to start the procedure by positioning the patient seated on a chair or supine on the contralateral side. The performing physician will approach the joint from the posterior aspect placing the probe parallel to the infraspinatus muscle fibers at the tendeno-muscular junction. The posterior recess is identified profound to infraspinatus muscle, in the labrum proximity. The needle will penetrate the skin at an approximately 2-cm distance, in plane, with an angulation of 45°, aiming at the superior margin of the labrum. In this way, the interaction with neurovascular structures (suprascapular nerve, circumflex artery) can be avoided and undesirable accidents eliminated. The penetration of the joint capsule is noticed as an obstacle and the puncture may be realized in particular in chronic inflammatory processes associated with higher capsule thickness and fibrotic changes. When final needle position is reached, arthrocentesis and/ or drug injection can be made. No resistance should be encountered while injecting, when the needle is properly positioned in the joint (Fig. 14.10).

Acromioclavicular Joint

The posterio-superior, "out-of-plane" approach is usually the preferred approach when injecting the acromioclavicular joint. The patient is sitting on a chair with the performing physician standing behind the patient (Fig. 14.11). After proper skin and

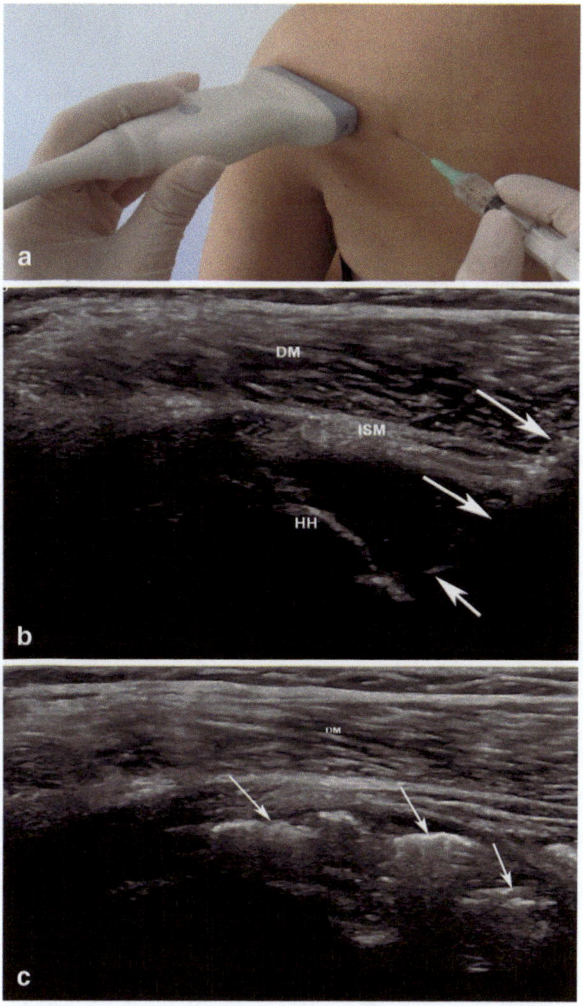

Fig. 14.10 **a** Patient and probe positioning for glenohumeral joint injection, "free-hand technique." **b** Needle (*arrows*) is penetrating with angulation of 45° inside the distended posterior glenohumeral recess. *DM* deltoid muscle, *HH* humeral head, *ISM* infraspinatus muscle. **c** Corticosteroid deposition (*arrows*, hyperechoic mass moving antigravitational generating acoustic shadow) inside the posterior glenohumeral recess

transducer disinfection, the probe will be positioned with one end on the clavicle and with the other on the acromion. The needle penetrates the skin approximately at a 1–2-cm distance from the probe having the angulation of 90°, entering below the acromioclavicular ligament [34].

The Subacromio-Subdeltoid Bursa

It is located deeper to the deltoid muscle and its approach can be easily made. Subacromio-subdeltoid bursitis usually occurs in association with other inflammatory, degenerative, or traumatic rotator cuff pathologies. Alone or in combination with tendon pathology, it represents a frequent cause of shoulder pain. The aspiration of

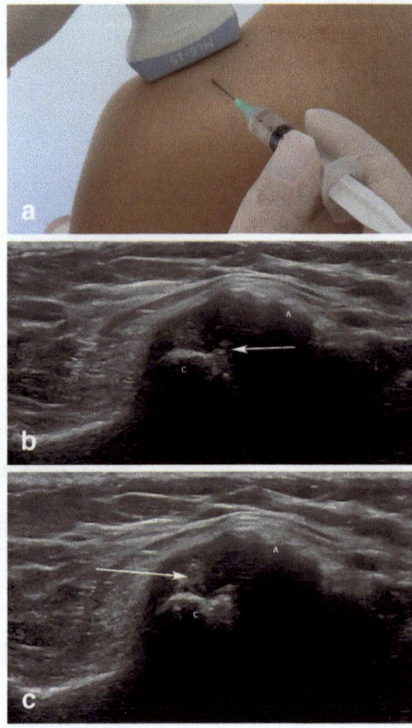

Fig. 14.11 **a** Patient and probe position for acromioclavicular joint injection, "free-hand technique," longitudinal approach, "out of plane" needle penetration. **b** Needle is depicted as a hyperechoic dot (*arrow*) inside the joint space, *C* clavicle, *A* acromion. **c** Hyperechoic mass (corticosteroid deposition, *arrow*) inside the acromioclavicular joint space distending the capsule

the bursal content represents an important emergency maneuvers leading to rapid therapeutic decompression as well as a diagnostic procedure (inflammatory, septic, hemorrhagic content, laboratory analysis—cytology, crystals, etc.) and can be done immediately after clinical examination and MSUS evaluation. The bursal content can be effusion, synovial hypertrophy or both, blood, etc. The MSUS-guided puncture accuracy is superior to the blind method in case of mixed or parenchymatous bursal content. Furthermore, the precise bursal CS injection can be carried safely avoiding rotator cuff tendon penetration or systemic intradeltoid CS administration [35–37] and subsequent better outcome is expected [38, 39]. In the last years, several attempts injecting intra-articular viscoelastic drug therapy showed some promising results [40].

Injection Technique and Positioning Skin and probe disinfection is made according to previous described technique. If possible, needle penetration will be "in plane" with angulation of 90°. In some cases of obese or very muscular subjects, it is necessary to angulate the needle more (60°–45°). After skin, subcutaneous tissue and deltoid muscle penetration, the needle tip reaches the bursa (Fig. 14.12). Real-time aspiration and/ or drug injection can be rapidly carried out and drug deposition and dispersion can be checked immediately after finalizing the invasive maneuver.

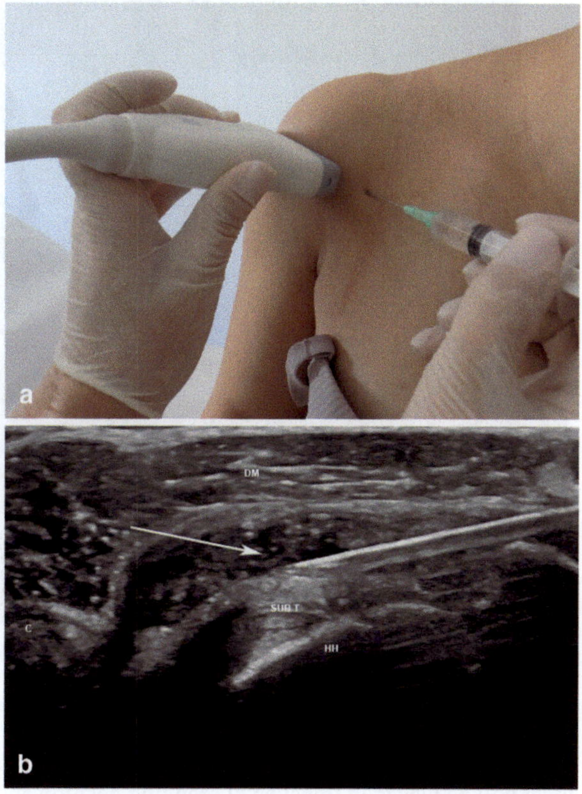

Fig. 14.12 **a** Patient and probe position for subacromio-subdeltoid bursitis, anterior approach. **b** Corticosteroid drug deposition (*arrow*) inside the subacromio-subdeltoid bursa, *C* coracoid bone, *HH* humeral head, *SUB T* subscapularis muscle tendon, *DM* deltoid muscle. The needle is depicted as a hyperechoic line creating reverberation artifact

BB Tenosynovitis

In case of true BB tenosynovitis or glenohumeral arthritis with BB tendon sheath effusion, anterior "in-plane" approach using a transversal scanning view is recommended (Fig. 14.13).

The Elbow

Elbow Joint

Elbow synovitis (humero-radial and humero-ulnar joints) may occur in several pathologies like RA, SpA, gout, septic, or traumatic circumstances. These joints can be approached either from the anterior or posterior aspect. In case of anterior approach, the patient can be seated or laid down on the bed with the elbow in maximal extension position. Preinterventional anterior joint aspect scanning will identify the neurovascular structures (median nerve and brachial artery, radial nerve and radial artery) and the distal BB tendon, structures that must be avoided while injecting.

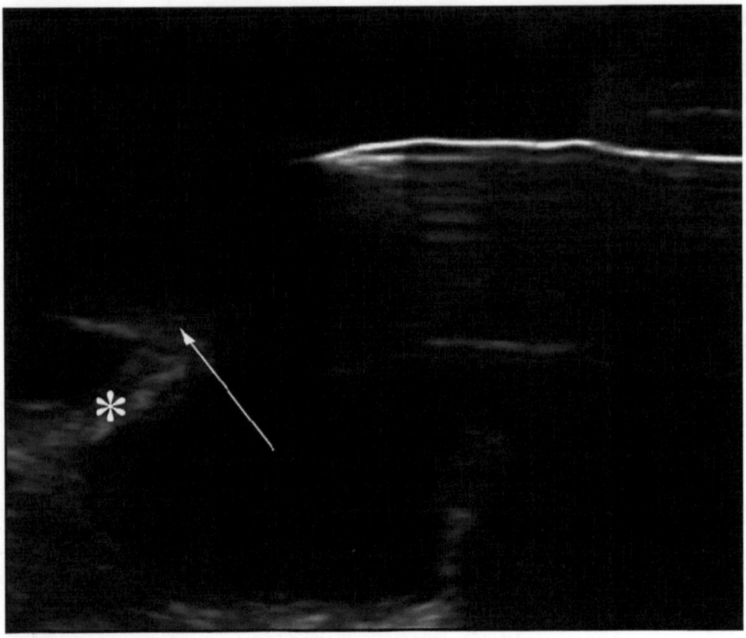

Fig. 14.13 Biceps brachial tenosynovitis. "In-plane," transversal scanning approach. *Arrow* biceps tendon, * mesotendon

Injection Technique and Positioning After skin and probe disinfection, the transducer will be placed in transversal scanning position and the needle will be inserted "in plane" at a 1–2.5-cm distance from the probe margin, with an angulation of 45°. The needle will penetrate the skin, subcutaneous tissue, the brachioradial muscle (located lateral and superficial to the brachial muscle) or the pronator muscle (located medial and superficial to the brachial muscle), depending on the puncture side, the brachial muscle, and lastly the joint capsule. The capsule penetration will generate a degree of resistance depending on its thickness and fibrotic changes. The needle tip will be visualized afterwards inside the articular space and aspiration and/or drug administration can be carried out under continuous visual control. The drug will be identified as a hyperechoic mass with antigravitational movement accompanied by capsule distension (Fig. 14.14). There are some limitations of the technique in cases of the reduction of the elbow extension due to chronic inflammatory pathology.

The posterior approach is the preferred due to a more comfortable position for both patient and physician and because of the lack of important neurovascular structures in the proximity, except the ulnar nerve. Its position in the proximal cubital channel must be checked before invasive maneuver commencement. The patient is seated, positioned for posterior elbow compartment scanning (Fig. 14.15). After olecranon recess visualization (transversal view), needle insertion is made profound to the triceps muscle tendon, angulation of 90°.

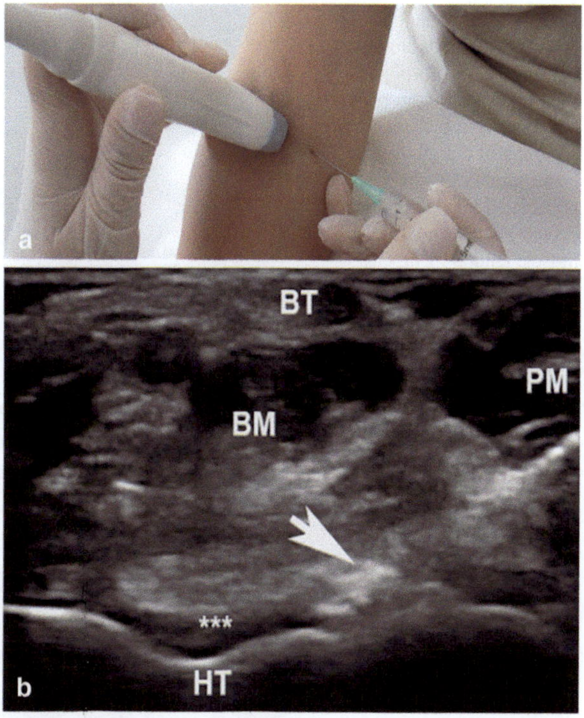

Fig. 14.14 **a** Patient and probe position for elbow joint puncture, "in-plane," transversal scanning, anterior approach. **b** Transversal scanning of the elbow joint—the needle tip is depicted inside the joint. *HT* humeral trochlea, *BM* brachial muscle, *PM* pronator muscle, *BT* biceps tendon, ** joint space, *arrow* needle

Olecranon Bursa

Pathological inflammatory/hemorrhagic affection of the olecranon bursa is a relatively common finding in patients with gout, RA, spondylarthritis or septic/traumatic conditions. The bursal puncture is easy to be performed because of its superficial location. The blind puncture is highly successful but in cases with synovial hypertrophy or blood clots, guided puncture is much more precise (Fig. 14.16). There is total freedom in needle insertion position "in and out of plane" which, in this case, depends more on the physician's comfortable position. After bursal content aspiration, in selected cases, CS deposition can be immediately made through the same needle [36, 41].

Lateral and Medial Enthesopathy

Lateral and medial enthesopathy is a common pathology, frequently seen in clinical practice. Several invasive techniques have been described tailored to the underlying pathology. The first treatment option is to deposit the CS outside the enthesis, at the level of the interface with the subcutaneous tissue. In this approach, enthesis needle injury and intratendinous drug injection are avoided (Fig. 14.17). The second option is the intralesional drug deposition. This technique is more specific for sport medi-

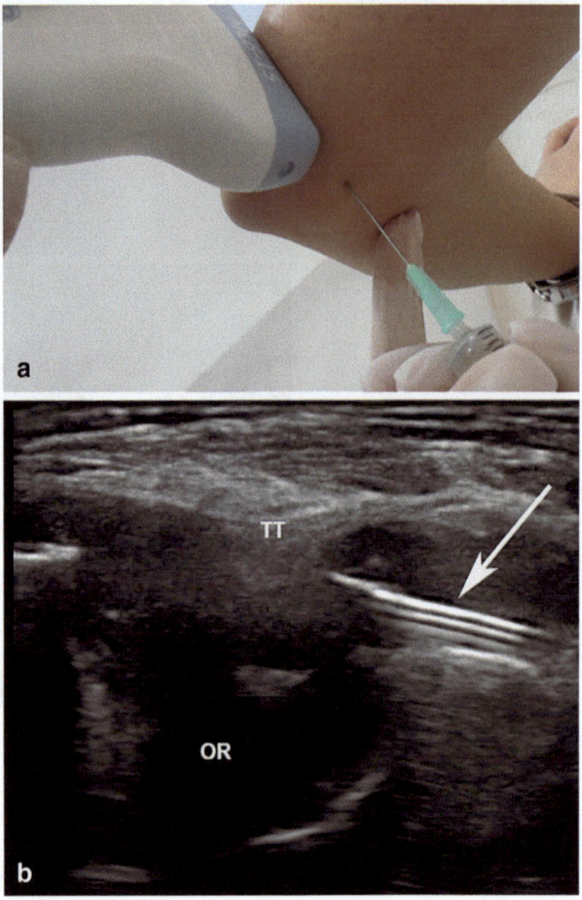

Fig. 14.15 **a** Patient and probe position for elbow joint punction, "in-plane," transversal scanning, posterior approach. **b** Transversal scanning of the elbow joint, posterior recess—the needle is depicted inside the joint (*arrow*) below the triceps muscle tendon (TT), OR olecranon recess

cine injuries and uses viscoelastic drugs. A third option refers to the dry needling technique with repetitive intralesional needle insertion which helps in stimulation of neovascularization and acceleration of the reparative process [42–44].

Injection Technique and Positioning The patient is sitting with the forearms placed on a table or pillow, in pronation (for lateral common extensor tendons enthesis) or supination position (for medial common flexor tendons enthesis). The enthesis is identified in a longitudinal or transversal approach and a short needle is inserted at a 0.5–1-cm distance with an angulation of 45° close to the interface with the subcutaneous tissue or at intralesional level as previously described (when hockey stick probe is used, needle angulation can be kept at 90°). Drug injection is made under continuous real-time visual control.

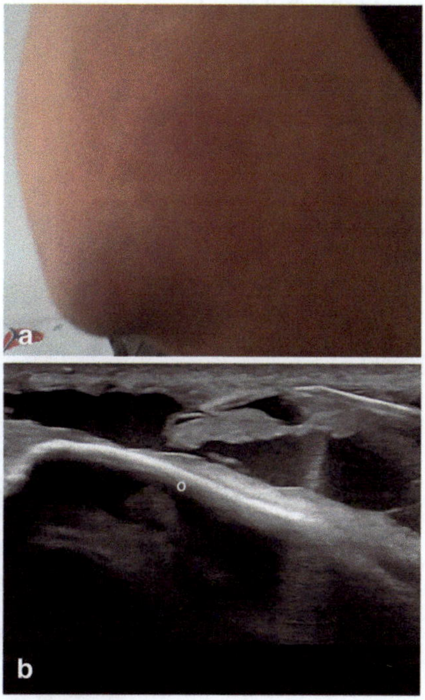

Fig. 14.16 **a** Olecranon bursitis in a patient with gout. **b** Posterior aspect of the elbow, *O* olecranon. The needle is identified inside the distended olecranon bursa

Wrist and Hand Region

Wrist and hand joints as well as periarticular structures may sustain inflammatory and/or degenerative lesions in several pathologic conditions as RA, SpA, microcrystalline arthropathies, OA or secondary to mechanical/ traumatic strain, etc.

Injection Technique and Positioning Interventional maneuvers in the wrist region require the following position: the patient sitting on a chair, facing the performing physician, with the hands on a table, in neutral position.

Radiocarpal, Intercarpal Joints

For radio-carpal joint interventions, the most convenient approach is in transversal dorsal scanning position, distal to Lister's tubercle. After region and probe disinfection, the needle penetrates at a distance of 0.5–1 cm from the probe margin, with an angulation of 60–45°. After capsule puncture, aspiration and/or drug deposition can be safely made under continuous visual control. The accuracy rate in injecting this area is high (93.5 %). The intra-articular drug is further seen as a hyperechoic cloudy mass moving antigravitational in the capsule proximity (Fig. 14.18). For the intercarpal joints, we recommend the scafo-lunate approach. The technique is similar with that described earlier, with the position of the probe more distally [44–47].

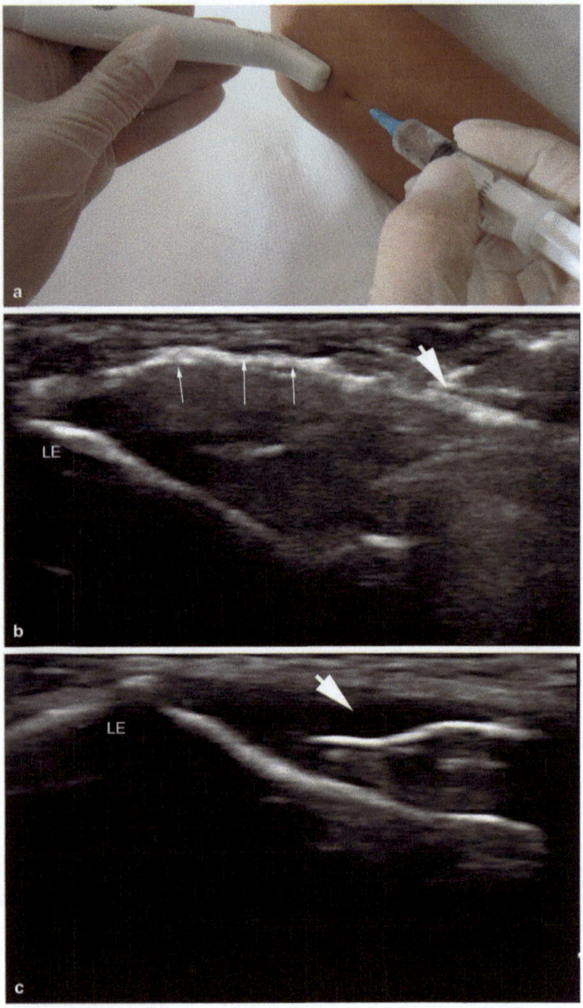

Fig. 14.17 **a** Patient and transducer position—longitudinal scanning of the common extensor tendons at the level of the lateral epicondyle. **b** Longitudinal scanning of the common extensor tendons; *LE* lateral epicondyle, needle penetrating with angulation of 45° (*arrow head*), corticosteroid drug deposition at the interface between the common enthesis and subcutaneous tissue (*small arrows*). **c** Hyaluronic acid drug deposition inside the enthesis lesion (*arrow head*), *LE* lateral epicondyle

MCP, PIP, and DIP Joints

Metacarpophalangeal (MCP), proximal interphalangeal (PIP), and distal interphalangeal (DIP) joints are injected more easily in a longitudinal approach. In order to avoid tendon perforation, an "out-of-plane" needle insertion is recommended (Fig. 14.19).

Wrist and Hand Tendons

Tenosynovitis is another good example for MSUS-guided injections with high accuracy of drug deposition in comparison to the blind technique [48–50]. Wrist and

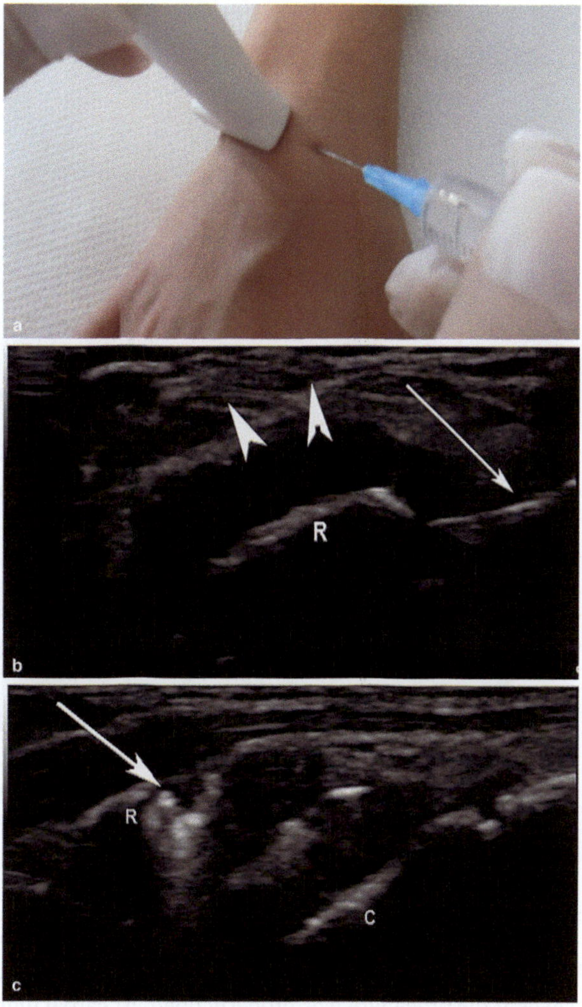

Fig. 14.18 **a** Patient and transducer position—transversal scanning of the radio-carpal joint. **b** Transversal scanning of the radio-carpal joint; *R* radius, capsular distension (*arrow heads*), needle (*arrow*). **c** Postprocedural longitudinal scanning of the radio-carpal joint; *R* radius, *C* carpal bone, corticosteroid drug deposition is identified intra-articular (*arrow*)

hand tendons are superficial structures and therefore can be approached easily. Medication deposition inside the tendons sheath follows the same technique. The tendons' scanning and injecting position depends strictly on the performing physician's choice and can be done in transversal or longitudinal view. The "in-plane" approach is usually advised as it offers continuous real-time visualization of the needle tip penetration. At wrist level, pathology can be encountered on either dorsal (extensor compartments I–VI) or volar aspects (carpal tunnel tendons—digit flexors I–V, flexor carpi radialis, flexor carpi ulnaris). At the digits level, the inflammatory changes presents usually on the volar side (superficial and profound flexor tendons) [51]. For digits, needles are preferred to be shorter and thinner.

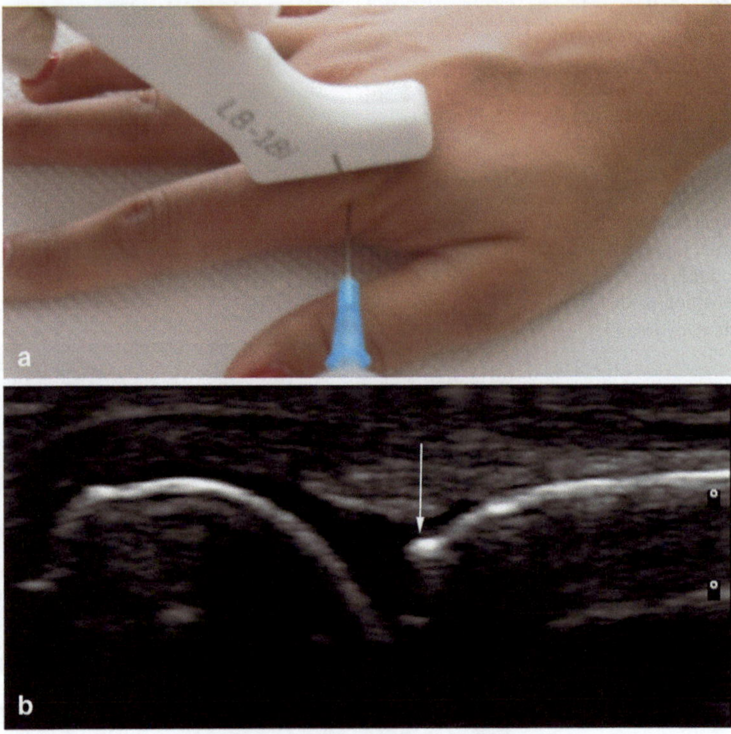

Fig. 14.19 **a** Patient and transducer position for MCP joint injection—longitudinal joint scanning and "out-of-plane" needle penetration. **b** Longitudinal scanning of the second MCP joint; the needle is visualized as a hyperechoic dot inside the joint (*arrow*). MCP metacarpophalangeal

Injection Technique and Positioning The hands will be placed in a comfortable, neutral position, on a table according to the area of interest (Figs. 14.20 and 14.21). Immediately after drug deposition, it is possible to scan again the target region and check if the procedure was properly done. At carpal tunnel level, CS deposition will be made in the proximity of the median nerve or, in case of flexor tendons tenosynovitis, anywhere inside the carpal tunnel. The needle will penetrate the disinfected skin at a 0.5–1-cm distance, angulation of 60–45°, avoiding neurovascular or tendon structures, reaching the target area (Fig. 14.22), [52, 53].

The Hip Region

Hip joint and periarticular hip structures may be affected in chronic inflammatory (RA, SpA, microcrystalline arthropathies), degenerative, congenital, and traumatic conditions (primary, secondary OA).

The hip joint is a deep organ and clinical examination (CE)-delivered information is limited. MSUS is useful not only for hip pathology evaluation but also for

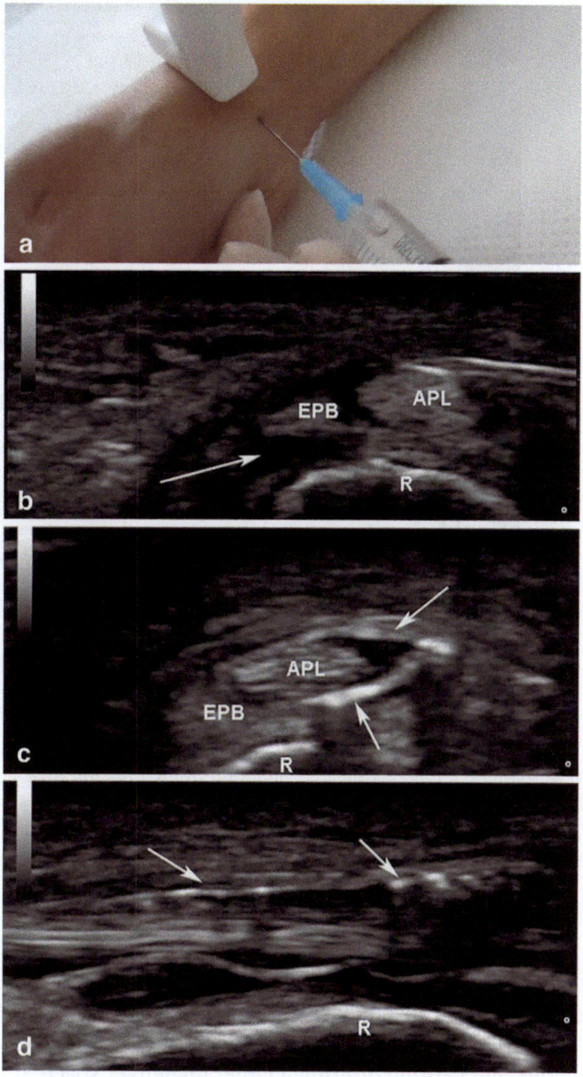

Fig. 14.20 **a** Patient and transducer position for extensors compartment I injection at the wrist level. **b** Transversal scanning at the level of the first extensors compartment (*APL* abductor pollicis longus, *EPB* extensor pollicis brevis), hypoechoic material distending the tendons sheath (*arrow*), *R* radius; the needle is identified as a hyperechoic line inside the tendons sheath. **c** Postprocedural corticosteroid drug deposition inside the tendons sheath (transversal scanning, *arrows*). **d** Postprocedural corticosteroid drug deposition inside the tendons sheath (longitudinal scanning, *arrows*)

guiding interventional maneuvers both for the hip joint itself and the surrounding bursae. Procedures can be carried out for diagnostic aspiration (microscopy—culture and Gram stain) or therapeutic injections [54–66]. The current OARSI and ASAS/ EULAR therapeutic guidelines include hip CS injection, which, unfortunately, is a rare therapeutic approach as standardized criteria for patients selection are still missing and the number of performing physicians with experience in the technique remains small [70]. The use of intra-articular hip viscoelastic drugs is an encouraging option but still under study because of the current low trial number [67–70]. MSUS-guided technique allows a correct needle penetration (avoiding neurovascular structures injury), joint aspiration and a more precise drug deposition which assures a higher efficacy and better outcome [59, 62, 70–72].

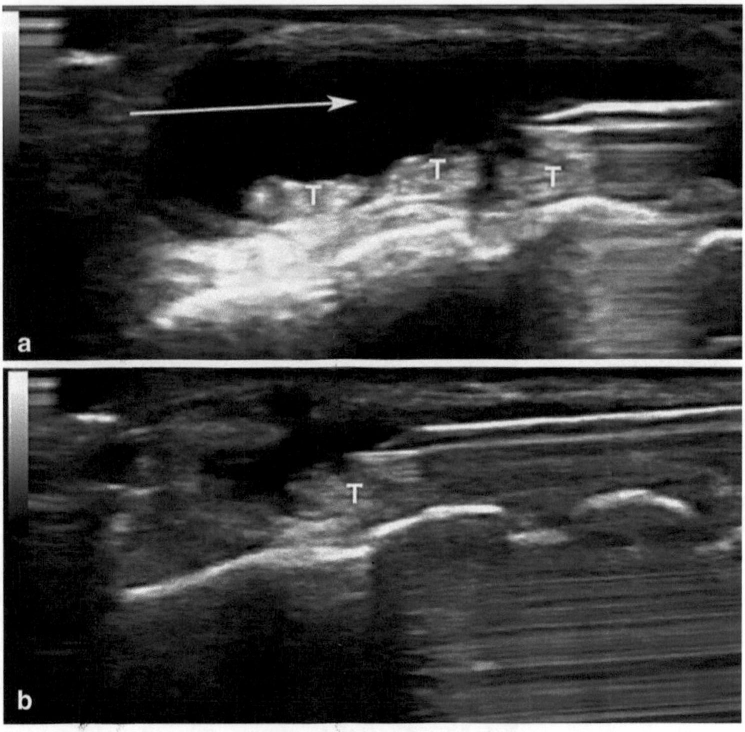

Fig. 14.21 **a** Transversal scanning at the level of extensors compartment IV (wrist). The needle is penetrating the tendons sheath, aspiration of the effusion (*arrow*) follows; *T* extensor tendons. **b** Aspiration of the effusion almost finalized

Hip Joint

Injection Technique and Positioning The first step for injection preparation is the positioning of the patient lying on the back, heels together, and leg in external rotation—the same position as for anterior hip examination—or with the leg in internal rotation, maneuver which will increase the distance between the bony cortex and capsule offering a larger maneuver space. It is recommended to use low-frequency probes—linear (3.5–7.5 MHz) or convex (2.5–5 MHz) for muscular or obese patients. After skin and probe disinfection, the probe is placed for the longitudinal anterior approach and visualization of the anterior joint recess. The needle is inserted at a distance of 2 cm, "in plane," in an inferior or superior approach, 45° angulation [55, 66]. It is recommended to use spinal puncture needles with different calibers 18–20 gauge for aspiration and viscoelastic drugs and thinner (22–24 gauge) for CS, anesthetic or saline injection. For better visualization, in order to increase the reflectivity, some performing physicians scratch the needle surface with a sterile scalpel and keep the mandrin inside it while inserting [55].

After skin penetration, needle advancement is followed penetrating the rectus femoris (superficial) and iliopsoas muscles (profound), the joint capsule, directing the tip to the head–neck junction. After bony cortex contact, needle retraction for

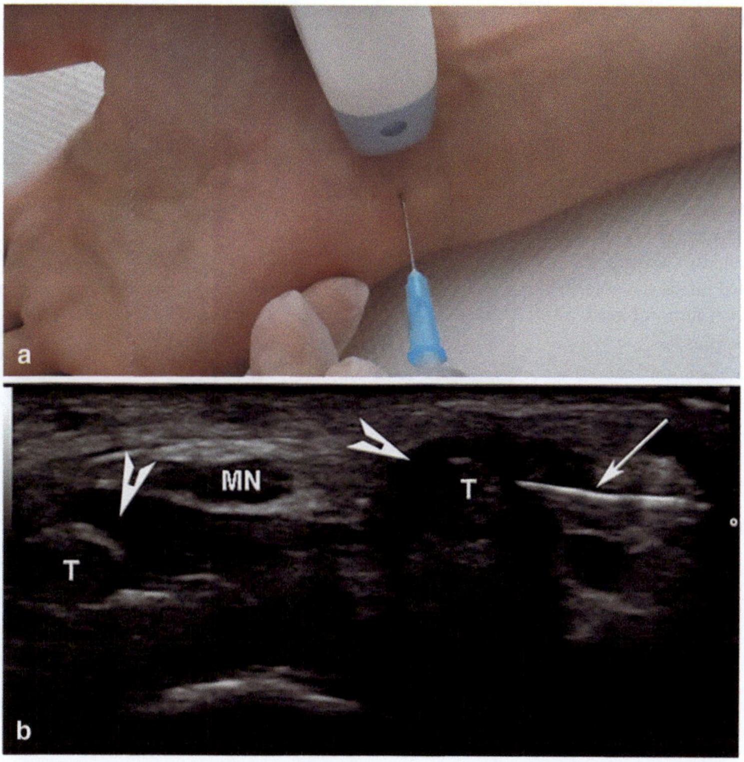

Fig. 14.22 **a** Patient and transducer position for carpal tunnel injection. **b** Transversal scanning at the level of the carpal tunnel; the needle is visualized as a hyperechoic line (*arrow*) penetrating close to a tendon (T), hypoechoic material is depicted around the tendons (effusion, *arrowheads*), *MN* median nerve

2–3 mm is mandatory in order to avoid the engagement of the tip inside the posterior synovial layer and facilitate its placement inside the anterior recess. Real-time visualization of the articular recess is possible during the aspiration process/ drug injection. Post-procedural drug distribution can be followed afterwards in multiplanar scanning sections (Fig. 14.23). Side effects due to intra-articular injection occur rarely if antiseptic rules are strictly respected [14, 69, 70, 73].

Iliopsoas Bursitis

Iliopsoas bursa is located deep to the iliopsoas muscle, in immediate contact with the capsule. It communicates with the hip joint and may be occupied by inflammatory (septic/ aseptic) content in several pathological conditions. The MSUS-guided injection follows the same steps as for the hip joint puncture. The needle tip will penetrate the skin, subcutaneous tissue, rectus femoris, and iliopsoas muscles and will enter the distended bursa. Sometimes it is necessary to puncture both joint and bursa for efficient aspiration [24].

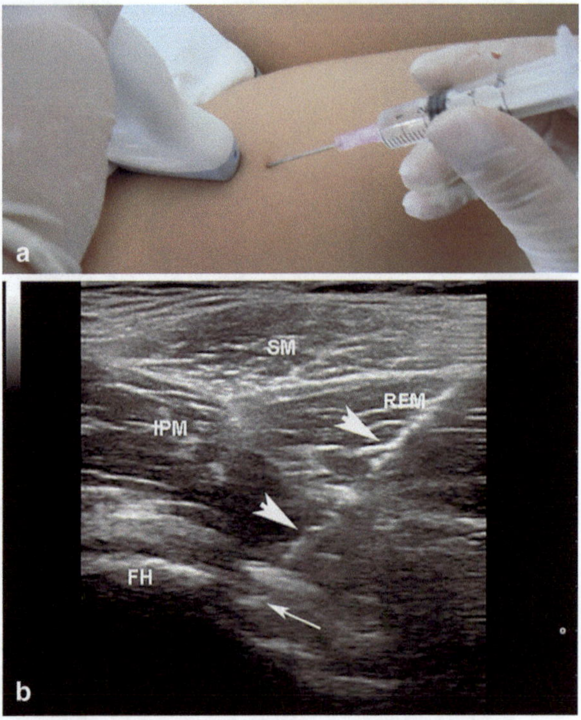

Fig. 14.23 **a** Patient and transductor position for "free-hand" technique injection of the hip joint. **b** Longitudinal scanning of the hip joint. The needle (*arrow heads*) is penetrating in angulation of 45° the rectus femoris (RFM) and iliopsoas (IPM) muscles, the needle tip (*arrow*) is identified inside the anterior recess; *FH* femoral head, *SM* sartorius muscle

The Trochanteric Bursae

Trochanteric bursitis occurs in inflammatory conditions (SpA, microcrystalline arthropathy), after trauma or friction, and is clinically highly symptomatic. Therefore, immediate therapeutic approach is needed. The patient is lying on the contralateral side and after trochanteric area scanning, a longitudinal "in-plane" injection approach is usually chosen. The bursae are superficial and the needle is inserted easily at a 1–2-cm distance from the proximal end of the transducer with angulation between 90° and 45° determined by the amount of subcutaneous tissue mass. After aspiration, in selected cases, CS deposition can be made immediately (Fig. 14.24), [74, 75].

The Ischiatic Bursa

Ischiatic bursitis occurs rare in clinical practice but usually it is highly symptomatic. The patient is in supine, flexed hip and knee joints. The optimal scanning plane is chosen and needle insertion is made with an angulation of 90–45°. After content aspiration, CS deposition may follow [75].

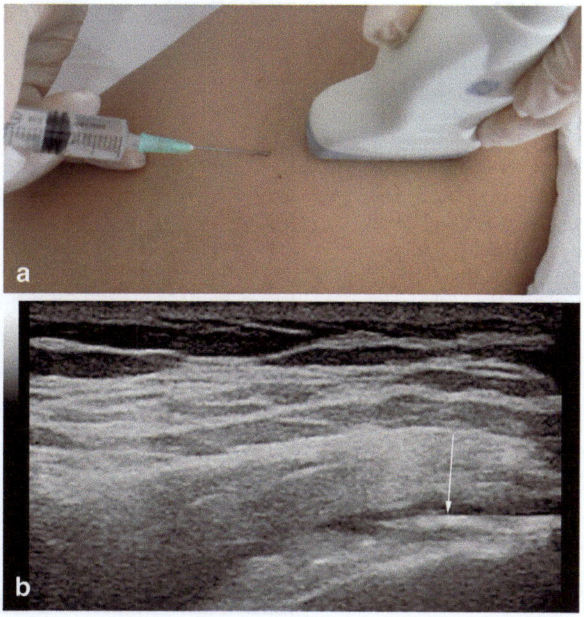

Fig. 14.24 **a** Patient and transducer position for "free-hand" technique injection of the trochanteric bursa. **b** Longitudinal scanning position at the level of the greater trochanter. The needle (*arrow*) is depicted inside the trochanteric bursa

The Knee Region

The Femoro-Tibial Joint and Suprapatellar Bursa

Knee joint effusion is very frequently encountered in clinical practice in all kind of inflammatory or degenerative pathologies and there is a lot of experience in practicing blind knee puncture for aspiration and/or drug deposition. Despite this, recent studies show low accuracy in case of blind joint needle penetration [9, 45, 76–79]. Indeed, viscoelastic or intra-articular drug administration in a "dry knee" may create problems when blindly performed. Independently of the goal of the puncture, nowadays, MSUS-guided knee injection is preferred because it is more precise, the amount of aspirating fluid is superior to blind maneuvers and the procedures are less painful [9, 80].

Injection Technique and Positioning The patient is lying with the knee in full extension/slightly flexed or sitting with a knee flexion at 90°. After the probe and skin area disinfection, the probe can be placed in transversal view at the level of the suprapatellar bursa or parapatellar joint recesses for joint aspiration of moderate to big effusions and at the level of the parapatellar recesses for mild effusion or viscoelastic drug deposition. The needle penetrates the skin at a distance of 2–2.5 cm, angulation of 90°. The accuracy for the lateral approach is usually high [24, 81]. In the case of skin pathology a different, unconventional approach, can be chosen (Figs. 14.25 and 14.26).

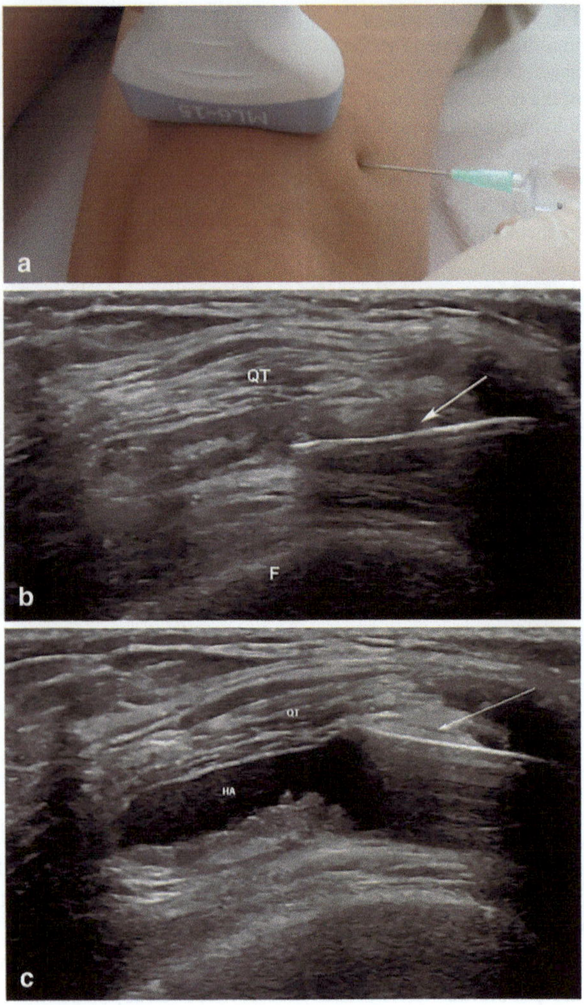

Fig. 14.25 **a** Patient and probe position for knee injection (superior joint recess). **b** Free-hand injection technique— viscosupplementation drug deposition in a "dry knee"; needle penetration (*arrow*), *QT* quadriceps tendon, *F* femoral cortex. **c** Postprocedural image—drug deposition (HA) identified as a hypoechoic mass profound to the quadriceps tendon

Knee Tendons

Quadricipital, patellar, femoral biceps or iliotibial band tendinopathy are rare events outside the world of sport medicine. The injection target area is represented by the peritendinous area, especially when peritendon pathology accompanies the tendon inflammation. The scanning section can be chosen in relation with the best target lesion visualization and the most comfortable procedure position [24, 82].

The Superficial Patellar Bursa

The superficial patellar bursitis (septic, mechanical, trauma) is recommended to be punctured rapidly. Being a superficial structure, the access to it is very easy and can

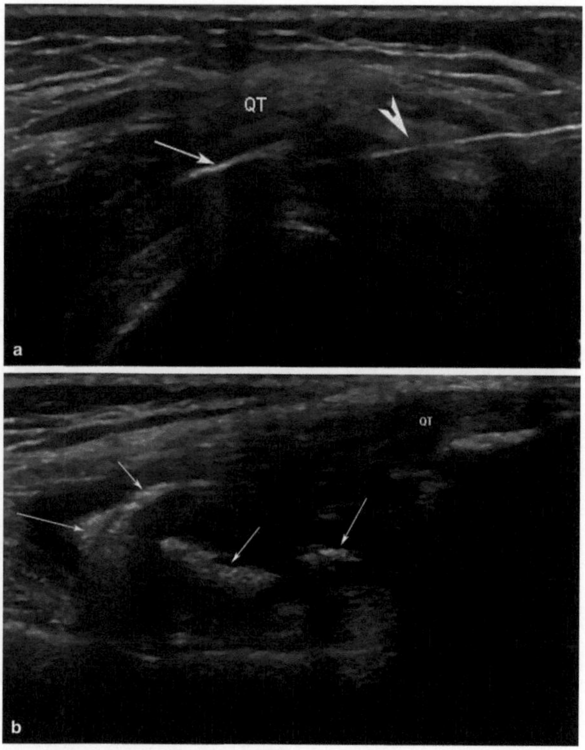

Fig. 14.26 **a** Free-hand technique—puncture of the suprapatellar bursa of the knee, transversal approach. Needle (*arrow head*) is penetrating the bursa profound to the quadriceps tendon (QT); deposition of the corticosteroid drug (*arrow*) is seen inside the bursa. **b** Postprocedural longitudinal scanning technique of the suprapatellar compartment of the knee; corticosteroid drug (*arrows*) is depicted inside the bursa

be done in any scanning plane. Keeping the security distance between needle and probe, content aspiration/ drug deposition can be continuously monitored.

The Profound Infrapatellar Bursitis

The approach is made in transversal plane, deep to the transducers footprint in order to avoid the injection of the patellar tendon, which is relatively a thick structure. The needle is inserted "in plane," with 90° angulation (Fig. 14.27).

Popliteal Cysts

Popliteal cysts (Baker's cyst, other popliteal cysts) generate pain in the posterior compartment of the knee, limitation of knee flexion, or compression of the popliteal neurovascular structures. The cyst content can be represented by fluid and/or synovial hypertrophy and accumulation under pressure may lead to cysts rupture and calf musculature dissection. Therefore, aspiration of the cyst content is usually advised, both for local decompression and fluid microscopy analysis. CS deposition is useful in case of local synovial hypertrophy. It is a very simple interventional maneuver which helps to avoid the surgical intervention most of the times.

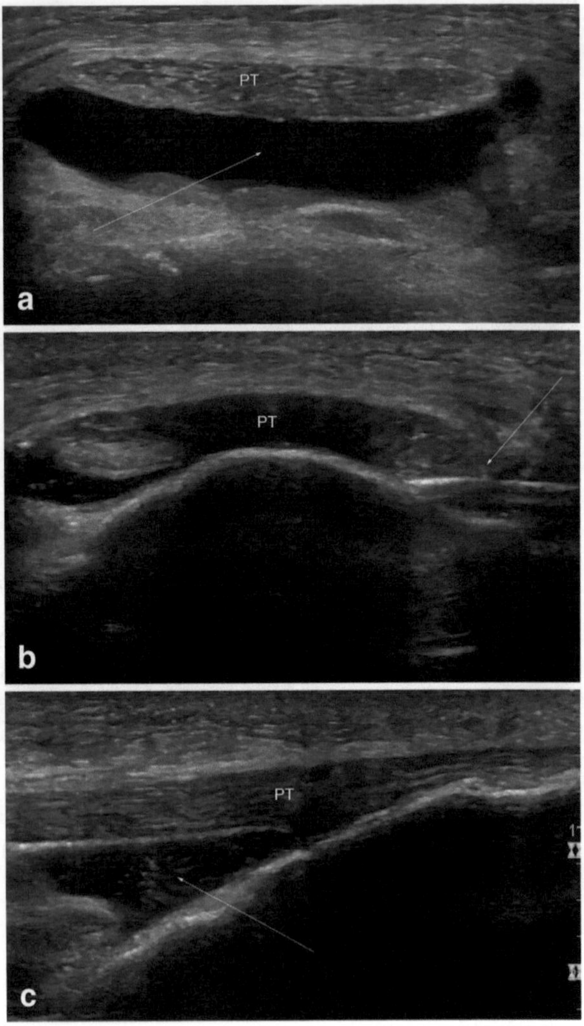

Fig. 14.27 **a** Transversal scanning at the level of the infrapatellar distended (*arrow*) bursa; *PT* patellar tendon. **b** Transversal scanning of the infrapatellar bursa (*arrow*), "free-hand" puncture technique, needle (*arrow*) is penetrating inside the bursa—partial aspiration of the bursal effusion; *PT* patellar tendon. **c** Postprocedural longitudinal scanning of the patellar tendon (PT) and bursa filled up with corticosteroid drug, hypoechoic mass with crystal aggregates inside

Injection Technique and Positioning The patient is in supine position, the transducer is placed at the level of the popliteal fossa and after cyst identification the needle is inserted with an angulation of 45°. It is recommended to puncture the skin at the distal level of the cyst in order to evacuate a maximal quantity of fluid. Needle repositioning is sometimes necessary in cases with multilocular cysts or in the case of fibrin or blood clots obliterating the needle tip. Occasionally, aspiration of small cystic spaces may be carried out, in particular when sepsis or crystal presence require confirmation (Fig. 14.28) [8, 24, 83].

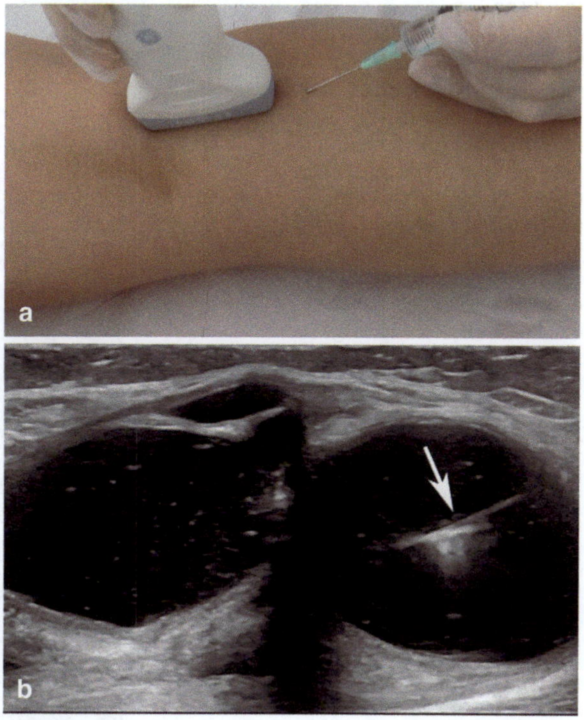

Fig. 14.28 **a** Patient and probe position for puncturing a popliteal cyst, "in-plane," longitudinal approach. **b** Longitudinal approach; the needle (*arrow*) is depicted inside the cyst

The Ankle and Foot Region

The ankle and foot regions are complex anatomical areas with multiple stratified layers such as joints, periarticular musculous, tendinous, ligamentar, and vasculo-nervous structures. Inflammatory, degenerative, as well as traumatic pathology may spontaneously coexist which may lead to the affection of a single or more structures at the same time. Preinterventional evaluation include a rigorous, systematic scanning protocol of the tibiotalar, subtalar, intertarsal, and MTP joints as well as of the anterior, medial, lateral, and posterior compartment.

Tibiotalar and Subtalar Joints

Injection Technique and Positioning The tibiotalar joint is the biggest joint in the ankle region and inflammatory or degenerative lesions may occur in RA, SpA, microcrystalline arthropathies, OA, and after traumatic events. The patient lies down with flexed knees, soles on the bed and with a slightly ankle extension. The joint puncture can be made in longitudinal or transversal approach, the choice depending more on the physician's experience. In longitudinal view, it is important to avoid the injection of the tendons, neuro-vascular structures located superficially to the joint. Therefore, it is recommended to use "out of plane" injection technique. Still, there is a certain risk to inject the tendons belly because of their tight location

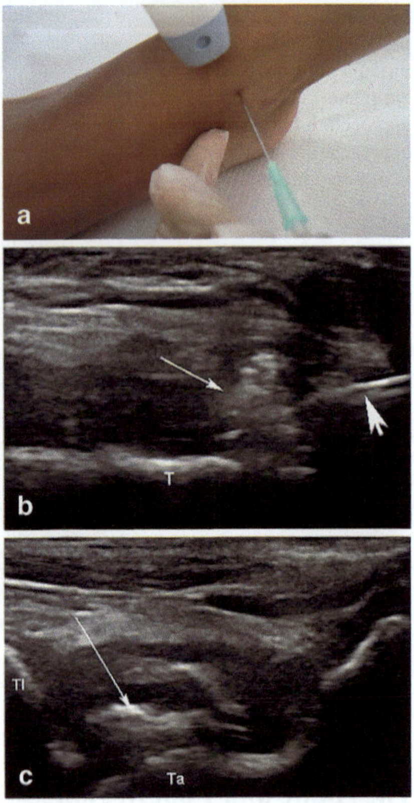

Fig. 14.29 **a** Patient and probe position for tibiotalar joint puncture. **b** Transversal scanning of the tibiotalar joint, T-talus, the needle is identified as a hyperechoic line (*arrow head*) and corticosteroid drug is depicted as a hyperechoic mass inside the joint (*arrow*). **c** Postprocedural longitudinal scanning of the tibiotalar joint, Ti-tibia, Ta-talus, corticosteroid drug inside the joint (*arrow*)

under the extensor retinaculum. The approach can be longitudinal- inferior or longitudinal superior with an angulation of 45°. In transversal view, the needle penetrates deep to the anterior tendons compartment, angulation 90°, "in plane" approach. Adopting this approach will help to avoid the tendons and neuro-vascular structures injury (Fig. 14.29), [24, 84, 85]. Other subtalar, intertarsal joints can be injected in the same way by moving the transducer distally (Fig. 14.30). MSUS-guided interventional maneuvers at the tibiotalar and subtalar joints have high accuracy [45, 86, 87].

Ankle Tendon Compartments

Injection Technique and Positioning Tenosynovitis at the anterior, medial, or lateral compartments of the ankle joint is a very frequent pathology. The technique is the same for all compartments, transversal or longitudinal scanning view, just the patient position varies. For the anterior compartment (tibialis anterior, extensor halucis longus, extensor digitorum longus tendons) we keep the same position as previously described for the tibiotalar joint. For the medial compartment (tibialis posterior tendon, flexor digitorum longus, flexor halucis longus tendons), the

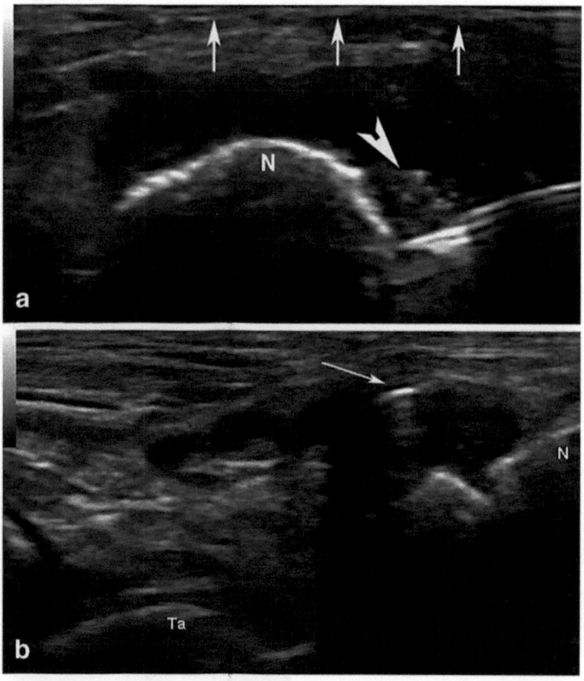

Fig. 14.30 **a** Talo-navicular joint puncture, transversal scanning approach; *N* navicular bone cortex. Hypoechoic material is distending the joint capsule (*arrows*); the needle is identified as a hyperechoic line; corticosteroid drug is visible as a hyperechoic mass at the needle type (*arrow head*). **b** Postprocedural longitudinal scanning of the talo-navicular joint, Ta-talus bone; corticosteroid drug is depicted inside the joint (*arrow*) with antigravitational movement

patient will be placed in "frog leg" position, whereas for the lateral compartment the leg will be in pronation with slight ankle extension and the sole on the bed (Fig. 14.31), [88].

The posterior compartment contains the Achille's tendon, the deep pre-Achillean and superficial retro-Achillean bursae, and the posterior aspect of the talo-calcaneal joint. The patient lies in supine position, with the feet hanging over the beds margin. For the tendon and bursae injection, it is recommended to use the transversal view. Pre-Achillian bursa is the main target for injections and needle insertion is made deep to the tendon, 90° angulation, "in plane" technique. Aspiration of the bursal content and/ or CS deposition is easily made without noticing any resistance (Fig. 14.32) [9, 15, 24, 88].

Talo-calcaneal joint can be visualized in the longitudinal view. After depth adjustment the needle is inserted "out of plane" at approximately 3 cm deep to the transducers footprint. The needle will be visualized as a hyperechoic dot inside the joint [89].

Intratendinous viscoelastic drug injection is used as an alternative therapeutic option mainly in sport medicine. Preinterventional MSUS tendon assessment identifies the degenerated tendon areas and needle is inserted in a longitudinal approach, 45° angulation, directly inside the lesion. Drug deposition and distribution inside the tendon is followed in real time. Follow up control is recommended to be made in approximately 3 weeks. In case of tendonitis associated with peritenonitis (SpA), peritendinous CS or biologic therapy injection at the interface of the tendon with the subcutaneous tissue can be administrated [24, 88].

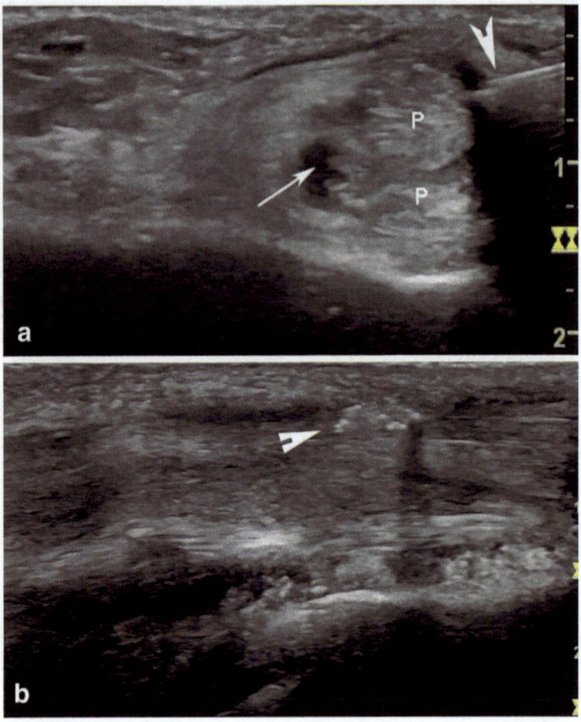

Fig. 14.31 **a** Transversal scanning at the level at the lateral ankle tendons compartment, *P* peroneus brevis and longus tendons; the needle (*arrowhead*) is penetrating inside the tendon sheath distended by effusion (*arrow*). **b** Postprocedural corticosteroid drug deposition (*arrowhead*) depicted as a hyperechoic mass inside the tendons sheath

Plantar Fascia

Injection Technique and Positioning Plantar fasciitis is a very common pathology seen in association with the SpA spectrum. However, it is also commonly reported by patients suffering from degenerative conditions. The therapeutic strategy is based on local CS injections and physiotherapy procedures. The patient's position is the same as for the posterior ankle compartment assessment but with the dorsal aspect of the foot on the bed in order to stabilize the foot when performing alone. When the injection is performed with feet hanging over the bed edge, a second person must stabilize the foot. The transducer is placed in a longitudinal or transversal view at the level of the plantar fascia insertion; the needle penetrates at 0.5–1 cm distance from the probe, angulation 90–45° up to the interface between the plantar fascia and subcutaneous tissue. Drug deposition is made and migration along the plantar fascia can be visualized in real time (Fig. 14.33), [90, 91]. Some authors recommend needle insertion up to the entheseal lesion followed by intralesional drug deposition [24], whereas others recommend CS deposition profound to the plantar fascia in order to avoid postprocedural fascia rupture [92].

Metatarsophalangeal Joints

Injection Technique and Positioning Inflammatory or degenerative changes in the metatarsophalangeal (MTP) joint may require local aspiration and/or drug depo-

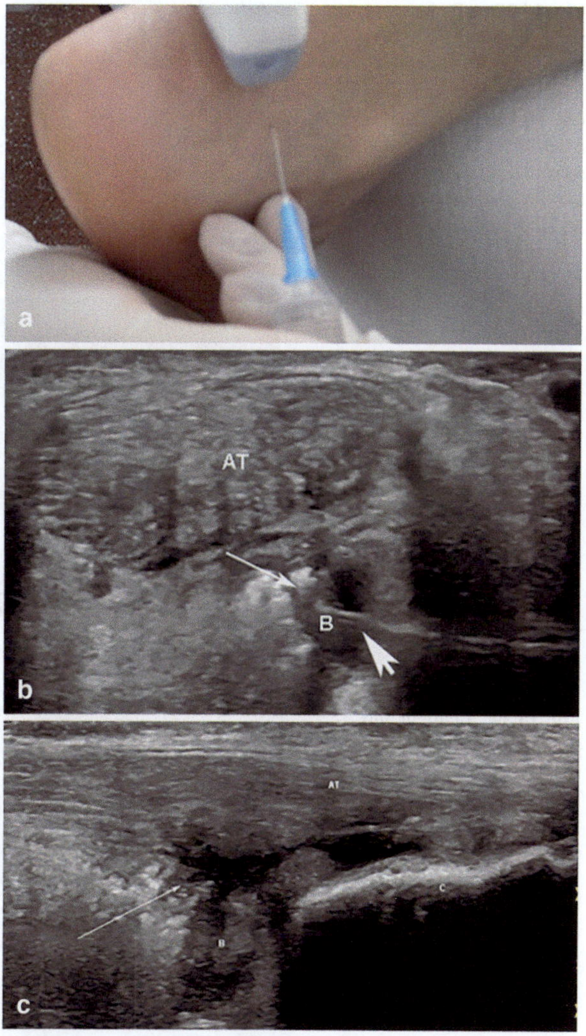

Fig. 14.32 **a** Patient and probe position for Achille's profound bursa injection. **b** Transversal scanning of the Achillian tendon and pre-Achillian bursa; the needle (*arrow head*) is penetrating the bursa; corticosteroid drug deposition (*arrow*) can be visualized inside the bursa (B). **c** Postprocedural longitudinal scanning of the Achille's tendon, pre-Achillian bursa; corticosteroid deposition inside the bursa (*arrow*), *C* calcaneus bone

sition. The patient's position is with the soles on the bed and flexed knees. A longitudinal or transversal scanning section is used to guide the needle inside the joint. In order to avoid the extensor tendon injection, needle insertion is made "out of plane," angulation of 45° until the penetration of the capsule is reached. Aspiration and/ or CS or viscoelastic drugs deposition can be made (Fig. 14.34, Fig. 14.35), [86, 87].

Temporomandibular Joint

Temporomandibular synovitis (RA, juvenile idiopathic arthritis (JIA), SpA) generates pain and functional distress with high impact on the patients' quality of life. Injection of the joint is made with the patient lying supine on the contralateral side

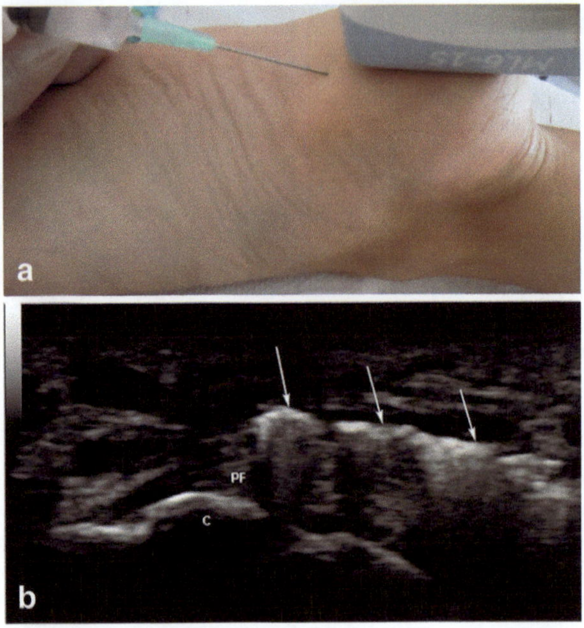

Fig. 14.33 **a** Patient and probe position for injection at the level of the plantar fascia. **b** Longitudinal scanning of the plantar fascia (PF). **c** calcaneus bone, corticosteroid drug deposition at the interface between the PF and the subcutaneous tissue (*arrows*)

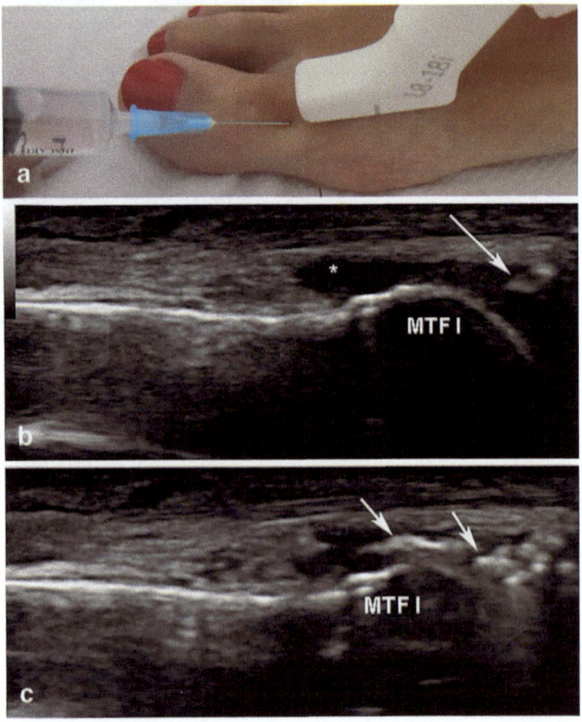

Fig. 14.34 **a** Patient and probe position for metatarsophalangeal I joint (MTP I). **b** Longitudinal approach, "in-plane" injection technique—intra-articular MTP I joint; * intra-articular effusion, needle (*arrow*). **c** Postprocedural corticosteroid drug detection inside the joint (*arrows*)

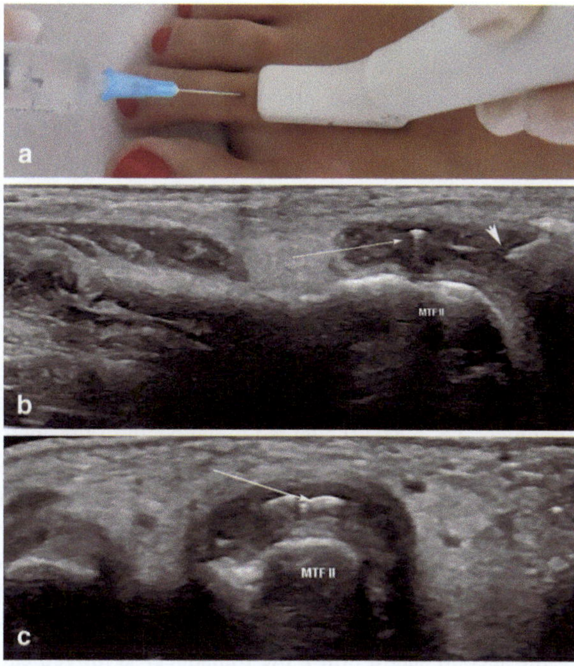

Fig. 14.35 **a** Patient and probe position for metatarsophalangeal II joint (MTPI I). **b** Longitudinal approach, "in-plane" injection technique—intra-articular MTP II joint; needle tip is identified inside the joint (*arrow head*), corticosteroid drug moving antigravitational inside the joint (*arrow*). **c** Postprocedural corticosteroid drug detection inside the joint (*arrow*)—transversal scanning view

with preauricular area exposure (joint's skin projection). After visualizing the joint in longitudinal view, the needle will penetrate the skin, subcutaneous tissue and capsule, 45° angulation and CS will be deposited (Fig. 14.36), [93, 94].

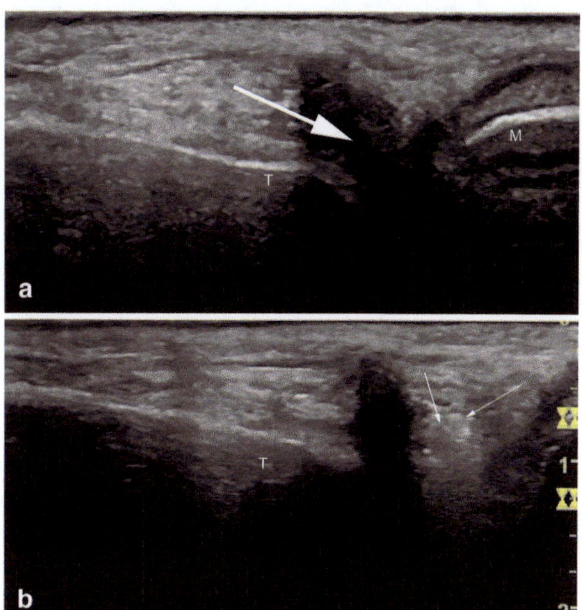

Fig. 14.36 Longitudinal scanning of the temporomandibular joint. *M* mandibular bone, *T* temporal bone; hypoechoic material (*arrow*) is distending the joint recess. Longitudinal scanning of the temporomandibular joint. *M* mandibular bone, *T* temporal bone; corticosteroid drug deposition inside the joint space—hyperechoic mass (*arrows*)

Sacroiliac Joints

Though sacroiliac joints interventional maneuvers gained some popularity recently, yet they have not attracted too many supporters because of: (1) the difficult injection technique (complex bone curves, angulation >45°, almost parallel to the US beam and subsequent needle visualization limits); (2) the contradictory study results regarding the success rate (22 % after computed tomography (CT) verification) and; (3) the long-term-post-injection outcome [95–100]. However, in cases where biologic treatment is not recommended or cannot be used, there is still a practical need to look for treatment alternatives in cases of sacroiliitis.

Injection Technique and Positioning The patient lies in supine position, arms close to the trunk. It is recommended to use lower frequency probes (linear <7 MHz or convex 2.5–5 MHz) adapted to the depth of the anatomical structures. Sacroiliac joint identification can be made at two levels—the superior one when placing the probe with one end on the iliac crest and the other on the L5 vertebra with subsequent caudal movement to the sacral region. The bony landmarks from medial to lateral are: the medial sacral crest, the lateral sacral crest, and the iliac bone. The first sacral hole is identified between the sacral crests. The sacroiliac joint is located between the lateral sacral crest and iliac bone. For the inferior level (at the level of the second sacral hole), the transducer is moved caudally. The joint orientation is vertical and anterior medial [95]. The superior compartment is predominantly fibrotic and the inferior one is synovial. Strong ligaments and capsule stabilize the joint.

The needle insertion technique is obligatory "free hand" and the sinusoidal, complex bony cortex may need frequent needle reorientation in order to penetrate the joint. The paraxial approach with angulation close to 0° creates difficulties in needle visualization. After skin penetration, needle advancement is made for approximately 1 cm to perforate the joint capsule. Drug deposition can be made under continuous visual control.

Anesthetic Nerve Blocks

Anesthetic troncular nerve blocks are used for regional anesthesia, especially in limbs for surgical purposes, and are accompanied by very low anesthetic risk and substantially low costs [14, 101, 102]. Other applications are the pain management in oncologic pathology with local neural invasion or in different nerve entrapment syndromes.

The identification of the nerve and proximity structures is possible by using MSUS, and guided perineural injection can be easily made with high precision and efficacy. It is recommended to use high-frequency probes and the same injection technique as for approaching the tendon sheaths. Depiction of the nerve structure will be made in transversal view and perineural drug deposition followed by the

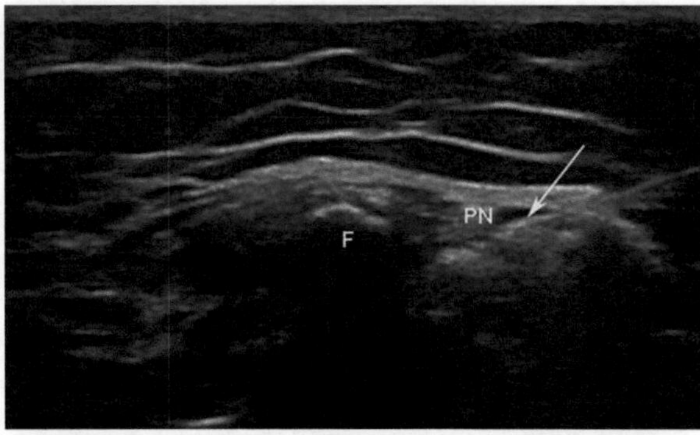

Fig. 14.37 Transversal scanning of the peroneal nerve (PN) at the level of the fibular head (F); needle tip is identified perineural (*arrow*)

formation of hypoechoic perineural hallo is visualized. Afterwards, the distribution along the nerve structure can be demonstrated. The method has a superior efficacy in comparison to the classic anatomic landmarks method or electroneurographic guiding method (Fig. 14.37), [101, 102].

Aspiration and Drainage of Different Fluid Collections, Abscesses

In the last years, aspiration and drainage of different cysts, collections, or abscesses were carried using ultrasonography as guiding tool. Apart from local decompression, it is important to collect biological material for different laboratory analysis—cytologic, bacteriologic, and crystals examination. The target area is identified as an anechoic/hypoechoic mass, sometimes with inhomogeneous floating content, with posterior signal enhancement. Puncture can be rapidly and easily made, avoiding the injury of neurovascular or tendinous structures. The needle (higher caliber, 16–20 gauge) is inserted under continuous real-time visual control, redirected if necessary when fibrin or blood clots or hypertrophic synovial tissue obstruct the tip. Depending on the local situation, anesthesia may be required along with catheter or drainage tubes' insertion, assuring supplementary drainage, in particular in case of big hematomas, where preventing relapses is highly recommended (Figs. 14.38 and 14.39), [24, 103].

MSUS-Guided Biopsy

Biopsy plays an important diagnostic role in various pathologic conditions. Recently, percutaneous US-guided biopsies gained high popularity due to the reduction of the procedure complexity, maneuver as well as recovery time, risks as well as costs. Its use was extended, in the last years, from tumor soft tissue and bony pathologies

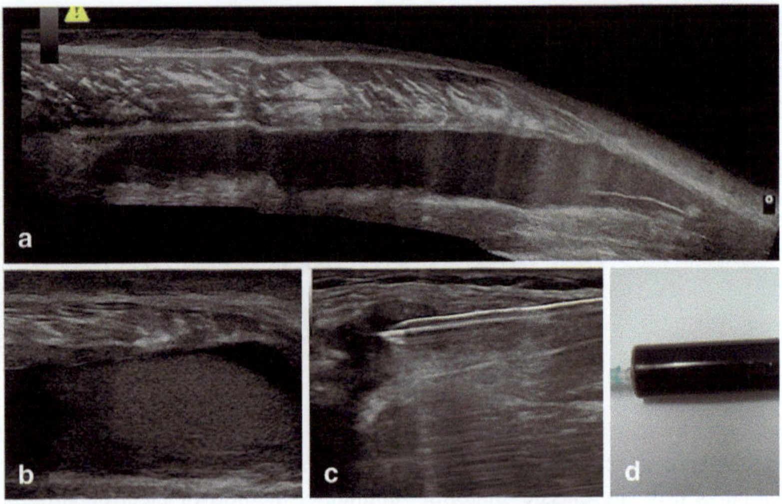

Fig. 14.38 **a** Calf hematoma visualized as a hypoechoic mass profound to muscle fibers: longitudinal scanning view. **b** transversal scanning view. **c** After hematoma aspiration. The needle is identified as a hyperechoic linear structure. **c** Aspect of the aspired fluid (blood)

to synovial tissue analysis too. The primary purpose is to obtain sufficient tissue for laboratory analysis and to directly visualize the sampling process with a nonradiant, less traumatizing, sensitive, high-resolution imaging technique [14, 104–106]. Apart from its diagnostic role in elucidating some cases with unprecise etiology, the analysis of the synovial tissue inside the joints, tendons sheaths, and bursae contributed to a better image of the pathogenetic processes, status, and prognosis in inflammatory arthritic conditions [14, 107–109].

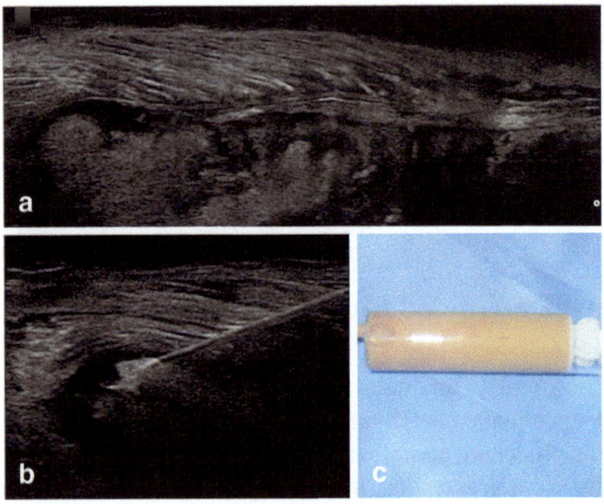

Fig. 14.39 **a** Calf abscess identified as a hyperechoic, inhomogeneous mass profound to the muscle. **b** During aspiration, the needle is identified as a hyperechoic linear structure; the tip is inside the abscess. **c** Aspect of the aspired content (pus)

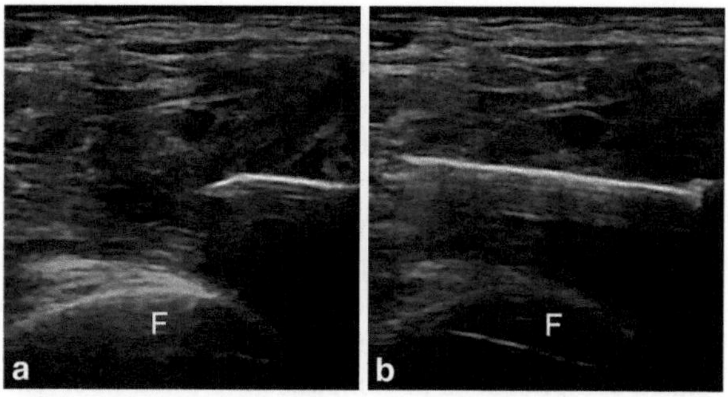

Fig. 14.40 **a** Knee biopsy—at the level of the lateral knee recess. The needle is placed in the optimal position for the biopsy. **b** Advancement of the needle after tissue sampling

The percutaneous synovial biopsy has major cost advantages as well as a better feasibility [24, 110]. In clinical practice, biopsies are recommended for synovial hypertrophy of unknown etiology. It is possible to perform biopsies in outpatient settings as well as hospitals in dedicated rooms without the need for patient's sedation.

Technically there are several steps. The first one is a systematic assessment of the target region and identification of those areas with synovial proliferation and positive Doppler signal. The MSUS scanning identifies then the neurovascular structures in the vicinity which should be avoided while the procedure is performed. The most convenient approach from skin to the target tissue should be considered, bearing in mind a comfortable position for the patient as well as the physician carrying out the procedure [108–110].

The devices used are Tru-Cut or automatic Bard needles or flexible forceps with 14–16 gauge caliber for large joints and 18 gauge for small joints and tendon sheaths. However, the needle choice may vary subject to the condition and the approach. After proper skin disinfection with bethadine, an incision of 3–5 mm is made at puncture site and the needle follows an "in-plane" approach to the target area. In case of automatic needles, the device should be armed prior to the joint penetration and released just outside the capsule or inside the joint, depending on the experience of the performing physician. Release of the needle inside the joint is less painful [24].

After capsule penetration, redirecting the needle is mandatory in order to obtain several samples followed by immediate proper preservation. The success rate in using automatic needles is 85–100 % for different anatomic areas (Fig. 14.40), [110–116].

Safety Profile

Safety profile analysis of MSUS-guided interventions highlights the two major advantages: the patients' exposure to the US beam during the maneuver and the placement of the transducer in the puncture area proximity. On the other hand, infection remains a possibility with all interventional maneuvers. Other potential drug-related side effects may incidentally occur. However, these are unrelated to the procedure itself, therefore it will not be discussed in this chapter.

Ultrasound is a nonradiant imaging tool and repeated evaluation, of any patient's category (children, pregnant patients, etc.) for diagnostic or interventional purposes, is safe without any radiation risk. In addition, recent studies showed a low risk for infection after US/ MSUS-guided interventions when rigorous cleaning and disinfection protocols are applied and safe distance between needle and transducer is kept [13, 14]. A very good safety profile is supported also by the reduction of incidental damage to adjacent neurovascular structures as well as other procedure-related complications in comparison to blind interventional maneuvers, especially in complex anatomical regions.

Conclusions

In conclusion, MSUS-guided maneuvers are ideal for clinical practice because they are not only highly efficient but also safe, rapid, and cheap. Introducing these maneuvers as part of our routine clinical practice will lead to better patients' outcome and procedure risk reduction and, at the end, will have a positive and important impact on the health-care resources.

References

1. Braun J, van den Berg RR, Baraliakos X, et al. 2010 update of the ASAS/ EULAR recommendations for the management of ankylosing spondylitis. Ann Rheum Dis. 2011;70:896–904.
2. Pendleton A, Arden N, Dougados M, et al. EULAR recommendations for the management of knee osteoarthritis: report of a task force of EULAR standing committee for International Clinical Studies Including Therapeutics (ESCISTS). Ann Rheum Dis. 2000;59:936–44.
3. Zhang W, Doherty M, Pascual E, et al. EULAR recommendations for calcium pyrophosphate deposition. Part II: management. Ann Rheum Dis. 2011;70:571–75.
4. Zhang W, Doherty M, Bardin T, et al. EULAR evidence based recommendations for gout. Part II: management. Report of a task force of the EULAR Standing Committee for International Clinical Studies Including Therapeutics (ESCISIT). Ann Rheum Dis. 2006;65:1312–24.
5. Zhang W, Doherty M, Arden N, et al. EULAR evidence based recommendations for the management of hip osteoarthritis: report of a task force of EULAR standing committee for International Clinical Studies Including Therapeutics (ESCISTS). Ann Rheum Dis. 2005;64(5):669–81.
6. Zhang W, Doherty M, Pascual E, et al. EULAR evidence based recommendations for gout. Part I: diagnosis. Report of a task force of EULAR standing committee for International Clinical Studies Including Therapeutics (ESCISTS). Ann Rheum Dis. 2006;65:1301–11.
7. Zhang W, Moskowitz RRW, Nuki G, et al. OARSI recommendations for the management of hip and knee osteoarthritis. Part II: OARSI evidence – based, expert consensus guidelines. Osteoarthr Cartil. 2008;16(12):137–62.
8. Hochberg MC, Altman RD, April KT, et al. American College of Rheumatology 2012 recommendations for the use of nonpharmacologic and pharmacologic therapies in osteoarthritis of the hand, hip, and knee. Arthritis Care Res. 2012;64(4):465–74.

9. Epis O, Bruschi E. Interventional ultrasound: a critical overview on ultrasound- guided injections and biopsies. Clin Exp Rheumatol. 2014;32(Suppl 80):S78–84.
10. Zhang W, Doherty M, Pascual E, et al. EULAR evidence based recommendations for gout. Part I: diagnosis. Report of a task force of EULAR standing committee for International Clinical Studies Including Therapeutics (ESCISTS). Ann Rheum Dis. 2006;65:1301–11.
11. Zhang W, Doherty M, Bardin T, et al. European League Against Rheumatism recommendations for calcium pyrophosphate deposition. Part I: terminology and diagnostis. Ann Rheum Dis. 2011;70:563–70.
12. Slawsky K, McInnis M, Goss TF, Lee DW. The clinical economics of ultrasound- guided procedures. Technical Report. Wauwatosa: General Electric; 2011.
13. Cervini P, Hesley GK, Thompson RL, Sampathkumar P, Knudsen JM. Incidence of infectious complications after an ultrasound- guided intervention. Am J Roentgenol. 2010;195:846–50.
14. Micu MC, Vlad MV, Bolboaca SD, et al. Musculoskeletal ultrasound guided manoeuvres-security profile. Med Ultrason. 2014;16(3):215–22.
15. D'agostino MA, Schmidt WA. Ultrasound-guided injections in rheumatology: actual knowledge on efficacy and procedures. Best Pract Res Clin Rheumatol. 2013;27:283–94.
16. Wakefield RJ, D'Agostino MA. Essential applications of musculoskeletal ultrasound in rheumatology. Philadelphia: Saunders Elsevier; 2010.
17. Mandl P, Naredo E, Conaghan PG, et al. Practice of ultrasound—guided arthrocentesis and joint injection, including training and implementation, in Europe: results of a survey of experts and scientific societies. Rheumatology. 2012;51:184–90.
18. Backhaus M, Burmester G, Gerber T, et al. Guidelines for musculoskeletal ultrasound in rheumatology. Ann Rheum Dis. 2001;60(7):641–9.
19. Klauser AS, Tagliafico A, Allen GM, et al. Clinical indications for musculoskeletal ultrasound: a Delphi-based consensus paper of the European Society of Musculoskeletal Radiology. Eur Radiol. 2012;22:1140–8.
20. McAlindon T, Kissin E, Nazarian L, Ranganath V, Prakash S, Taylor M, et al. American college of rheumatology report on reasonable use of musculoskeletal ultrasonography in rheumatology clinical practice. Arthritis Care Res (Hoboken). 2012;64:1625–40.
21. Karim Z, Wakefield RJ, Conaghan PG, et al. The impact of ultrasonography on diagnosis and management of patients with musculoskeletal conditions. Arthritis Rheum. 2000;44:2932–3.
22. D'Agostino MA, Ayral X, Baron G, Ravaud P, Breban M, Dougados M. Impact of ultrasound imaging on local corticosteroid injections of symptomatic ankle, hind-,and mid-foot in chronic inflammatory diseases. Arthritis Rheum. 2005;53:284–92.
23. Micu MC, Alcalde M, Saenz JI, Crespo M, Collado P, Bolboaca SD, et al. Impact of musculoskeletal ultrasound in an outpatient rheumatology clinic. Arthritis Care Res. 2013;65(4):615–21.
24. Fodor D. Aplicatii ale ecografiei musculoscheletale in reumatologie. Bucuresti: Ed. Med; 2013.
25. Stephens MB, Beutler AI, O'Connor FG. Musculoskeletal injections: a review if the evidence. Am Fam Physician. 2008;78(8):971–6.
26. Bloc S, Mercadal L, Garnier T, et al. Evaluation of a new disinfection method for ultrasound probes used for regional anesthesia. Ulraviolet C light. J Ultrasond Med. 2011;30(6):785–8.
27. Epis O, Iagnocco A, Meenagh G, Riente L, Delle Sedie A, Filippucci E, et al. Ultrasound imaging for the rheumatologist XVI. Ultrasound – guided procedures. Clin Exp Rheumatol. 2008;26(4):515–8.
28. Koibuchi H, Fujii Y, Kotani K, et al. Degradation of ultrasound probes caused by disinfection with alcohol. J Med Ultrasonics. 2011;38:97–100.
29. Sethi PM, El Attrache N. Accuracy of intra-articular injection of the gleno- humeral joint: a cadaveric study. Orthopedics. 2006;29:149–52.
30. Rutten MJCM, Collins JMP, Maresch BJ. Glenohumeral joint injection: a comparative study of ultrasound and fluoroscopically guided techniques before MR arthrography. Eur Radiol. 2009;19:722–30.
31. Naredo E, Cabero F, Cruz A, Uson J, Palop MJ, Crespo M. Ultrasound guided musculoskeletal injections. Ann Rheum Dis. 2005;64(2):341. (author reply 341).

32. Zwar B, Read W, Noakes B. Sonographycally guided gleno-humeral injection. Am Journal Rev. 2004;184:48–50.
33. Hulstyn MJ, Fadale PD. Arthroscopic anatomy of the shoulder. Orthop Clin North Am. 1995;26:597–612.
34. Sabeti-Aschraf M, Ochsner A, Schueller-Wedekamm C, et al. The infiltration of the AC joint performed by one specialist: Ultrasound versus palpation a prospective randomized pilot study. Eur J Radiol. 2010;75:e37–40.
35. Naredo E, Cabero F, Beneyto P, et al. Ultrasound versus blind injection of the subacromial bursa. J Rheumatol. 2004;31:308–14.
36. Bruyn GAW, Schmidt WA. How to perform ultrasound guided injections. Best Pract Res Clin Rheumatol. 2009;23:269–79.
37. Micu MC, Iagnocco A. Musculoskeletal ultrasound guiding shoulder injection when "one picture is worth ten thousand landmarks". Med Ultrason. 2013;15(1):3–5.
38. Naredo E, Cabero F, Cruz A, Uson J, Palop MJ, Crespo M. Ultrasound guided musculoskeletal injections. Ann Rheum Dis. 2005;64(2):341. (author reply 341).
39. Pickup L, Smith TO, Denton ERE, Toms AP. The clinical and functional outcomes of ultrasoundguided vs landmark-guided injections for adults with shoulder pathology—a systematic review and metaanalysis. Rheumatology. 2013;52:743–51.
40. Tagliafico A, Russo G, Boccakini S, et al. Ultrasound-guided interventional procedures around the shoulder. Radiol Med. 2014;119:318–26.
41. Radunovic G, Vlad V, Micu MC, et al. Ultrasound assessment of the elbow. Med Ultrason. 2012;14(2):141–6.
42. Petrella RJ, Cogliano A, Decaria J, Mohamed N, Lee R. Management of tennis elbow with sodium hyaluronate periarticular injections. Sports Med Arthrosc Rehabil Ther Technol. 2010;2:4.
43. Gaujoux-Viala C, Dougados M, Gossec L. Efficacy and safety of steroid injections for shoulder and elbw tendonitis: a meta-analysis of randomized controlled trials. Ann Rheum Dis. 2009;68:1843–9.
44. Grassi W, Lammana G, Farina A, Cervini C. Synovitis of small joints sonographyc guided diagnostic and therapeutic approach. Ann Rheum Dis. 1999;58:595–7.
45. Cunnington J, Marshall N, Hide G, et al. A randomized, double-blind, controlled study of ultrasound-guided corticosteroid Injection into the joint of patients with inflammatory arthritis. Arthritis Rheum. 2010;63:1862–9.
46. Raza K, Lee CY, Pilling D, et al. Ultrasound guidance allows needle placement and aspiration from small joints in patients with early inflammatory arthritis. Rheumatology. 2003;42:976–9.
47. Lohman M, Vasenius J, Nieminen O. Ultrasound guidance for puncture and injection in the radiocarpal joint. Acta Radiol. 2007;48:744–7.
48. Zingas C, Failla JM, Van Holsbeeck M. Injection accuracy and clinical relief of de Quervain' s tendonitis. J Hand Surg. 1998;23:89–96.
49. Mirzanli C, Ozturk K, Ayanoglu S, Imren Y, Aliustaoglu S. Accuracy of intrasheath injection techniques for de Quervain's disease: a cadaveric study. J Hand Surg Eur. 2012;37:155–60.
50. Mc Dermott JD, Ilyas AM, Nazarian LN, Leinberg CF. Ultrasound-guided injections for de Quervain's tenosynovitis. Clin Orthop Relat Res. 2012;470:1925–31.
51. Bodor M, Flossman T. Ultrasound-guided first annular pulley injection for trigger finger. J Ultrasound Med. 2009;28:737–43.
52. Grassi W, Farina A, Filipucci E, Cervini C. Intralesional therapy in carpal tunnel syndrome: a sonographic-guided approach. Clin Exp Rheumatol. 2002;20(1):73–6.
53. Smith J, Wisniewski SJ, Finnoff JT, Payne JM. Sonographically guided carpal tunnel injections. The ulnar approach. J Ultrasound Med. 2008;27:1485–90.
54. Kruse DW. Intraarticular cortisone injection for osteoarthritis of the hip. Is it effective? Is it safe? Curr Rev Musculoskeletal Med. 2008;1:227–33.
55. Sofka CM, Saboeiro G, Adler RS. Ultrasound-guided adult hip injections. J Vasc Interv Radiol. 2005;16:1121–3.
56. Bradley MJ. An in-vitro study to understand successful free-hand ultrasound guided intervention. Clin Radiol. 2001;56(6):495–8.

57. Qvistgaard E, Kristoffersen H, Terslev L, Danneskiold-Samsøe B, Torp-Pedersen S, Bliddal H. Guidance by ultrasound of intra-articular injections in the knee and hip joints. Osteoarthr Cartil. 2001;9:512–7.
58. Qvistgaard E, Christensen R, Torp-Pedersen S, et al. Intra-articular treatment of hip osteoarthritis: a randomized trial of hyaluronic acid, corticosteroid and isotonic saline. Osteoarthr Cartil. 2006;14:163–70.
59. Micu MC, Bogdan G, Fodor D. Steroid injection for hip osteoarthritis: efficacy under ultrasound guidance. Rheumatology. 2010;49:1940–4.
60. Flanagan J, Casale FF, Thomas TL, Desai KN. Intra-articular injection for pain relief in patients awaiting hip replacement. Ann R Coll Surg Engl. 1988;70:156–7.
61. Kullenberg B, Runesson R, Tuvhag R, Olsson C, Resch S. Intraarticular corticosteroid injection: pain relief in osteoarthritis of the hip? J Rheumatol. 2004;31:2265–8.
62. Robinson P, Keenan AM, Conaghan G. Clinical effectiveness and dose response of image-guided intra-articular corticosteroid injection for hip osteoarthritis. Rheumatology. 2007;46:285–91.
63. Margules KR. Fluoroscopically directed steroid instillation in the treatment of hip osteoarthritis: safety and efficacy in 510 cases. Arthritis Rheum. 2001;44:2449–50.
64. Lambert RGW, Hutchings EJ, Grace MGA, et al. Steroid Injection for osteoarthritis of the hip: a randomized, double-blind, placebo-controlled trial. Arthritis Rheum. 2007;56:2278–87.
65. Atchia I, Kane D, Reed M, Isaacs JD, Birrell FN, et al. Efficacy of a single ultrasound guided injection for the treatment of hip osteoarthritis. Ann Rheum Dis. 2011;70:110–6.
66. Micu MC. Ultrasound guided hip injection techniques. Osteoarthr Diagn Treat Surg. In Tech 2012.
67. Migliore A, Massafra U, Bizzi E, Vacca F, Martin-Martin S, Granata M, et al. Comparative, double-blind, controlled study of intra-articular hyaluronic acid (Hyalubrix®) injections versus local anesthetic in osteoarthritis of the hip. Arthritis Res Ther. 2009;11(6):1–8.
68. Migliore A, Tormenta S, Martin Martin LS, et al. Safety profile of 185 ultrasound guided intra-articular injections for treatment of rheumatic diseases in the hip. Reumatismo. 2004;56:104–9.
69. Migliore A, Giovannangeli F, Granata M, Lagana B. Hyalgan G-F 20: review of its safety and efficacy in the management of joint pain in osteoarthritis. Clin Med insights Arthritis Musculoskelet Disord. 2010;3:55–68.
70. Levi DS. Intra-articular hip injections using ultrasound guidance: accuracy using a linear array transducer. Phys Med Rehabil. 2013;5:129–34.
71. Pourbagher MA, Ozalay M, Pourbagher A. Accuracy and outcome of sono- graphically guided intra-articular sodium hyaluronate injections in patients with osteo- arthritis of the hip. J Ultrasound Med. 2005;24:1391–5.
72. Smith J, Hurdle MF, Weingarten TN. Accuracy of sonographically guided intra-articular injections in the native adult hip. J Ultrasound Med. 2009;28:329–35.
73. Kruse DW. Intraarticular cortisone injection for osteoarthritis of the hip. Is it effective? Is it safe? Curr Rev Musculoskeletal Med. 2008;1:227–33.
74. Labrosse JM, Cardinal E, Leduc BE, et al. Effectiveness of ultrasound-guided corticosteroid Injection for the treatment of gluteus medius tendinopathy. Am J Roentgenol. 2010;194:202–6.
75. Rowbotham EL, Grainger AJ. Ultrasound-guided intervention around the hip joint. Am J Roentgenol. 2011;197:W122–7.
76. Curtiss HM, Finnoff JT, Peck E, Hollman J, Muir J, Smith J. Accuracy of ultrasound-guided and palpation-guided knee injections by an experienced and less-experienced injector using a superolateral approach: a cadaveric study. Phys Med Rehabil. 2011;3(6):507–15.
77. Balint PV, Kane D, Hunter J, McInnes LB, Field M, Sturrock RD. Ultrasound guided versus conventional joint and soft tissue fluid aspiration in rheumatology practice: a pilot study. J Rheumatol. 2002;29(10):2209–13.
78. Jackson DW, Evans NA, Thomas BM. Accuracy of needle placement into the intra- articular space of the knee. J Bone Joint Surg Am. 2002;84:1522–7.

79. Jang SH, Lee SC, Lee JH, Nam SH, Cho KR, Park Y. Comparison of ultrasound (US)-guided intra-articular injections by in-plain and out-of-plain on medial portal of the knee. Rheumatol Int. 2013;33:1951–9.
80. Bum Park Y, Ah Choi W, Kim YK, Chul Lee S, Hae Lee J. Accuracy of blind versus ultrasound-guided suprapatellar bursal injection. J Clin Ultrasound. 2012;40:20–5.
81. Esenyel C, Demirhan M, Esenyel M, et al. Comparison of four different intra-articular sites in the knee: a cadaver study. Knee Surg Sport Traumatol Arthrosc. 2007;15:573–7.
82. Adler RS. Percutaneous ultrasound-guided peritendinous injections. 1st Edition. Informa Healthcare; 2007.
83. Koski JM. Ultrasound guided injections in rheumatology. J Rheumatol. 2000;27:2131–8.
84. Micu MC, Nestorova R, Petranova T, Porta F, Radunovic G, Vlad V, et al. Ultrasound of the ankle and foot in rheumatology. Med Ultrason. 2012;14(1):34–41.
85. Wisniewski SJ, Smith J, Patterson DG, Carmichael SW, Pawlina W. Ultrasound-guided versus nonguided tibiotalar joint and sinus tarsi injections: a cadaveric study. Phys Med Rehabil. 2010;2(4):277–81.
86. Reach JS, Easley ME, Chuckpaiwong B, Nunley JA. 2nd. Accuracy of ultrasoud guided injections of the foot and ankle. Foot Ankle Int. 2009;30:239–42.
87. Khosla S, Thiele R, Bamhauer JF. Ultrasound guidance for intra- articular injections of the foot and ankle. Foot Ankle Int. 2009;30:886–90.
88. Di Geso L, Filipucci E, Meenagh G, Gutierrez M, Ciapetti A, Salaffi F, et al. CS injection of tenosynovitis in patients with chronic inflammatory arthritis: the role of US. Rheumatology (Oxford). 2012;51(7):1299–303.
89. Fredberg U, Ostgaard R. Effect of ultrasound-guided, peritendinous injections of adalimumab and anakinra in chronic achilles tendinopathy: a pilot study. Scand J Med Sci Sports. 2009 Jun;19(3):338–44.
90. Smith J, Finnoff JT, Henning PT, Turner NS. Accuracy of sonographically guided posterior subtalar joint injections: comparison of 3 techniques. Ultrasound Med. 2009 Nov 1;28(11):1549–57.
91. Kane D, Greaney T, Bresnihan B, Gibney R, Fitzgerald O. Ultrasound guided injection of recalcitrant plantar fasciitis. Ann Rheum Dis. 1998;57(6):383–4.
92. McMillan AM, Landorf KB, Gilheany MF, Bird AR, Morrow AD, Menz HB. Ultrasound guided corticosteroid injection for plantar fasciitis: randomised controlled trial. Br Med J. 2012;344:e3260.
93. Maide E, Presley ChB, Murthy N, Pawlina W, Smith J. Sonographically guided deep plantar fascia injections. Where does the injectate go? J Ultrasound Med. 2013;32:1451–9.
94. Young CM, Shiels WE, Coley BD, et al. Ultrasound-guided corticosteroid injection therapy for juvenile idiopathic arthritis: 12-year care experience. Pediatr Radiol. 2012 Dec;42(12):1481–9.
95. Habibi S, Ellis J, Strike H, Ramanan AV. Safety and efficacy of US-guided CS injection into temporomandibular joints in children with active JIA. Rheumatology (Oxford). 2012;51(5):874–7.
96. Klauser A, De Zordo T, Feuchtner G, et al. Feasibility of ultrasound-guided sacroiliac joint injection considering sonoanatomic landmarks at two different levels in cadavers and patients. Arthritis Rheum. 2008;59(11):1618–24.
97. Hartung W, Ross CJ, Straub R, et al. Ultrasound-guided sacroiliac joint injection in patients with established sacroiliitis: precise IA injection verified by MRI scanning does not predict clinical outcome. Rheumatology. 2010;49:1479–82.
98. Iagnocco A and Naredo E. Ultrasound-guided corticosteroid injection in rheumatology: accuracy or efficacy? Rheumatology. 2010;49(8):1427–8.
99. Canoso JJ. Aspiration and injection of joints and periarticular tissues (including articular and intralesional therapy). In: Hochberg MC, Silman AJ, Smolen JS, Weinblatt ME, Weisman MMH, editors. Rheumatology. Mosby Elsevier; 2008. P. 421–32.
100. Zochling J, van der Heijde D, Burgos-Vargas R, Collantes E, Davis JC Jr, Dijkmans B, et al. ASAS/EULAR recommendations for the management of ankylosing spondylitis. Ann Rheum Dis. 2006;65:442–52.

101. Rosenberg JM, Quint TJ, de Rosayro AM. Computerized tomographic localisation of clinically guided sacroiliac joint injections. Clin J Pain. 2000;16:18–21.
102. Peterson MK. Ultrasound guided nerve blocks. Br J Anaesth. 2002;88(5):621–3.
103. Perlas A, Bruli R, Chan V, et al. Ultrasound guidance improves the success of sciatic nerve block at the poplitea fossa. Reg Anesth Pain Med. 2008;33:259–65.
104. Punzzi L, Cimmino MA, Frizziero V, et al. Recomandayioni della Societa Italiana di Reumatologia per l' esecutione dell'artrocentesi. Reumatismo. 2007;59:227–34.
105. Royall NA, Farrin E, Bahner DP, Stawicki SPA. Ultrasound- assisted musculoscheletal procedures: a practical overview of current literature. World J Orthop. 2011;2(7):57–66.
106. Torriani M, Echtebehre M, Amstalden E. Sonographically guided core needle biopsy of bone and soft tissue tumors. J Ultrasound Med. 2002;21:275–81.
107. Yeow KM, Tan CF, Chen JS, Hsueh C. Diagnostic sensitivity of ultrasound—guided core needle biopsy in soft tissue masses about superficial bone lesions. J Ultrasound Med. 2000;19:849–55.
108. Koski JM, Helle M. Ultrasound guided synovial biopsy using portal and forceps. Ann Rheum Dis. 2005;64:926–9.
109. Felea I, Fodor D, Schiotis R, Georgiu C, Bojan A, Rednic S. Ultrasound findings in AL musculoskeletal amyloidosis. Med Ultrason. 2011 Mar;13(1):76–9.
110. Schumacher HR. Synovial fluid analysis and synovial biopsy. In: Kelley WN, Harris ED, Ruddy S, Sledge CB, editors. Textbook of rheumatology. 5th ed. Phyladelphia: Saunders; 1997. P. 609–25.
111. Skrzynski MC, Biermann JS, Montag A, Simon MA. Diagnostic accuracz and charge savings of outpatient core needle biopsy compared with open biopsy of musculoskeletal tumors. J Bone Joint Surg Am. 1996;78:644–9.
112. Peer S, Freuis T, Loiyides A, Gruber H. Ultrasound guided core needle biopsy of soft tissue tumors; a fool froof technique? Med Ultrason. 2011;13(3):187–94.
113. Rubens DJ, Fultz PJ, Gottlieb RH, Rubin SJ. Effective ultrasonographically guided intervention for diagnosis of musculoskeletal lesions. J Ultrasound Med. 1997;16:831–42.
114. Le HBQ, Lee ST, Munk PL. Image-guided musculoskeletal biopsies. Semin Intervent Radiol. 2010;27:191–8.
115. Battaglia M, Pollastri P, Ferrao A, Betoni F, Bacci G, Galletti S. The role of ultrasound-guided needle biopsy in the diagnosis of soft-tissue tumors. J Ultrasound. 2007;10:59–62.
116. Bickels J, Jelinek JS, Shmookler BM, Neff RS, Malawer MM. Biopsy of musculoskeletal tumors: current concepts. Clin Ortop Related Res. 1999;368:212–9.

Index

Y. El Miedany (ed.), *Musculoskeletal Ultrasonography in Rheumatic Diseases*,
DOI 10.1007/978-3-319-15723-8

MIX
Papier aus verantwortungsvollen Quellen
Paper from responsible sources
FSC® C105338

If you have any concerns about our products,
you can contact us on
ProductSafety@springernature.com

In case Publisher is established outside the EU,
the EU authorized representative is:
Springer Nature Customer Service Center GmbH
Europaplatz 3, 69115 Heidelberg, Germany

Printed by Libri Plureos GmbH
in Hamburg, Germany